Introduction to Chemical Engineering

Published by John Wiley & Sons, Inc., Hoboken, New Jersey.
Published simultaneously in Canada.

Library of Congress Cataloging-in-Publication Data

Names: C.M. van 't Land, 1937– author.
Title: Introduction to chemical engineering / C.M. van 't Land.
Description: 1st edition. | Hoboken, NJ : Wiley, 2024. | Includes bibliographical references and index.
Identifiers: LCCN 2021035195 (print) | LCCN 2021035196 (ebook) | ISBN 9781119634089 (hardback) | ISBN 9781119634096 (adobe pdf) | ISBN 9781119634126 (epub)
Subjects: LCSH: Chemical engineering.
Classification: LCC TP155 .L26 2024 (print) | LCC TP155 (ebook) | DDC 660–dc23
LC record available at https://lccn.loc.gov/2021035195
LC ebook record available at https://lccn.loc.gov/2021035196

Cover Design: Wiley
Cover Image: Courtesy of C.M. van 't Land

Set in 9.5/12.5pt STIXTwoText by Straive, Pondicherry, India

Contents

Preface

After graduation as a chemical engineer at the University of Twente in The Netherlands in 1971 I worked in research for the multinational company Akzo Nobel. My retirement started in 2000. In 2005, I proposed the Dutch organization PAO Techniek en Management to organize a seminar "Introduction to Chemical Engineering". The goal was to enable chemists in the process industry to acquire knowledge of chemical engineering. The seminar was given successfully in The Netherlands between 2005 and 2015 and its scope was gradually extended. Material from standard textbooks was combined with material obtained while at work. The seminar appeared to be attractive also for, e.g., mechanical engineers and physicists. The seminar was written in English as that is the "lingua franca" of the process industry. Then, in 2015, the interest declined. It is possible that internet information is the cause. I then proposed John Wiley & Sons to publish a book containing the material of the seminar. An agreement was made and the present book is the result. I am grateful for the permissions obtained for the incorporation of material from various sources. The sources are mentioned in the book. I wish to thank especially the following publishers:

Delft Academic Press/VSSD at Delft in The Netherlands,
Elsevier at Amsterdam in The Netherlands,
McGraw-Hill at New York in the U.S.A.,
John Wiley & Sons at Hoboken in the U.S.A.,
Taylor and Francis Group at Abingdon in the U.K., and
Wolters Kluwer at Alphen aan den Rijn in The Netherlands.

I am greatly indebted to my wife, Annechien, for her constant encouragement and patience.

C.M. van 't Land

Prologue

1. A Typical Chemical Production System

Introduction

The production system for crystalline ethylene diamine tetra-acetic acid (EDTA) within Nouryon Industrial Chemicals will be described. Nouryon is a multinational company manufacturing chemicals. The company employs at present approximately 10,000 employees and is active in more than 80 countries. The activities of Nouryon were part of the multinational company AkzoNobel till 2019. EDTA is a chelate, it is also called a sequestering agent. The commercial name of the product is Dissolvine Z.

Chelates

The word "chela" means claw in Greek. Chelates have the ability to seize metal ions and control them, making it difficult for a different substance to liberate them. EDTA and compounds derived from EDTA are chelates. For example, the calcium disodium salt of EDTA is applied to deactivate undesirable heavy metal ions that catalyze (promote) the degradation of vegetable oils and fats. By preventing this degradation, which makes food and beverages rancid, food quality is preserved and shelf life is increased. The application of EDTA in boilers, heat exchangers, and other water circulation systems present in power, brewing, sugar, and dairy industries is a further example. The compound forms stable, water soluble metal complexes with all potentially harmful metal ions, dissolving existing metal complexes and preventing new ones to form.

The other side of the coin is that sequestering agents can also be applied to deliberately administer a metal. For example, a Nouryon iron chelate has been approved by the World Health Organization and the Food and Agricultural Organization to be applied as an iron supplement in food. The iron chelate is soluble in water.

The Chemistry of the EDTA-Na_4 Synthesis

The tetra sodium salt of EDTA is the intermediate for the manufacture of EDTA.

The compound is made by the alkaline cyanomethylation of ethylene diamine (EDA) by means of sodium cyanide and formaldehyde:

$$4NaCN + 4HCOH + EDA + 4H_2O \rightarrow EDTA\text{-}Na_4 + 4NH_3$$

This is an overall equation. The alkaline cyanomethylation comprises in fact a series of consecutive reactions.

EDA: 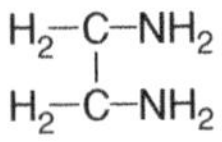

EDTA—Na_4:

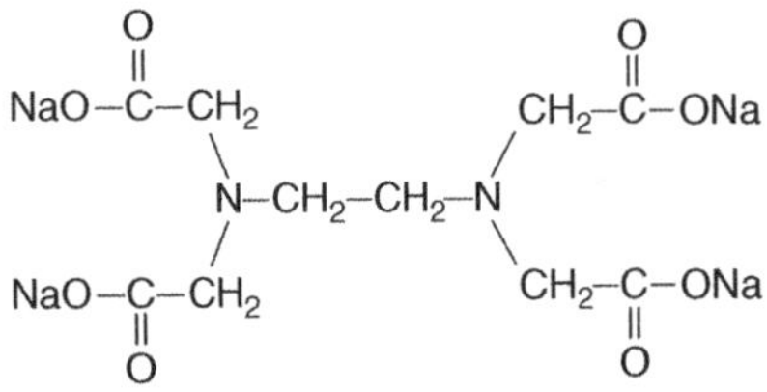

The principal by-product is NH_3, ammonia, which is continuously boiled off during the reaction. The reaction is carried out batchwise. The reaction is exothermic. The reaction conditions are 100–105 °C and the pressure is atmospheric. The pH of the reaction is higher than 12, some free caustic soda is added to the reactor. The reaction rate is high, the reaction is brought to completion in seconds. The reaction does not need a catalyst. An excess of NaCN causes the formation of sodium formate. An excess of formaldehyde causes the formation of oligomeric compounds. With respect to EDA, the yield is 95–100%. For selectivity reasons, it is not possible to carry out the reaction continuously.

The removal of ammonia is key as ammonia reacts with sodium cyanide, formaldehyde, and water to the trisodium salt of nitrilo tri-acetic acid:

$$3NaCN + 3HCOH + NH_3 + 3H_2O \rightarrow NTA\text{-}Na_3 + NH_3$$

NTA-Na_3:

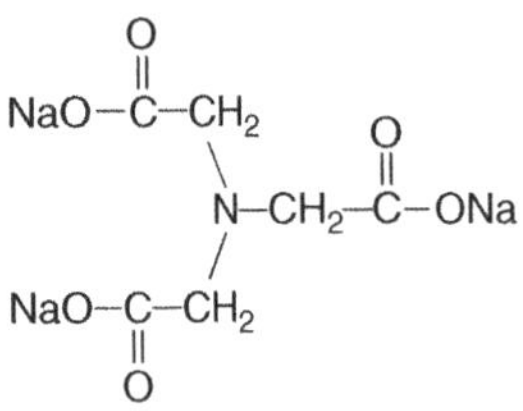

The impurity NTA-Na_3 and other impurities (like glycolic acid salt) are not detrimental to most applications of chelating agents.

The Industrial Reaction to Produce EDTA-Na_4

See Figure 1. First, the reactor will be described. Its total volume is approximately 25 m^3. The reactants are mixed by means of a proprietary stirrer. Heat is transferred indirectly to the reactor contents by means of a coil in the lower part of the reactor. The coil is heated by means of steam. The reaction is an exothermic one; however, extra heat is needed for the evaporation of water and ammonia. There are baffles in the lower part of the reactor, their height

Figure 1 EDTA-Na_4 reactor.

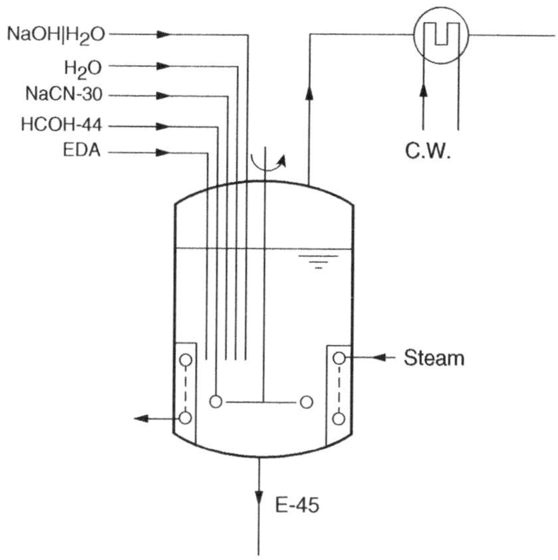

equals the height of the coil. Formaldehyde is dosed by means of a sparger around the agitator. Formaldehyde is the most critical component. An aqueous solution of sodium cyanide and EDA are dosed by means of dip-pipes. Gaseous ammonia and water are removed overhead and are condensed in a series of two surface condensers. In the first condenser, water is condensed mainly whereas NH_3-25 is obtained in the second condenser. The saturated vapor pressure of NH_3-25 is 1 bara at 25 °C.

Second, the manufacturing procedure will be described. A certain amount of EDA is added to the reactor. Small amounts of water and caustic soda (NaOH) are also added to adjust the alkalinity. Next, the stirrer is activated and the reactor contents are heated to 100 °C. The dosing of the aqueous solutions of sodium cyanide, formaldehyde, and EDA is started and the reaction proceeds. Ammonia and water are removed overhead. The batch time is 4–5 h. Each batch yields 15 t of a 45% by weight solution of the tetra sodium salt of EDTA in water and 4 t of a 25% by weight solution of ammonia in water. Next, the reactor contents are pumped to a second reactor to bring the reaction to completion under moderate boiling conditions. Bleaching the contents of the second reactor is the next step. The intermediate is termed Dissolvine E-45. It can be sold as such, it can be spray-dried, or it can be converted into the acid. The conversion into the acid will be described in the next section.

The aqueous solution of sodium cyanide is highly toxic to humans. Contact with an acid converts sodium cyanide to hydrogen cyanide which is a hazardous respiratory poison. Hydrogen cyanide dissolves in water and is transported via the bloodstream to the individual cells of the body where it blocks oxygen uptake by combining with the enzymes which control cellular oxidation. Oxygen uptake at the cellular level is blocked as long as the cyanide is present. Normal cellular oxygen uptake can resume if death of the cells has not already occurred. The cyanides are particularly hazardous because of their low threshold toxicity level coupled with the fact that they are odorless. Several operators are trained to be able to give an injection with an antidote if need be.

The Conversion into EDTA

See Figure 2. EDTA-H_4 (the acid) is made by means of reaction crystallization, liquid/solid-separation on a belt filter, and drying in a thin-film dryer, a Solidaire. A Solidaire is a horizontal cylinder equipped with a stirrer rotating at high speed.

The reaction crystallization occurs continuously in a cascade of two well-mixed reactors. The process occurs at atmospheric pressure and approximately 80 °C. The reaction occurs between the 45% by weight solution of the tetra sodium salt of EDTA and an aqueous 30% by weight solution of hydrochloric acid. The pH in the first reactor is 2.3 whereas the pH in the second reactor is 1.45. At a pH of 2.3, relatively large crystals are formed. The reaction is brought to completion at a pH of 1.45. Reacting at a pH of 1.45 would lead to a good yield rightaway; however, the filter cake would be relatively wet. Accomplishing the bulk of the crystallization at a pH of 2.3 and finishing the reaction crystallization at a pH of 1.45 results in a filter cake having a relatively low moisture content. That means that the drying effort is relatively modest. Liquid/solid-separation is carried out on an Outotec Larox RT belt filter. See Figure 3. The stages on the belt filter are mother liquid removal, washing, and steaming. The latter step lowers the moisture content of the cake, probably through a viscosity decrease. 18% water by weight is a typical cake moisture content.

Next, hot air (155 °C) and product travel concurrently through a dryer that is depicted in Figure 4. The air cools down to 70 °C and thereby accomplishes convective drying. At the same time, a jacket achieves contact drying. The jacket temperature is 160–170 °C (condensing steam). The Solidaire dryer has been selected because the product tends to form incrustations during the drying process. The vigorously rotating agitator prevents this from happening. Solidaire is a Bépex (Hosokawa) trade name.

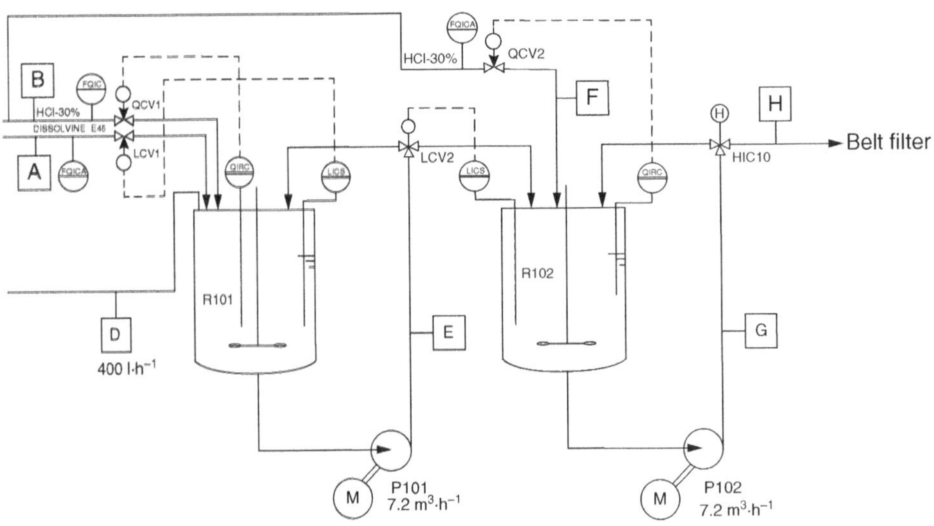

Figure 2 EDTA-H_4 reactors.

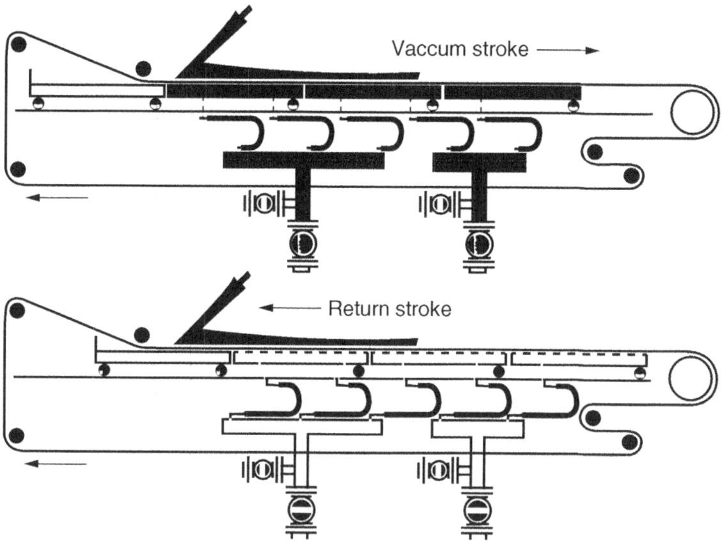

Figure 3 Belt filter. *Source:* Courtesy of Outotec Oyj, Espoo, Finland.

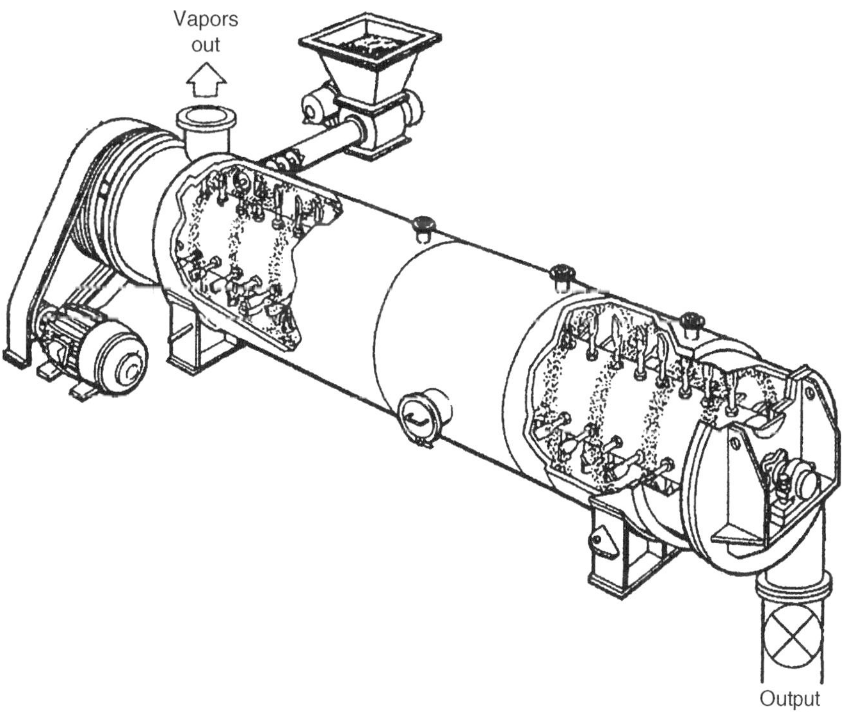

Figure 4 Thin-film dryer. *Source:* Courtesy of Bepex International LCC, Minneapolis, U.S.A.

The Chemical Plant

The Prologue started with a description of the manufacture of EDTA-Na_4. The chemistry, mixing, heat transfer, and stripping appeared to be important. Next, the manufacture of EDTA was treated. Mixing occurs at the reaction crystallization. The acidification is followed by liquid/solid separation, leaching (washing), and drying.

2. Chemical Reactors and Unit Operations

A distinction is made between chemical reactors and unit operations. Unit operations are steps common to most industrial processes such as heat transfer, distillation, fluid flow, filtration, crushing and grinding, and crystallization.

In response to the rapid rise of commercial manufacturing of chemicals used in industry, Lewis Mills Norton founded a course of study in chemical engineering in the chemistry department of MIT in 1888 [1]. The Report to the President of MIT at the time stated: "The chemical engineer is a mechanical engineer who has given special attention to the problems of chemical manufacture." This is a prophetic statement of chemical engineering's duality between the "plumbers" and the "chefs". In 1888, the course of instruction focused on descriptive industrial chemistry with an extended study of mechanical engineering. The educational method used at the time was the case method. One wonders, however, how the study of the manufacture of soda ash would help to, for example, separate gasoline and fuel oil from crude oil.

Arthur D. Little stated in 1915 in a Visiting Committee Report to the President of MIT that the core of the chemical engineering education should be centered on the "unit operations." Walker, Lewis, and McAdams wrote the textbook "Principles of Chemical Engineering" in 1923. The authors said: "We have selected for treatment basic operations common to all chemical industries, rather than details of specific processes, and so far as is now possible, the treatment is mathematically quantitative as well as qualitatively descriptive."

The results were dramatic. All chemical engineering educators in the United States learned unit operations from this textbook, and they taught it to generations of students.

In 1945, it became clear that technological advances were made largely by scientists who had not studied engineering. The development of atomic energy, laser, and radar are examples. A fundamental approach appeared to be successful. The textbook "Transport Phenomena" was published in 1960 [2]. The authors remarked: "Herein we present the subjects of momentum transport (viscous flow), energy transport (heat conduction, convection, and radiation), and mass transport (diffusion). . . . Because of the current demand in engineering education to put more emphasis on understanding basic physical principles than on the blind use of empiricism, we feel there is a very definite need for a book of this kind."

This publication and other aspects caused a dramatic change in the chemical engineering curriculum: a new standard of research based on more advanced mathematics and fundamental physics and physical chemistry. The latter approach became a great success.

References

1. Wei, J. (1996). A century of changing paradigms in chemical engineering. *CHEMTECH*, 26 (5), 16–18.
2. Bird, R. B., Stewart, W. E., and Lightfoot, E. N. (1960). *Transport Phenomena*. Hoboken, NJ: Wiley.

Part I

Transport Phenomena

Part I Content

1

Mass Balances

1.1 Introduction

A mass balance is an accounting for mass flows and changes in inventory of mass for a system. When there is no generation (or usage) of material within the system: accumulation = input − output.

Lavoisier formulated the conservation law of mass for chemical reactions for the first time in 1775: the sum of the masses of the substances entering into a reaction equals the sum of the masses of the products of the reaction. He formulated his law after studying the formation of mercuric oxide and its decomposition. On oxidizing mercury to orange mercuric oxide, the mass increased because of the reaction with oxygen. On decomposing the compound, the original mass of mercury was obtained again.

Mass balance calculations are almost invariably a prerequisite to all other calculations in the solution of chemical engineering problems.

Balances can be made on many quantities different from mass, for example, on pounds sterling. Most of the principles are based on common sense and logical thinking.

In plants, on making a mass balance for a mixing point, a piece of equipment, a part of a plant, or a plant, it is important that the entering and the leaving flows are measured independently from each other. It is not a good practice to calculate a leaving flow from an entering flow or vice versa. If, still and all, this is done, it is advised to indicate this clearly.

In chemical plants, sometimes, there are parts which operate continuously while other parts function batchwise. For mass balances, both for design and for an analysis of a functioning plant, it is important to choose a unit of time and establish both the entering and leaving masses.

The law of conservation of mass is valid for process plants. The law is not valid for nuclear reactions.

Introduction to Chemical Engineering, First Edition. C.M. van 't Land.

Example 1.1

John is skating on the Dutch canals. After a while, he wants to take a rest. He enters a coffee-house and asks for a mug with hot milk. He is being served but then changes his mind. Could he give the milk back and get a mug with hot chocolate instead? It will give him calories for the remainder of the tour. No problem and he gets his hot chocolate. John enjoys the drink and soon after that gets on his feet again and is ready to leave. However, the proprietor addresses him: "Sir, could you pay please?"

JOHN: "Pay for what?"
PROPRIETOR: "For the hot chocolate!"
JOHN: "But I gave you the milk back for the chocolate."
PROPRIETOR: "Yes, but you did not pay the milk either!"
JOHN: "Of course I did not pay the milk because I did not drink it."

The proprietor is confused and John leaves the coffee-house. Who is right?

Approach 1

Make a mass balance.

John enters with an empty stomach. He leaves with chocolate in his stomach. He must pay for the chocolate. The discussion is all eyewash.

Approach 2

Find the mistake in the reasoning.

John uses the milk as a means of exchange. However, the milk does not belong to him. He must pay for the hot chocolate.

Example 1.2

This example comes from the United States.

Three guys rent a hotel room and are charged $30 ($10 each). It occurred a while ago. They walk across the street and while there the hotel manager decides he charged them too much. The manager gives the bell boy $5 and tells him to return it to the three men. The bell boy, being practical and a bit dishonest, thinks: how can they split $5. The bell boy decides to keep $2 and give them each $1 back. Now the dilemma: the men each receives back $1 meaning they have now paid $9 each. Three times nine is $27 plus the $2 the bell boy has makes $29. What happened to the other $1? See Figure 1.1.

Considerations relevant for Example 1.2

Approach 1
Make a dollar balance for each of the three parties concerned.
Men: have spent 27.
Manager: received 25.
Bell boy: received 2.
27 = 25 + 2
There is no one-dollar issue. The discussion is idle.

Approach 2
Find the mistake in the reasoning.
Make a block scheme.

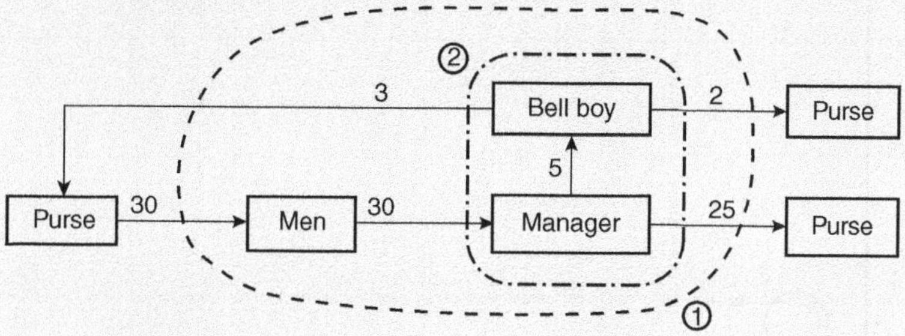

Circumscription ①: 30 = 3 + 2 + 25
Circumscription ②: 30 = 3 + 2 + 25

It is impossible to make a circumscription where 27 and 2 appear at the same side of the equation.

Figure 1.1 Sleeping in a hotel in the United States.

1.2 Theory

In a steady-state process, neither accumulation nor depletion occurs.

When a chemical reaction does not occur, an independent equation can be written to balance each compound. If a chemical reaction occurs, the component material balances have to be based on the elements themselves (H, O, S, etc.) or their equivalents (H_2, O_2, S_4, etc.).

A material balance problem has a unique solution only if the number of independent equations and the number of unknown variables are equal.

Example 1.3

A mixture of benzene, toluene, and xylene is fed to a distillation column. The feed rate is 10,000 kg · h^{-1}. The feed contains 60% by weight of benzene, 30% by weight of toluene, and 10% by weight of xylene. See Figure 1.2.

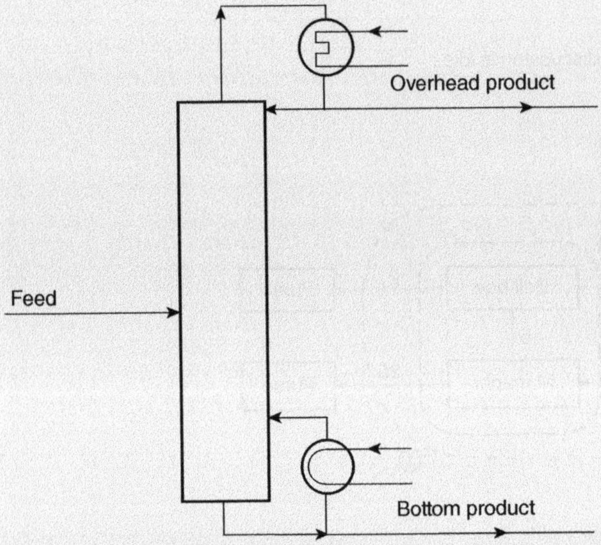

Figure 1.2 A distillation column.

Number of variables (3 flows and 3 components)	: 9
Number of independent mass balance equations	: 3
Number of fixed variables for a unique solution	: 9 − 3 = 6
Number of variables already specified	: 3
Number of variables to be specified by the designer	: 6 − 3 = 3

Specified by the designer: 5 kg · h^{-1} of toluene and 3 kg · h^{-1} of xylene in the overheads and 5 kg · h^{-1} of benzene in the bottoms.

Benzene in feed − benzene in bottoms = benzene in overheads

$0.6 \cdot 10{,}000 - 5 = 5{,}995 \text{ kg} \cdot \text{h}^{-1}$

Toluene in feed − toluene in overheads = toluene in bottoms

$0.3 \cdot 10{,}000 - 5 = 2{,}995 \text{ kg} \cdot \text{h}^{-1}$

Xylene in feed − xylene in overheads = xylene in bottoms

$0.1 \cdot 10{,}000 - 3 = 997 \text{ kg} \cdot \text{h}^{-1}$

Example 1.4

See Figure 1.3. The example concerns the air oxidation of cumene in a 0.75-l glass laboratory reactor. The oxidation proceeds at atmospheric pressure and 112 °C. When the air oxidation is carried out in the presence of an aqueous solution of sodium hydroxide, the reaction product is phenyl iso-propanol (cumyl alcohol).

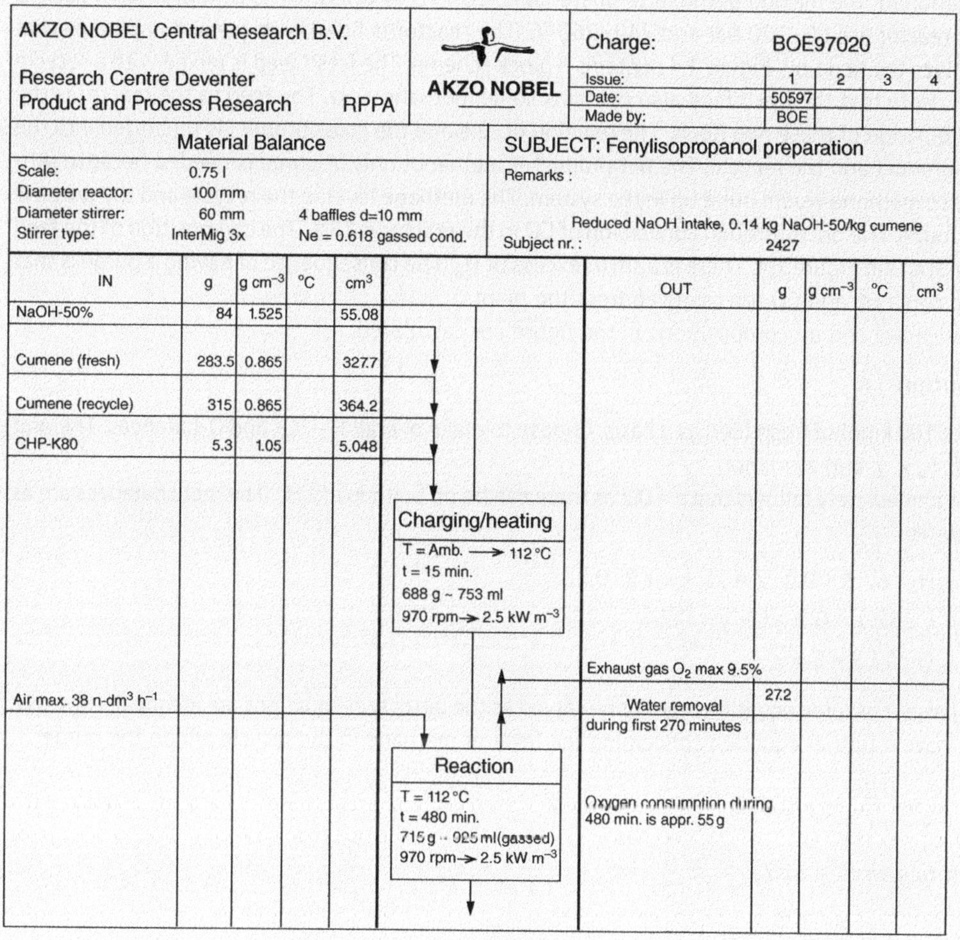

Figure 1.3 Laboratory batch air oxidation of a hydrocarbon.

Example 1.5

Coal can be converted into more convenient liquid products. H_2 and CO are the main gases that can be generated from in situ (in the ground) coal combustion in the presence of steam (as occurs naturally in the presence of ground water) [1]. The gaseous mixture is called synthesis gas because it can be used for the manufacture of various products, for example, methanol. The equation for the making of the latter material is: $CO + H_2 \rightarrow CH_3OH$. The conditions in a packed bed reactor are 50–100 bar and 240–260 °C. The reactor is filled with metal compounds that catalyze the reaction. Figure 1.4 contains a block scheme. The fresh feed is mixed with a recycle flow. Note that the fresh feed also contains some methane, CH_4. The feed to the reactor is the combination of these two flows. The reaction occurs and the gross product is separated into the net product and the recycle. The net product is methanol only. A purge is needed because otherwise methane would build up in the system. The methane level in the recycle and the purge is 3.2 mol%. The once-through conversion of CO in the reactor is 18%. The composition of the feed in mol% is in Figure 1.4. There is a small excess of H_2. The consequence of having a purge is that also some H_2 and CO are removed from the plant.

All flows and all compositions in the figure are calculated.

Solution

Take 100 kmol of fresh feed as a basis. Choose to make overall H_2-, C-, and O-balances. The unit of E, R, x, y, and z is kmol.

It immediately follows that $z = 0.2$ as there can be no buildup of CH_4. The mole balances are as follows.

$$H_2 : 67.3 + 0.2 \cdot 2 = 2E + x + 2 \cdot 0.2.$$

$$C : 32.5 + 0.2 = E + y + 0.2.$$

$$O : 32.5 = E + y.$$

This set of three equations cannot be solved as the latter two equations are identical. However, we have a further equation concerning the level of methane in the recycle and in the purge.

$$\frac{0.2}{x + y + 0.2} = 0.032.$$

It follows: $E = 31.25$.

$x = 4.8$.

$y = 1.25$.

The once-through balance is used to determine R. The mole fraction of CO in R is:

$$\frac{1.25}{4.8 + 1.25 + 0.2} = 0.2.$$

And an equation for the conversion in the reactor is:

$$(32.5 + 0.2 \cdot R)(1 - 0.18) = 1.25 + 0.2 \cdot R,$$

$$R = 705,$$

$$P = 6.25.$$

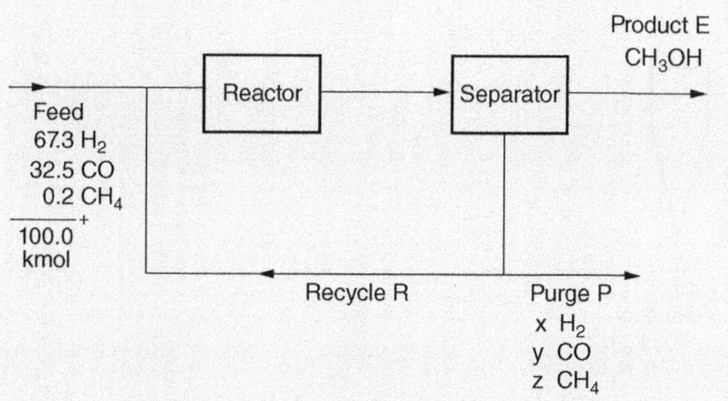

Figure 1.4 Continuous methanol manufacture.

1.3 Additional Material

In a plant, making a mass balance was used to establish what was wrong in the plant. It concerned a drying operation in a flash dryer. The name flash dryer was given to the dryer because the drying occurs in approximately 1 s. See Figure 1.5. Hot air and wet product travel concurrently through the dryer and drying is accomplished during the transport of the material. Air is heated indirectly in a heat exchanger. At the top of the dryer, product and spent air are separated in a cyclone. 1,000 $kg \cdot h^{-1}$ of wet filter cake could be dried in this dryer. However, gradually the capacity fell to 850 $kg \cdot h^{-1}$. To attain the final moisture content, it appeared necessary to raise the gas exit temperature from 65 to 80 °C. Because of product quality reasons, it was not possible to raise the air inlet temperature. First, the moisture content of the filter cake was checked. It appeared that it had not changed. It was speculated that the cake properties had changed, however, that could not be confirmed. The temperature and the amount of fresh air had not changed either. It was decided to make mass balances in $kg \cdot h^{-1}$ for both dry air and water vapor.

Entering flows			
	Air flow in	Evaporation	Sum
Dry air	10,627.8	0.0	10,627.8
Water vapor	42.5 +	295.0 +	337.5 +
	10,670.3	295.0	10,965.3
Leaving flow			
Dry air	11,005.3		
Water vapor	737.5 +		
	11,742.5		

The dry air flow leaving the dryer is slightly larger than the dry air flow entering the dryer. This can be explained by air ingress through the product inlet. The water flow leaving the dryer is more than twice the water vapor flow entering the dryer. This can be explained by the leakage of steam into the hot air flow in the heat exchanger. The heater was repaired and the dryer performance improved accordingly.

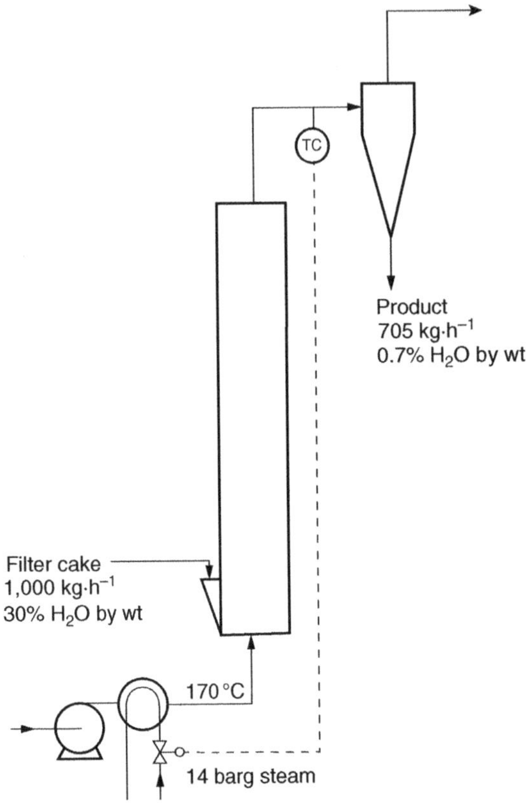

Figure 1.5 Flash drying.

Reference

1. Himmelblau, D. M., and Riggs, J. B. (2004). *Basic Principles and Calculations in Chemical Engineering* (pp. 369–371). Upper Saddle River, NJ: Prentice Hall.

2

Energy Balances

2.1 Definitions

Energy balances calculations are, like mass balance calculations, usually a prerequisite to other calculations in the solution of chemical engineering problems.

Heat

Heat can be supplied to a system by a chemical reaction. It is also that part of the total energy flow across a system boundary that is caused by a temperature difference between the system and the surroundings.

Work

Work is done whenever a force acts through a distance. Work per unit of time (usually s) is power.

Joule placed water in an insulated container, and agitated the water with a rotating stirrer. He found that a definite amount of work was required per unit mass of water for every degree of temperature rise caused by the stirring. It is possible to completely convert work (mechanical energy) into heat. It is also possible to convert heat into mechanical energy, however, not completely, as we know from power stations.

Internal Energy

Internal energy refers to the energy of the elementary particles making up the substance. At Joule's experiment, the energy added as work is contained as internal energy.

Kinetic Energy

Kinetic energy is the energy a substance possesses because of its motion as a whole.

Potential Energy

Potential energy is the energy a substance possesses because of its position as a whole.

Pressure Energy

Pressure energy is the energy a substance possesses because of its pressure as a whole.

Quantitatively, heat is more important than the other forms of energy. This will be illustrated by means of two examples.

Introduction to Chemical Engineering, First Edition. C.M. van 't Land.

Example 2.1

377,100 J are needed for the heating of 1 kg of water from 10 to 100 °C. Something we do routinely in the morning in the kitchen. With this amount of energy, it is possible to lift a human of 75 kg 512 m.

Example 2.2

The driver of a car does not pay the required attention while the car heads for a tree. Just in time, the driver realizes what is going to happen if he does not interfere. He activates the brakes of the car and manages to stop the vehicle before the tree. The kinetic energy of the car has been converted into sensible heat of the brake linings.

2.2 The General Energy Balance

A steady-state process is depicted in Figure 2.1.

$$\left(\Delta U + \Delta\left(\frac{p}{\rho}\right) + g\cdot\Delta h + \frac{\Delta v^2}{2}\right)\Phi_m = Q - W \quad W. \tag{2.1}$$

Φ_m : mass flow in $kg \cdot s^{-1}$.

Differences of the flows

$\Delta U \cdot \Phi_m$: internal energy
$\Delta(p/\rho)\Phi_m$: pressure energy
$g \cdot \Delta h \cdot \Phi_m$: potential energy
$(\Delta v^2/2)\Phi_m$: kinetic energy

Q : heat flow added to the system
W : power performed by the system

$$H = U + \frac{p}{\rho} \quad J \cdot kg^{-1}.$$

H : enthalpy

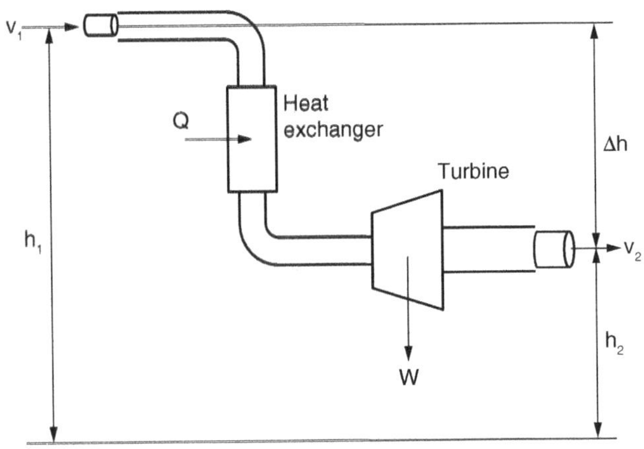

Figure 2.1 The general energy balance.

For a flow process, it is the enthalpy change rather than the internal-energy change that is of importance. Enthalpy data are tabulated.

For many of the applications considered, the kinetic- and potential-energy terms are small compared with other terms and can be neglected. Equation 2.1 then reduces to:

$$\Delta H \cdot \Phi_m = Q - W \quad W.$$

In a heat exchanger, it can be stated: $\Delta H = c_p \cdot \Delta T$
c_p : specific heat at constant pressure in $J \cdot kg^{-1} \cdot K^{-1}$
Leading to:

$$\Phi_m \cdot c_p \cdot \Delta T = Q \quad W.$$

2.3 Applications of the General Energy Balance

2.3.1 Pump

A pump raises the pressure of a liquid flow Φ_m from pressure p_1 to a higher pressure p_2. It is assumed that neither the potential energy nor the kinetic energy are affected. The pumping process is adiabatic (no heat exchange with the surroundings). By internal friction, some of the power is converted into heat. Equation 2.1 reduces to:

$$\left((U_2 - U_1) + \left(\frac{1}{\rho}\right)(p_2 - p_1)\right)\Phi_m = -W.$$

If there is no friction:

$$\left(\frac{1}{\rho}\right)(p_2 - p_1)\Phi_m = -W$$

or:

$$(p_2 - p_1)\Phi_v = -W.$$

Example 2.3

A liquid is transferred by means of a pump. The pump motor is sized.

$p_1 = 1$ bar.

$p_2 = 5$ bar.

$\Phi_v = 0.01\ m^3 \cdot s^{-1}$,

$(5 \cdot 10^5 - 10^5)0.01 = 4{,}000$ W.

It is thus relatively simple to calculate the power to pump an incompressible fluid. Theoretically, this equals $\Phi_v \cdot \Delta p$ W. In practice, because of internal friction, double this power will be required, that is, 8 kW. A 10-kW motor should be installed.

Water is an incompressible fluid. Then, the internal energy is a function of the temperature only. Any friction raises the temperature of the fluid.

2.3.2 Air Oxidation of Cumene

Introduction

The air oxidation of cumene was mentioned in Example 1.4 in Section 1.2. The chemical reaction was carried out batchwise. The mass balance for the laboratory reaction on 2-l scale was given there. The air oxidation was also carried out successfully on pilot-plant scale. The reason for carrying out the work of this section was the plan to carry out the reaction industrially in a modified 25-m^3 reactor. Work carried out in a flow calorimeter to measure the reaction heat is dealt with. The latter value was required for the design of the reactor modification. The interpretation of the measurements in terms of the General Energy Equation will be shown in this section.

General

The reaction equation for the oxidation of cumene in the presence of a solution of sodium hydroxide into phenyl iso-propanol (also called cumyl alcohol) is:

$$2C_6H_5C_3H_7 + O_2 \rightarrow 2C_6H_5C_3OH_7.$$

The reaction is carried out at atmospheric pressure and the temperature is 112 °C. The reaction is exothermic. A small amount of cumyl hydroperoxide is added to the reactor contents as a catalyst. First, cumene, sodium hydroxide solution, and the catalyst are added to the reactor. Next, the stirrer is activated. Subsequently, the reactor contents are heated to the reaction temperature and the air flow through the reactor contents is started. The air flow remains constant during the oxidation. The stirrer homogenizes the reactor contents, distributes air, and promotes heat transfer from the reactor contents to the reactor jacket. In the beginning, the composition of the gas flow leaving the reactor is almost equal to the air composition. After a short time, the reaction starts and the oxygen content of the leaving gas flow decreases. After approximately 3 h, the oxygen content of the leaving gas flow reaches a minimum, that is, about 2% by volume. Next, the oxygen content of the leaving air gradually increases. The reaction is stopped when the oxygen content of the leaving air reaches 9.5% by volume after approximately 8 h. The reason is that for a slightly higher oxygen concentration, a gas explosion is possible.

The Flow Calorimeter

See Figure 2.2. The reaction is carried out on 2-l scale. The situation after air has been passed through for 3 h is depicted. The reaction heat in W is maximum at this point in time. The heat balance of the reactor is dealt with as follows. Heat is generated by the chemical reaction. There is also heat input by the stirrer. Heat is absorbed by air entering the reactor and by the reactor jacket. There is also reflux cooling which proceeds as follows. Gas leaving the reactor contains cumene vapor and water vapor. These two components are condensed overhead and returned to the reactor as liquids.

By measuring the cooling water flow through the reactor jacket and its temperature increase, it is possible to calculate the jacket cooling load in W. By measuring the cumene and water flows coming from the reflux condenser, it is possible to calculate the reflux condenser cooling load.

The calculation of the reaction heat proceeds as follows. After air has been passed through for 3 h, the air flow to the reactor is replaced abruptly by a nitrogen flow. The nitrogen flow in $nl \cdot h^{-1}$ equals the previous air flow in $nl \cdot h^{-1}$. On replacing the air flow by a nitrogen flow, the chemical reaction stops. Now, to keep the reactor temperature at 112 °C, the reactor jacket must be heated instead of cooled. The difference in heat flow is measured and this measurement, in combination with the measurements of the reflux flows, enables the calculation of the reaction heat.

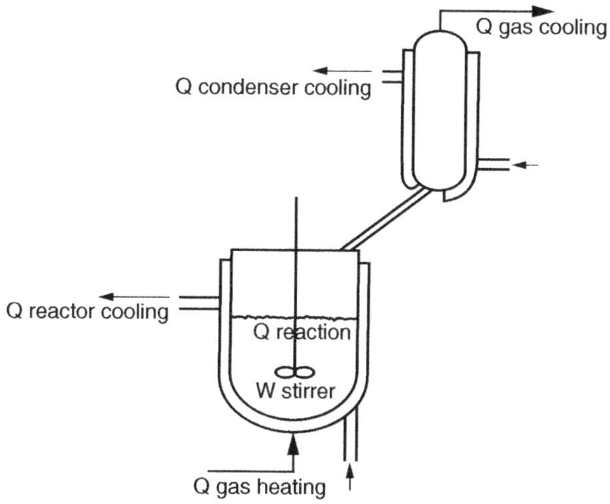

Figure 2.2 Air oxidation of a HC in a reaction calorimeter.

Physical properties

Air

ρ at 0 °C and atmospheric pressure 1.29 $kg \cdot m^{-3}$

$c_p = 1{,}000\, J \cdot kg^{-1} \cdot K^{-1}$

Nitrogen

ρ at 0 °C and atmospheric pressure 1.26 $kg \cdot m^{-3}$

$c_p = 1{,}000\, J \cdot kg^{-1} \cdot K^{-1}$

Water

ΔH of evaporation at 112 °C $2.23 \cdot 10^6\, J \cdot kg^{-1}$

$c_p = 4{,}190\, J \cdot kg^{-1} \cdot K^{-1}$

Liquid cumene

ΔH of evaporation at 112 °C $3.4 \cdot 10^5\, J \cdot kg^{-1}$

$c_p = 1{,}930\, J \cdot kg^{-1} \cdot K^{-1}$

Data of the flow calorimeter experiment
Reaction temperature 112 °C
Reaction pressure: atmospheric
Initial charge 753 ml
Stirrer power input 2.4 $kW \cdot m^{-3}$ initial reaction mixture
Air flow 38 $nl \cdot h^{-1}$ at 20 °C
Nitrogen flow 38 $nl \cdot h^{-1}$ at 20 °C
Both gas flows are dry.
Oxygen absorption rate 7 $nl \cdot h^{-1}$

Per nl of gas leaving the reactor, 0.5 g of water and 3.0 g of cumene are evaporated (stripped), condensed, and returned to the reactor at 50 °C.

On switching from air to nitrogen, the difference between the two jacket heat loads is 38.1 W.

Interpretation of the Flow Calorimeter Data

A first thought could be that 38.1 W represents the reaction heat. However, a correction is needed. The reasoning is as follows. On oxidizing the reactor contents, 38 $nl \cdot h^{-1}$ are fed to the reactor and 31 $nl \cdot h^{-1}$ leave the reactor. On passing nitrogen through, 38 $nl \cdot h^{-1}$ are passed on to the reactor and 38 $nl \cdot h^{-1}$ leave the reactor. The latter gas flow entrains more water and cumene from the reactor than the former gas flow. Moreover, the latter gas flow is larger and absorbs more heat from the reactor contents than the former gas flow. Therefore, the 38.1 W must be corrected. When passing nitrogen through the reactor, more heat is removed by the reflux condenser than when the reactor contents are oxidized and that is compensated by the reactor jacket. Thus, the reaction heat will be smaller than 38.1 W. On calculating the correction, it is assumed that the gas flow leaving the reflux condenser contains neither water nor cumene. That is, it is assumed that all water and cumene leaving the reactor by evaporation are returned to the reactor after condensation in the reflux cooler.

The correction comprises three calculations. First, the extra sensible heat for the nitrogen flow is calculated. Second, the extra heat for the larger amount of entrained water is calculated. Third, the extra heat for the larger amount of entrained cumene is calculated.

$$
\begin{array}{ll}
\text{Nitrogen:} & 7 \cdot 10^{-3} \cdot 1.26 \cdot 1{,}000(112-20)/3{,}600 = 0.2 \\
\text{Water} & :7 \cdot 10^{-3} \cdot 0.5(2.23 \cdot 10^{6} + (112-50)4{,}190)/3{,}600 = 2.4 \\
\text{Cumene:} & 7 \cdot 10^{-3} \cdot 3.0(3.4 \cdot 10^{5} + (112-50)1{,}930)/3{,}600 = \underline{2.7} + \\
 & 5.3
\end{array}
$$

The reaction heat is thus 38.1 − 5.3 = 32.8 W.

Per l of initial reaction mixture: 32.8/0.753 = <u>43.6 W</u>.

Note 2.1

The gas leaving the reactor is cooled to 50 °C. The cooling causes the condensation of water and cumene flowing back into the reactor. However, the gas leaving the calorimeter system still contains some water vapor and some cumene vapor. That has not been taken into account.

Note 2.2

The heat of evaporation of water in $J \cdot kg^{-1}$ is about 6.5 times greater than the heat of evaporation of cumene in $J \cdot kg^{-1}$. The two heats of evaporation are approximately equal if expressed as $J \cdot kmol^{-1}$. The molecular weight of water is 18 $kg \cdot kmol^{-1}$ whereas the molecular weight of cumene is 120 $kg \cdot kmol^{-1}$. The background is that, on evaporation, per kmol the same amounts of bonds between molecules in the liquid phase must be broken.

Scaling-up

The reactor jacket heat load is calculated.

The initial charge of the industrial reactor is 15.8 m^3.

The air flow is $0.038 \cdot 1.58 \cdot 10^4/0.753 = 797\ nm^3 \cdot h^{-1}$, say $800\ nm^3 \cdot h^{-1}$.

Reaction heat when the air oxidation has been carried out for 3 h:

$$43.6 \cdot 10^{-3} \cdot 1.58 \cdot 10^4 = 689\ kW.$$

Heat losses to the surroundings are neglected.

The reference temperature is 0 °C.

The air flow entering the reactor is $(800/3{,}600)1.29 = 0.287\ kg \cdot s^{-1}$. Of this gas flow, $(31/38)(8 \cdot 10^5/3{,}600) = 181.3\ nl \cdot s^{-1}$ leave the reactor.

Heat input, kW	
Reaction	689
Agitator $15.8 \cdot 2.4$	38
Air flow $0.287 \cdot 1{,}000 \cdot 20 \cdot 10^{-3}$	6 +
	733
Heat extraction, kW	
Overhead condenser	
Water $181.3 \cdot 0.5 \cdot 10^{-3}(2.23 \cdot 10^6 + (112 - 50)4{,}190)10^{-3}$	226
Cumene $181.3 \cdot 3.0 \cdot 10^{-3}(3.4 \cdot 10^5 + (112 - 50)1{,}930)10^{-3}$	250
Gas flow $181.3 \cdot 10^{-3} \cdot 1.26 \cdot 1{,}000 \cdot 112 \cdot 10^{-3}$	26
Jacket (balance)	231 +
	733

It should be possible to extract 231 kW from the reactor cooling jacket.

The General Energy Balance is considered. The gas flow through the calorimeter system is taken as the flowing medium. Differences concerning potential energy and kinetic energy are neglected. The following equation is valid:

$\Phi_{m2} \cdot H_2 - \Phi_{m1} \cdot H_1 = Q - W \quad W$. In numbers:

$$26 - 6 = 689 - 226 - 250 - 231 + 38 \quad W.$$

2.4 The Mechanical Energy Equation

The Mechanical Energy Equation is also called Bernoulli's equation.

$$\left(\Delta\left(\frac{p}{\rho}\right) + g \cdot \Delta h + \Delta\left(\frac{v^2}{2}\right)\right)\Phi_m = -W - W_{fr} \quad W. \tag{2.2}$$

Pressure energy, potential energy, and kinetic energy are forms of mechanical energy. The equation expresses the concept that mechanical energy in a flow system is conserved, except for that part converted into other forms of energy as a result of friction, that is, W_{fr}. It is applicable for cases at which there is neither deliberate heat supply nor extraction.

On subtracting Equation 2.2 from 2.1, one obtains:

$$\Phi_m \cdot \Delta U - Q = W_{fr}.$$

This relationship shows that the energy transferred because of internal friction appears either as internal energy or as heat transferred to the surroundings. Frictional energy is transferred as heat to the surroundings when the flow is isothermal. The frictional energy appears as internal energy when heat is not exchanged with the surroundings. A process at which heat is not exchanged with the surroundings is called an adiabatic process.

The following equation is also called Bernoulli's equation:

$$p + \rho g h + \rho \cdot \frac{v^2}{2} = a \quad N \cdot m^{-2}. \tag{2.3}$$

This equation is applicable for an incompressible flowing medium not performing work and not experiencing friction. The expression shows the fact that the sum of the heads related to pressure, elevation, and velocity is constant. A certain type of head can be converted into a different type of head.

2.5 Applications of the Mechanical Energy Balance

The Single-Fluid Nozzle

See Figure 2.3. The liquid leaves the nozzle with a given velocity. The liquid is assumed to be incompressible and the flow is assumed to occur without friction. Bernoulli's law (Equation 2.3) is then applicable:

$$p_1 + \rho g h + \rho \cdot \frac{v_1^2}{2} = p_2 + \rho g h + \rho \cdot \frac{v_2^2}{2}.$$

v_1 is much smaller than v_2 and can be neglected. It follows:

$$v_2 = \sqrt{2 \cdot \frac{\Delta p}{\rho}}.$$

Because of friction losses, the actual velocity is smaller:

$$v_2 = C\sqrt{2 \cdot \frac{\Delta p}{\rho}}.$$

This simple expression indicates that the velocity is proportional to the square root of the pressure difference and inversely proportional to the square root of the specific mass.

C is 0.82 if the flow is turbulent, the hole length is equal to the hole diameter, and the hole is sharp and has a circular cross-section [1].

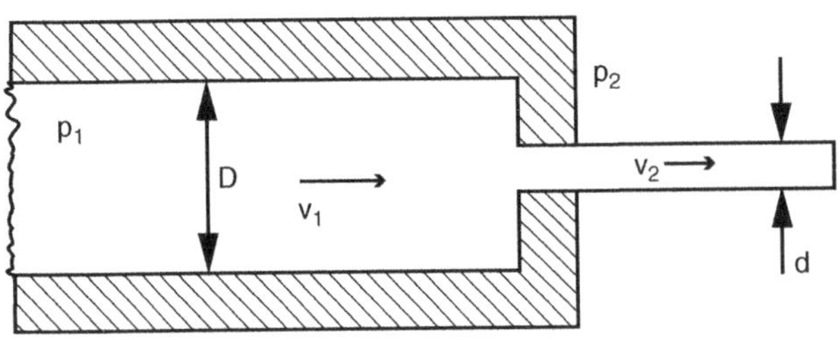

Figure 2.3 The single-fluid nozzle.

Example 2.4

Water flows with a pressure difference of 8 bar through a circular nozzle having a diameter of 3 mm and a length of 3 mm. The hole is sharp. Calculate the flow through the nozzle.

$$\rho = 1{,}000\ \text{kg}\cdot\text{m}^{-3}.$$

$$v_2 = 0.82\sqrt{2\cdot 8\cdot 10^5/1{,}000} = 32.8\ \text{m}\cdot\text{s}^{-1}.$$

The outcome of the calculation is approximately correct if the Reynolds Number is greater than 10^4. Then, the flow through the nozzle is turbulent. Re, the Reynolds Number, equals $\rho vd/\mu$. μ, the dynamic viscosity of water, is 10^{-3} $\text{N}\cdot\text{s}\cdot\text{m}^{-2}$ at 20 °C. $\text{Re} = 1{,}000\cdot 32.8\cdot 0.003/10^{-3} = 9.84\cdot 10^4$. The flow through the nozzle is turbulent.

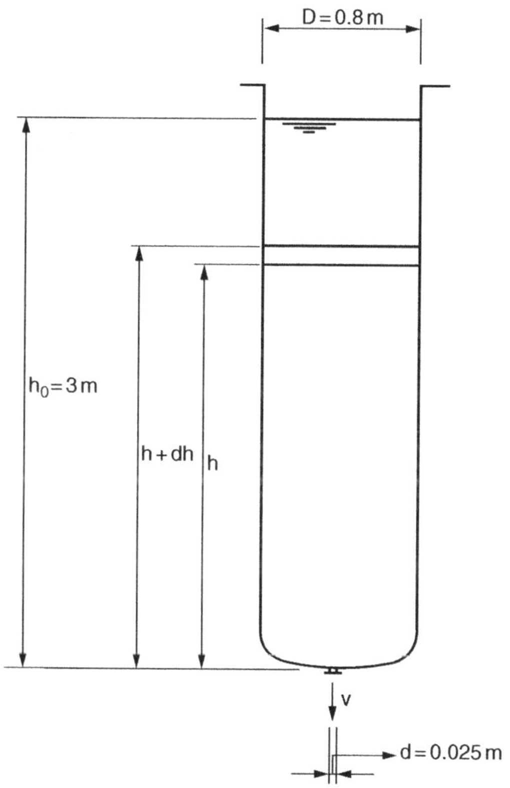

Figure 2.4 Emptying a vessel.

Emptying a Vessel

See Figure 2.4. The static pressure at the bottom of the vessel is ρgh. The velocity of the leaving water is v. v is a function of the height of the liquid and the height of the liquid is a function of time. Elevation (static pressure) head is converted into velocity head. Equation 2.3 gives $(\rho\cdot v^2/2) = \rho gh$ leading to $v = \sqrt{2gh}$. $v = C\sqrt{2gh}$ because of contraction and friction losses. We have here a variable which is a function of time. In such a case, the calculus can help us solve problems. A small time span dt is defined. The liquid level decreases with dh in that small time span.

F_1 and F_2 are the cross-sectional areas of, respectively, the vessel and the hole in the bottom. The mass balance in the small time span dt:

$$-F_1 \cdot dh = F_2 \cdot v \cdot dt.$$

The minus-sign must be used because the height decreases when time elapses.

Next, v is replaced by $C\sqrt{2gh}$. The differential equation can be rearranged as follows:

$$dt = -\frac{F_1}{F_2} \cdot \frac{dh}{C\sqrt{2gh}}.$$

This differential equation can be integrated. The integration proceeds as follows:

$$t = -\frac{F_1}{F_2} \cdot \frac{1}{C\sqrt{2g}} \cdot \int dh/\sqrt{h},$$

$$t = -\frac{F_1}{F_2} \cdot \frac{\sqrt{2}}{C\sqrt{g}} \cdot \sqrt{h} + C_1 \text{ s}.$$

C_1 is an integration constant. Its value can be found by realizing that $h = h_o$ when $t = 0$. It follows:

$$C_1 = \frac{F_1}{F_2} \cdot \frac{\sqrt{2}}{C\sqrt{g}} \cdot \sqrt{h_o}.$$

The time to reach a certain height is then given by:

$$t = \frac{F_1}{F_2} \cdot \frac{\sqrt{2}}{C\sqrt{g}} \left(\sqrt{h_o} - \sqrt{h}\right) \text{ s}.$$

$h = 0$ when the vessel is empty.

The flow coefficient C is assumed to be independent of the water height. $C = 0.62$ is applicable for circular sharp holes having a length smaller than one quarter of the hole diameter [1].

Example 2.5

The time to empty a vessel filled with water is calculated. The characteristics of the vessel are:

$F_1 = 0.5\ m^2$.

$F_2 = 5 \cdot 10^{-4}\ m^2$.

$h_o = 3$ m.

$$t = \frac{F_1}{F_2} \cdot \frac{\sqrt{2h_o}}{C\sqrt{g}} = \frac{0.5}{5 \cdot 10^{-4}} \cdot \frac{\sqrt{6}}{0.62\sqrt{9.81}} = 1{,}261 \text{ s} = 21 \text{ min}.$$

Flow Through a Weir

See Figure 2.5. The design is used to measure a flow. It is possible to arrive at the flow by measuring the height h_o. The derivation proceeds as follows. The liquid velocity is a function of the height. At

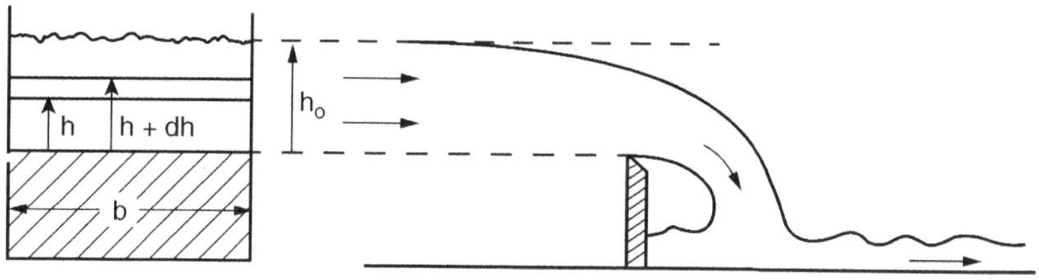

Figure 2.5 Flow through a weir. *Source:* Courtesy of Delft Academic Press/VSSD, Delft, The Netherlands.

height h_o, the velocity may be considered negligibly low. The velocity of the streamline coming from height h is v_h. Application of Bernoulli's law leads to the expression:

$$\rho gh + \frac{1}{2}\rho \cdot v_h^2 = \rho g \cdot h_o \rightarrow v_h = \sqrt{2g(h_o - h)}.$$

The calculus can also assist us in this case. Unlike at the previous example, the situation is steady-state, that is, it does not change with time. However, the liquid velocity is a function of the height. A small flow $d\Phi_v$ is a function of the local velocity.

$$d\Phi_v = v_h \cdot b \cdot dh,$$

$$d\Phi_v = b\sqrt{2g(h_o - h)} \cdot dh.$$

This differential equation can be integrated. The integration proceeds as follows:

$$\Phi_v = b\sqrt{2g} \int_{h=0}^{h=h_o} \sqrt{h_o - h} \cdot dh,$$

$$\Phi_v = -b\sqrt{2g} \int_{h=0}^{h=h_o} \sqrt{h_o - h} \cdot d(h_o - h),$$

$$\Phi_v = -b\sqrt{2g} \int_{h=0}^{h=h_o} d(h_o - h)^{3/2},$$

$$\Phi_v = \frac{2}{3} \cdot b\sqrt{2g \cdot h_o^3}.$$

In practice, there is contraction and there are friction losses. The practical formula becomes for $b \gg h_o$ and $h_o > 0.01$ m:

$$\Phi_v = 0.62 \cdot \frac{2}{3} \cdot b\sqrt{2g \cdot h_o^3} = 0.58 \cdot b\sqrt{g \cdot h_o^3} \text{ [2]}.$$

Example 2.6

The flow over a weir is calculated. The data concerning the weir and the flow are:

$h_o = 0.1$ m.

$b = 2$ m.

$\Phi_v = 0.58 \cdot 2\sqrt{9.81 \cdot 0.1^3} = 0.116\ m^3 \cdot s^{-1} = 417.6\ m^3 \cdot h^{-1}$.

References

1. Beek, W. J., Muttzall, K. M. K., and van Heuven, J. W. (1999). *Transport Phenomena* (p. 89). Chichester, UK: Wiley.
2. Beek, W. J., Muttzall, K. M. K., and van Heuven, J. W. (1999). *Transport Phenomena* (p. 95). Chichester, UK: Wiley.

3
Viscosity

3.1 Definition

See Figure 3.1. A parallel-sided slab of fluid is considered which is sandwiched between a stationary base and a moving plate with area A. A force F is applied to the plate to cause it to move with a constant velocity. The resistance offered by the liquid then equals this force F, but acts in the opposite direction. Provided the flow is laminar, the following equation is valid for many fluids:

$$\frac{F}{A} = \tau = -\mu \cdot \frac{dv_y}{dx}. \tag{3.1}$$

τ, the shear stress in $N \cdot m^{-2}$, is proportional to the velocity gradient. The velocity gradient is also called the shear rate. μ, the proportionality constant, is called the dynamic viscosity in $N \cdot s \cdot m^{-2}$. The equation is called Newton's law of friction (1723). Fluids behaving in this fashion are termed Newtonian fluids. All gases and most simple liquids obey Equation 3.1. Primarily pastes, slurries, and high polymers do not follow this simple law. They are called non-Newtonian fluids and are discussed in Section 3.3. The minus sign is used in the equation because the velocity decreases in the positive x-direction.

It is also possible to express the dynamic viscosity in cP. A viscosity of 1 cP equals a viscosity of $10^{-3}\ N \cdot s \cdot m^{-2}$.

The shear stress τ appears in Equation 3.1. τ is sometimes called the momentum flux. The background is as follows. The fluid acquires a certain amount of momentum in the y-direction in the neighborhood of the moving surface. Fluid having this momentum, imparts some of it to an adjacent fluid layer. This transmission is repeated in the x-direction till the stationary layer next to the base is reached. Thus, τ may also be interpreted as the viscous flux of momentum. It is possible to consider the velocity gradient as the driving force for momentum transport. There is an analogy with the flow of heat from a hot region to a colder one. The temperature difference is the driving force in that case.

The kinematic viscosity ν is obtained by dividing the dynamic viscosity by the specific mass ρ in $kg \cdot m^{-3}$. The unit is $m^2 \cdot s^{-1}$.

Introduction to Chemical Engineering, First Edition. C.M. van 't Land.

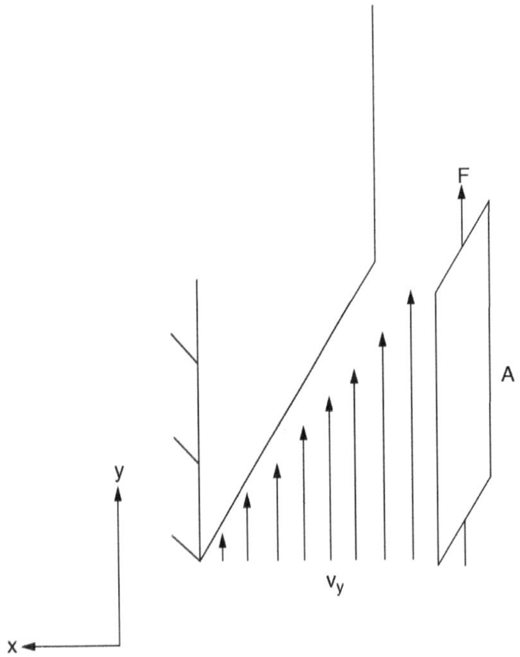

Figure 3.1 Moving plate.

Example 3.1

In Figure 3.1, a plate moves with a velocity of 0.3 $m \cdot s^{-1}$ in the positive y-direction. There is water between the moving plate and the stationary base. The water viscosity is $0.7 \cdot 10^{-3}$ $N \cdot s \cdot m^{-2}$. The distance between the stationary base and the plate is $3 \cdot 10^{-4}$ m. The shear stress experienced by the moving plate is:

$$\tau = 0.7 \cdot 10^{-3} \cdot \frac{0.3}{3 \cdot 10^{-4}} = 0.7\,\mathrm{N} \cdot \mathrm{m}^{-2}.$$

The dynamic water and air viscosities at room temperature are, respectively, about 10^{-3} and $20 \cdot 10^{-6}\,\mathrm{N} \cdot \mathrm{s} \cdot \mathrm{m}^{-2}$.

The viscosities of, for example, water and sugar syrup at ambient temperature differ by a factor of $5 \cdot 10^4$. There is a great variation in liquid viscosities.

The flow of gases and many liquids can be described by means of Equation 3.1. The elastic deformation of solid bodies, for example, a steel bar, can be described by means of Hooke's law. That law states that the relative elongation of a solid body is proportional to the applied tensile stress. The solid body regains its original form on relieving the stress if the deformation has been elastic. That is not the case if the solid body has been deformed plastically.

3.2 Newtonian Fluids

See Figure 3.2. The behavior of a Newtonian fluid can be represented by Line 1. For such fluids, the dynamic and the kinematic viscosities are not dependent on the shear stress but dependent on the temperature only. There is no time-dependency either.

3.3 Non-Newtonian Fluids

3.3.1 The Viscosity is a Function of the Temperature and the Shear Rate

In this section, the flowing medium is no longer indicated as a fluid, but as a liquid. First, shear-thinning (pseudoplastic) liquids are discussed. They form the largest group of non-Newtonian liquids. Their behavior is characterized by Line 2 in Figure 3.2. The viscosity decreases on increasing the shear rate. Figure 3.3 shows various possibilities for slurries and pastes to behave in this way.

If a straight line is obtained on plotting $^{10}\log \tau$ as a function of $^{10}\log(dv_y/dx)$ with a slope $m < 1$, the liquid is termed an Ostwald-De Waele liquid. The relevant equation is:

$$\tau = -K\left(\frac{dv_y}{dx}\right)^m. \tag{3.2}$$

It follows: $\mu_{eff} = K(dv_y/dx)^{m-1}$.

K is the consistency index, m is the flow index, and μ_{eff} is the effective viscosity. The liquid obeys a "power law."

Blood is a liquid exhibiting shear-thinning behavior. On flowing through a capillary in, for example, the brain, the red blood corpuscles undergo deformation as indicated in Figure 3.3.

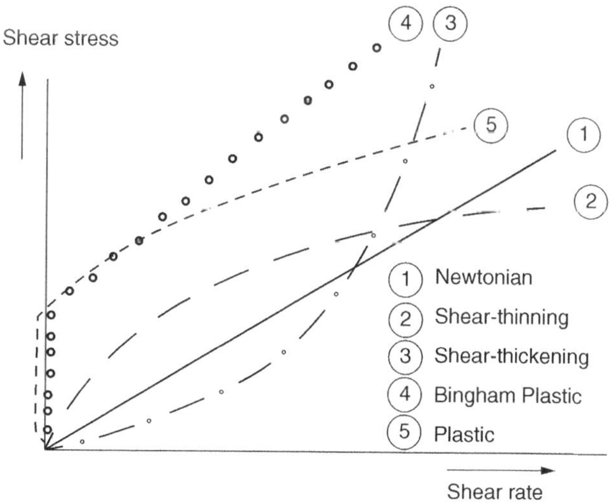

Figure 3.2 Shear stress as a function of shear rate. *Source:* Courtesy of Wolters Kluwer, Alphen aan den Rijn, The Netherlands.

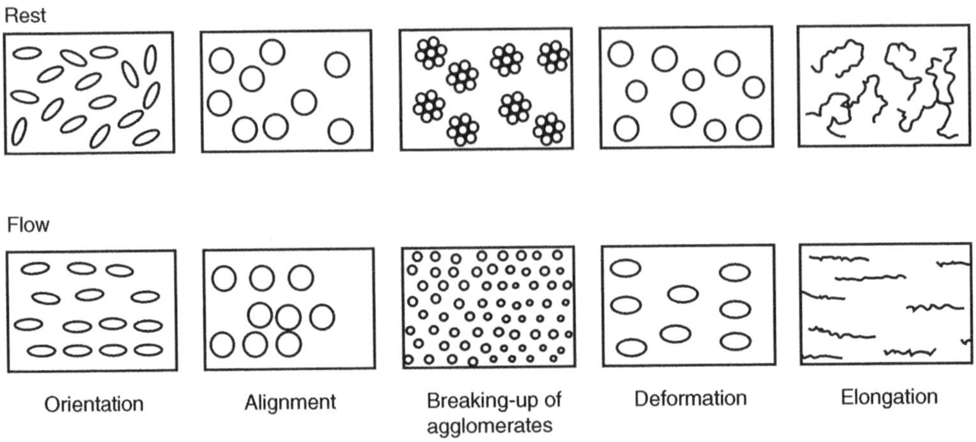

Figure 3.3 Shear-thinning. *Source:* Courtesy of Wolters Kluwer, Alphen aan den Rijn, The Netherlands.

That causes a decrease of the viscosity. Pressure to pass blood through the brain is generated by the heart. With the latter pressure as driving force, the flow of blood increases when the viscosity of blood decreases. That is a characteristic of laminar flow as will be discussed in the next chapter.

Aqueous solutions of carboxymethyl cellulose also exhibit shear-thinning behavior.

Second, shear-thickening (dilatant) liquids are discussed. The viscosity increases on increasing the shear rate. Their behavior is characterized by Line 3 in Figure 3.2. The phenomenon is found for concentrated dispersions and suspensions. Figure 3.4 shows the reason for this behavior. The number of collisions between the particles per unit of time rises when the shear rate increases, the particles push each other away. Next, there will be a shortage of continuous phase to fill the gaps between the particles completely. This phenomenon can be noticed when you make a walk on the beach near the tide-line. The sand around your feet "dries" out. The seawater sinks into the sand.

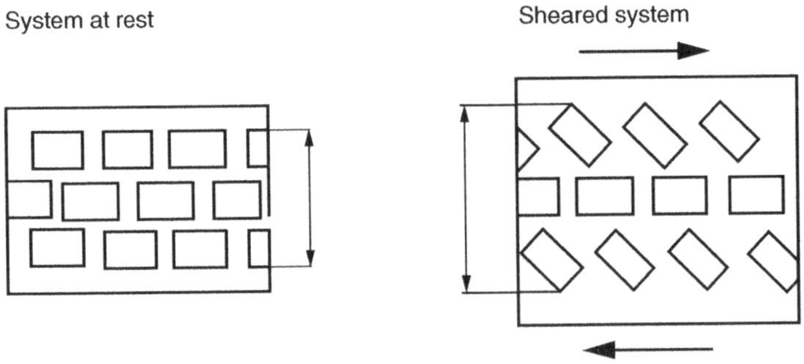

Figure 3.4 Shear-thickening. *Source:* Courtesy of Wolters Kluwer, Alphen aan den Rijn, The Netherlands.

Example 3.2

A slurry is prepared containing 1 part by weight of water and 1–2 parts by weight of corn starch. On moving a finger slowly through this suspension, little resistance is met. However, on increasing the speed with which the finger is moved through the suspension, the resistance increases. At low speeds, there is little resistance and vice versa.

d3o is a second example of a material exhibiting shear-thickening behavior. It is a pasty-like material which is used for sportswear. When the deformations are relatively slow, the material gives in without much resistance. When the rate of deformation is high, the material stiffens. Thus, sportsmen and -women are protected against injuries on falling. See Figure 3.5.

Silly Putty is a further example of a pasty-like material showing this behavior. A ball of Silly Putty bounces like a tennis ball. Putting the ball on the table causes it to flow out into the form of a small pancake. Tensile stress causes a bar to break as if it were a solid material. Kneading two differently colored balls produces a larger ball having a third color, showing that is possible to mix them as liquids.

Third, liquids are discussed that require a yield value for the shear stress in order to make the material flow. When the shear stress is lower than a certain value, the material behaves like an elastic solid material obeying Hooke's law. When the shear stress exceeds this value, the material is a liquid. The behavior of these materials is characterized by Lines 4 and 5 in Figure 3.2. Line 4 is representative for Bingham Plastics. When the material starts to flow, the behavior is Newtonian. The relevant equation is:

$$\tau - \tau_0 = -\mu \cdot \frac{dv_y}{dx}. \tag{3.3}$$

Line 5 characterizes Plastics. Beyond the yield shear stress, the material is shear-thinning. See Figure 3.6 showing that agglomerates can break up causing material to flow. Examples of materials belonging to the third category are pastes, paints that can easily be applied, and yet do not form "tears," and ice-cream. Further examples are landslides after heavy rainfall and avalanches in the mountains. In The Netherlands, a dike collapsed after a dry period at Wilnis in 2003. This incident will be discussed after the next paragraph.

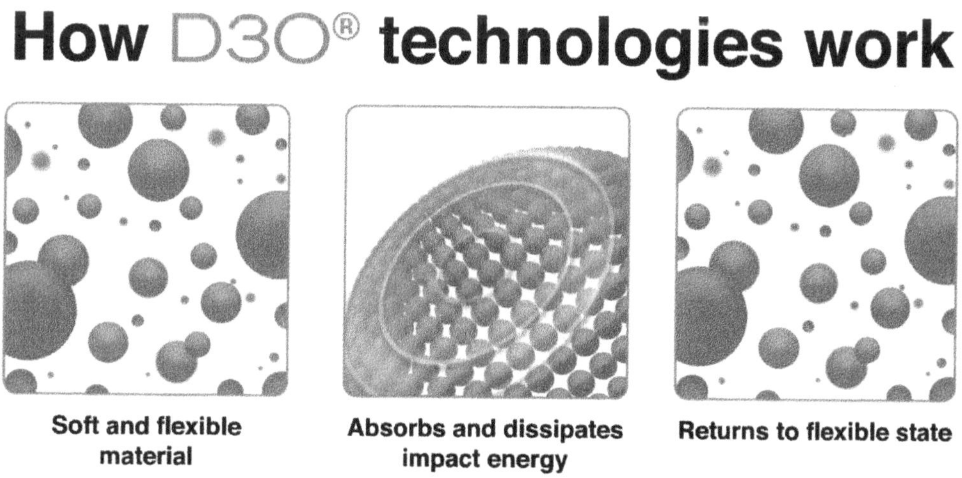

Figure 3.5 An application of a viscous shear-thickening liquid. *Source:* Courtesy of D3O, The Lab, Croydon, UK.

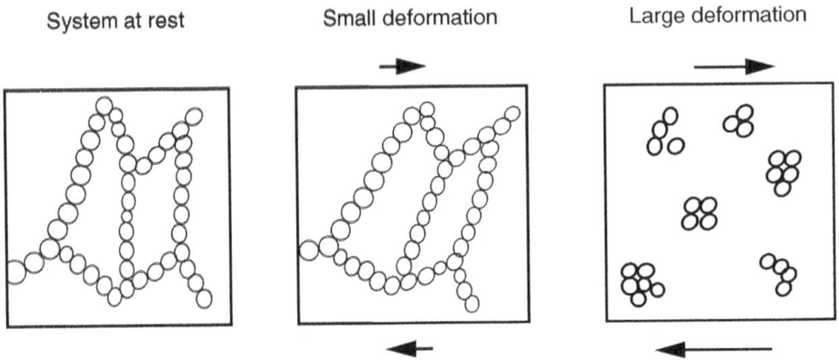

Figure 3.6 Bingham Plastic or Plastic. *Source:* Courtesy of Wolters Kluwer, Alphen aan den Rijn, The Netherlands.

In this paragraph, the viscosity of mixtures of water and a wettable powder will be discussed. The powder is not soluble in water and the weight average particle size is, for example, 30 μm. On adding a small amount of powder to water, for example, 5% by weight, the suspension can be stirred easily. The viscosity of the suspension increases when more powder is added to the suspension. On continuing the addition of powder, a point is reached where the viscosity suddenly rises substantially. The suspension has turned into a paste. On adding more powder, the paste breaks up and it again becomes relatively easy to stir the mixture. The point where the viscosity rises substantially is the point where water is just still the continuous phase. However, on adding more powder, water no longer is the continuous phase. In contact dryers, sometimes pastes are stirred. Not considering the possibility of the described phenomenon has led to breaking of shafts in these dryers.

It is now possible to consider the collapse of the Dutch dike in 2003 in some detail. When the dike contains a normal amount of water, it can be compared to a paste of the former paragraph. However, if the dike dries out, it loses strength and that is what happened in 2003. Water in the canal the dike was supposed to contain, pushed the dike aside. A dike also loses strength when it becomes wet after heavy rainfall. But that is not what happened at Wilnis in 2003.

3.3.2 The Viscosity is a Function of Time

Thixotropy is the phenomenon that the viscosity, on maintaining a shear rate, decreases as a function of time. See Figure 3.7. Gradually, crosslinking is undone.

Rheopecticity is the phenomenon that the viscosity, on maintaining a shear rate, increases as a function of time. Some lattices exhibit this behavior.

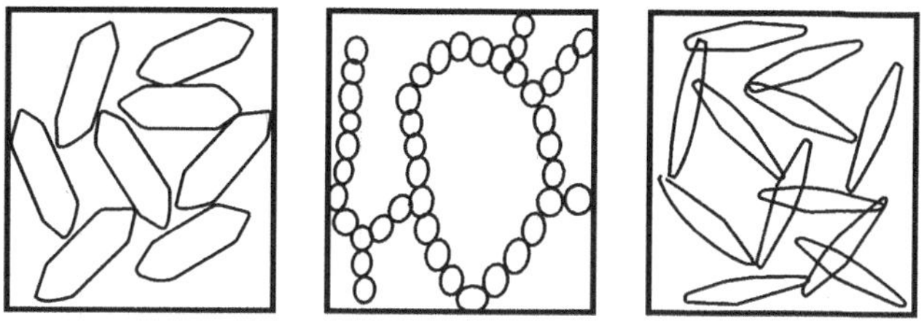

Figure 3.7 Cross-linked structures cause thixotropy. *Source:* Courtesy of Wolters Kluwer, Alphen aan den Rijn, The Netherlands.

3.4 Viscoelasticity

Elastic behavior is characterized by Hooke's law. Viscous behavior is characterized by Newton's law. Practically, all materials exhibit both elastic and viscous behavior. First, we will take a look at liquids and reference is made to Figure 3.1. The plate with area A moves with a constant velocity. Suddenly, the plate is stopped. The shear stress will not disappear instantaneously, but gradually. The shear stress has decreased to 1/e of its original value in the relaxation time. Flat stones will bounce if thrown at high velocity onto the surface of water in a pond. Water cannot yield immediately if the shear rate is high. Second, we will take a look at glaciers. They consist of a solid material; however, they flow when the shear stress is high.

3.5 Viscosity of Newtonian Fluids

3.5.1 Gases

The following expression is valid according to kinetic theory: $\mu \propto \rho \cdot v_m \cdot l$. The meaning of the symbols is as follows:

ρ: the specific mass of the gas in $kg \cdot m^{-3}$

v_m: the average molecular velocity in $m \cdot s^{-1}$

l: the mean free path length of the molecules in m

The average free path length of the molecules is inversely proportional to the specific mass: $l \propto 1/\rho$.

It follows that the viscosity is independent from the pressure as the effects of pressure on ρ and l cancel each other out. This prediction agrees well with experimental data up to about 10 at [1]. According to kinetic theory, the average molecular velocity is proportional to the root of the absolute temperature: $v_m \propto \sqrt{T}$. This predicted temperature dependency is less satisfactory than the predicted pressure dependency. Experimental data indicate that the exponent of T is 0.65–0.8 rather than 0.5.

See Figure 3.8. The viscosities of some well-known gases are, up to approximately 10 at, in the range $10^{-5} - 3 \cdot 10^{-5}\ N \cdot s \cdot m^{-2}$.

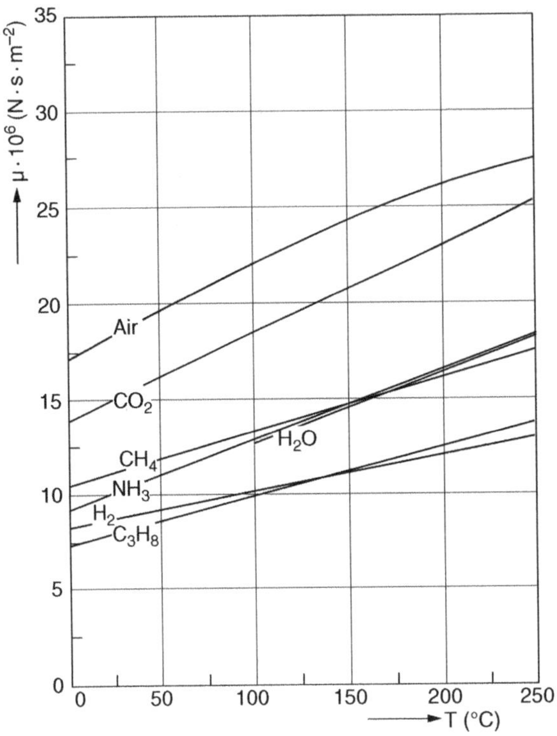

Figure 3.8 Dynamic viscosity of a few gases as a function of temperature at atmospheric pressure. *Source:* Courtesy of Delft Academic Press/VSSD, Delft, The Netherlands.

3.5.2 Liquids

Liquid viscosities vary more than gas viscosities. See Figure 3.9. Liquid viscosities decrease when the temperature increases and are, like gas viscosities, hardly dependent on the pressure. Andrade formulated a simple relationship between liquid viscosity and the absolute temperature:

$$\ln \mu_l = A + \frac{B}{T}. \tag{3.4}$$

A and B are constants. On plotting log μ as a function of 1/T, often straight lines are obtained. The Orrick and Erbar method for the estimation of low-temperature liquid viscosity is mentioned [2]. This method employs a group contribution method to estimate A and B. These constants are different from the constants A and B to be used in Equation 3.4. Their expression is:

$$\ln \frac{\mu_l}{\rho_l \cdot M} = A + \frac{B}{T}. \tag{3.5}$$

The meaning of the symbols is as follows:
μ_l: viscosity in cP
ρ_l: specific mass at 20 °C in $g \cdot cm^{-3}$
M: molecular weight in $g \cdot mol^{-1}$
T: absolute temperature in K

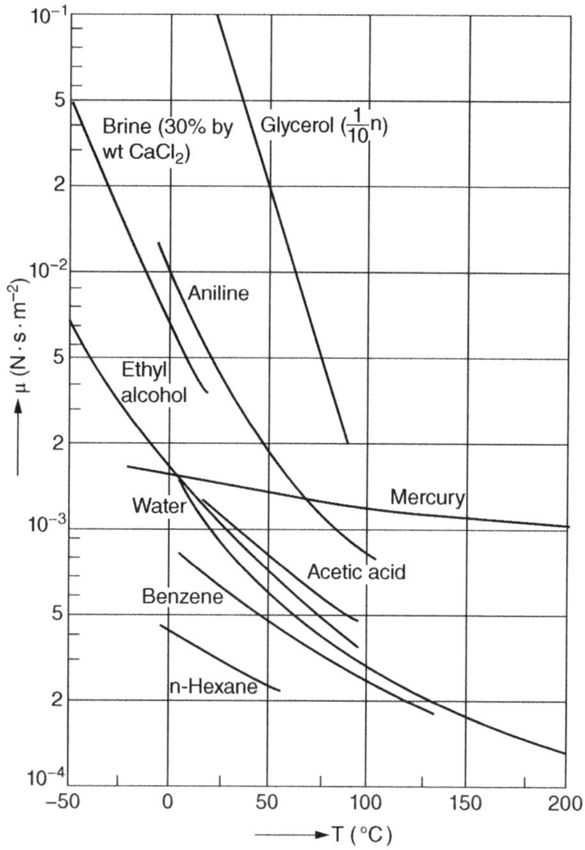

Figure 3.9 Dynamic viscosity of a few liquids as a function of temperature. *Source:* Courtesy of Delft Academic Press/VSSD, Delft, The Netherlands.

Example 3.3

Estimate the viscosity of liquid 1-butanol at 120 °C. The liquid is under pressure as the boiling point is 117.7 °C at atmospheric pressure. The experimental value is 0.394 cP.

$A = -10.79.$

$B = 2{,}271.$

ρ_l at 20 °C : $0.8096\ \mathrm{g\cdot cm^{-3}}.$

$M = 74.123\ \mathrm{g\cdot mol^{-1}}.$

$$\ln\frac{\mu_l}{0.8096\cdot 74.123} = -10.79 + \frac{2{,}271}{393},$$

$\mu_l = 0.400$ cP at 120 °C.

Remark
Figures 3.2, 3.3, 3.4, 3.6, and 3.7 appeared originally in [3].

References

1. Bird, R. B., Stewart, E. S., and Lightfoot, E. N. (1960). *Transport Phenomena* (p. 21). New York, NY: Wiley.
2. Poling, B. E., Prausnitz, J. M., and O'Connell, J. P. (2001). *The Properties of Liquids and Gases* (pp. 9.59–9.61). New York, NY: McGraw-Hill.
3. Blom, C., Jongschaap, R. J. J., and Mellema, J. (1988). *Introduction to Rheology*. Deventer, The Netherlands: Kluwer Technische Boeken B.V. (in Dutch).

4

Laminar Flow

4.1 Steady-state Flow Through a Circular Tube

Laminar flow of a Newtonian fluid through a circular tube is caused by a pressure gradient. For laminar flow, all velocity vectors are in the x-direction. There is no exchange of mass between the moving layers. Laminar flow is different from turbulent flow. In Chapter 5, a criterion to distinguish between laminar flow and turbulent flow will be discussed. It is desired to arrive at the velocity distribution in the tube for a Newtonian fluid. See Figure 4.1. The approach is to write down the balance of forces of a control volume moving along with the flow.

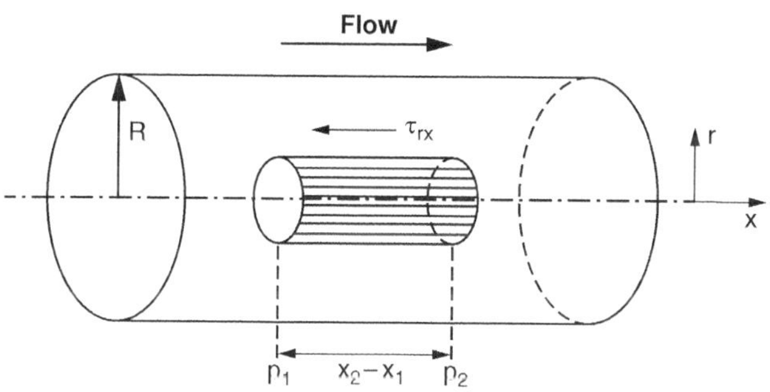

Figure 4.1 Flow through a circular tube. *Source:* Transport Phenomena by Beek, W. J. et al. (1999). Reproduced with permission of John Wiley & Sons Inc.

Pressure forces act on the cross-sections of the control volume and shear forces act at the cylindrical surface. The force balance is:

$$\pi \cdot r^2(p_1 - p_2) = \tau_{rx} \cdot 2\pi r(x_2 - x_1).$$

The notation τ_{rx} means that the variable τ is a function of r. However, it is not a function of x. Being a vector, it is in the x-direction.

It follows:

$$\tau_{rx} = \frac{r}{2} \cdot \frac{p_1 - p_2}{x_2 - x_1} = \frac{r}{2} \cdot -\frac{dp}{dx}.$$

Introduction to Chemical Engineering, First Edition. C.M. van 't Land.

The shear stress has a maximum value at the wall and is zero in the center. See Figure 4.2. The equation of Newton's law of friction is:

$$\tau_{rx} = -\mu \cdot \frac{dv_{rx}}{dr}.$$

The remark made for τ_{rx} also applies for v_{rx}.

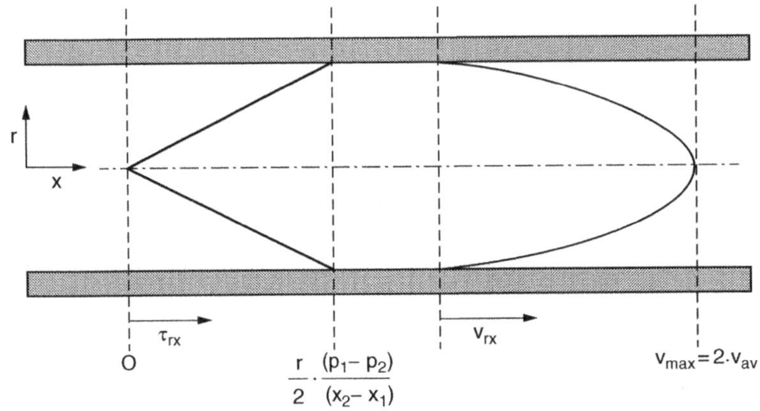

Figure 4.2 Shear distribution and velocity distribution for laminar flow in a circular tube. *Source:* Transport Phenomena by Beek, W. J. et al. (1999). Reproduced with permission of John Wiley & Sons Inc.

Combining the latter equation with the former one leads to:

$$-\mu \cdot \frac{dv_{rx}}{dr} = \frac{r}{2} \cdot -\frac{dp}{dx}.$$

This is the equation of motion which can be integrated. The pressure gradient is a constant. The integration proceeds as follows:

$$dv_{rx} = -\left(-\frac{dp}{dx}\right)\frac{1}{2\mu} \cdot r \cdot dr,$$

$$v_{rx} = -\left(-\frac{dp}{dx}\right)\frac{1}{2\mu}\int r \cdot dr,$$

$$v_{rx} = -\left(-\frac{dp}{dx}\right)\frac{1}{4\mu} \cdot r^2 + C.$$

C is an integration constant. Its value can be found by realizing that $v_{rx} = 0$ for $r = R$:

$$0 = -\left(-\frac{dp}{dx}\right)\frac{1}{4\mu} \cdot R^2 + C \rightarrow C = \left(-\frac{dp}{dx}\right)\frac{1}{4\mu} \cdot R^2.$$

And, finally, the following equation is obtained:

$$v_{rx} = \frac{1}{4\mu} \cdot -\frac{dp}{dx}\left(R^2 - r^2\right). \tag{4.1}$$

Thus, the velocity distribution is parabolic, with maximum flow velocity at $r = 0$:

$$v_{max} = \frac{R^2}{4\mu} \cdot -\frac{dp}{dx}. \tag{4.2}$$

Formally, we can state (see Figure 4.3):

$$d\Phi_v = 2\pi r \cdot v_{rx} \cdot dr.$$

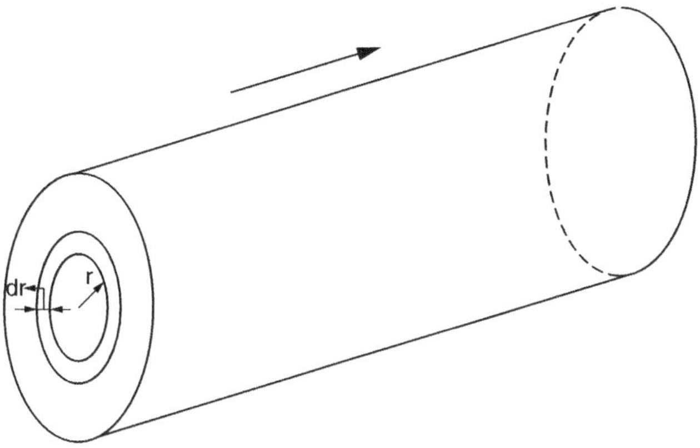

Figure 4.3 Laminar flow through a tube.

This is a differential equation that can be integrated. First, v_{rx} is substituted by the right-hand side of Equation 4.1:

$$d\Phi_v = 2\pi r \cdot \frac{1}{4\mu}\left(-\frac{dp}{dx}\right)\left(R^2 - r^2\right)dr.$$

The integration then proceeds as follows:

$$\Phi_v = \frac{2\pi}{4\mu}\left(-\frac{dp}{dx}\right) \cdot \int_{r=0}^{r=R} r\left(R^2 - r^2\right)dr,$$

$$\Phi_v = \frac{\pi}{2\mu}\left(-\frac{dp}{dx}\right)\left(\frac{R^4}{2} - \frac{R^4}{4}\right),$$

$$\Phi_v = \frac{\pi \cdot R^4}{8\mu}\left(-\frac{dp}{dx}\right). \qquad (4.3)$$

A further formal statement is:

$$\Phi_v = \pi \cdot R^2 \cdot v_{av}.$$

It follows:

$$v_{av} = \frac{R^2}{8\mu} \cdot \left(-\frac{dp}{dx}\right). \qquad (4.4)$$

It follows from Equations 4.2 and 4.4:

$$v_{max} = 2v_{av}.$$

Equation 4.3 is the Hagen–Poiseuille equation which is used at various methods for the measurement of viscosities of Newtonian fluids. Two simple viscosimeters for liquids, based on this equation, are shown in Figure 4.4. A further method for liquids, based on a different treatment, is discussed in Section 4.2. The viscosity of gases can be calculated by passing a known gas flow through a long capillary tube and measuring the pressure loss. The viscosity can then be obtained from Equation 4.3.

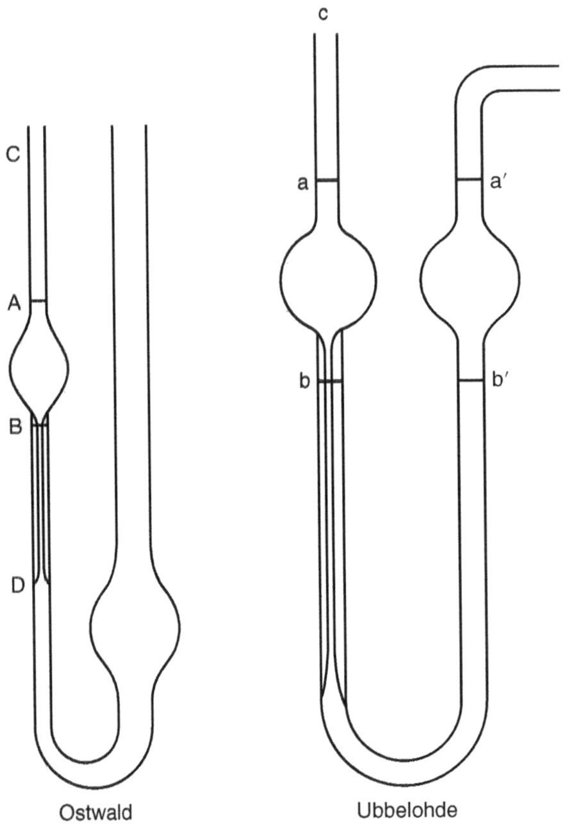

Figure 4.4 Two simple viscosimeters.

Example 4.1

The viscosity of an organic liquid is measured with an Ostwald viscosimeter. The viscosimeter is thermostatted. The capillary diameter is 1 mm. The initial upper liquid level is A and the time to reach B is measured. The volume between A and B is 10 ml.

$\rho = 850\ \text{kg} \cdot \text{m}^{-3}$.

$h = 0.3\ \text{m}$.

$R = 0.5 \cdot 10^{-3}\ \text{m}$.

10 ml flows through the capillary in 127.4 s.

The driving force for the flow is the pressure difference between B and D. The latter is the elevation head (static head) and it equals ρgh.

$$\mu = \frac{R^2}{8 \cdot v_{av}} \cdot -\frac{dp}{dx} = \frac{0.0005^2}{8 \cdot \dfrac{10.10^{-6}}{\frac{\pi}{4}(0.001)^2 \cdot 127.4}} \cdot \frac{850 \cdot 9.81 \cdot 0.3}{0.3} = 2.6 \cdot 10^{-3}\ \text{N} \cdot \text{s} \cdot \text{m}^{-2}.$$

The dynamic viscosity can also be expressed as 2.6 cP.

Usually, the viscosity thus found must be corrected. The correction factor is, for example, 0.9. Thus, the viscosity becomes $0.9 \cdot 2.6 = 2.3$ cP.

Note that in the equation, the specific mass of the liquid and the time are the only variable factors. Thus, it is possible to state: $\mu = C\rho t$. C is for a given Ostwald viscosimeter a known constant. It can be determined by testing a reference liquid of which the viscosity is known. Then, it is sufficient to measure the time for and the specific mass of a test liquid.

The second viscosimeter shown in Figure 4.4 is the Ubbelohde viscosimeter. The instrument functions as follows. a and b′ are the initial liquid levels. A test is started by applying air pressure at c. The time is measured to reach the levels a′ and b. The bulbs have the same size and are at the same elevation. In this instrument, the air pressure is the driving force for passing the liquid through the capillary. So, it is no longer necessary to know the specific mass of the liquids tested. One might realize that, initially, the vertical distance between a and b′ adds to the driving force. That is true, however, at the end of the liquid transfer, the opposite effect, that is, the vertical distance between a′ and b is active. These two effects cancel each other out. Moreover, as this method is used for viscous liquids, hydrostatic heads are relatively small effects. If a liquid having a known viscosity μ_1 can be transferred in time t_1 and the time to transfer the test liquid having a viscosity μ_2 is t_2, the following equation applies:

$$\mu_2 = \frac{t_2}{t_1} \cdot \mu_1.$$

The two liquids must be transferred by means of the same air pressure.

Example 4.2

The viscosity of air is determined. Air flows through a capillary having an internal diameter of 0.5 mm and a length of 10 m. The temperature is 20 °C. The measured flow is 10^{-8} $m^3 \cdot s^{-1}$. The measured pressure loss is 1,100 $N \cdot s \cdot m^{-2}$.

$$v_{av} = \frac{10^{-8}}{\frac{\pi}{4}(5 \cdot 10^{-4})^2} = 0.0509 \text{ m} \cdot \text{s}^{-1},$$

$$\mu = \frac{R^2}{8 \cdot v_{av}} \cdot -\frac{dp}{dx} = \frac{(5 \cdot 10^{-4})^2}{4 \cdot 8 \cdot 0.0509} \cdot \frac{1{,}100}{10} = 1.688 \cdot 10^{-5} \text{ N} \cdot \text{s} \cdot \text{m}^{-2}.$$

One might realize that the volumetric flow changes when the pressure changes. And the pressure changes, as we measure a pressure loss. Thus, the indicated flow is an approximate flow and the measured viscosity is not entirely accurate. That is true, however, for the present discussion the accuracy is adequate.

4.2 Rotational Viscosimeters

Rotational viscosimeters are applied for both Newtonian and non-Newtonian liquids. The laminar flow is caused by a surface moving through a liquid. For example, a liquid is confined in an annulus and the inner cylinder rotates whereas the outer cylinder is stationary. These viscosimeters are

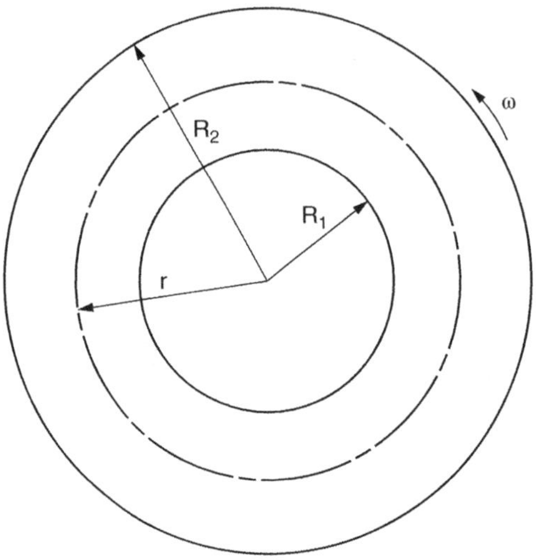

Figure 4.5 Coaxial cylinder geometry.

indicated to be of the Searle type. It is also possible to do it the other way round and those meters are of the Couette type. The Searle type is used more often because it is relatively easy to thermostat the instrument. Also in this case, it is possible to calculate the velocity distribution in the annulus. See Figure 4.5. In that figure, the outer cylinder rotates and the inner cylinder is stationary. The calculations proceed as follows. The momentum on the inner cylinder having a length L is:

$$T = 2\pi \cdot R_1 \cdot L \cdot \tau_1 \cdot R_1 = 2\pi \cdot R_1^2 \cdot L \cdot \tau_1.$$

Note: area times stress equals force and force times arm is momentum. τ_2 is the shear stress on the outer cylinder. The shear stress at radius r is:

$$\tau = \frac{T}{2\pi \cdot r^2 \cdot L}. \tag{4.5}$$

The momentum T is constant as it is passed on from the outer cylinder to the inner cylinder. The liquid is assumed to be Newtonian. The expression for Newton's law of friction is:

$$\tau = \mu r \cdot \frac{d\omega}{dr}. \tag{4.6}$$

Combining Equations 4.5 and 4.6 results in:

$$d\omega = \frac{T}{2\pi\mu L} \cdot \frac{dr}{r^3}.$$

The integration of this differential equation proceeds as follows:

$$\Omega = -\frac{T}{4\pi\mu L} \int_{r=R_1}^{r=R_2} dr^{-2}.$$

Ω is the angular velocity of the outer cylinder.

$$\Omega = \frac{T}{4\pi\mu L}\left(\frac{1}{R_1^2} - \frac{1}{R_2^2}\right).$$

This equation can be written explicit in μ:

$$\mu = \frac{T}{4\pi\Omega L}\left(\frac{1}{R_1^2} - \frac{1}{R_2^2}\right). \tag{4.7}$$

It follows from Equation 4.7 that the dynamic viscosity is directly proportional to the measured momentum. There is a proportionality constant for each rotational speed. Note that it is not necessary to know the specific mass of the liquid.

The rotational viscosimeters can also be used to establish, for example, thixotropy.

4.3 Additional Remarks

Two relatively simple cases concerning the velocity distribution for steady-state laminar flow have been treated. The four steps to arrive at a velocity distribution are:

- writing down the balance of forces or the momentum balance,
- stating Newton's law of friction,
- substituting the balance of forces or the momentum balance into Newton's law of friction, and
- integrating the resulting expression.

5

Turbulent Flow

5.1 Velocity Distribution

The velocity distribution for a fluid flowing through a tube is parabolic for laminar flow. Reynolds injected a dye stream into water flowing through a tube. At low velocities, the dye stream formed a smooth, straight streak. The dye completely dispersed at high velocities due to radial mixing. Such flow is called turbulent. See Figure 5.1. The velocity distribution for turbulent flow is blunter than for laminar flow. See Figure 5.2.

For tubular reactors, turbulent flow is preferred over laminar flow as the residence time is almost the same for all molecules. Still, the flow near the wall is laminar and that fact has led to the concept of the laminar boundary layer.

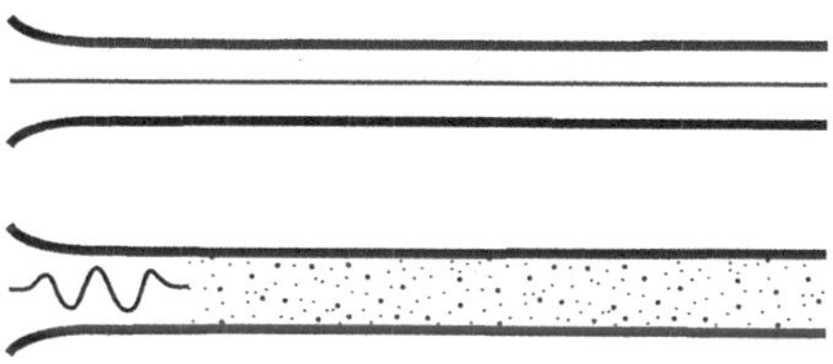

Figure 5.1 Laminar flow and turbulent flow.

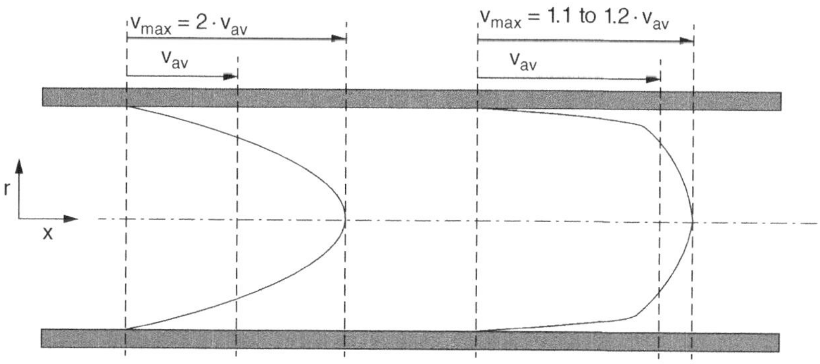

Figure 5.2 Laminar and turbulent velocity profiles in a tube. *Source:* Transport Phenomena by Beek, W. J. et al. (1999). Reproduced with permission of John Wiley & Sons Inc.

Introduction to Chemical Engineering, First Edition. C.M. van 't Land.

5.2 The Reynolds Number

The flow pattern in a tube or other conduit is characterized by Re, the Reynolds number.

For a tube:

$$Re = \frac{\rho v D}{\mu}. \tag{5.1}$$

ρ: specific mass in $kg \cdot m^{-3}$
v: average velocity in $m \cdot s^{-1}$
D: tube diameter in m
μ: dynamic viscosity in $N \cdot s \cdot m^{-2}$

Two flow fields having the same Re are dynamically similar. The flow through a straight tube is laminar if Re < 2,100. The flow becomes turbulent for Re > 4,000. The region 2,100 < Re < 4,000 is a transition region. Re gives the ratio between the inertial force and the viscous force. This will be elucidated.

First, the inertial force will be discussed. The velocity head of a flowing fluid is $0.5 \cdot \rho \cdot v^2$ $N \cdot m^{-2}$. The inertial force is the product of the velocity head and an area, for example, $0.5 \cdot \rho \cdot v^2 \cdot D^2$ N.

Second, the viscous force will be mentioned. The shear stress is proportional to $\mu \cdot (v/D)$. The viscous force is the product of shear stress and area, for example, $(\mu v \cdot D^2/D) = \mu v D$.

Dividing the inertial force by the viscous force results in $\frac{\rho \cdot v^2 \cdot D^2}{2\mu v D} = \frac{\rho v D}{2\mu} \infty \frac{\rho v D}{\mu}$.

5.3 Pressure Drop in Horizontal Conduits

The steady-state flow of a Newtonian fluid through a horizontal conduit is considered. See Figure 5.3. The shear stress at the wall can be given as:

$$\tau_w = -\mu \cdot \frac{dv_x}{dy}. \tag{5.2}$$

This equation is valid because there is no turbulence at the wall. The velocity distribution is required for the calculation of τ_w. For turbulent flow, it is not possible to assess the velocity distribution. However, on using similarity considerations, it is possible to derive that the quotient

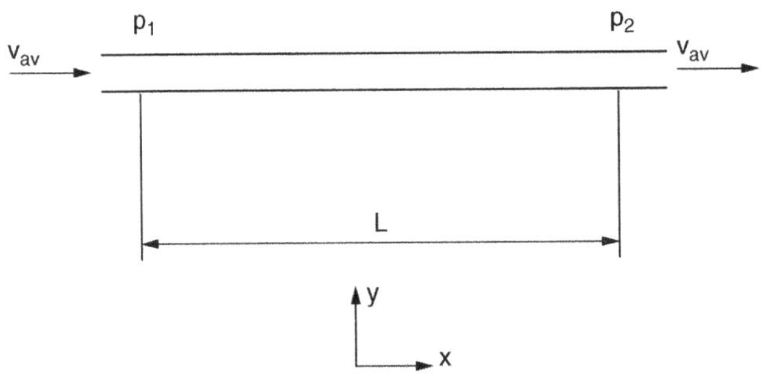

Figure 5.3 Steady-state flow of a Newtonian fluid through a horizontal conduit.

of τ_w and $\rho \cdot v_{av}^2$ in geometrical similar situations depends on Re only. Conventionally, this is stated as follows:

$$\tau_w = f \cdot 0.5 \cdot \rho \cdot v_{av}^2. \tag{5.3}$$

This equation is made plausible as follows. Think of a line in a refinery. Through the line flows, for example, gasoline. The liquid has an average velocity of 1.5 m · s^{-1}. While the gasoline flows through the line, the liquid tries to entrain the line. To prevent this to happen, the line is secured. That is, the line support exerts a force on the line to prevent the entrainment of the line. The other side of the coin is that the line tries to slow down the gasoline. Equation 5.3 expresses that the shear force at the wall is a function of f and the velocity head $0.5 \cdot \rho \cdot v_{av}^2$. f is the Fanning friction factor and is a function of Re and the geometry.

The balance of forces for a control volume moving along with the flow is written down: $F(p_1 - p_2) = SL \cdot \tau_w$. Rewriting this equation results in:

$$\tau_w = \frac{F(p_1 - p_2)}{SL}. \tag{5.4}$$

F: cross-sectional area of the conduit in m^2
S: wetted perimeter of the conduit in m
L: length of the conduit in m

Combining Equations 5.3 and 5.4 results in:

$$(p_1 - p_2) = \Delta p = 4f \cdot \frac{SL}{4F} \cdot 0.5 \cdot \rho \cdot v_{av}^2. \tag{5.5}$$

This is the Fanning Equation for the pressure loss by friction. By definition, F/S is termed the hydraulic radius m_h.

Its dimension is m. D_h, the hydraulic diameter, equals $4 \cdot m_h$.

For tubes, Equation 5.5 becomes:

$$\Delta p = 4f \cdot \frac{L}{D} \cdot 0.5 \cdot \rho \cdot v_{av}^2. \tag{5.6}$$

Figure 5.4 gives the hydraulic diameter of various conduits.

Equation 5.6 is generally used to calculate pressure loss due to flow through a conduit. A tube is a typical conduit.

On replacing − dp by Δp and dx by L in Equation 4.4 and equating Equations 4.4 and 5.6, it follows:

$$4f = \frac{64}{Re}. \tag{5.7}$$

Thus, for $Re < 2{,}100$, 4f is inversely proportional to Re. For $2{,}000 < Re < 10^5$ and for smooth tubes:

$$4f = 0.316 \cdot Re^{-0.25}. \tag{5.8}$$

This the Blasius Equation.

In practice, conduits are not smooth. Figure 5.5 illustrates the roughness of conduits. Figure 5.6 is the Moody Diagram. Basically, it is applicable for tubes [1]. It enables us to determine 4f as a function of Re. The tube roughness x/D is a parameter.

It is also possible to use Figure 5.6 for other conduits. Then, in Equation 5.6, the hydraulic diameter must be used. Figure 5.4 contains the hydraulic diameter of various conduits. The results of

Flow situation		Hydraulic diameter $D_h = 4F/S$	F
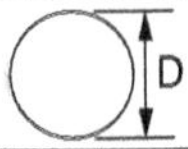	Circular pipe	D	$\frac{\pi}{4} D^2$
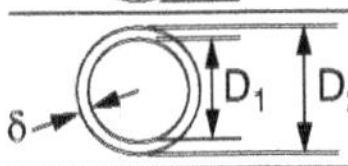	Concentric pipe or slit	$D_2 - D_1 = 2\delta$	$\frac{\pi}{4} (D_2^2 - D_1^2)$
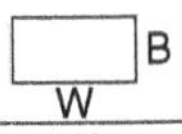	Rectangular pipe	$\frac{2WB}{W + B}$	WB
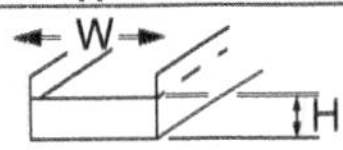	Open channel	$\frac{4WH}{W + 2H}$	WH
	Open channel	$H\sqrt{2}$	H^2
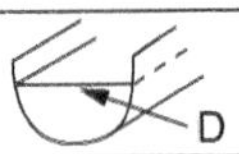	Half - filled	D	$\frac{\pi}{8} D^2$
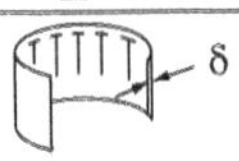	Liquid film	4δ	$\delta \pi D$

Figure 5.4 Hydraulic diameter of various conduits. *Source:* Transport Phenomena by Beek, W. J. et al. (1999). Reproduced with permission of John Wiley & Sons Inc.

calculations using Equation 5.6 for other conduits than tubes are approximately correct. Thus, Figure 5.6 is a powerful generalization for Newtonian fluids flowing turbulently through conduits having various types of cross-section.

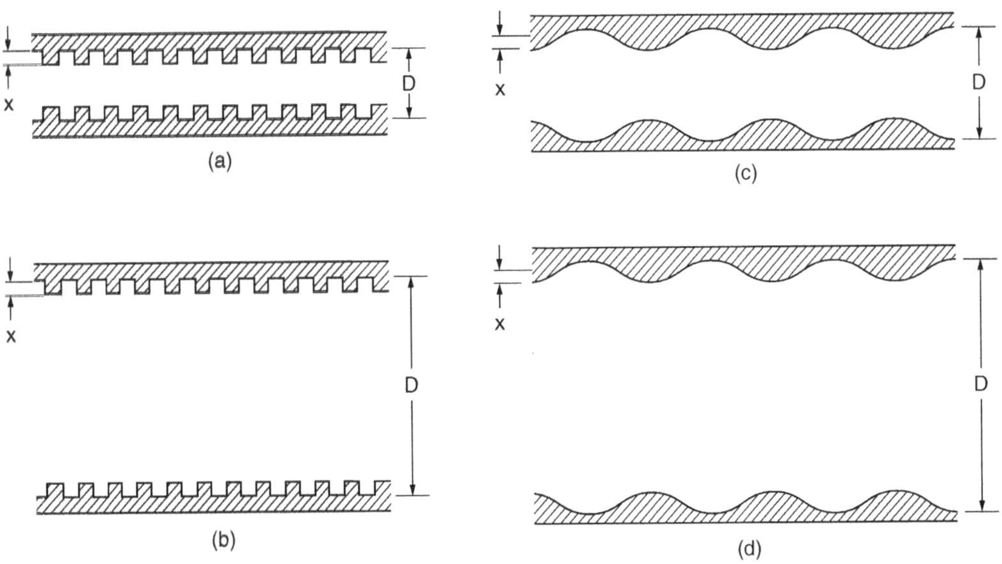

Figure 5.5 Tube roughness.

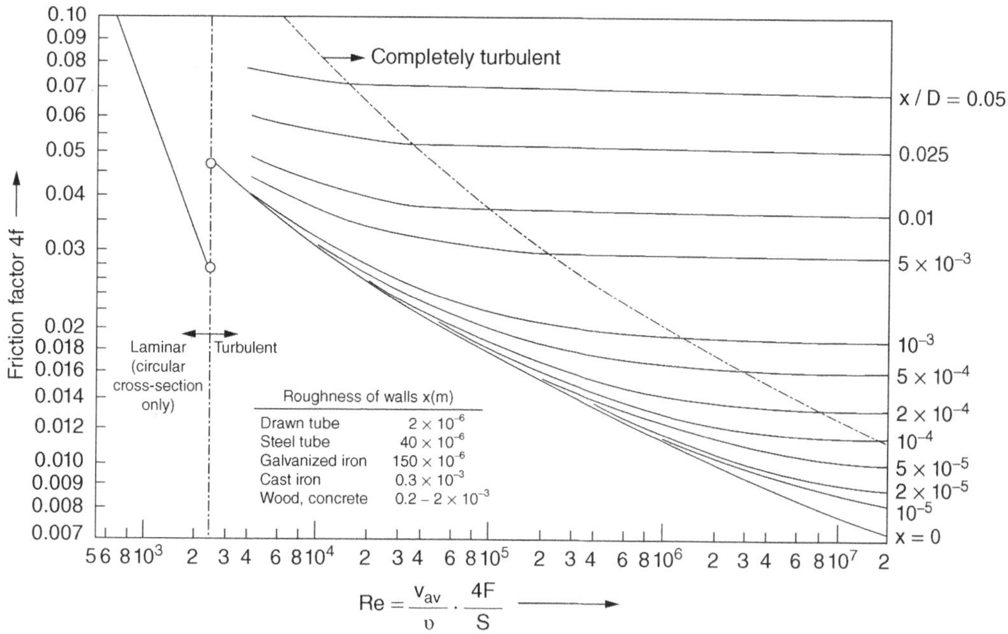

Figure 5.6 Friction factor for flow through tubes. *Source:* Transport Phenomena by Beek, W. J. et al. (1999). Reproduced with permission of John Wiley & Sons Inc.

For laminar flow, Figure 5.6 is applicable only for tubes. The proportionality constant (64 for tubes) depends on the form of the cross-section.

For high Re-values, 4f is hardly dependent on Re. Doubling the velocity in this range means a multiplication of the pressure loss with a factor 4: $\Delta p \propto v^2$.

Tube velocities for low-viscosity liquids are usually in the range 1–2 $m \cdot s^{-1}$. For gas velocities, the usual range is 10–20 $m \cdot s^{-1}$.

5.4 Pressure Drop in Tube Systems

Tube systems contain, for example, valves, bends, and constrictions. The pressure loss in a fitting is usually described by means of the following equation:

$$\Delta p = K_W \cdot 0.5 \cdot \rho \cdot v^2. \tag{5.9}$$

K_W is a dimensionless friction factor. It is a function of the local geometry and is, provided $Re > 10^5$, independent from Re. The downstream velocity is taken for the calculation of Re. A number of K_W-values are listed in Figure 5.7.

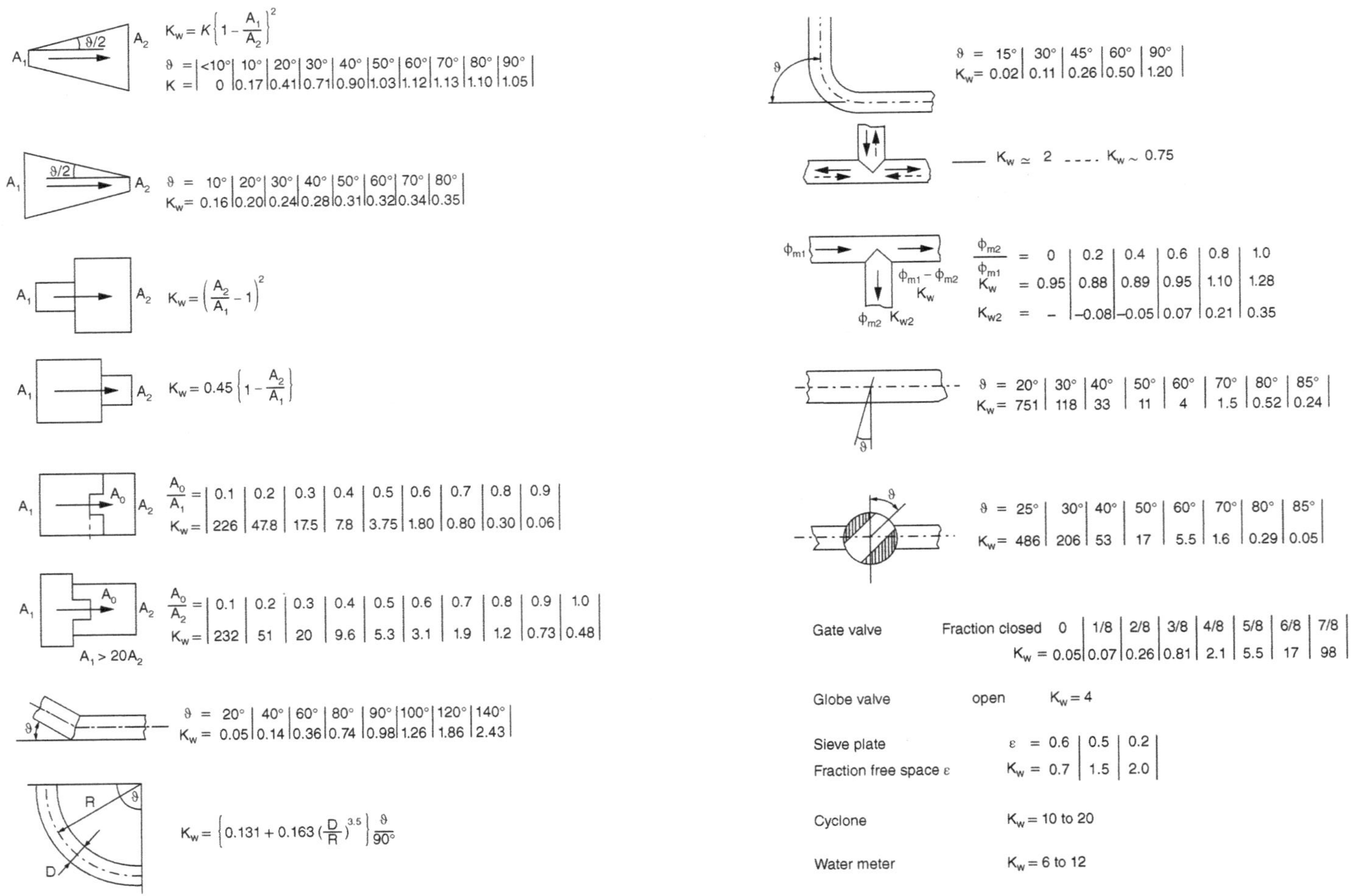

Figure 5.7 Some values of the friction loss factor K_w when $Re > 10^5$ for downstream flow. *Source:* Transport Phenomena by Beek, W. J. et al. (1999). Reproduced with permission of John Wiley & Sons Inc.

5.5 Flow Around Obstacles

5.5.1 Introduction

Flow around obstacles exerts a force F on an obstacle. Two different mechanisms can be distinguished, that is, friction resistance and form resistance.

Friction Resistance

The force is caused by friction resistance only in the special case of flow along a flat plate. See Figure 5.8. The figure shows that the disturbance of the relative flow velocity v_r increases in the y-direction when we move from $x = 0$ to higher x-values. According to boundary layer theory, the shear stress at the wall can be expressed as follows [2]:

$$\tau_w = -0.667\sqrt{\frac{\mu}{\rho \cdot v_r \cdot x}} \cdot 0.5 \cdot \rho \cdot v_r^2 \quad \mathrm{N \cdot m^{-2}}.$$

Note that $\mu/\rho \cdot v_r \cdot x$ is the inverse of Re.

The force F on the plate can be found by integrating the product of shear stress and area from $x = 0$ to $x = L$. The width of the plate is B. The integration proceeds as follows:

$$F = 2 \cdot 0.667\sqrt{\frac{\mu}{\rho \cdot v_r}} \cdot B \cdot 0.5 \cdot \rho \cdot v_r^2 \cdot \int_{x=0}^{x=L} x^{-1/2} dx,$$

$$F = 2 \cdot 2 \cdot 0.667\sqrt{\frac{\mu}{\rho \cdot v_r}} \cdot B \cdot 0.5 \cdot \rho \cdot {v_r}^2 \cdot \int_{x=0}^{x=L} dx^{1/2},$$

$$F = 2.67\sqrt{\frac{\mu}{\rho \cdot v_r \cdot L}} \cdot BL \cdot 0.5 \cdot \rho \cdot v_r^2 \ \mathrm{N}. \tag{5.10}$$

The factor 2 in the first equation for F is because the flat plate has two sides.

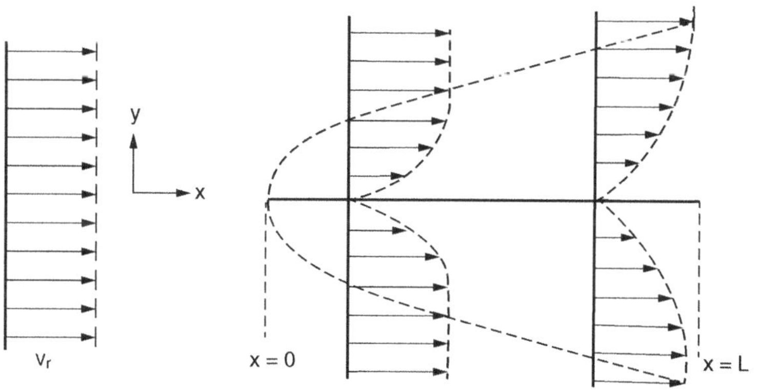

Figure 5.8 Flow along a plate. *Source:* Courtesy of Delft Academic Press/VSSD, Delft, The Netherlands.

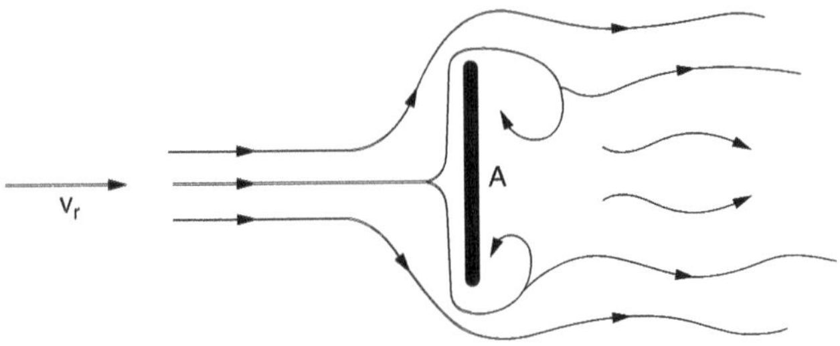

Figure 5.9 Eddy formation behind an obstacle in the flow. *Source:* Courtesy of Delft Academic Press/VSSD, Delft, The Netherlands.

Form Resistance

A flat plate perpendicular to the flow represents a pronounced case of form resistance. See Figure 5.9. The upfront pressure is approximately $0.5 \cdot \rho \cdot v_r^2$, the velocity head. At the edge of the plate, the pressure has been converted into extra kinetic energy which is only partly converted into pressure again. The cause is eddying in the "dead water" behind the plate. As a result, the order of magnitude of the force on the plate is given by:

$$F \approx A \cdot 0.5 \cdot \rho \cdot v_r^2.$$

Generally, the force on an obstacle in the flow is described with a drag coefficient C_W:

$$F = C_W \cdot A \cdot 0.5 \cdot \rho \cdot v_r^2. \tag{5.11}$$

A is the largest cross-section of the obstacle perpendicular to v_r. C_W is a function of $Re = \rho \cdot v_r \cdot D/\mu$ in which D is a characteristic dimension of the obstacle. At high Re-values, form resistance exceeds friction resistance.

Combination

In practice, both friction resistance and form resistance are met. However, the two mechanisms are not treated separately. The approach used has been adapted to form resistance. Equation 5.11 is generally used for design purposes.

5.5.2 Dispersed Spherical Particles

Often, solid particles or droplets are dispersed in a fluid phase. They move under the influence of an external force. The force can be caused by gravity, centrifugation, a magnetic field, or an electric field. The resistance to flow determines the relative velocity of the particles or droplets. The flow field around one single sphere depends on Re, which is defined as $\rho \cdot v_r \cdot D/\mu$, D being the diameter of the sphere. See Figure 5.10.

At highly viscous flow ($Re < 1$), inertia effects do not occur and the streamlines converge again behind the sphere. Then, Stokes' law is valid:

$$F = 3\pi\mu D \cdot v_r. \tag{5.12}$$

C_W in Equation 5.11 is then $24/Re$. We deal with friction resistance mainly.

At higher Re-values, both friction resistance and form resistance play a role. The influence of the former becomes smaller when Re increases. The form resistance is largely determined by the place where the boundary layer is released from the surface. For laminar flow ($1 < Re < 10^3$), $C_W = 18.5 \cdot Re^{-0.6}$ (empirical formula).

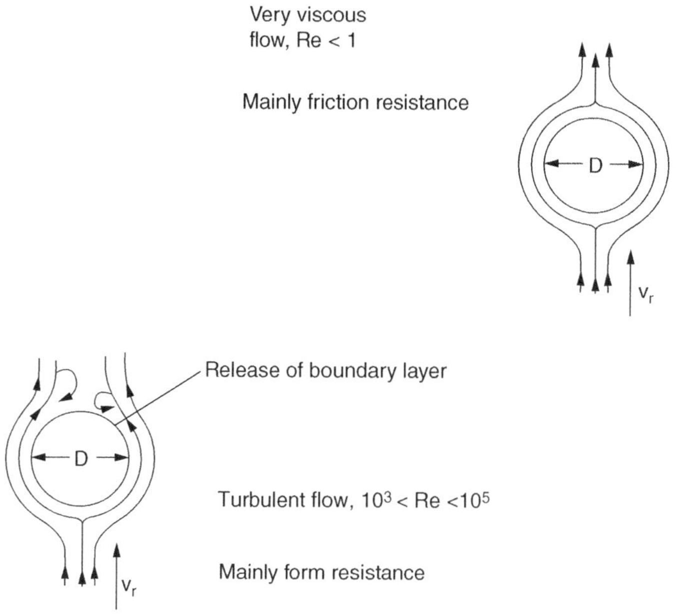

Figure 5.10 Flow pattern around a sphere. *Source:* Courtesy of Delft Academic Press/VSSD, Delft, The Netherlands.

For $10^3 < Re < 10^5$, said release occurs at a short distance behind the center of the sphere. C_w is then constant and about 0.43 in this range of Re numbers. This fact is known as Newton's law of resistance.

The friction boundary layer becomes turbulent for $Re > 2 \cdot 10^5$. In that range, the point where the boundary layer is released is shifted back causing a rapid fall of C_w. For $Re > 2 \cdot 10^5$, $C_w = 0.13$.

Figure 5.11 shows C_w for a sphere as a function of Re. Data concerning the drag coefficients of various objects can be found in the literature [3].

When a sphere starts to fall through a stagnant medium, the movement is initially accelerated. The acceleration stops when the terminal velocity is reached. There is a balance of forces when that velocity is reached. The two forces are the gravity force pulling the sphere down and the force exerted by the medium through which the sphere falls. The terminal velocity of the falling sphere can be calculated from the equation:

$$\frac{\pi}{6} \cdot D^3(\rho_s - \rho_f)g = C_w \cdot \frac{\pi}{4} \cdot D^2 \cdot 0.5 \cdot \rho_f \cdot v_r^2. \tag{5.13}$$

For $Re < 1$: $v_r = \dfrac{(\rho_s - \rho_f)}{18 \cdot \mu} \cdot D^2$.

For $1 < Re < 10^3$, v_f can be found by trial and error by means of Figure 5.11. In that range, $C_w = 18.5 \cdot Re^{-0.6}$.

For $10^3 < Re < 2 \cdot 10^5$: $v_r = 1.76\sqrt{\left(\frac{\rho_s - \rho_f}{\rho_f}\right)gD}$.

For $Re > 2 \cdot 10^5$: $v_r = 3.20\sqrt{\left(\frac{\rho_s - \rho_f}{\rho_f}\right)gD}$.

A review for C_w is as follows:

$Re < 1$	$24/Re$
$1 < Re < 10^3$	$18.5 \cdot Re^{-0.6}$

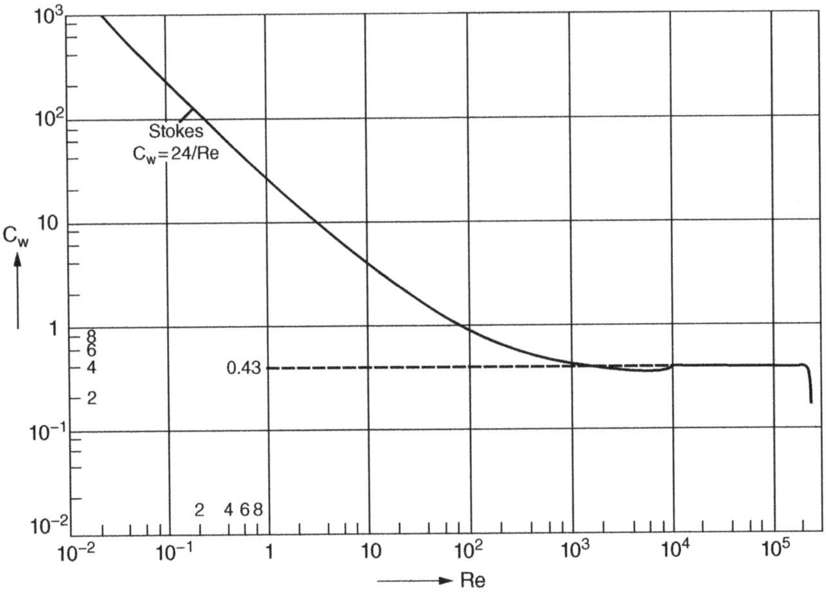

Figure 5.11 C_w = f(Re) for a sphere. *Source:* Courtesy of Delft Academic Press/VSSD, Delft, The Netherlands.

$10^3 < Re < 2 \cdot 10^5$	0.43
$Re > 2 \cdot 10^5$	0.13

Example 5.1

It is desired to prill stearic acid. Prilling is the process of spray crystallization. Liquid stearic acid is sprayed to produce droplets that fall through air, the cooling medium, and crystallize to particles. Figure 5.12 shows a prilling plant schematically. The melt has a temperature of 80 °C whereas stearic acid's melting point is 70 °C. A pump transfers the liquid to the top of the prilling tower. Here, the melt is sprayed, the droplets fall down and crystallize. A particle size of 1.2 mm is required. The particles leave the tower at 30 °C. The cooling air enters at 15 °C and leaves at 25 °C. The terminal velocity of the stearic acid particles is calculated.

$\rho_s = 1{,}009\ kg \cdot m^{-3}$ at 50 °C.

$\rho_g = 1.205\ kg \cdot m^{-3}$ at 20 °C.

$\mu_g = 1.82 \cdot 10^{-5}\ N \cdot s \cdot m^{-2}$ at 20 °C.

The calculation starts with the choice of a terminal velocity and a check of its correctness. Table 5.1 [4] is used to select $v_r = 4.5\ m \cdot s^{-1}$.

$$Re = \frac{\rho_g \cdot v_r \cdot d_p}{\mu_g} = \frac{1.205 \cdot 4.5 \cdot 1.2 \cdot 10^{-3}}{1.82 \cdot 10^{-5}} = 357.5.$$

Equation 5.13: $\frac{\pi}{6} \cdot D^3 \left(\rho_s - \rho_g\right) g = C_W \cdot \frac{\pi}{4} \cdot D^2 \cdot 0.5 \cdot \rho_g \cdot v_r^2.$

Left-hand side: $\frac{\pi}{6}(1.2 \cdot 10^{-3})^3 (1{,}009 - 1.205) 9.81 = 8.95 \cdot 10^{-6}$ N.

Right-hand side: $0.65 \cdot \frac{\pi}{4}(1.2 \cdot 10^{-3})^2 \cdot 0.5 \cdot 1.205 \cdot 4.5^2 = 8.97 \cdot 10^{-6}$ N.

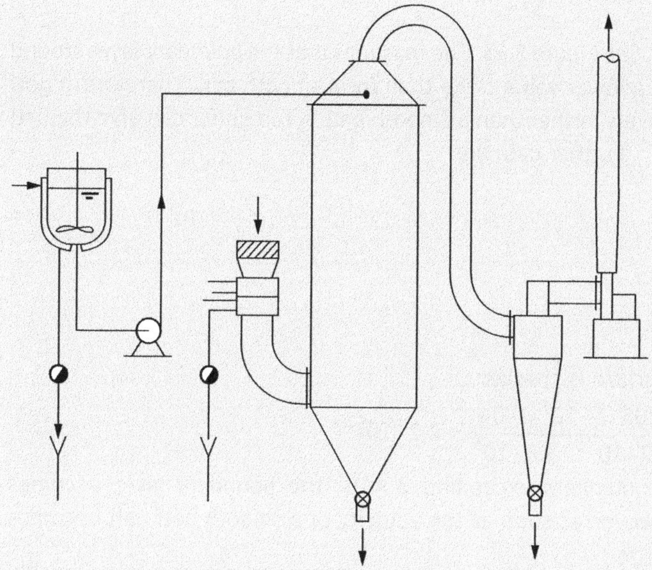

Figure 5.12 A prill tower.

C_w is read in Figure 5.11 as a function of Re. The left-hand side of the equation equals the right-hand side, so the choice of 4.5 m · s^{-1} is correct. If that should not have been the case, a new value of v_r should be taken. The correct value of v_r is thus found by iteration.

Table 5.1 Terminal velocity of spherical particles in air in m · s^{-1}.

	$\Delta\rho = (\rho_S - \rho_F)$, kg · m^{-3}					
	1,000			3,000		
	°C			°C		
Diameter, μm	**15**	**90**	**205**	**15**	**90**	**205**
50	0.074	0.062	0.052	0.14	0.12	0.10
100	0.25	0.22	0.19	0.46	0.41	0.37
250	0.91	0.87	0.83	1.51	1.48	1.43
500	1.98	2.01	1.99	3.17	3.20	3.26
600	2.39	2.43	2.46	3.78	3.84	3.94
700	2.79	2.87	2.91	4.33	4.51	4.70
800	3.14	3.23	3.35	4.85	5.07	5.27
900	3.48	3.66	3.79	5.42	5.64	5.91
1,000	3.84	4.00	4.21	5.97	6.19	6.50
1,250	4.73	4.91	5.12	7.10	7.53	7.86
1,500	5.37	5.76	5.98	7.78	8.65	9.40
2,000	6.71	7.18	7.71	9.72	10.6	11.7
2,500	7.71	8.45	9.20	11.3	12.5	13.7
3,000	8.75	9.56	10.5	14.8	14.0	15.5
5,000	11.6	13.1	14.6	16.7	18.6	21.2

Example 5.2

Golf balls have a dimpled surface. See Figure 5.13. The reason is that the boundary layer around a golf ball becomes turbulent at a lower value of Re than for a smooth ball. Therefore, a golf ball having a dimpled surface travels farther than a smooth ball. A top golfer can give the ball an initial velocity of 300 km · h^{-1}. Further data are:

$D = 0.04$ m.

$\rho_g = 1.205\ \text{kg} \cdot \text{m}^{-3}$.

$\mu_g = 1.82 \cdot 10^{-5}\ \text{N} \cdot \text{s} \cdot \text{m}^{-2}$.

The effectivity of the dimpled surface is checked.

Initially, $\text{Re} = \dfrac{\rho_g \cdot v_r \cdot D}{\mu_g} = \dfrac{1.205 \cdot 3 \cdot 10^5 \cdot 0.04}{1.82 \cdot 10^{-5} \cdot 3.6 \cdot 10^3} = 2.2 \cdot 10^5$.

Figure 5.11 shows that for Re numbers exceeding $2 \cdot 10^5$ the boundary layer becomes turbulent for a smooth ball. However, as soon as the velocity of a smooth golf ball becomes smaller, for instance, $300 \cdot \dfrac{2 \cdot 10^5}{2.2 \cdot 10^5} = 273\ \text{km} \cdot \text{h}^{-1}$, the air resistance increases considerably as C_w rises from 0.13 to 0.43. The velocity at which this transition occurs is considerably lower than 273 km · h^{-1} for a golf ball having a dimpled surface. Thus, yes, it pays off to give golf balls a dimpled surface.

Figure 5.13 Golf ball with a dimpled surface.

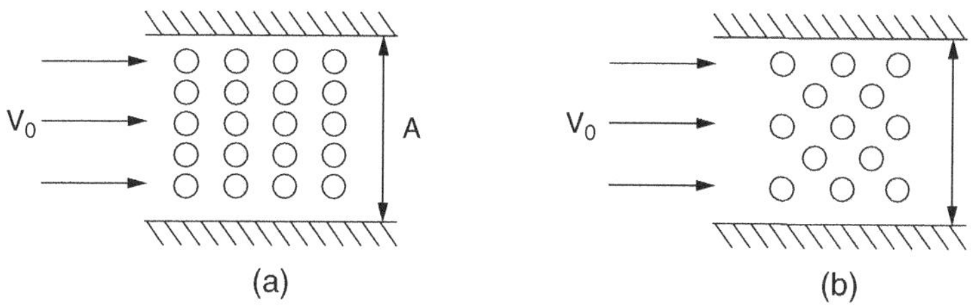

Figure 5.14 Flow perpendicular to a bank of pipes: (a) in line, (b) staggered. *Source:* Transport Phenomena by Beek, W. J. et al. (1999). Reproduced with permission of John Wiley & Sons Inc.

5.6 Terminal Velocity of a Swarm of Particles

The settling rate of a swarm of monodisperse spherical particles in a liquid is smaller than the settling rate of an individual spherical particle. Monodisperse means that all particles have the same size. The equation of Richardson and Zaki [5] can be used:

$$(v_r)_s = v_r \cdot \varepsilon^n. \tag{5.14}$$

ε is the volume fraction of the continuous phase whereas n is a number which depends on Re of a particle settling alone (not in a swarm) in the liquid. n is approximately 4.5 for Re = 1 whereas n = 2.4 when Re > 500.

5.7 Flow Resistance of Heat Exchangers with Tubes

It is assumed that the direction of the flow is perpendicular to the tubes. First, a single tube is considered. For $10^3 < Re < 10^5$, $K_w = 1.2$ [6]. See Figure 5.14. In practice, banks of tubes are often used as heating (or cooling) elements in a liquid or gas flow. Two ways of arranging the tubes can be distinguished: (a) in line and (b) staggered.

In case (a), most tubes are in the "wake" of their predecessor, whereas in case (b) they are not. Due to the shielding in case (a), the resistance of arrangement (a) is lower than that of arrangement (b). The heat transfer coefficient for arrangement (a) is also lower than the heat transfer coefficient for arrangement (b). For case (b), as a rule of thumb, we can use the fact that the flow resistance coefficient of a tube is about 1.5 times higher than for a single tube if the distance between two adjacent tubes equals the tube diameter. The latter statement is valid for $30 < Re < 2{,}000$ [6].

Example 5.3

A horizontal air duct of galvanized iron with rectangular cross-section ($0.3 \cdot 0.3\ m^2$) having a length of 10 m contains a heat exchanger. The heat exchanger consists of 15 rows of 6 horizontal tubes placed transversely to the flow direction. The rows are installed in a staggered position. The external tube diameter is 0.02 m. The horizontal distance between the centers of the rows is 0.02 m. The mean duct temperature is 80 °C and the mean duct pressure is $1.3 \cdot 10^5\ N \cdot m^{-2}$. The line velocity is $12\ m \cdot s^{-1}$. The pressure loss due to the heat exchanger and the power for the flow through the line and the heat exchanger are estimated.

Air data

The air specific mass is 1.29 kg · m^{-3} at 0 °C and atmospheric pressure.

$$\rho_g = 1.29 \cdot \frac{273}{273+80} \cdot 1.3 = 1.3 \text{ kg} \cdot \text{m}^{-3}.$$

$$\mu = 2.2 \cdot 10^{-5} \text{ N} \cdot \text{s} \cdot \text{m}^{-2} \text{ at } 20\,°\text{C}.$$

Line pressure loss

$$D_h = \frac{4F}{S} = \frac{4 \cdot 0.3 \cdot 0.3}{4 \cdot 0.3} = 0.3 \text{ m}.$$

$$Re = \frac{1.3 \cdot 12 \cdot 0.3}{2.2 \cdot 10^{-5}} = 2.13 \cdot 10^5.$$

Figure 5.5: $x = 150 \cdot 10^{-6}$ m, $\frac{x}{D_h} = 5 \cdot 10^{-4}$, and $4f = 0.0185$

$$\Delta p = 4f \cdot \frac{L}{D_h} \cdot 0.5 \cdot \rho_g \cdot v^2 = 0.0185 \cdot \frac{10}{0.3} \cdot 0.5 \cdot 1.3 \cdot 12^2 = 58 \text{ N} \cdot \text{m}^{-2}.$$

Heat exchanger pressure loss

Single tube

The air velocity is for one tube: $\frac{300}{300-6\cdot 20} \cdot 12 = 20 \text{ m} \cdot \text{s}^{-1}$

$$Re = \frac{\rho_g \cdot v \cdot D_t}{\mu} = \frac{1.3 \cdot 20 \cdot 0.02}{2.2 \cdot 10^{-5}} = 2.4 \cdot 10^4.$$

$$K_w = 1.2 \ (10^3 < Re < 10^5).$$

$$K'_w = 1.5 \cdot 1.2 = 1.8.$$

Strictly speaking, Re must be between 30 and 2,000 for this approximation. It is assumed that it is also true for $Re = 2.4 \cdot 10^4$.

The force on one tube is: $F = 1.8 \cdot 0.5 \cdot 1.3 \cdot 20^2 \cdot 0.3 \cdot 0.02 = 2.81$ N.

The force on the bank is: $6 \cdot 15 \cdot 2.81 = 253$ N.

$$\Delta p = \frac{253}{0.3 \cdot 0.3} = 2{,}811 \text{ N} \cdot \text{m}^{-2}.$$

The power required is: $\Phi_v \cdot \Delta p = 0.3 \cdot 0.3 \cdot 12(2{,}811 + 58) = 3{,}100$ W.

References

1. Beek, W. J., Muttzall, K. M. K., and van Heuven, J. W. (1999). *Transport Phenomena* (p. 69). Chichester, UK: Wiley.
2. Beek, W. J., Muttzall, K. M. K., and van Heuven, J. W. (1999). *Transport Phenomena* (p. 105). Chichester, UK: Wiley.
3. Beek, W. J., Muttzall, K. M. K., and van Heuven, J. W. (1999). *Transport Phenomena* (p. 107). Chichester, UK: Wiley.

4. Nonhebel, G., and Moss, A. A. H. (1971). *Drying of Solids in the Chemical Industry* (p. 244). London, UK: Butterworth.
5. Richardson, J. F., and Zaki, W. N. (1954). Sedimentation and fluidisation: Part 1. *Transactions of the Institution of Chemical Engineers*, 32, 35–53.
6. Beek, W. J., Muttzall, K. M. K., and van Heuven, J. W. (1999). *Transport Phenomena* (p. 114). Chichester, UK: Wiley.

6

Flow Meters

6.1 Introduction

Two groups of flow meters can be distinguished. The first group consists of meters that generate a signal by using some of the energy of the flowing fluid. They are called fluid-energy activated flow meters. The second group of flow meters comprises those meters that derive their basic signal from the interaction of the flow and an external stimulus. They are called external stimulus flow meters. Several meters of both categories will be discussed.

6.2 Fluid-energy Activated Flow Meters

6.2.1 Oval-gear Flow Meter

See Figure 6.1. Oval-shaped, gear-toothed rotors rotate within a chamber of specified geometry. As these rotors turn, they sweep out and trap a very precise volume of fluid between the outer oval shape of the gears and the inner chamber walls. Magnets are embedded in the rotors. The meter is unsuitable for gases.

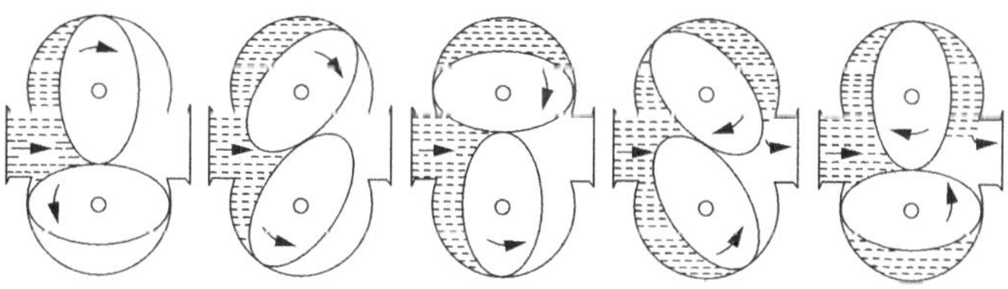

Figure 6.1 The oval-gear flow meter. *Source:* Courtesy of Bopp & Reuther Messtechnik GmbH, Speyer, Germany.

6.2.2 Orifice Meter

See Figure 6.2. An orifice is a restricted opening through which fluid flows. An orifice meter consists of a plate with a sharp-edged circular hole that is inserted between two flanges of a piping system for the purpose of establishing the flow rate from pressure-difference measurements across the orifice. The flow pattern indicates that the streamlines tend to converge a short distance downstream from the plane of the orifice; therefore, the minimum-flow area is smaller than the area of the opening in the orifice plate. This minimum-flow area is known as the "contracted area of the jet." The ratio of

Introduction to Chemical Engineering, First Edition. C.M. van 't Land.

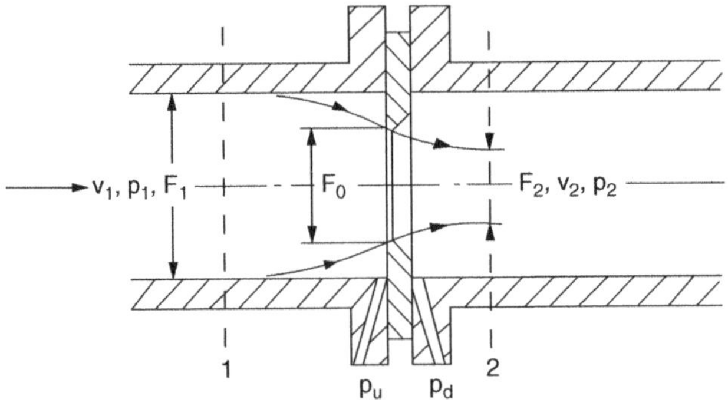

Figure 6.2 The orifice plate. *Source:* Courtesy of Delft Academic Press/VSSD, Delft, The Netherlands.

the areas F_2/F_0 is called the contraction coefficient μ whereas the ratio of the areas F_0/F_1 is termed m. Both μ and m are smaller than 1. When the line is horizontal, Bernoulli's expression becomes:

$$0.5 \cdot \rho \cdot v_1^2 + p_1 = 0.5 \cdot \rho \cdot v_2^2 + p_2.$$

v_2 can be replaced by $v_1/\mu m$ in this equation.
It is then possible to express v_1 as a function of p_1, p_2, ρ, μ, and m.
The mass flow is: $\Phi_m = \rho \cdot v_1 \cdot F_1$.
On substituting the expression for v_1 in the latter expression, one obtains:

$$\Phi_m = \frac{\mu \cdot F_0}{\sqrt{1 - \mu^2 \cdot m^2}} \sqrt{2\rho(p_1 - p_2)}.$$

Instead of p_1 and p_2, p_u and p_d are measured in the orifice meter of Figure 6.2. The instrument in the figure is called an orifice meter with corner taps. There are several possibilities concerning the location of the taps. To cope with the locations of the pressure measurements and to correct for friction, an empirical factor z′ is introduced. Furthermore, at the transport of gases, the upstream specific mass is different from the downstream specific mass. To cope with that effect, an additional factor ε is introduced. ε = 1 for liquids and $0.9 < \varepsilon < 1.0$ for gases.

$$\Phi_m = \frac{\mu \cdot z' \cdot \varepsilon \cdot F_0}{\sqrt{1 - \mu^2 \cdot m^2}} \sqrt{2\rho(p_u - p_d)}.$$

The usual presentation of the previous equation is [1]:

$$\Phi_m = \frac{C}{\sqrt{1 - \beta^4}} \cdot \varepsilon \cdot \frac{\pi}{4} \cdot d^2 \sqrt{2\rho(p_u - p_d)}. \quad (6.1)$$

d in Equation 6.1 is the diameter of the orifice whereas β is the ratio between d and the line diameter. C is the discharge coefficient and is tabulated. C is a function of:

- $Re_1 = \frac{\rho v D}{\mu}$,
- $\frac{d}{D}$, and
- the design of the orifice and the taps.

The design of an orifice meter is specified in the mentioned reference. An orifice meter needs both upstream and downstream of the instrument lengths of straight pipe. The lengths required are specified as well. For the measurement of gas flows, tabulated data for the establishment of ε are given. The reference also contains methods to calculate the pressure loss of the instrument.

The orifice meter is simple and reliable. The pressure loss is a drawback.

Example 6.1

It is desired to dose 45–60 t · h^{-1} of cyclohexane. The hydrocarbon has a temperature of 20 °C. Select an orifice meter.

Cyclohexane data

$$\rho = 779 \text{ kg} \cdot \text{m}^{-3} \text{ at } 20\,°\text{C}.$$

$$\mu = 0.98 \cdot 10^{-3} \text{ N} \cdot \text{s} \cdot \text{m}^{-2} \text{ at } 20\,°\text{C}.$$

Select a line diameter of 0.1 m and an orifice meter with β = 0.7. An orifice meter with corner taps is in focus.

$$\Phi_m = 45 \text{ t} \cdot \text{h}^{-1}.$$

$$v_1 = \frac{45{,}000}{3{,}600 \cdot \frac{\pi}{4} \cdot 0.1^2 \cdot 779} = 2.04 \text{ m} \cdot \text{s}^{-1}.$$

$$Re_1 = \frac{779 \cdot 2.04 \cdot 0.1}{0.98 \cdot 10^{-3}} = 1.62 \cdot 10^5.$$

$$C = 0.6086 \text{ and } \frac{C}{\sqrt{1 - 0.7^4}} = 0.698.$$

$$p_u - p_d = \frac{45{,}000^2}{3{,}600^2 \cdot 0.698^2 \cdot 1^2 \left(\frac{\pi}{4} \cdot 0.07^2\right)^2 \cdot 779 \cdot 2} = 13{,}898 \text{ N} \cdot \text{m}^{-2}.$$

$$\Phi_m = 60 \text{ t} \cdot \text{h}^{-1}.$$

$$v_2 = \frac{60{,}000}{3{,}600 \cdot \frac{\pi}{4} \cdot 0.1^2 \cdot 779} = 2.72 \text{ m} \cdot \text{s}^{-1}.$$

$$Re_2 = \frac{779 \cdot 2.72 \cdot 0.1}{0.98 \cdot 10^{-3}} = 2.16 \cdot 10^5.$$

$$C = 0.6072 \text{ and } \frac{C}{\sqrt{1 - 0.7^4}} = 0.697.$$

$$p_u - p_d = \left(\frac{60}{45}\right)^2 \left(\frac{0.698}{0.697}\right)^2 13{,}898 = 24{,}779 \text{ N} \cdot \text{m}^{-2}.$$

Example 6.2

The pressure loss of the orifice meter of Example 6.1 is calculated for a flow of 60 t · h^{-1}.

The pressure loss is calculated as a fraction of ($p_u - p_d$). The fraction is $1 - \beta^{1.9} = 0.492$ and 0.492 · 24,779 = 12,191 N · m^{-2}.

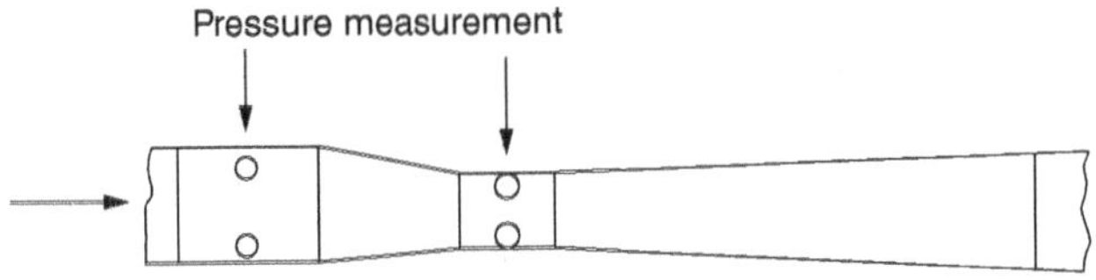

Figure 6.3 Classical venturi meter.

6.2.3 Venturi Meter

A venturi meter is, like the orifice meter, used to establish a flow rate from a pressure-difference measurement. A typical venturi meter consists of a cylindrical inlet section, a smooth entrance cone (acceleration cone) having an angle between 20 and 22°, a short cylindrical throat section, and a diffuser cone (deceleration cone) having an angle in the range 7–15° [2]. See Figure 6.3. Small diffuser angles tend to minimize head loss from pipe friction, flow separation, and increased turbulence. Pressures are measured upstream the throat and at the throat. The difference of the pressures is representative for the flow. The treatment of the venturi meter resembles the treatment of the orifice meter and Equation 6.1 is also applicable in this case. The pressure loss of a venturi meter is smaller than the pressure loss of an orifice meter.

6.2.4 Rotameter

The rotameter is a common flow measurement device whose operation is based on drag principles. The rotameter, shown schematically in Figure 6.4, consists of a tapered tube and a solid float (bob) that is free to move vertically in the tube. At any flow rate within the range of the meter, fluid entering the bottom raises the float (thereby increasing the area between the float and the tube) until the drag and buoyancy forces are balanced by the mass of the float. The relevant equation is:

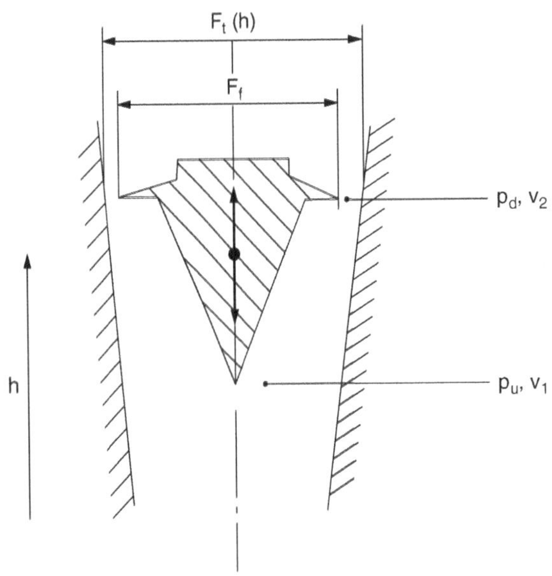

Figure 6.4 Rotameter. *Source:* Courtesy of Delft Academic Press/VSSD, Delft, The Netherlands.

$$(p_u - p_d)F_1 = V_f(\rho_f - \rho).$$

F_f is the float area, V_f is the float volume, and ρ_f is the specific mass of the float material. Substitution of $(p_u - p_d)$ according to the latter equation into Equation 6.1 yields the rotameter formula:

$$\Phi_m = \alpha \cdot \varepsilon \cdot F_0 \sqrt{\frac{2\rho \cdot V_f(\rho_f - \rho)}{F_1}}. \tag{6.2}$$

F_0 is the area through which the fluid flows between the solid float and the wall and α is termed the flow coefficient. It replaces the quotient $C/\sqrt{1 - \beta^4}$ in Equation 6.1.

α can be read from graphs. F_0 varies linearly with the rotameter height. Note: the rotameter tube is not a truncated cone.

The rotameter can be considered an orifice meter with a fixed pressure difference and a variable passage.

Example 6.3

On dosing a gas, the system pressure is increased with a factor 3. The temperature is kept constant. A rotameter float is kept at the same position. How much larger is the mass flow?

$\Phi_m \infty \sqrt{p}$, hence the mass flow is $\sqrt{3}$ larger, that is, 1.73.

Next, the temperature is increased from 20 to 70 °C. Again, $\Phi_m \infty \sqrt{\rho}$ and the mass flow is $\sqrt{\frac{273 + 20}{273 + 70}}$, that is, 0.92 smaller.

$$1.73 \cdot 0.92 = 1.59.$$

For the dosing of gases, temperature and pressure must be defined. For this reason, gas mass flow meters are often preferred.

6.3 External Stimulus Flow Meters

6.3.1 Thermal Flow Meter

A way to measure mass flow rates of fluids (mostly gases) is offered by transferring some heat to them. The meters have a downstream temperature sensor and an upstream one. The downstream unheated sensor measures the fluid's temperature. The upstream sensor is heated and also indicates a temperature. If the fluid would be stagnant, the temperature of the upstream sensor would be maximum. As soon as fluid starts to flow, the temperature of the upstream sensor falls. The flowing fluid has a cooling effect on that sensor. That cooling effect is used as the measuring principle. Thermal mass flow meters come in two different versions. In one version, the temperature difference between the two sensors is kept constant and the required power to achieve that result is measured. That power is representative for the mass flow rate. In the second version, the power to heat the upstream sensor is kept constant and the temperature difference between the two sensors is measured. That temperature difference is representative for the mass flow rate. Thermal mass flow meters do not have moving parts and provide a relatively unobstructed flow path.

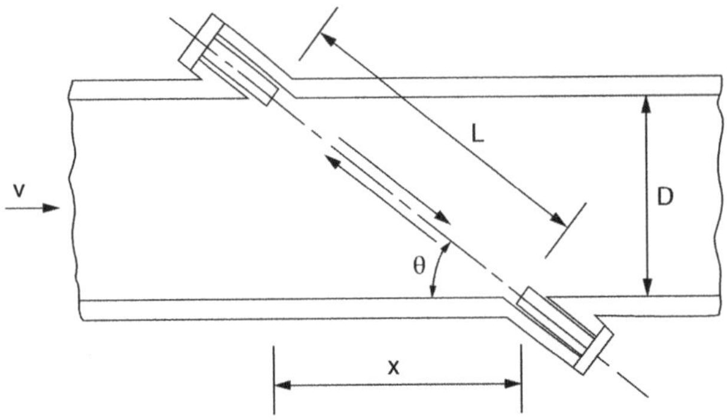

Figure 6.5 Geometry of the transit-time flow meter. *Source:* Courtesy of Cambridge University Press, Cambridge, UK.

6.3.2 Ultrasonic Flow Meters

The description will be limited to the transit-time flow meter. Its functioning depends on the slight difference in time needed for an ultrasound wave to travel upstream rather than downstream a measured flow. Waves are launched each way, their time of transit is measured, and the difference is representative for the velocity of the flow. It is usual to send the beams across the flow but not at right angles to the flow, so that there is a component of the fluid velocity along the path of the acoustic beams. See Figure 6.5. Ultrasonic flow meters can be used for both liquids and gases. They do not have moving parts and provide a relatively unobstructed flow path.

References

1. DIN Deutsches Institut für Normierung e.V. Measurement of fluid flow by means of pressure differential devices inserted in circular cross-section conduits running full – Part 2: Orifice plates (ISO 5167-2:2003); German version EN ISO 516-2:2003. Berlin, Germany: Beuth Verlag GmbH.
2. DIN Deutsches Institut für Normierung e.V. Measurement of fluid flow by means of pressure differential devices inserted in circular cross-section conduits running full – Part 4: Venturi tubes (ISO 5167-4:2003); German Version EN ISO 5167-4:2003. Berlin, Germany: Beuth Verlag GmbH.

7

Case Studies Flow Phenomena

7.1 Energy Consumption: Calculation of the Power Potential of a High Artificial Lake

An artificial lake with a capacity of $6.0 \cdot 10^6$ m^3 of water lies on the Mons St. Nicholas in Luxemburg. The lake is, mainly during the night, filled with water from the river Our in 12 h. The artificial lake is 230 m higher than the river. Turbines on the river bank are, to generate electric power, driven with water from the lake during the next 12 h. Water from the lake is supplied to the turbines through two parallel concrete tubes having a length of 650 m and a diameter of 6 m.

a) What power could this station deliver, apart from losses, if it is in continuous operation for 12 out of 24 h?
b) What is the required pump power with a pump efficiency of 75%?
c) What are the friction losses expressed as a percentage of the theoretically possible power? Let the resistance value for the entry and exit losses, each related to the velocity in the tubes, be 2.5.

The water specific mass is 1,000 kg $\cdot$ m^{-3} and the water dynamic viscosity is $1.2 \cdot 10^{-3}$ N $\cdot$ s $\cdot$ m^{-2}.

a) Water flow $\Phi_v =$ m$^3 \cdot$ s^{-1}.
   Hydrostatic head $\Delta p = \rho g h =$ N $\cdot$ m^{-2}.
   Power $P = \Phi_v \cdot \Delta p =$ W.
b) Pump power $= \underline{P} =$ W.

c) Water velocity v = m$\cdot$ s^{-1}.

Pressure loss entry and exit $\Delta p = 2 \cdot K_W \cdot 0.5 \cdot \rho \cdot v^2 =$ N $\cdot$ m^{-2}.

$$\text{Re} = \frac{\rho v D}{\mu} =$$

Turbulent flow
Figure 5.5, concrete: 4f =

$$\Delta p = 4f \cdot \frac{L}{D} \cdot 0.5 \cdot \rho \cdot v^2 = \quad \text{N} \cdot \text{m}^{-2}.$$

Total pressure loss + = N $\cdot$ m^{-2}.

Friction power $P' = \Phi_v \cdot \Delta p =$ W.
Friction power as a percentage of the power calculated in (a):

Introduction to Chemical Engineering, First Edition. C.M. van 't Land.

Solution

a) $\Phi_v = 138.9\ m^3 \cdot s^{-1}$,

$$\Delta p = 1{,}000 \cdot 9.81 \cdot 230 = 2.26 \cdot 10^6\ N \cdot m^{-2},$$

$$P = 138.9 \cdot 2.26 \cdot 10^6 = 3.14 \cdot 10^8\ W = 314\ MW.$$

b) Pump power $\frac{314}{0.75} = 419\ MW.$

c) $v = 2.5\ m \cdot s^{-1}$.

Entry and exit: $\Delta p = 2 \cdot 2.5 \cdot 0.5 \cdot 1{,}000 \cdot 2.5^2 = 1.56 \cdot 10^4\ N \cdot m^{-2}$.

Line

$$Re = \frac{1{,}000 \cdot 2.5 \cdot 6}{1.2 \cdot 10^{-3}} = 1.25 \cdot 10^7,$$

$$4f = 0.013,$$

$$\Delta p = 0.013 \cdot \frac{650}{6} \cdot 0.5 \cdot 1{,}000 \cdot 2.5^2 = 4{,}401\ N \cdot m^{-2},$$

$$1.56 \cdot 10^4 + 4{,}401 = 2.00 \cdot 10^4\ N \cdot m^{-2},$$

$$P' = 2.00 \cdot 10^4 \cdot 138.9 = 2.78 \cdot 10^6\ W = 2.78\ MW,$$

$$\frac{2.8}{313} \cdot 100 = 0.89\%.$$

7.2 Estimation of the Size of a Pump Motor

An aqueous soda solution is pumped from a vessel at atmospheric pressure to the top of an absorption tower working at a pressure of 1.2 bara. See Figure 7.1. It is requested to estimate the size of the pump motor.

$$\Phi_V = 42\ m^3 \cdot h^{-1}.$$

$$L = 25\ m.$$

$$D = 0.095\ m.$$

$$\rho = 1{,}100\ kg \cdot m^{-3}.$$

$$\mu = 1.1\ cP\ \left(1.1 \cdot 10^{-3}\ N \cdot s \cdot m^{-2}\right).$$

$$\Delta p\ (\text{spray nozzle}) = 0.35\ bar\ \left(0.35 \cdot 10^5\ N \cdot m^{-2}\right).$$

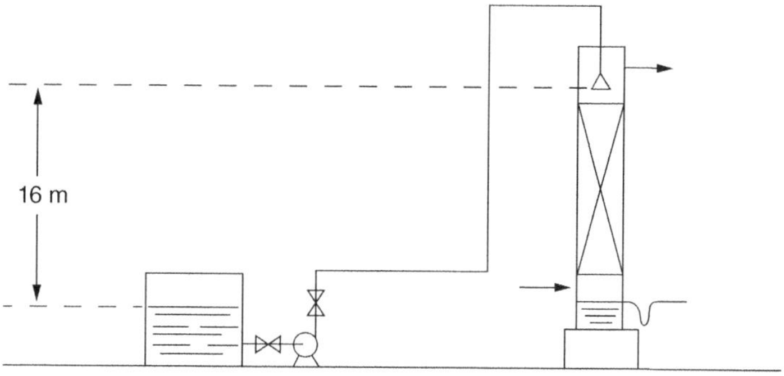

Figure 7.1 Estimation of the size of a pump motor.

Calculation Scheme

Bernoulli's Equation (Equation 2.2):

$$\left(\Delta\frac{p}{\rho} + g\cdot\Delta h + 0.5\cdot\Delta v^2\right)\Phi_m = -W - W_{fr} \quad W.$$

W is the power to be exerted by the pump. The calculation of W is the goal.

$$(\Delta p + \rho g\cdot\Delta h + 0.5\cdot\rho\cdot\Delta v^2)\Phi_v + W_{fr} = -W \quad W,$$

$$\Delta p\cdot\Phi_v = \qquad W,$$

$$\rho g\cdot\Delta h\cdot\Phi_v = \qquad W,$$

$$v = \qquad m\cdot s^{-1},$$

$$0.5\cdot\rho\cdot\Delta v^2\cdot\Phi_v = \qquad W.$$

W_{fr}
Line

$$Re = \frac{\rho v D}{\mu} =$$

Turbulent flow
Figure 5.5, drawn tube: 4f =

$$\Delta p = 4f\cdot\frac{L}{D}\cdot 0.5\cdot\rho\cdot v^2 = \qquad N\cdot m^{-2}.$$

See Figure 5.7 concerning four bends and two valves ($\delta = 70°$).

$$\Delta p = (\qquad + \qquad)0.5\cdot\rho\cdot v^2 = \qquad N\cdot m^{-2}.$$

$$W_{fr} = (\qquad + \qquad)\Phi_v = \qquad W.$$

$$-W = \qquad + \qquad + \qquad + \qquad = \qquad W.$$

Divide by 0.5 because of:

- pump hydraulic efficiency,
- gear efficiency, and
- motor efficiency.

The size of the pump motor is   kW.

Solution

$$\Delta p \cdot \Phi_v = (0.2 + 0.35)10^5 \cdot \frac{42}{3{,}600} = 642\ \mathrm{W}.$$

$$\rho g \cdot \Delta h \cdot \Phi_v = 1{,}100 \cdot 9.81 \cdot 16 \cdot \frac{42}{3{,}600} = 2{,}014\ \mathrm{W}.$$

$$v = \frac{42}{3{,}600 \cdot \frac{\pi}{4} \cdot 0.095^2} = 1.65\ \mathrm{m \cdot s^{-1}}.$$

$$0.5 \cdot \rho \cdot \Delta v^2 \cdot \Phi_v = 0.5 \cdot 1{,}100 \cdot 1.65^2 \cdot \frac{42}{3{,}600} = 17\ \mathrm{W}.$$

W_{fr}

Line

$$Re = \frac{\rho v D}{\mu} = \frac{1{,}100 \cdot 1.65 \cdot 0.095}{1.1 \cdot 10^{-3}} = 1.57 \cdot 10^5.$$

Turbulent flow

Figure 5.5, drawn tube: $4f = 0.016$.

$$\Delta p = 4f \cdot \frac{L}{D} \cdot 0.5 \cdot \rho \cdot v^2 = 0.016 \cdot \frac{25}{0.095} \cdot 0.5 \cdot 1{,}100 \cdot 1.65^2 = 6{,}305\ \mathrm{N \cdot m^{-2}}.$$

Figure 5.7, four bends and two valves ($\delta = 70°$):

$$\Delta p = (4 \cdot 1.20 + 2 \cdot 1.6)0.5 \cdot \rho \cdot v^2 = 8 \cdot 0.5 \cdot 1{,}100 \cdot 1.65^2 = 1.20 \cdot 10^4\ \mathrm{N \cdot m^{-2}}.$$

$$W_{fr} = \left(6{,}305 + 1.20 \cdot 10^4\right)\frac{42}{3{,}600} = 214\ \mathrm{W}.$$

$$-W = 642 + 2{,}014 + 17 + 214 = 2{,}887\ \mathrm{W}.$$

$$\frac{2{,}887}{0.5} = 5{,}774\ \mathrm{W}.$$

Take a 6-kW motor.

8

Heat Conduction

8.1 Introduction

Heat conduction is a heat transport mechanism. The other transport modes are convection and radiation. Heat transport is important. The two main objectives are:

- The transfer of heat from a medium to a second medium, for example, the removal or supply of the heat generated or required for a chemical reaction, and
- The prevention by insulation of heat transfer from, for example, steam lines.

Heat transport by means of conduction and convection are the two main transport mechanisms. Often, both modes play a role. The removal of heat generated by a chemical reaction by a jacket of a stirred reactor is an example. It is assumed that the temperature of the reactor contents is kept constant by means of stirring causing forced convection. The heat diffuses through the thermal laminar boundary layer by conduction. Heat transport through the reactor wall is also by conduction. In the jacket, the heat diffuses through the thermal laminar boundary layer and is carried away by the flowing medium.

The thermal conductivity is an important property for the heat conduction. This transport coefficient is defined in Figure 8.1. The thermal conductivity is also important for convective heat transport because a flowing medium can only absorb heat from a warm wall if it is conducting. The property will be discussed in Section 8.2. Steady-state heat conduction is attempted in Section 8.3. A typical case of unsteady-state heat transfer by conduction is treated in Section 8.4.

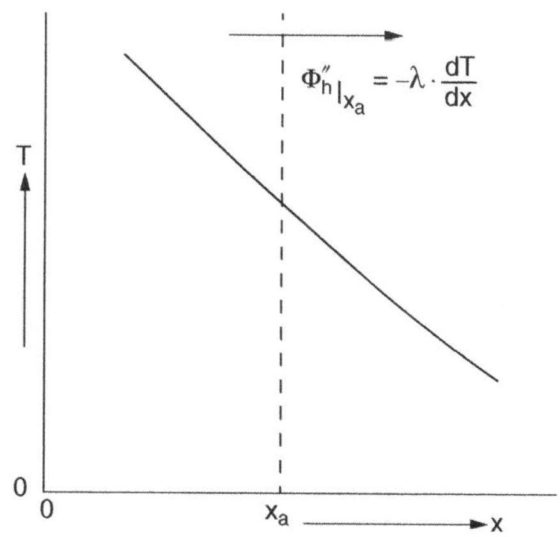

Figure 8.1 The definition of the thermal conductivity.

Introduction to Chemical Engineering, First Edition. C.M. van 't Land.

8.2 Thermal Conductivity

Gases

The following expression is valid according to kinetic theory for monatomic gases (like neon):

$\lambda \propto \rho \cdot v_m \cdot l \cdot c_v$ $W \cdot m^{-1} \cdot K^{-1}$ [1]. The meaning of the symbols is as follows:

ρ: specific mass of the gas in $kg \cdot m^{-3}$

v_m: average molecular velocity in $m \cdot s^{-1} (\propto \sqrt{T})$

l: mean free path in m ($\propto 1/\rho$)

c_v: specific heat at constant volume in $J \cdot kg^{-1} \cdot K^{-1}$

It follows that the thermal conductivity is independent of pressure as $l \propto 1/\rho$.

For this statement, the pressure should be neither too high nor too low.

The product $\rho \cdot l \cdot c_v$ does not vary strongly with temperature. Hence, λ is, like v_m, approximately proportional to $\sqrt{T}$.

Concerning the viscosity of gases, it was stated in Section 3.5.1 that, also according to kinetic theory, $\mu \propto \rho \cdot v_m \cdot l$ $N \cdot s \cdot m^{-2}$. The latter expression is not restricted to monatomic gases.

The expressions for the viscosity and the thermal conductivity can be combined to $\lambda = B \cdot c_v \cdot \mu$. The latter equation can be used to estimate λ when c_v and μ are known. An approximation for B is:

$$B = 2.25\left(\frac{R}{c_v}\right) + 1.$$

R is the gas constant in $J \cdot kmol^{-1} \cdot K^{-1}$ and c_v is in the latter expression in $J \cdot kmol^{-1} \cdot K^{-1}$. B equals 2.5 for ideal monatomic gases like neon because $c_v = (3/2)R$ in that case [2].

Figure 8.2 shows the variation of λ with temperature for a few gases at basically atmospheric pressure. The thermal conductivity of air is in the range 0.02–0.03 $W \cdot m^{-1} \cdot K^{-1}$.

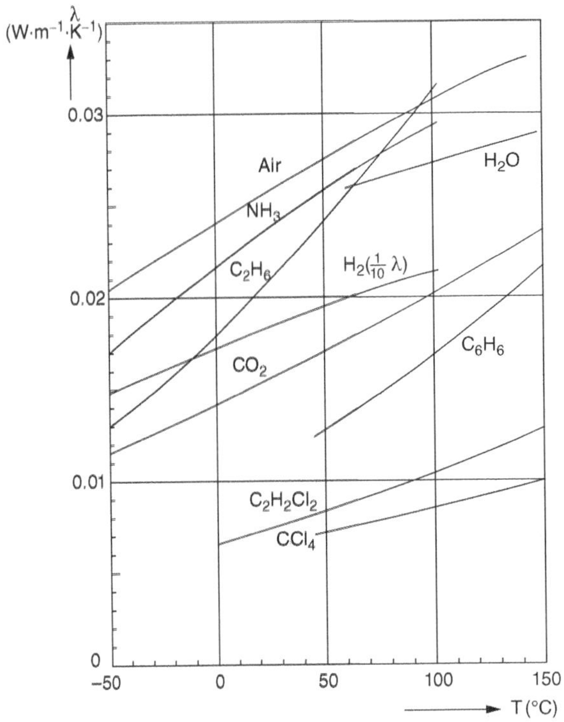

Figure 8.2 The thermal conductivity of a few gases as a function of the temperature. *Source:* Courtesy of Delft Academic Press/VSSD, Delft, The Netherlands.

Liquids

Practical rules do not exist. There is no relationship with viscosity. Usually, the thermal conductivity decreases slightly with temperature. However, for water, λ increases from 0 to 100 °C.

Most organic liquids have λ-values in the range 0.1–0.15 $W \cdot m^{-1} \cdot K^{-1}$. Associated compounds like alcohols have higher λ's, water being an extreme with $\lambda = 0.6\ W \cdot m^{-1} \cdot K^{-1}$ at ambient temperature. Mercury is a liquid metal.

Figure 8.3 shows the variation of λ with temperature for a few liquids.

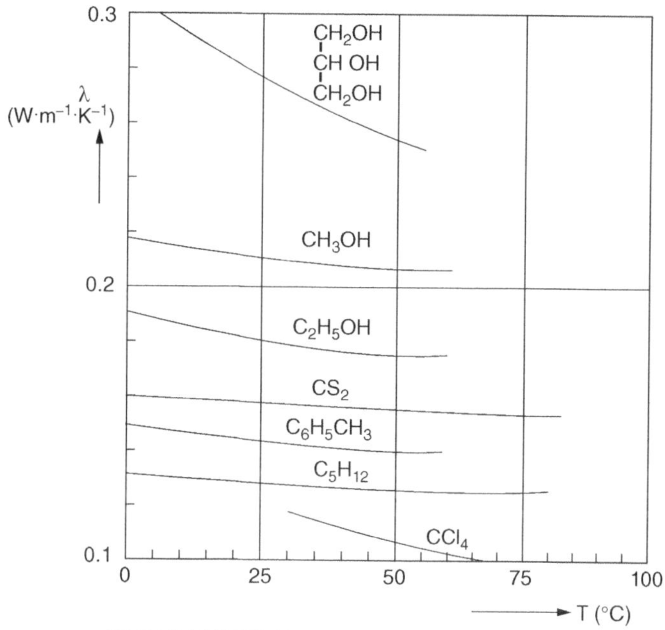

	Ice				Water							
T	−100	−50	−25	0	0	25	50	75	125	200	300	375
λ	3.48	2.76	2.49	2.23	0.56	0.61	0.64	0.67	0.68	0.66	0.54	0.21

Figure 8.3 The thermal conductivity of a few liquids as a function of the temperature. *Source:* Courtesy of Delft Academic Press/VSSD, Delft, The Netherlands.

Solids

A distinction is made between crystalline and amorphous materials. In crystals, the elementary particles are arranged regularly in space. A measurement method to assess the presence of a crystalline lattice is X-ray diffraction. A regular pattern is obtained for crystalline materials. A regular pattern cannot be established for amorphous materials. Four different types of crystals can be distinguished:

- metal,
- ionic,
- molecular, and
- atomic.

Metals are good conductors. See Figure 8.4. Like electricity, heat is conducted by free electrons. Wiedemann–Franz–Lorenz' law is valid: $\lambda \propto T\sigma$ in which σ is the electrical conductivity. The proportionality constant is strongly dependent on the purity of the metal. Thermal conductivity of metals is in the range 20–200 $W \cdot m^{-1} \cdot K^{-1}$. The conductivity of alloys is worse than the conductivity of the metals forming the alloys.

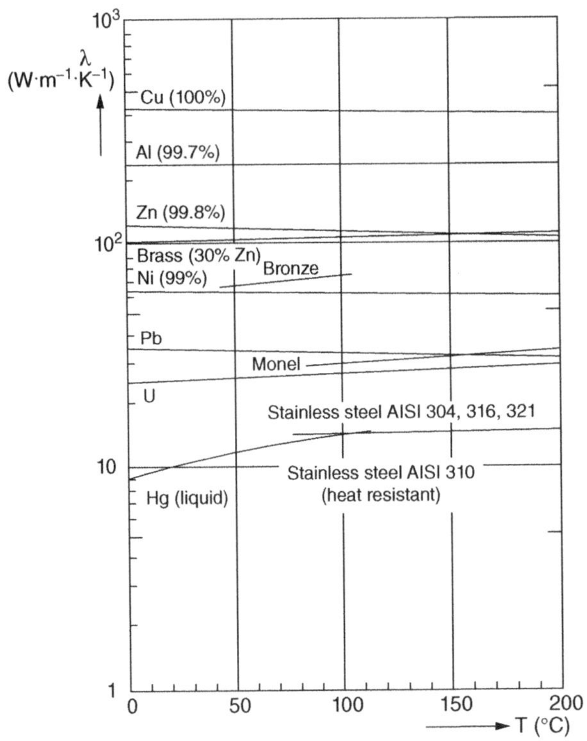

Figure 8.4 The thermal conductivity of a few metals and alloys as a function of the temperature. *Source:* Courtesy of Delft Academic Press/VSSD, Delft, The Netherlands.

Ionic crystals are relatively good conductors. For example, $\lambda = 5.8\ W \cdot m^{-1} \cdot K^{-1}$ for vacuum pan salt at 30 °C.

Molecular crystals are held together by Van der Waals forces. Thermal conductivity is low. Thermal conductivities in $W \cdot m^{-1} \cdot K^{-1}$ are given for three compounds:

- 0.175 for stearic acid,
- 0.37 for naphthalene, and
- 0.26 for mono-chloro acetic acid.

Atomic crystals are crystals made up of atoms. Bonds between atoms can, like in diamond, be covalent. Bonds can also be caused by Van der Waals forces, like in solid argon.

Heat is transported by lattice vibrations in ionic, atomic, and molecular crystals. These vibrations are damped more strongly at higher than at lower temperatures. Hence, for pure crystals, $\lambda \propto 1/T$.

For disturbed crystal lattices: $\lambda \propto (T + \theta)^{-1}$. θ is large for amorphous materials and λ is then hardly dependent on the temperature.

λ of an amorphous material is lower than λ of the corresponding crystalline material. For example, λ is 6–10 $W \cdot m^{-1} \cdot K^{-1}$ for crystalline silica at 0 °C whereas λ is 0.8–1.3 $W \cdot m^{-1} \cdot K^{-1}$ for glass at 20 °C.

Anisotropy can play a role in crystals. That means that the physical property is a function of the direction in which the property is measured. For example, the thermal conductivity of fibrous materials depends on whether the property is measured in the direction of the fibers or perpendicular to the fibers.

Heterogeneous Systems

Refractory and insulating materials are examples. Usually, solid/gas systems are met. Either the solid phase or the gaseous phase is continuous. The former option is preferred when the materials must be solid structures like in furnaces. The thermal conductivity is between 0.1 and 1.0 of the solid's thermal conductivity.

Figure 8.5 shows the thermal conductivity of systems of which the gaseous phase is continuous. Heat transfer by free convection cannot take place because the air cavities are small. Mainly, the air takes care of the particle-to-particle heat transfer by conduction.

8.3 Steady-state Heat Conduction

Heat conduction through a wall is depicted in Figure 8.6. The following equation is valid:

$$\Phi_h'' = \frac{\lambda}{d}(T_1 - T_0)\ W \cdot m^{-2}. \tag{8.1}$$

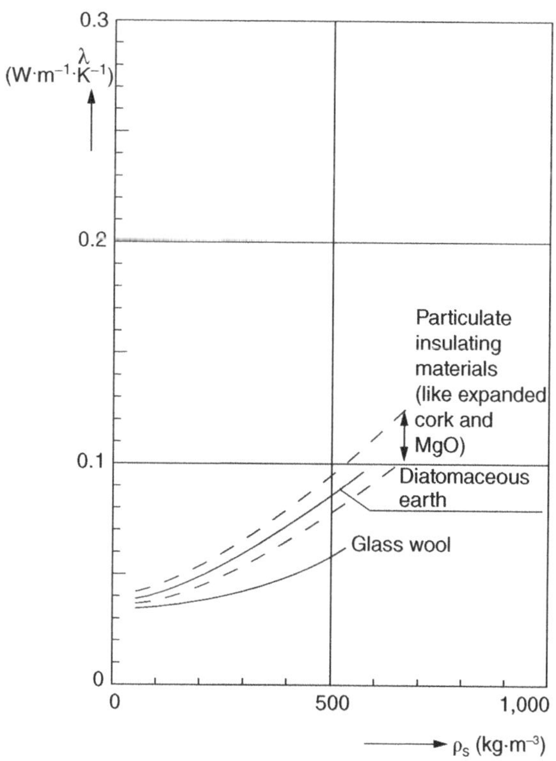

Figure 8.5 The thermal conductivity of some insulating materials at 20 °C as a function of the bulk density. *Source:* Courtesy of Delft Academic Press/VSSD, Delft, The Netherlands.

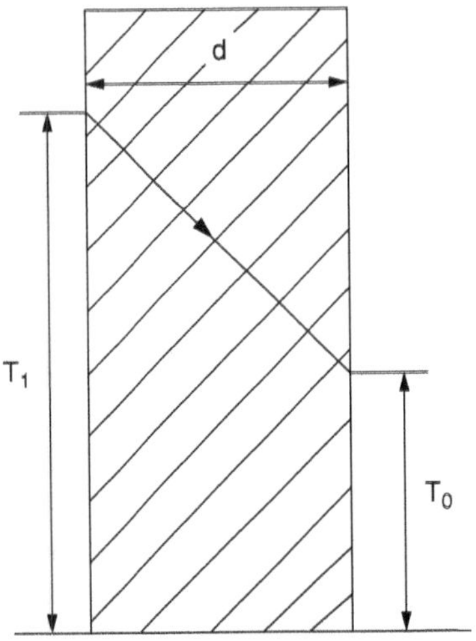

Figure 8.6 Heat conduction through a wall.

Heat conduction through a wall consisting of three different layers is shown in Figure 8.7. Now, the following equations are valid:

$$\Phi_h'' = \frac{\lambda_1}{x_2 - x_1}(T_1 - T_2) = \frac{\lambda_2}{x_3 - x_2}(T_2 - T_3) = \frac{\lambda_3}{x_4 - x_3}(T_3 - T_4)\ \mathrm{W \cdot m^{-2}}. \tag{8.2}$$

It is possible to write:

$$T_1 - T_2 = \frac{\Phi_h''(x_2 - x_1)}{\lambda_1},$$

$$T_2 - T_3 = \frac{\Phi_h''(x_3 - x_2)}{\lambda_2},$$

$$T_3 - T_4 = \frac{\Phi_h''(x_4 - x_3)}{\lambda_3}.$$

$+$

$$T_1 - T_4 = \Phi_h''\left(\frac{(x_2 - x_1)}{\lambda_1} + \frac{(x_3 - x_2)}{\lambda_2} + \frac{(x_4 - x_3)}{\lambda_3}\right). \tag{8.3}$$

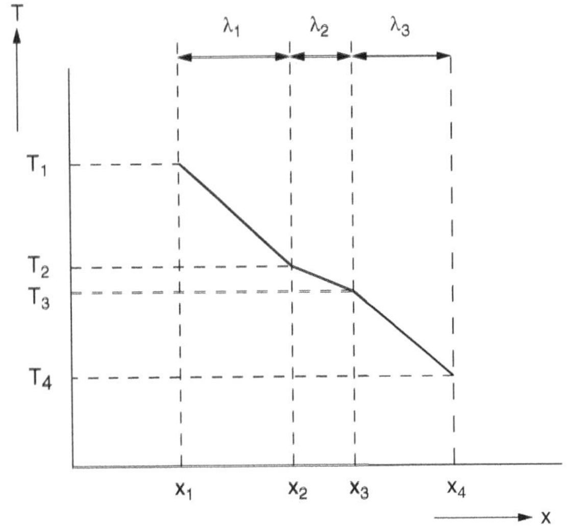

Figure 8.7 Heat conduction through a wall consisting of three layers. *Source:* Courtesy of Delft Academic Press/VSSD, Delft, The Netherlands.

The expression between the large brackets is the sum of the resistances to heat flow. The equation states that the overall temperature difference equals the heat flow multiplied by the sum of the resistances. This is analogous to Ohm's law for three resistances in series: $V = i(R_1 + R_2 + R_3)$ V.

Figure 8.8 shows the conduction of heat through a cylindrical wall. The heat flow through the surface at $r = R_1$, at $r = R_2$, and at all surfaces in between is constant. On going from the surface at $r = R_1$ to the surface at $r = R_2$, the surface through which the heat flows increases. Such a case can be treated by means of the calculus. The basic equation for a hollow cylinder having a length L m is:

$$\Phi_h = -\frac{\lambda}{dr} \cdot dT \cdot 2\pi r L \text{ W},$$

$$dT = -\frac{\Phi_h}{2\pi\lambda L} \cdot \frac{dr}{r} \text{ K}.$$

The integration of this differential equation proceeds as follows:

$$\int_{T_2}^{T_1} dT = -\frac{\Phi_h}{2\pi\lambda L} \int_{R_2}^{R_1} \frac{dr}{r},$$

$$T_1 - T_2 = -\frac{\Phi_h}{2\pi\lambda L} \cdot \ln(R_1/R_2),$$

$$\Phi_h = \frac{2\pi\lambda L(T_1 - T_2)}{\ln(R_2/R_1)}. \quad (8.4)$$

Integrating between T_1 and T and between r and R_1 leads to:

$$\Phi_h = \frac{2\pi\lambda L(T_1 - T)}{\ln(r/R_1)}.$$

Figure 8.8 Steady-state heat conduction through a cylindrical wall. *Source:* Courtesy of Delft Academic Press/VSSD, Delft, The Netherlands.

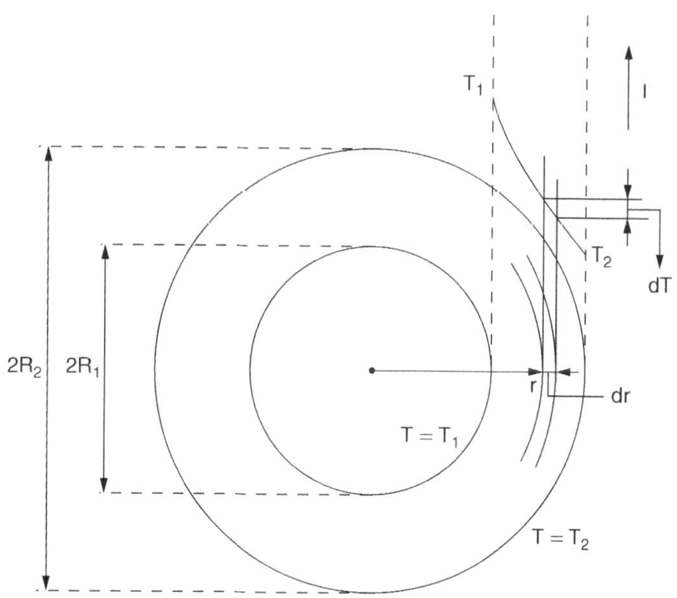

On dividing that expression by Equation 8.4, the temperature distribution appears:

$$\frac{T_1 - T}{T_1 - T_2} = \frac{\ln(r/R_1)}{\ln(R_2/R_1)}. \tag{8.5}$$

It is possible to calculate the temperature as a function of the radius.

Example 8.1

Superheated steam having a pressure of 15 bar and a temperature of 350 °C flows through a line having an outside diameter of 108 mm and an inside diameter of 100 mm. The line is 100 m long and is insulated with glass wool having a thickness of 50 mm. The heat loss of the line in W is calculated.

According to Figure 8.5, the thermal conductivity of glass wool is 0.05 $W \cdot m^{-1} \cdot K^{-1}$. It is estimated that the temperature of the glass wool decreases from 350 to 40 °C. In first instance, it is assumed that the steam temperature remains at 350 °C.

$$\Phi_h = \frac{2\pi\lambda L(T_1 - T_2)}{\ln R_2/R_1} = \frac{2\pi \cdot 0.05 \cdot 100(350 - 40)}{\ln 104/54} = 1.49 \cdot 10^4\ W.$$

The heat loss from this line is 14.9 kW.

Notes

Dividing the heat flow by the area and by the difference between the surface temperature and the ambient temperature gives the coefficient for the transfer of heat from the surface to the atmosphere by convection.

$$\frac{1.49 \cdot 10^4}{100 \cdot 2\pi \cdot 0.104(40 - 15)} = 9.1\ W \cdot m^{-2} \cdot K^{-1}.$$

This is a normal figure. The assumption of 40 °C is correct. Radiation does not yet contribute at 40 °C.

The heat loss causes a temperature decrease of the steam. A decrease of 12.5 K can be estimated when the steam velocity is 15 $m \cdot s^{-1}$.

Next, the heat conductivity through a hollow sphere is treated.

This type of heat conduction occurs at, for example, the raising of nuclear energy inside spheres. Such spheres are present in a pebble bed reactor (PBR). They have a diameter of, for example, 60 mm. A 5-mm outer shell of the pebbles consists of graphite. The outside radius of the sphere is R_2 whereas the outside radius of the spherical cavity is R_1. The temperatures are, respectively, T_2 and T_1. Figure 8.8 can be consulted.

The heat flow through the surface at $r = R_1$, at $r = R_2$, and through all surfaces in between is constant. Also, in this case, the application of the calculus is useful. The basic equation is:

$$\Phi_h = -\frac{\lambda}{dr} \cdot dT \cdot 4\pi \cdot r^2\ W,$$

$$dT = -\frac{\Phi_h}{4\pi\lambda} \cdot \frac{dr}{r^2}.$$

The integration of this differential equation proceeds as follows:

$$\int_{T_2}^{T_1} dT = -\frac{\Phi_h}{4\pi\lambda} \int_{R_2}^{R_1} \frac{dr}{r^2} \text{ K},$$

$$T_1 - T_2 = \frac{\Phi_h}{4\pi\lambda} \int_{R_2}^{R_1} dr^{-1},$$

$$T_1 - T_2 = \frac{\Phi_h}{4\pi\lambda} \left(\frac{1}{R_1} - \frac{1}{R_2} \right),$$

$$\Phi_h = \frac{4\pi\lambda(T_1 - T_2)}{1/R_1 - 1/R_2} \text{ W}. \quad (8.5)$$

Integrating between T_1 and T and between R_1 and r leads to:

$$\Phi_h = \frac{4\pi\lambda(T_1 - T)}{1/R_1 + 1/r}.$$

Dividing the latter equation by Equation 8.5 results in the temperature distribution:

$$\frac{T_1 - T_2}{T_1 - T} = \frac{1/R_1 - 1/R_2}{1/R_1 - 1/r}.$$

It is possible to calculate the temperature as a function of the radius.

The heat release of a sphere having a radius of R_1 and a surface temperature T_1 to a stagnant, infinitely extended surrounding having a constant λ and a temperature T_2 at an infinitely large distance is:

$$\Phi_h = 4\pi\lambda \cdot R_1(T_1 - T_2) \text{ W}.$$

R_2 is very large in Equation 8.5 in that case and $1/R_2$ becomes 0.

8.4 Heating or Cooling of a Solid Body

A slab having a thickness D is taken as an example. At time $t = 0$, the temperature of both walls is raised from T_0 to T_1. The temperature in the slab increases as shown in Figure 8.9. The temperature is a function of both time and location. We see a case of unsteady-state heat transfer by conduction. Fourier (1768–1830) was a pioneer in heat transfer research and also studied this type of heat transfer. His work concerning this subject resulted in graphs shown in Figures 8.10 and 8.11. The graphs contain data applicable to slabs. There are also data for other geometries. Figure 8.10 is valid for the temperature T_m of the center of the geometries whereas Figure 8.11 is applicable for the mean temperature T_{av} of the geometries.

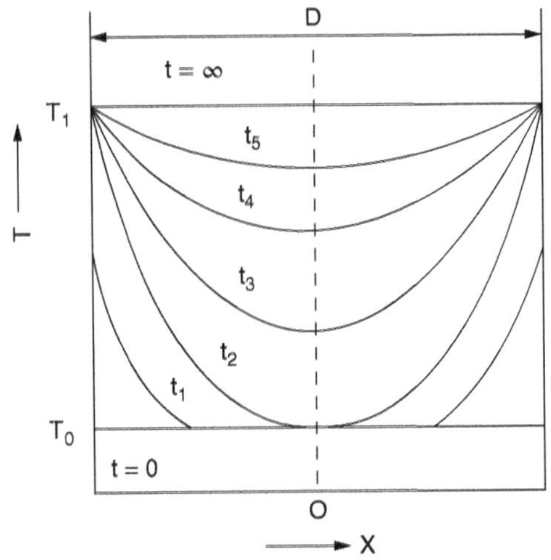

Figure 8.9 Heat penetration into a slab having a thickness D. *Source:* Transport Phenomena by Beek, W. J. et al. (1999). Reproduced with permission of John Wiley & Sons Inc.

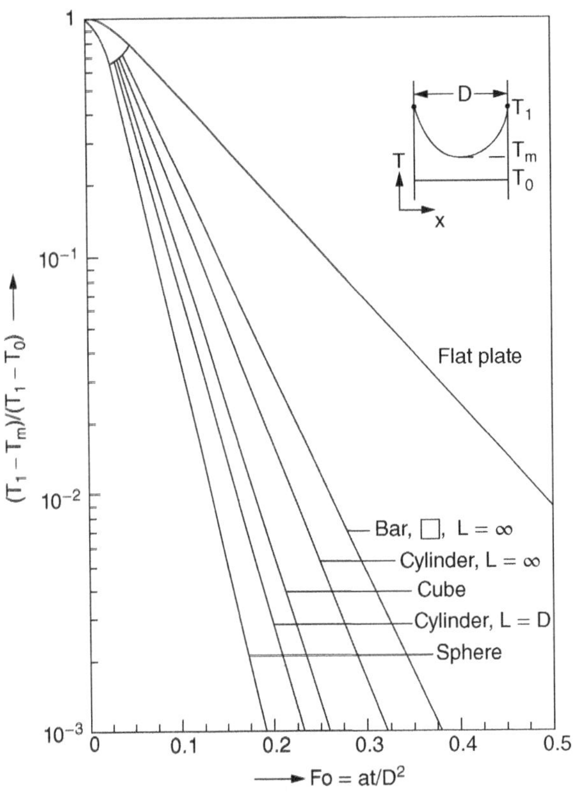

Figure 8.10 Central temperature during unsteady-state heat conduction. *Source:* Transport Phenomena by Beek, W. J. et al. (1999). Reproduced with permission of John Wiley & Sons Inc.

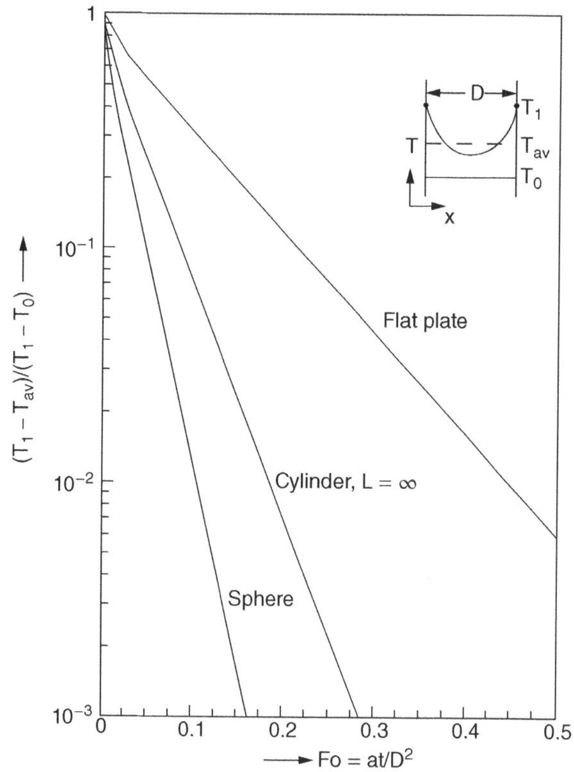

Figure 8.11 Mean temperature during unsteady-state heat conduction. *Source:* Transport Phenomena by Beek, W. J. et al. (1999). Reproduced with permission of John Wiley & Sons Inc.

Example 8.2

A stainless steel slab (flat plate) has a thickness D of 0.1 m. Its temperature T_0 is initially 10 °C. The temperature at the boundaries T_1 is raised to 100 °C at t = 0. The time required to reach T_m = 99 °C at the center of the slab is calculated.

$$\frac{T_1 - T_m}{T_1 - T_0} = \frac{100 - 99}{100 - 10} = \frac{1}{90} = 0.011.$$

Figure 8.10: $\frac{at}{D^2} = 0.48$

$\frac{at}{D^2}$ is the Fourier number (Fo) appearing in all problems of unsteady-state heat conduction. a is called the heat diffusivity. It is a measure for the rate at which temperature differences in a medium are smoothed out by heat conduction.

The relevant properties of stainless steel are:

$\lambda = 16\ \mathrm{W \cdot m^{-1} \cdot K^{-1}}$ (Figure 8.4).

$c_p = 500\ \mathrm{J \cdot kg^{-1} \cdot K^{-1}}$.

$\rho = 8{,}000\ \mathrm{kg \cdot m^{-3}}$.

$$a = \frac{\lambda}{c_p \cdot \rho} = 4 \cdot 10^{-6}\ \mathrm{m^2 \cdot s^{-1}}.$$

t = 1,200 s.

References

1. Bird, R. B., Stewart, W. E., and Lightfoot, E. N. (1960). *Transport Phenomena* (p. 255). New York, NY: Wiley.
2. Bird, R. B., Stewart, W. E., and Lightfoot, E. N. (1960). *Transport Phenomena* (p. 254). New York, NY: Wiley.

9

Convective Heat Transfer

9.1 Heat Exchangers

In first instance, a simple model of a heat exchanger is considered. See Figure 9.1. It concerns two concentric tubes. Heat is transferred from Liquid′ to Liquid″. The two liquids can flow concurrently or countercurrently. The area required for heat transfer can be calculated from the equation:

$$\Phi_h = UA \cdot \Delta T_m \text{ W}. \tag{9.1}$$

U is the overall heat transfer coefficient in $W \cdot m^{-2} \cdot K^{-1}$. The overall heat transfer resistance 1/U equals the sum of the partial resistances.

$$\frac{1}{U} = \frac{1}{\alpha'} + \frac{d_w}{\lambda_w} + \frac{d_d}{\lambda_d} + \frac{1}{\alpha''}. \tag{9.2}$$

See Figure 9.2.

A is the area in m^2 which is available for heat transfer.

ΔT_m is the logarithmic mean temperature difference:

$$\Delta T_m = \frac{\Delta T_0 - \Delta T_L}{\ln\left(\Delta T_0/\Delta T_L\right)}. \tag{9.3}$$

ΔT_0 and ΔT_L are the temperature differences between the process flow and the medium at $x = 0$ and $x = L$. Equation 9.3 is valid for both concurrent and countercurrent flow. It will be derived for countercurrent flow. See Figure 9.1. The temperatures of both media and the temperature difference of both media are functions of the location x. Liquid′ is cooled whereas Liquid″ is heated. The application of the calculus is useful for such a case.

The energy balance for the heat exchanger is:

$$\Phi_h = \Phi'_m \cdot c'_p\left(T'_0 - T'_L\right) = \Phi''_m \cdot c''_p\left(T''_0 - T''_L\right).$$

A small length dx of the heat exchanger is defined. The temperatures of both media change with, respectively, dT' and dT'' on passing dx. The energy balance for the small length dx is:

$$d\Phi_h = -\Phi'_m \cdot c'_p \cdot dT' = -\Phi''_m \cdot c''_p \cdot dT''.$$

The minus signs in the latter equations are required because the temperatures of both media decrease when x increases. The small heat flow from Liquid′ to Liquid″ can also be expressed kinetically in a form analogous to the form of Equation 9.1:

$$d\Phi_h = US \cdot dx(T' - T'') = US \cdot dx \cdot \Delta T. \tag{9.4}$$

Introduction to Chemical Engineering, First Edition. C.M. van 't Land.

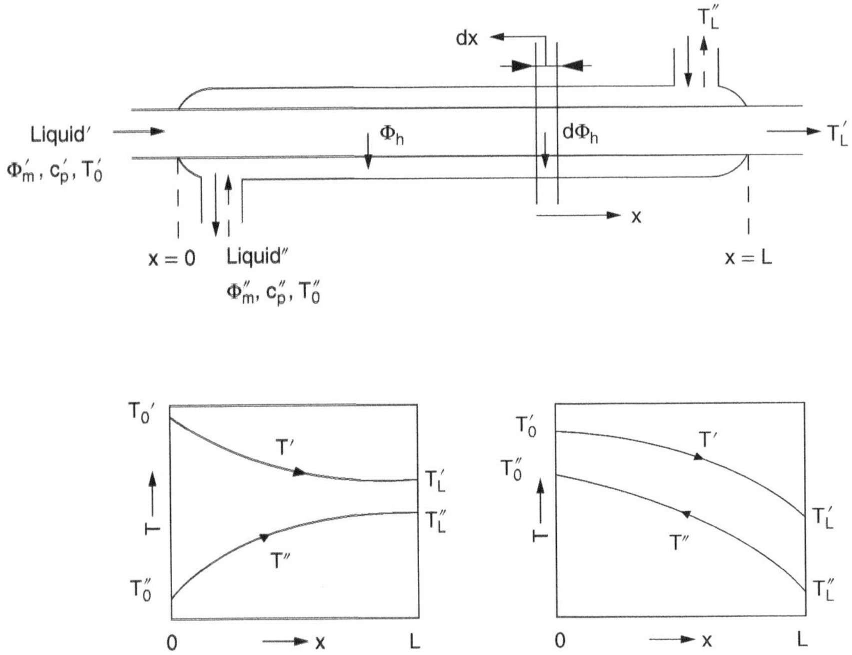

Figure 9.1 A heat exchanger built as two concentric tubes. The flow can be concurrent or countercurrent. *Source:* Courtesy of Delft Academic Press/VSSD, Delft, The Netherlands.

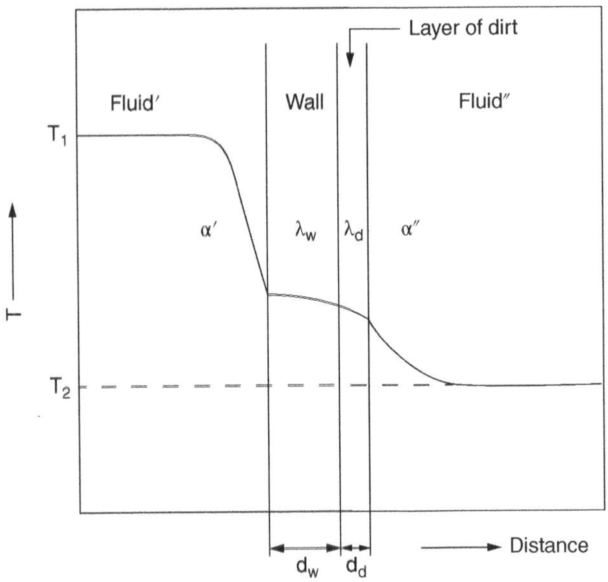

Figure 9.2 Temperature profile at heat transfer from Fluid′ to Fluid″. *Source:* Transport Phenomena by Beek, W. J. et al. (1999). Reproduced with permission of John Wiley & Sons Inc.

S is the length of the periphery of the tube through which the heat transfer occurs in m.

The energy balance of the heat exchanger gives directly:

$$\frac{\Phi_h}{\Phi'_m . c'_p} = T'_0 - T'_L \text{ and } \frac{\Phi_h}{\Phi''_m \cdot c''_p} = T''_0 - T''_L.$$

Subtraction of the first of these two equations from the second one leads to:

$$\Phi_h \left(\frac{1}{\Phi''_m \cdot c''_p} - \frac{1}{\Phi'_m \cdot c'_p} \right) = T''_0 - T''_L - T'_0 + T'_L = \Delta T_L - \Delta T_0. \tag{9.5}$$

The energy balance for a small length dx gives directly:

$$\frac{d\Phi_h}{\Phi'_m \cdot c'_p} = -dT' \text{ and } \frac{d\Phi_h}{\Phi''_m \cdot c''_p} = -dT''.$$

Subtraction of the first of these two equations from the second one leads to:

$$d\Phi_h \left(\frac{1}{\Phi''_m \cdot c''_p} - \frac{1}{\Phi'_m \cdot c'_p} \right) = dT' - dT'' = d\Delta T. \tag{9.6}$$

Dividing Equation 9.6 by Equation 9.5 results in:

$$\frac{d\Phi_h}{\Phi_h} = \frac{d\Delta T}{\Delta T_L - \Delta T_0}. \tag{9.7}$$

Substitution of Equation 9.4 into Equation 9.7 gives:

$$\frac{US \cdot dx \cdot \Delta T}{\Phi_h} = \frac{d\Delta T}{\Delta T_L - \Delta T_0} \text{ and } \frac{d\Delta T}{\Delta T} = U \cdot \frac{\Delta T_L - \Delta T_0}{\Phi_h} \cdot S \cdot dx.$$

This is a differential equation which can be integrated. The integration proceeds as follows:

$$\int_{\Delta T_0}^{\Delta T_L} \frac{d\Delta T}{\Delta T} = \frac{U(\Delta T_L - \Delta T_0)S}{\Phi_h} \int_0^L dx,$$

$$\ln \frac{\Delta T_L}{\Delta T_0} = \frac{U(\Delta T_L - \Delta T_0)SL}{\Phi_h} \text{ and } \Phi_h = \frac{U(\Delta T_L - \Delta T_0)A}{\ln (\Delta T_L / \Delta T_0)} - UA \cdot \Delta T_m.$$

Table 9.1 contains a summary of heat transfer coefficients. The calculation of U-values will be treated in Section 9.2. In practice, most heat exchangers do not belong to the simple "single pass" type but to the "multiple pass" type. Equation 9.3 is still used; however, a correction factor F is added:

$$\Delta T_m = F \cdot \frac{\Delta T_L - \Delta T_0}{\ln (\Delta T_L / \Delta T_0)}. \tag{9.8}$$

Figure 9.3 gives F as a function of X whereas Y is a parameter. The graph is valid for a heat exchanger with two passes at the tube side and one pass at the shell side. More than one pass at either side of the heat exchanger are designed to use the available space in plants as economically as possible. The heat exchanger of Figure 9.4 is called a heat exchanger with hairpin tubes.

Table 9.1 Summary of heat transfer coefficients ($W \cdot m^{-2} \cdot K^{-1}$).

Heat flow → to: ↓ from:	Gas (free convection) α = 5–15	Gas (flowing) α = 10–100	Liquid (free convection) α = 50–1,000	Liquid (flowing): water α = 3,000–10,000, other liquids α = 500–2,000	Boiling liquid: water α = 3,500–60,000, other liquids α = 1,000–20,000
Gas (free convection) α = 5–15	Room/outside air through glass U = 1–2	Superheaters U = 3–10			
Gas (flowing) α = 10–100		Heat exchangers for gases U = 10–30	Gas boiler U = 10–50	Oven U = 10–40 + radiation	Boiler U = 10–40 + radiation
Liquid (free convection) α = 50–10,000	Radiator central heating U = 5–15		Oil bath for heating U = 25–500	Cooling coil U = 500–1,500 if stirred	
Liquid (flowing): water α = 3,000–10,000, other liquids α = 500–3,000		Gas coolers U = 10–50	Heating coil in vessel: water/water without stirring U = 50–250, with stirring U = 500–2,000	Heat exchanger: water/water U = 900–2,500, water/other liquids U = 200–1,000	Evaporators of cooling units, brine coolers U = 300–1,000
Condensing vapor: water α = 5,000–30,000, other liquids α = 1,000–4,000	Steam radiators U = 5–20	Air heaters U = 10–50	Steam jackets around vessels with stirrers, water U = 300–1,000, other liquids U = 150–500	Condensers: steam/water U = 1,000–4,000, other vapor/water U = 300–1,000	Evaporators: steam/water U = 1,500–6,000, steam/other liquids U = 300–2,000

Source: Transport Phenomena by Beek, W. J. et al. (1999). Reproduced with permission of John Wiley & Sons Inc.

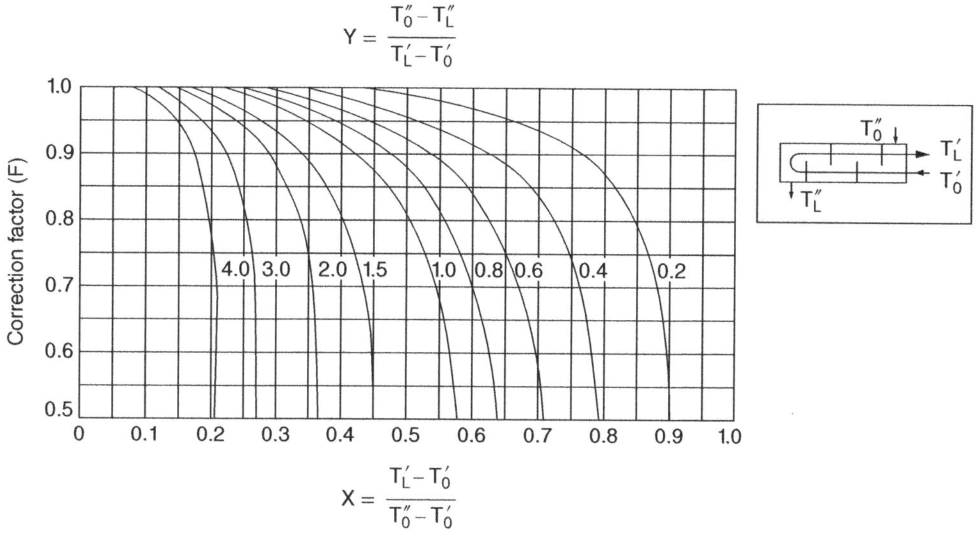

Figure 9.3 Correction for ΔT_m for multiple-pass heat exchanger (one shell pass and two tube passes). *Source:* Transport Phenomena by Beek, W. J. et al. (1999). Reproduced with permission of John Wiley & Sons Inc.

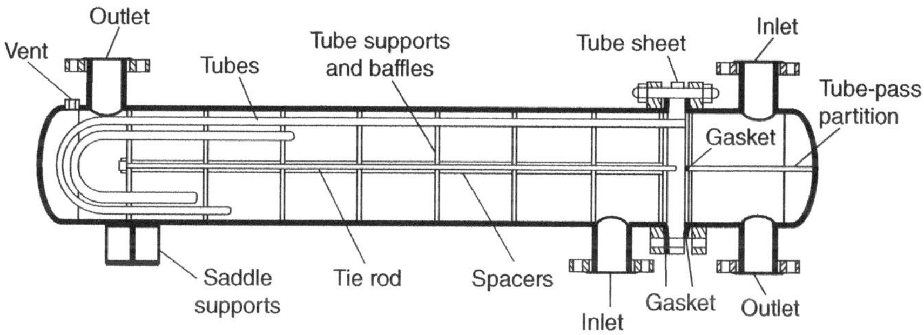

Figure 9.4 Heat exchanger with hairpin tubes.

Example 9.1

It is required to cool 10^4 kg · h^{-1} of benzene from 80 to 40 °C in a heat exchanger. The cooling medium is water having a temperature of 15 °C. The cooling water flow is $1.5 \cdot 10^4$ kg · h^{-1}. The area of the heat exchanger is calculated for three cases: (a) a countercurrent concentric tube heat exchanger, (b) a concurrent concentric tube heat exchanger, and (c) a heat exchanger with hairpin tubes according to Figure 9.4. The cooling water flows through the tubes. The specific heats of benzene and water are, respectively, 1,730 and 4,190 J · kg^{-1} · K^{-1}.

The heat load is: $\Phi_h = \dfrac{10^4 \cdot 1{,}730(80 - 40)}{3{,}600} = 1.92 \cdot 10^5$ W.

The cooling water temperature rise is: $\dfrac{1.92 \cdot 10^5 \cdot 3{,}600}{1.5 \cdot 10^4 \cdot 4{,}190} = 11.0$ K.

The cooling water outlet temperature is 15 + 11 = 26 °C.
An overall heat transfer coefficient of 600 $W \cdot m^{-2} \cdot K^{-1}$ is selected from Table 9.1.
Countercurrent concentric tube heat exchanger

$$\Delta T_m = \frac{(40-15)-(80-26)}{\ln(40-15)/(80-26)} = 37.7\ K.$$

$$A = \frac{1.92 \cdot 10^5}{600 \cdot 37.7} = 8.5\ m^2.$$

Concurrent concentric tube heat exchanger

$$\Delta T_m = \frac{(80-15)-(40-26)}{\ln(80-15)/(40-26)} = 33.2\ K.$$

$$A = \frac{1.92 \cdot 10^5}{600 \cdot 33.2} = 9.6\ m^2.$$

Thus, a countercurrent heat exchanger is more efficient than a concurrent heat exchanger. The driving force of the former is greater leading to a smaller required area.
Heat exchanger with hairpin tubes according to Figure 9.4.

$$X = \frac{26-15}{80-15} = 0.17.$$

$$Y = \frac{80-40}{26-15} = 3.6.$$

F = 0.95 according to Figure 9.4 and $A = \frac{1.92 \cdot 10^5}{600 \cdot 0.95 \cdot 37.7} = 8.9\ m^2$.

Note that ΔT_m is calculated as if the heat exchanger with hairpin tubes were a countercurrent tube heat exchanger. The calculated area of 8.9 m^2 is in between the areas calculated for the countercurrent and concurrent versions. The flow in the heat exchanger with hairpin tubes is in actual fact partly countercurrent and partly concurrent.

9.2 Heat Transfer Correlations

A reflux condenser on top of a reactor and a coil in a reactor are typical heat exchangers in the process industry. The area of a heat exchanger can be calculated by means of Equation 9.1. The calculation of Φ_h, the heat load, was discussed in Section 9.1. The calculation of ΔT_m, the logarithmic mean temperature difference, was also treated in Section 9.1. The calculation of heat transfer coefficients will now be dealt with.

Heat Exchange Between a Fixed Wall and a Fluid

Many measurements have been carried out in the first half of the previous century. The results could be summarized by equations of the following type:

$$Nu = f\left(\frac{X}{D}, Re, Pr\right). \tag{9.9}$$

Nu: the Nusselt number, that is, $\alpha D/\lambda$
x: the location, for example, the path traveled in a tube
D: a characteristic dimension, for example, a tube diameter
Re: the Reynolds number characterizing the flow
Pr: the Prandtl number containing fluid properties, $Pr = v/a$ with $a = \lambda/(c_p \cdot \rho)$

Turbulent flow of gases and liquids through tubes
The relevant equation is [1]:

$$Nu = 0.027 \cdot Re^{0.80} \cdot Pr^{1/3} \left(\frac{\mu}{\mu_w}\right)^{1/7}. \tag{9.10}$$

The conditions concerning the applicability of this equation are:

- the fluid properties must be determined at the average fluid temperature,
- the flow should be adjusted,
- $2 \cdot 10^3 < Re < 10^5$, and
- $Pr \geq 0.7$.

μ/μ_w is a viscosity correction, μ is the fluid viscosity at the average fluid temperature whereas μ_w is the fluid viscosity at the average maximum boundary layer temperature.

Example 9.2

Water flows with a velocity of 2 $m \cdot s^{-1}$ through a 70/30 Cu/Ni tube having an internal diameter of 29 mm. Steam condenses on the outside of the tube and maintains a temperature of 150 °C. The water is heated from 15 to 80 °C. The heat transfer coefficient for the flowing water is calculated.

The average water bulk temperature is $\frac{80 + 15}{2} = 47.5$ °C.

The water properties at 47.5 °C are:

$\rho = 989\ kg \cdot m^{-3}$.

$c_p = 4{,}170\ J \cdot kg^{-1} \cdot K^{-1}$.

$\mu = 5.72 \cdot 10^{-4}\ N \cdot s \cdot m^{-2}$.

$\lambda = 0.638\ W \cdot m^{-1} \cdot K^{-1}$.

$$Re = \frac{\rho v \cdot D_t}{\mu} = \frac{989 \cdot 2 \cdot 0.029}{5.72 \cdot 10^{-4}} = 1.00 \cdot 10^5.$$

The flow is turbulent.

$Pr = \frac{v}{a}$ whereas $v = \frac{\mu}{\rho}$ and $a = \frac{\lambda}{c_p \cdot \rho}$. Hence:

$$Pr = \frac{\mu}{\rho} \cdot \frac{c_p \cdot \rho}{\lambda} = \frac{\mu \cdot c_p}{\lambda} = \frac{5.72 \cdot 10^{-4} \cdot 4{,}170}{0.638} = 3.74.$$

It is assumed that the average maximum boundary temperature is $\frac{150 + 47.5}{2} = 98.75\ °C$. At the latter temperature, $\mu_w = 2.86 \cdot 10^{-4}\ N \cdot s \cdot m^{-2}$.

$$Nu = 0.027\left(1.00 \cdot 10^5\right)^{0.80} \cdot 3.74^{1/3}\left(5.72 \cdot 10^{-4}/2.86 \cdot 10^{-4}\right)^{1/7} = 462.7,$$

$$Nu = \frac{\alpha \cdot D}{\lambda} = 462.7 \rightarrow \alpha = \frac{0.638 \cdot 462.7}{0.029} = 1.02 \cdot 10^4\ W \cdot m^{-2} \cdot K^{-1}.$$

The heat transfer coefficient at the process side was calculated. To obtain the overall heat transfer coefficient, we would also need data concerning the transfer of heat through the wall and at the medium side. See Equation 9.2. The transfer of heat through the wall is considered by d_w/λ_w. Regarding the heat transfer coefficient at the medium side, we would need a heat transfer correlation. It is advised to consult the VDI Heat Atlas concerning heat transfer correlations [2]. The VDI Heat Atlas also contains methods to calculate the pressure loss of heat exchangers, both at the process side and at the medium side.

The Nusselt Number in a Stagnant Medium

The heat release of a sphere having a radius R_1 and a surface temperature T_1 to a stagnant, infinitely extended surrounding having a constant λ and a temperature T_∞ at an infinitely large distance is: $\Phi_h = \alpha 4\pi \cdot R_1^2(T_1 - T_\infty)$. The heat transfer is, in the latter equation, described as if it were convective heat transfer. It can also be described in terms of heat conduction. According to Section 8.3, the relevant equation is: $\Phi_h = 4\pi\lambda \cdot R_1(T_1 - T_\infty)$.

$1 = \frac{\alpha \cdot R_1}{\lambda}$ is obtained on dividing the first equation by the second equation. Thus, $Nu = \frac{2\alpha \cdot R_1}{\lambda} = \frac{\alpha \cdot D_1}{\lambda} = 2.$

References

1. Beek, W. J., Muttzall, K. M. K., and van Heuven, J. W. (1999). *Transport Phenomena* (p. 183). Chichester, UK: Wiley.
2. VDI e.V. (2010). *VDI Heat Atlas*. Berlin Heidelberg, Germany: Springer-Verlag.

10

Heat Transfer by Radiation

10.1 Introduction

Technically, heat transfer by radiation is accomplished by electromagnetic waves. Ultrasonic waves can also transfer heat; however, that is not considered in this chapter. A distinction is made between infrared radiation (IR), radio waves (RF, that is, radio frequency), and microwaves (MW).

The wavelengths and frequencies of electromagnetic waves are coupled by means of the formula $v = f\lambda$, in which v is the propagation velocity of electromagnetic waves in a vacuum, that is, $2.9978 \cdot 10^6\ m \cdot s^{-1}$, f is the frequency in Hz, that is, s^{-1}, and λ is the wavelength in m. The wavelengths of IR are in the range 1–10 μm, whereas characteristic wavelengths of microwaves and radio waves are 10 cm and 100 m, respectively.

IR can be generated in one out of two ways: electrically, by passing a current through an electrical resistance which emits, or by burning gas that heats an emitting ceramic plate. The other two types of radiation are generated by means of electrical energy.

10.2 IR

The thermal agitation of molecules of a body causes IR radiation. Radiation exchange between two surfaces may take place if the interspace is more or less diathermanous for the entire or part of the wavelength range. Heat transport then occurs in two directions and, if the surfaces have different temperatures, the difference between the two opposite heat flows is not equal to zero.

Generally, the wavelength in μm of IR radiation is used for classification: (a) short 0.76–2.0, (b) medium 2.0–4.0, and (c) long 4.0–10.0.

The wavelengths of visible light vary from 0.4 μm (blue) to 0.8 μm (red). When an apple is red, incident light is absorbed but for the red variety. The latter type of radiation is reflected. A black body absorbs all incident radiation, whereas a white body reflects all incident radiation. These qualifications for visible light are also used to classify bodies receiving IR radiation. A black body absorbs all radiation received (100% absorption, no reflection, no transmission). A white body does not absorb IR radiation (100% reflection, no absorption, no transmission). A gray body absorbs part of the IR radiation and reflects the remaining part. A transparent body (for example, a gas) does not interact with IR radiation.

First, black bodies are treated. A black body emits a characteristic radiation. The spectral distribution of the energy is dependent on the temperature only. See Figure 10.1. The amount of energy

Introduction to Chemical Engineering, First Edition. C.M. van 't Land.

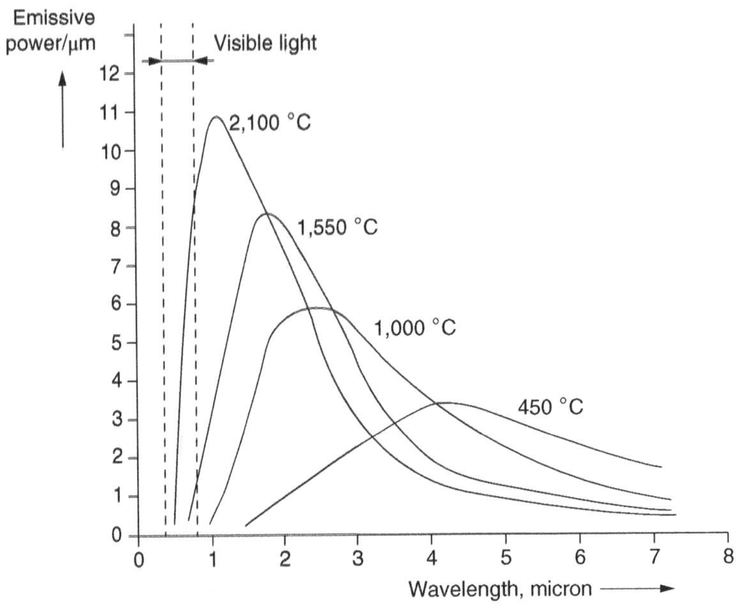

Figure 10.1 Emission spectra of a black body.

increases when the temperature increases. For each temperature, the energy is represented by the areas under the curved lines. Stefan–Boltzmann's law expresses this quantitatively:

$$\Phi_h'' = \sigma \cdot T^4 \text{ W} \cdot \text{m}^{-2}. \tag{10.1}$$

σ is Stefan–Boltzmann's constant: $5.675 \cdot 10^{-8}$ W · m^{-2} · K^{-4}. The Planck radiation law gives the energy distribution over the wave length range. The wavelength at which the maximum energy prevails decreases when the temperature increases. This is expressed by Wien's displacement law:

$$\lambda_{max} \cdot T = 2{,}897 \text{ μm} \cdot \text{K}. \tag{10.2}$$

λ_{max} is 10 μm at ambient temperature. 10 μm is in the far infrared.

Now, four different cases of energy exchange between two bodies will be treated. When the bodies are parallel, they are supposed to be large to be able to neglect the effect of escaping radiation. Furthermore, any reflection is diffusely, that is, in all directions. Finally, the gas between two bodies is transparent.

The energy exchange between two parallel black bodies can be calculated by means of Equation 10.1.

Example 10.1

The temperatures of two parallel black bodies are 800 and 75 °C. The heat flow between the two bodies is: $\Phi_h'' = \sigma(T_1^4 - T_2^4) = 5.675 \cdot 10^{-8}((800 + 273)^4 - (75 + 273)^4) = 74{,}393$ W · m^{-2}.

Next, gray bodies are discussed. For these bodies, a is the absorption coefficient and e is the emission coefficient. a and e are both equal to 1 for black bodies. Like for visible light, a = 1 means

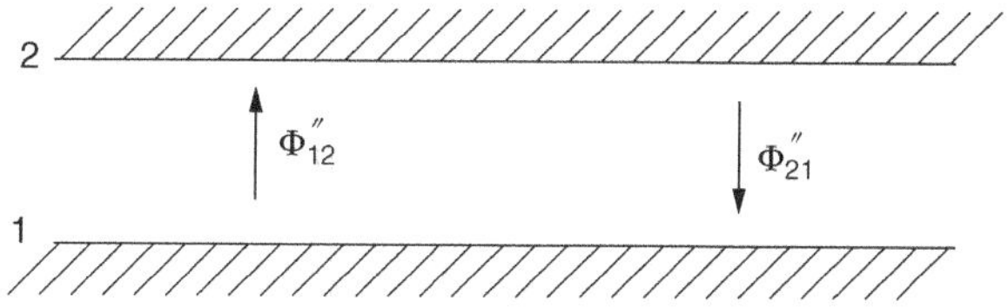

Figure 10.2 Two planes exchanging infrared radiation. *Source:* Courtesy of Delft Academic Press/VSSD, Delft, The Netherlands.

that all incident radiation is absorbed. In practice, most nonmetallic surfaces have absorption coefficients exceeding 0.8. Thus, they are, for IR radiation, almost black. Polished and clean metallic surfaces have absorption coefficients in the range 0.2–0.5. Thus, they are, for IR radiation, almost white. $a = 0$ is valid for a white body whereas $0 < a < 1$ is valid for gray bodies. Kirchhoff's law states that, for a specific body, $a = e$. The latter law can be proven as follows. See Figure 10.2.

Two parallel planes are, respectively, black and gray. They have the same temperature T. We have $\Phi''_{12} = \sigma \cdot T^4$ and $\Phi''_{21} = e_2 \cdot \sigma \cdot T^4 + (1 - a_2)\Phi''_{12}$. The two specific heat flows are equal because the temperatures of the two planes are equal. It then follows: $e_2 = a_2$.

It can be proven that the heat transfer between two parallel planes which are gray bodies can be expressed as follows:

$$\Phi''_h = \frac{\sigma\left(T_1^4 - T_2^4\right)}{1/e_1 + 1/e_2 - 1}\ \mathrm{W \cdot m^{-2}}. \tag{10.3}$$

The proof of Equation 10.3 proceeds as follows. See Figure 10.2.

$$\Phi''_{12} = e_1 \cdot \sigma \cdot T_1^4 + \Phi''_{21}(1 - a_1)\ \mathrm{W \cdot m^{-2}}.$$

$$\Phi''_{21} = e_2 \cdot \sigma \cdot T_2^4 + \Phi''_{12}(1 - a_2)\ \mathrm{W \cdot m^{-2}}.$$

$$\Phi''_h = e_1 \cdot \sigma \cdot T_1^4 - a_1 \cdot \Phi''_{21}\ \mathrm{W \cdot m^{-2}}.$$

The first step is to replace a_1 by e_1 and a_2 by e_2. The second step is to replace Φ''_{12} in the second equation by the first equation. The third step is then writing the equation obtained explicit in Φ''_{21}. The final step is the replacement of Φ''_{21} in the third equation by the obtained expression. When both e_1 and e_2 are equal to 1 (black bodies), the application of Equation 10.1 is obtained again.

A gray body can be convex and completely surrounded by a second gray body. Then, the following expression applies for the smaller body:

$$\Phi''_h = \frac{\sigma\left(T_1^4 - T_2^4\right)}{1/e_1 + A_1/A_2(1/e_2 - 1)}\ \mathrm{W \cdot m^{-2}}. \tag{10.4}$$

The proof of Equation 10.4 proceeds as follows. See Figure 10.3.

$$\Phi_1 = A_1 \cdot e_1 \cdot \sigma \cdot T_1^4 + \frac{A_1}{A_2}(1 - a_1)\Phi_2\ \mathrm{W}. \tag{10.5}$$

Φ_1 is the radiation in W emitted by Body 1. The first part of the right hand side (RHS) of the equation is the radiation emitted by the body per se. The second part of the RHS of the equation represents the radiation received by Body 1 from Body 2 and reflected by Body 1. Bodies 1 and 2

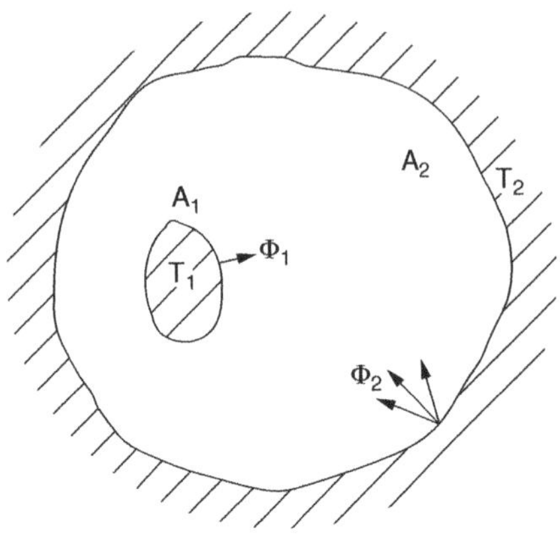

Figure 10.3 A body completely surrounding a second body while the two exchange infrared radiation. *Source:* Courtesy of Delft Academic Press/VSSD, Delft, The Netherlands.

compete concerning the receipt of radiation emitted by Body 2. The emission Body 1 receives from Body 2 is $\Phi_2 \cdot A_1/(A_1 + A_2)$. On multiplying the latter expression by $(1 - a_1)$, the reflected radiation is obtained: $(1 - a_1)\Phi_2 \cdot A_1/(A_1 + A_2)$. That expression is simplified to $\Phi_2(1 - a_1)A_1/A_2$. The simplification enables to arrive at Equation 10.4. The simplification is permitted if the area of Body 2 is much larger than the area of Body 1.

$$\Phi_2 = A_2 \cdot e_2 \cdot \sigma \cdot T_2^4 + \left(\Phi_1 + \left(1 - \frac{A_1}{A_2}\right)\Phi_2\right)(1 - a_2)\ \text{W}. \tag{10.6}$$

Φ_2 is the radiation emitted by Body 2. The first part of the RHS of the equation is the radiation emitted by the body per se. The second part of the RHS consists of two parts. First, there is $\Phi_1(1 - a_2)$. That is the radiation received by Body 2 from Body 1 and reflected by Body 2. Note that all radiation from Body 1 reaches Body 2 whereas only part of the radiation from Body 2 reaches Body 1. The second part of the second part of the RHS of the equation concerns the radiation emitted by Body 2 and received and reflected by Body 2. That emission is $\Phi_2(1 - a_2)A_2/(A_1 + A_2)$. The latter expression is replaced by $\Phi_2(1 - a_2)(1 - A_1/A_2)$. That replacement enables to arrive at Equation 10.4. It is also in line with the simplification introduced in Equation 10.5. If the fraction of the radiation emitted by Body 2 and arriving at Body 1 is A_1/A_2, then the fraction of the radiation emitted by Body 2 and arriving at Body 2 is $1 - A_1/A_2$.

The net transport is given by:

$$\Phi_h = A_1 \cdot e_1 \cdot \sigma \cdot T_1^4 - \frac{A_1}{A_2} \cdot a_1 \cdot \Phi_2\ \text{W}. \tag{10.7}$$

The first part of the RHS of the equation concerns radiation emitted by Body 1. The second part regards radiation received by Body 1 from Body 2 and absorbed. Again, instead of the fraction $A_1/(A_1 + A_2)$, the fraction A_1/A_2 is included.

The first step to arrive at Equation 10.4 is to replace a_1 by e_1 and a_2 by e_2. We have three equations enabling us to arrive at Equation 10.4. The second step is to replace Φ_1 in the second equation by the first equation. The third step is to write the equation thus obtained explicit in Φ_2. The final step is to replace Φ_2 in the third equation by the obtained expression. Note that the heat flow in Equation 10.4 is the specific heat flow for Body 1, that is, the heat flow expressed per m^2 area of Body 1.

Strictly speaking, Equation 10.4 is valid for concentric spheres and cylinders only.

If $A_2 \gg A_1$ and if e_2 is close to 1:

$$\Phi''_h = e_1 \cdot \sigma\left(T_1^4 - T_2^4\right) \, W \cdot m^{-2}. \tag{10.8}$$

Equation 10.8 is used frequently to calculate heat losses of bodies. It is common practice to equate T_2 to the ambient temperature. It is also possible to define for such a case:

$$\Phi''_h = \alpha_r(T_1 - T_2) \, W \cdot m^{-2}.$$

On dividing the latter equation by Equation 10.8, the result is:

$$\alpha_r = \frac{e_1 \cdot \sigma\left(T_1^4 - T_2^4\right)}{(T_1 - T_2)} \, W \cdot m^{-2} \cdot K^{-1}. \tag{10.9}$$

An approximate expression is:

$$\alpha_r = 4 \cdot e_1 \cdot \sigma \cdot T^3. \tag{10.10}$$

T is the arithmetic average of T_1 and T_2 in the latter equation. The approximation is valid only if the temperature difference is small compared to both T_1 and T_2.

IR radiation already plays a role at ambient temperature.

10.3 Dielectric Heating

10.3.1 General Aspects

Dielectric heating can be accomplished by means of radio waves or microwaves. Microwaves received their name because the length of the waves is small compared to the length of radio waves. The tempering of frozen food is a typical successful area of microwave processing. A further successful area of both processing by means of radio waves and processing by means of microwaves is drying. Dielectric heating offers advantages when the material processed is not particulate (for instance, textiles) or is particulate with a large particle size (centimeters or decimeters). Heat input by convection, conduction, or IR radiation is rather inefficient in these cases. Dielectric heating has the unique ability to generate heat within the product.

Neither radio waves nor microwaves have the potential to ionize human or animal tissues. γ-rays do have this potential as their frequencies are much higher than the frequencies used at dielectric heating. An aspect is that microwaves have the potential of heating human and animal tissues. Hence, regulations exist concerning microwave field strengths outside microwave equipment.

Materials processed must be susceptible to dielectric heating. That means that the material must be a semiconductor. At RF heating, the translation (moving as a whole) of ions is often important. At microwave heating, the oscillation of dipoles often plays a role. A molecule possesses a dipole when the center of positive electricity does not coincide with the center of negative electricity. Then, a molecule oscillates in an AC field and heat is thereby generated. The water molecule is a well-known example. Polyethylene is not susceptible to dielectric heating because its molecules do not possess a dipole.

The power equation for dielectric heating is [1]:

$$P = 2\pi f \cdot \varepsilon_o \cdot \varepsilon'_r \cdot \tan\varphi \cdot E^2 \, W \cdot m^{-3}. \tag{10.11}$$

f: the frequency in Hz, that is, s^{-1},
ε_o: the dielectric permittivity of free space, that is, $8.854 \cdot 10^{-12}$ $F \cdot m^{-1}$,

Table 10.1 Dielectric loss factors of various materials at two frequencies.

Compound	Temperature, °C	10^7 Hz	$3 \cdot 10^9$ Hz
Ice	−12	0.067	0.0029
Water	25	0.36	12.0
Methanol	25	0.81	15.3
Ethanol	25	0.80	1.63
Polyethylene	25	<0.00045	0.0007
Paper 78 $g \cdot m^{-2}$	22	0.25[a]	0.5[b]

[a] At $2.712 \cdot 10^7$ Hz.
[b] At $2.45 \cdot 10^9$ Hz.
Sources: Metaxas and Meredith [2] for paper and von Hippel [3] for the other compounds.

ε_r': the relative dielectric constant of the material processed, that is, the ratio, at a specified frequency, of the capacity of a capacitor, with the material as dielectric, to the capacity of the same capacitor in vacuo,
tan φ: the loss tangent, that is, the tangent of the phase angle between the field in the material and the applied field, and
E: the field strength in $V \cdot m^{-1}$.

The product $\varepsilon_r' \cdot \tan \varphi$ is termed the loss factor. The symbol ε_r'' is generally used for the loss factor. Table 10.1 gives the dielectric loss factors of some common materials. Water is receptive to both radio waves and microwaves, however, ice is not. The reason is that ice molecules can neither translate nor oscillate. The molecules of both liquid methanol and liquid ethanol have dipoles and are hence receptive. The loss factor of methanol is greater than the loss factor of ethanol because the alkane character of ethanol is more pronounced than methanol's alkane character.

Radio waves and microwaves are compared concerning maximum specific power input in $W \cdot m^{-3}$ and heating rate in $K \cdot s^{-1}$ in Table 10.2. The table is based on Equation 10.11. A solid material having a specific mass of 2,000 $kg \cdot m^{-3}$ and a specific heat of 2,000 $J \cdot kg^{-1} \cdot K^{-1}$ is considered. The field strength is 10^5 $V \cdot m^{-1}$. Two typical loss factors are 0.1 and 1.0. 10^7 Hz and $3 \cdot 10^9$ Hz are considered typical radio wave and microwave frequencies, respectively. The implicit assumption is that the field strength remains constant when the radiation penetrates the material. That is usually not the case, as the radiation, on penetrating the material, starts to heat the material and, hence, the field loses strength. Keeping this in mind, it is possible to draw qualitatively two conclusions.

Table 10.2 Maximum specific power and maximum heating rate.

Frequency, Hz	Specific power, $W \cdot m^{-3}$		Heating rate, $K \cdot s^{-1}$	
	Loss factor		Loss factor	
	0.1	1.0	0.1	1.0
10^7	$5.56 \cdot 10^5$	$5.56 \cdot 10^6$	0.14	1.4
$3 \cdot 10^9$	$1.67 \cdot 10^8$	$1.67 \cdot 10^9$	41.8	418

Microwaves are more powerful than radio waves and the loss factor is important for the specific power input.

A few words follow concerning the efficiency of dielectric heating. The efficiency of today's power plants is approximately 40%. The efficiency of the conversion of electric energy into microwaves having a frequency of $2.45 \cdot 10^9$ Hz and into radio waves is 70%. The efficiency of the acceptance of the electromagnetic radiation by the dielectric material is 85%. Thus, the overall efficiency is approximately 25%. Dielectric heating is a technique for special applications.

10.3.2 RF Heating

Conventional electric equipment is used for RF heating. That is, cables, switches, and capacitor plates are system components. The applicator is that part of the installation in which the product is heated, which can take one of three different forms: (a) two flat metal plates, (b) a stray electrode system, and (c) a staggered through-field electrode system. See Figure 10.4. Option (a) can be used for large objects. The clearance between the top of the object and the upper capacitor plate should be as small as possible. This design is used for, for example, the drying of hanks of yarn in the textile industry. In option (b), a horizontal nonuniform field passes through (usually) thin webs (up to 10 mm). The post-drying of paper to obtain a uniform low moisture content is an example. In option (c), electrodes are arranged both above and below the material. Relatively thick sheet material can then be processed. See Figure 10.4.

Figure 10.4 Heating by means of radio waves.

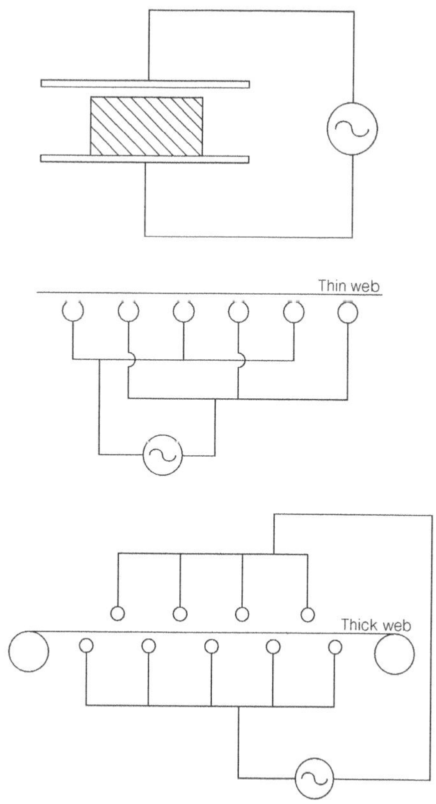

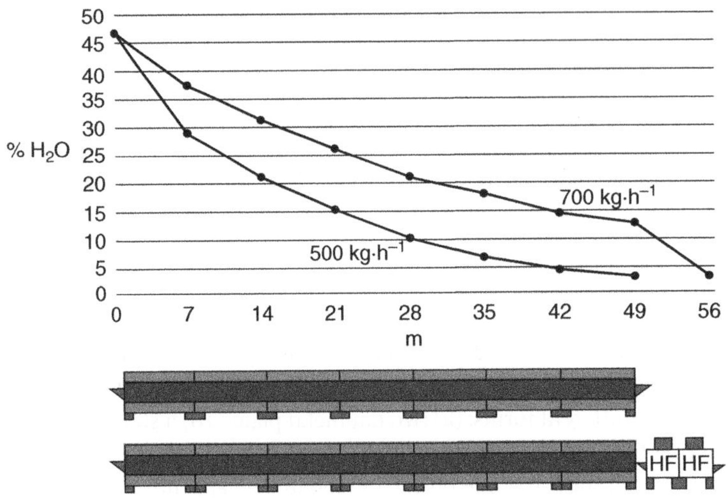

Figure 10.5 Typical capacity extension of a baking oven by the addition of an RF post baking oven. *Source:* Courtesy of STALAM S.p.A., Nove, Italy.

A typical example of RF heating is for the capacity extension of existing conveyor dryers for bakery products. See Figure 10.5.

10.3.3 Microwave Heating

Microwave heating utilizes unconventional electric equipment. After generation by means of a magnetron or klystron, the microwave energy is usually transported through wave guides. A waveguide is a hollow rectangular metal conduit, the dimensions of which match the nature of the transported microwaves. The parts taking care of the generation of microwaves within a magnetron or klystron are called tubes. Figure 10.6 shows various designs for microwave heating. A cavity applicator, typically used for microwave ovens, is at the top of Figure 10.6. A web is contacted with microwaves leaving slots in the middle drawing. A traveling wave applicator is shown at the bottom. A thin sheet of material is passed through the slots. Figure 10.7 shows continuous microwave heating. At microwave heating, a decrease of the heat consumption by the material processed does not automatically lead to a decrease of the tubes' power output. Microwave heating is in this respect different from RF heating. Measures must be taken to avoid overheating the tubes in case the material processed changes its susceptibility. First, the generator and the heating cavity are often physically separated to protect the generator from reflection. Second, provisions are designed between the generator and the heating cavity to absorb reflected power.

For a continuous heater, leakages of microwaves at the product inlet and outlet are absorbed by reject filter traps using quarter wave length choke elements or by dissipative traps through which an absorbing liquid flows. The latter provision can be a combination of an upper and a lower plastic pipe through which water flows. Safetywise, it is more important to contain microwaves than radio waves. The reason is that microwaves can provide more energy than radio waves because of the

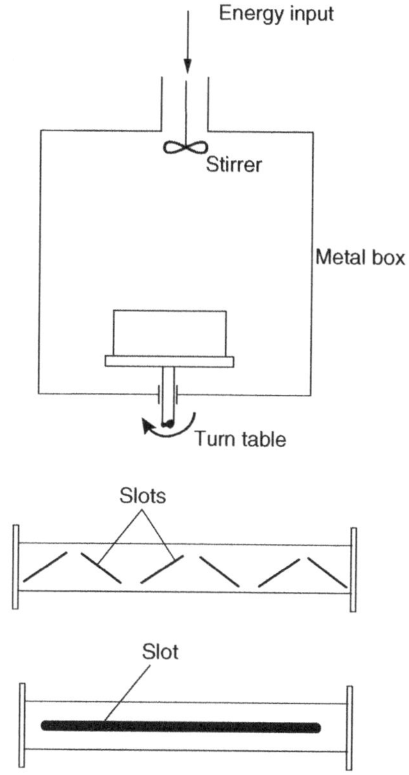

Figure 10.6 Heating by means of microwaves.

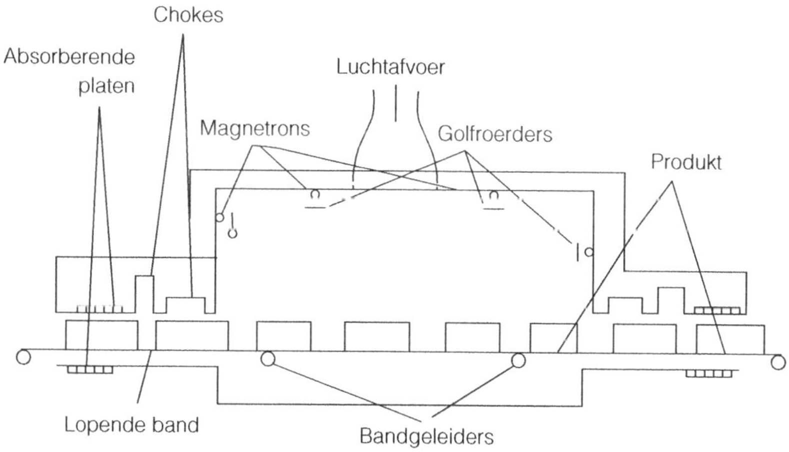

Figure 10.7 Continuous microwave heating. A translation of the words in the figure follows: absorberende platen, absorbing plates; chokes, chokes; magnetrons, magnetrons; luchtafvoer, air outlet; golfroerders, wave reflectors; produkt, product; lopende band, conveyor-belt; bandgeleiders, belt guides.

frequency difference. Containing microwaves is more difficult than containing radio waves because the wavelength of the former is smaller than the wavelength of the latter.

The German company Püschner at Schwanewede indicates that the largest module for the generation of microwaves having a frequency of $9.15 \cdot 10^8$ Hz is a 100-kW one. Their largest power module for the generation of microwaves having a frequency of $2.45 \cdot 10^9$ Hz is a 20–30-kW one. The reason for this difference is the control of the heat transfer. The generation of microwaves is accompanied by heat development. A magnetron is, in principle, a vacuum diode containing a cylindrical anode and a central cathode. A small wavelength (and a high frequency) requires a small distance between the two electrodes and vice versa. Thus, the dimensions of a magnetron generating microwaves having a frequency of $9.15 \cdot 10^8$ Hz are larger than the dimensions of a magnetron generating microwaves having a frequency of $2.45 \cdot 10^9$ Hz. A relatively large magnetron has relatively much area for heat transfer. Thus, the maximum size of a magnetron increases when the frequency decreases. At $2.45 \cdot 10^9$ Hz, a capacity of, for example, 75 kW can be obtained by having three magnetrons in parallel.

Microwave drying is found less often than RF drying. The two reasons are that microwave drying is investmentwise more expensive than RF drying and that microwave drying requires more precautions than RF drying. However, it is attractive that reasonable microwave powers can be combined with modest field strengths. That minimizes the risk of a breakdown.

The center of a spherical object is, at microwave heating, heated more strongly than the periphery. This is called the lens effect. The heating of a potato is an example.

Runaway

In many instances, the product stops absorbing microwave energy when the product gets dry. However, this is not always the case. Potato starch still absorbs microwave energy when the product is dry [4]. The cause is that the product contains 8% by weight of bound water when the product is dry. That water is not bound in a crystal lattice, it has a certain mobility and is susceptible to microwaves. See Table 10.3. That table contains data of drying experiments carried out at 200 mbar and 60 °C. On applying microwaves ($2.45 \cdot 10^9$ Hz) to this product at 0% water by weight, the temperature rises to 100 °C and higher. The loss factors of powder and compacted powder remain approximately constant when the moisture content varies from 0 to 15% by weight. Note: the product still contains 8% by weight of bound water when the moisture content is 0% water by weight.

Product Damage

Drying wood and clay by means of microwaves can cause product damage. This is the case when the steam raised within the material cannot escape and there is a pressure buildup. That can cause cracking of material. That does not occur readily at the drying of textile as it has an open structure.

Table 10.3 Microwave drying of wet potato starch.

			$\varepsilon_r' \cdot \tan \varphi$	$\varepsilon_r' \cdot \tan \varphi$
H_2O, % by weight	ε_r'	$\tan \varphi$	Powder	Densified powder[a]
0	2.86	0.3	0.85	1.30
3	3.42	0.34	1.15	1.54
7.5	3.13	0.28	0.89	1.50
15	3.58	0.29	1.05	1.83

[a] Bulk density 1,000 $kg \cdot m^{-3}$.

Product Discoloration

A case was reported of discoloration of a pharmaceutical product at batchwise combined contact drying and microwave drying. The phenomenon did not occur at indirect drying without microwaves. The measured temperature and pressure conditions were equal at the two tests. The probable explanation is that microwaves have caused so-called hot spots in the material.

References

1. Metaxas, A.C., and Meredith, R. J. (1983). *Industrial Microwave Heating* (p. 72). London, UK: Peter Peregrinus Ltd.
2. Metaxas, A. C., and Meredith, R. J. (1983). *Industrial Microwave Heating* (p. 28). London, UK: Peter Peregrinus Ltd.
3. von Hippel, A. R. (1954). *Dielectric Materials and Applications* (pp. 301, 327, 361, 362). Cambridge, MA: M.I.T. Press.
4. Vromans, H. (1994). Microwave drying of pharmaceutical excipients; comparison with conventional conductive drying. *European Journal of Pharmacy and Biopharmacy*, 40 (5), 333–336.

11

Case Studies Heat Transfer

11.1 Bulk Materials Heat Exchanger

In the process industry, many particulate materials are dried. Salt and sugar are typical examples. It is often necessary to cool these materials after the drying step. The reason is that warm particulate materials can cause condensation of water on silo walls. Tests concerning the cooling of sugar crystals are described [1]. Cooling of sugar crystals occurs in a shell-and-tube heat exchanger. The sugar crystals flow through 40 tubes from the top to the bottom of the exchanger. Cooling water flows through the shell from the bottom to the top of the exchanger. Thus, there is countercurrent flow. In a characteristic test, 1,037 $kg \cdot h^{-1}$ of sugar crystals are cooled from 46 to 30.5 °C. The cooling is accomplished by a cooling water flow of 4,519 $kg \cdot h^{-1}$ of which the temperature increases from 22.8 to 23.9 °C. It can be assumed that the length of the tubes is 3 m and that the inside and outside diameter of the tubes are, respectively, 29 and 32 mm. It is requested to calculate:

- the rate at which sugar crystals descend through the tubes in $cm \cdot s^{-1}$,
- the process heat flow in W,
- the medium heat flow in W,
- the heat exchanging area in m^2 based on the outside tube diameter,
- the logarithmic mean temperature difference in K, and
- the heat transfer coefficient in $W \cdot m^{-2} \cdot K^{-1}$.

Physical data

Sugar

$$\rho_b = 850\ kg \cdot m^{-3}.$$

$$c_p = 1{,}320\ J \cdot kg^{-1} \cdot K^{-1}.$$

Water

$$c_p = 4{,}180\ J \cdot kg^{-1} \cdot K^{-1}.$$

Introduction to Chemical Engineering, First Edition. C.M. van 't Land.

Solution

- Rate at which sugar crystals descend through the tubes:

$$\frac{1{,}037 \cdot 100}{3{,}600 \cdot 850 \cdot 40 \cdot \pi/4 \cdot 0.029^2} = 1.28\ \text{cm} \cdot \text{s}^{-1}.$$

- Process heat flow:

$$\frac{1{,}037 \cdot 1{,}320(46 - 30.5)}{3{,}600} = 5{,}894\ \text{W}.$$

- Medium heat flow:

$$\frac{4{,}519 \cdot 4{,}180(23.9 - 22.8)}{3{,}600} = 5{,}772\ \text{W}.$$

- Outer tube area for heat transfer:

$$40 \cdot \pi \cdot 0.032 \cdot 3 = 12.1\ \text{m}^2.$$

- ΔT_m:

$$\frac{(46 - 23.9) - (30.5 - 22.8)}{\ln (46 - 23.9)/(30.5 - 22.8)} = 13.7\ \text{K}.$$

- Heat transfer coefficient:

$$\frac{5{,}894}{12.1 \cdot 13.7} = 35.6\ \text{W} \cdot \text{m}^{-2} \cdot \text{K}^{-1}.$$

11.2 Heat Exchanger

A liquid flow enters a straight pipe having a wall temperature of 100 °C [2]. The liquid enters at 25 °C. The liquid flow in the pipe is turbulent. It has a temperature of 50 °C at 3 m from the entrance.

a) Calculate the distance at which the liquid temperature is 90 °C.
b) Calculate the distance at which the liquid temperature is 90 °C after the liquid flow has been doubled.

The first calculation
The basic equation for heat transfer in a heat exchanger is Equation 9.1. Write down two equations.

U is constant.

Calculate ΔT_{m1} and ΔT_{m2} by using Equation 9.3. Divide the first equation by the second one. The ratio of the two heat flows equals the ratio of the two temperature increases of the flows. The ratio of the two areas for heat transfer can now be calculated and equals the ratio of the two lengths.

The second calculation

The basic equation for heat transfer in a heat exchanger is Equation 9.1. Write down two equations.

$$\Delta T_{m1} = \Delta T_{m2}.$$

$$\Phi_{h2} = 2 \cdot \Phi_{h1}.$$

Note that U_2 is now not equal to U_1 as the liquid velocity has been raised. Equation 9.10 gives the correlation for heat transfer in tubes. U is proportional to $v^{0.8}$ according to this correlation. So, $U_2/U_1 = 2^{0.8}$ as $v_2/v_1 = 2$. Divide the first equation by the second one. The ratio of the two areas for heat transfer can now be calculated. The ratio of the lengths equals the latter ratio.

Solution

First calculation

$$\Phi_{h1} = U \cdot A_1 \cdot \Delta T_{m1} \text{ W}.$$

$$\Phi_{h2} = U \cdot A_2 \cdot \Delta T_{m2} \text{ W}.$$

$$\Delta T_{m1} = 61.7 \text{ K}.$$

$$\Delta T_{m2} = 32.3 \text{ K}.$$

$$\frac{\Phi_{h2}}{\Phi_{h1}} = \frac{U \cdot A_2 \cdot \Delta T_{m2}}{U \cdot A_1 \cdot \Delta T_{m1}}.$$

$$\frac{65}{25} = \frac{A_2 \cdot 32.3}{A_1 \cdot 61.7}.$$

$$\frac{A_2}{A_1} = 4.97.$$

$$L_2 = 4.97 \cdot L_1 = 4.97 \cdot 3 = 14.9 \text{ m}.$$

Second calculation

$$\Phi_{h1} = U_1 \cdot A_1 \cdot \Delta T_{m1}.$$

$$\Phi_{h2} = U_2 \cdot A_2 \cdot \Delta T_{m2}.$$

$$\Delta T_{m1} = \Delta T_{m2}.$$

$$\Phi_{h2} = 2 \cdot \Phi_{h1}.$$

$$U_2 = 2^{0.8} \cdot U_1.$$

$$\frac{\Phi_{h2}}{\Phi_{h1}} = \frac{U_2 \cdot A_2 \cdot \Delta T_{m2}}{U_1 \cdot A_1 \cdot \Delta T_{m1}}.$$

$$2 = 2^{0.8} \cdot \frac{A_2}{A_1}.$$

$$\frac{A_2}{A_1} = 1.15.$$

$$L_2 = 1.15 \cdot 14.9 = 17.1\ \text{m}.$$

11.3 Surface Temperature of the Sun

The maximum intensity of sunlight occurs at a wavelength of approximately 0.5 μm [3]. Calculate the surface temperature of the sun by using Equation 10.2.

Solution

Equation 10.2:

$$\lambda_{max} \cdot T = 2{,}897\ \mu\text{m} \cdot \text{K}.$$

$$T = \frac{2{,}897}{0.5} = 5{,}794\ \text{K} = 5{,}521\ °\text{C}.$$

11.4 Gas IR Textile Drying

The company GoGas at Dortmund in Germany developed, together with University of Erlangen, the porous combustion technology. Gas burns with air in a ceramic foam. If natural gas is used, the gas mass flow is approximately 100 $\text{nm}^3 \cdot \text{h}^{-1} \cdot \text{m}^{-2}$, which corresponds with 889 $\text{kW} \cdot \text{m}^{-2}$ (32,000 $\text{kJ} \cdot \text{nm}^{-3}$). When this burner type is used for, for example, the drying of textile, the burner surface temperature obtained is approximately 1,450 °C.

The emission coefficients of the burner surface and the textile are, respectively, 0.8 and 0.9. Calculate:

- the heat flow in $\text{kW} \cdot \text{m}^{-2}$ due to IR radiation by using Equation 10.3,
- the burner efficiency, and
- the evaporated water flow in $\text{kg} \cdot \text{h}^{-1} \cdot \text{m}^{-2}$ when the water evaporation takes 2,600 $\text{kJ} \cdot \text{kg}^{-1}$.

Solution

Equation 10.3: $$\Phi''_h = \frac{\sigma(T_1^4 - T_2^4)}{1/e_1 + 1/e_2 - 1}\ \text{W} \cdot \text{m}^{-2},$$

$$\Phi''_h = \frac{5.675 \cdot 10^{-8}(1,723^4 - 373^4)}{1/0.8 + 1/0.9 - 1} = 3.67 \cdot 10^5\ \text{W} \cdot \text{m}^{-2}.$$

Burner efficiency: $$\frac{3.67 \cdot 10^5}{8.89 \cdot 10^5} \cdot 100 = 41.3\%.$$

Evaporated water flow: $$\frac{3.67 \cdot 10^5 \cdot 3{,}600}{2{,}600{,}000} = 508.2\ \text{kg} \cdot \text{m}^{-2} \cdot \text{h}^{-1}.$$

11.5 Heat Loss by IR Radiation

A long rod (D = 1.1 cm, e = 0.8) having a temperature of 327 °C is installed vertically in quiet air of 27 °C [4]. The heat transfer coefficient of free convection is 15 $W \cdot m^{-2} \cdot K^{-1}$. Calculate:

- the IR radiation heat flow by using Equation 10.8,
- the heat transfer coefficient of IR radiation by using Equation 10.9,
- the heat transfer coefficient of IR radiation by using Equation 10.10,
- the ratio of the two coefficients, and
- the percentages heat lost by the rod due to radiation and convection.

Solution

IR radiation heat flow according to Equation 10.8:

$$\Phi''_{h} = 0.8 \cdot 5.675 \cdot 10^{-8}\left(600^4 - 300^4\right) = 5{,}516\ \mathrm{W \cdot m^{-2}}.$$

Heat transfer coefficient due to IR radiation according to Equation 10.9:

$$\alpha_r = \frac{5{,}516.1}{600 - 300} = 18.39\ \mathrm{W \cdot m^{-2} \cdot K^{-1}}.$$

Heat transfer coefficient due to IR radiation according to Equation 10.10:

$$\alpha_r = 4 \cdot 0.8 \cdot 5.675 \cdot 10^{-8} \cdot 450^3 = 16.55\ \mathrm{W \cdot m^{-2} \cdot K^{-1}}.$$

The difference between the two coefficients is approximately 10%. This relatively high percentage is caused by the rather large temperature difference.

The percentages loss due to IR radiation and free convection are, if $\alpha_r = 18.39\ \mathrm{W \cdot m^{-2} \cdot K^{-1}}$, respectively, 55 and 45%.

11.6 Microwave Drying of a Pharmaceutical Product

Table 11.1 shows experimental results of the drying of a pharmaceutical product [5, 6]. The evaporated liquid is an alcohol. What conclusion can be drawn from the experimental results?

Table 11.1 Microwave drying tests at Laboratoires Glaxo.

Dryer volume, l	75	75	300	300	300	600	600
Product[a]	p	p	p	p	a	p	a
Drying method[b]	c	c/mw	c	c/mw	c/mw	c/mw	c/mw
Batch size, kg	30	15	150	150	150	200–300	200–300
Drying time, min	60	30	135	30	40–45	35–50	35–50

[a] p stands for placebo and a for active material.
[b] c stands for contact drying and mw for microwave drying.
Sources: Robin et al. [5] and Pearlswig et al. [6].

Solution

The conclusion is that the drying time can, on scaling up, be kept approximately constant by combining microwave power input with indirect heating. The authors report that they kept the microwave power input in $W \cdot kg^{-1}$ constant.

References

1. Stark, B., Hustert, O., Schrevel, G., Philippart, L., Issle, F., and Frenzel, S. (2008). From plastics to sugar. *Process Worldwide*, 11 (4), 44–45.
2. Beek, W. J., Muttzall, K. M. K., and van Heuven, J. W. (1999). *Transport Phenomena* (pp. 200–201). Chichester, UK: Wiley.
3. Beek, W. J., Muttzall, K. M. K., and van Heuven, J. W. (1999). *Transport Phenomena* (p. 235). Chichester, UK: Wiley.
4. Beek, W. J., Muttzall, K. M. K., and van Heuven, J. W. (1999). *Transport Phenomena* (p. 234). Chichester, UK: Wiley.
5. Robin, P., Lucisano, L. J., and Pearlswig, D. M. (1994). Rationale for selection of a single-pot manufacturing process using microwave/vacuum drying. *Pharmaceutical Technology*, 18 (5), 28–36.
6. Pearlswig, D. M., Robin, P., and Lucisano, L. J. (1994). Simulation modeling applied to the development of a single-pot process using microwave/vacuum drying. *Pharmaceutical Technology*, 18 (6), 44–60.

12

Steady-state Diffusion

12.1 Introduction and Definition of the Diffusion Coefficient

Diffusion is the transport of mass by random Brownian movement of molecules. The flux is proportional to the concentration gradient:

$$\Phi''_{m} = -Đ \cdot \frac{d\rho}{dx}\ kg \cdot m^{-2} \cdot s^{-1}\ \text{or}$$

$$\Phi''_{mole} = -Đ \cdot \frac{dc}{dx}\ kmol \cdot m^{-2} \cdot s^{-1}.$$

This is called Fick's First Law. See Figure 12.1.

Figure 12.1 Fick's First Law.

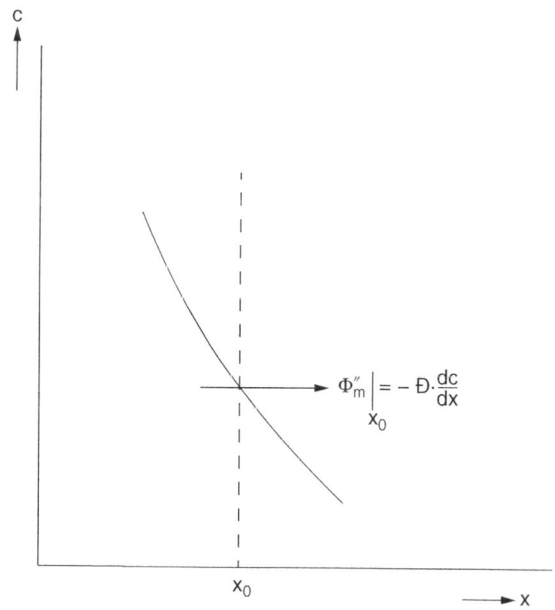

In flowing media, mass transport by diffusion is combined with mass transport by free or forced convection.

Both proportionality constants in the previous equations are called the diffusion coefficient of which the symbol is Đ. In principle, the proportionality constants are different. However, in practice, they are often almost equal.

Introduction to Chemical Engineering, First Edition. C.M. van 't Land.

The diffusion coefficient is defined in such a way that the net transport through a fixed plane is zero. If we restrict ourselves to binary systems (with components A and B), two zero-transport cases can be distinguished:

a) for every unit of mass A diffusing through the plane, a unit of mass B moves in the opposite direction, and
b) for every mole A, a mole B moves in the other direction through the plane.

These two cases will be taken one step further.

a) Net mass flow is nil, ρ in $kg \cdot m^{-3}$ is constant.

$$\Phi''_{m,A} = -Đ_{AB} \cdot \frac{d\rho_A}{dx}.$$

$$\Phi''_{m,B} = -Đ_{BA} \cdot \frac{d\rho_B}{dx}.$$

$$\Phi''_{m,A} + \Phi''_{m,B} = 0.$$

$$\rho_A + \rho_B = \rho \rightarrow d\rho_A + d\rho_B = 0.$$

Thus, $Đ_{AB} = Đ_{BA}$.

This case is applicable to liquids and solids. In the case of small concentrations, of, for example, A in B, the specific mass is approximately constant.

b) Net moles flux is nil, c (in $kmol \cdot m^{-3}$) is constant.

$$\Phi''_{mole,A} = -Đ_{AB} \cdot \frac{dc_A}{dx}.$$

$$\Phi''_{mole,B} = -Đ_{BA} \cdot \frac{dc_B}{dx}.$$

$$\Phi''_{mole,A} + \Phi''_{mole,B} = 0.$$

$$c_A + c_B = c \rightarrow dc_A + dc_B = 0.$$

Thus, $Đ_{AB} = Đ_{BA}$.

This case is suitable for the description of the diffusion transport in gases.

12.2 The Diffusion Coefficient

Gases

According to the kinetic theory, the following statement can be made for the diffusion coefficient of molecules A in gas A:

$$Đ \approx \frac{1}{3} \cdot v_m \cdot l.$$

v_m is the mean velocity of the molecules $(\propto \sqrt{T})$ whereas l is the mean free path of the molecules $(\propto 1/p)$. It follows:

$$Đ \propto \frac{T\sqrt{T}}{p}.$$

The latter statement predicts that the diffusion coefficient varies inversely with pressure; this prediction agrees well with experimental data for many gas mixtures up to about 10 at. The predicted temperature dependence is too weak, just as it was for viscosity and thermal conductivity, but it is qualitatively correct [1].

For binary mixtures, Đ is hardly dependent on the concentration. At normal temperature and pressure, $Đ \approx 10^{-5}\ m^2 \cdot s^{-1}$. For example, for water vapor/air: $Đ = 2.5 \cdot 10^{-5}\ m^2 \cdot s^{-1}$ at atmospheric pressure and 25 °C.

Liquids

The diffusion coefficients of liquids at ambient temperature are in the range $10^{-6} - 10^{-9}\ m^2 \cdot s^{-1}$. The diffusion coefficient at a given temperature can be estimated from the diffusion coefficient at ambient temperature by means of the Einstein–Nernst–Eyring relationship, that is, $Đ\mu/T$ is constant. μ is the dynamic viscosity of the liquid.

For concentrated solutions, the mass transport by diffusion is proportional to the activity gradient rather than the concentration gradient. Measurements based on concentrations may then lead to a strong dependency of Đ on concentration. For CO_2/water, $Đ = 1.95 \cdot 10^{-6}\ m^2 \cdot s^{-1}$ and for 5% ethanol by weight in water, $Đ = 1.13 \cdot 10^{-6}\ m^2 \cdot s^{-1}$ at 25 °C.

Solids

The diffusion coefficients of solid materials at ambient temperature are in the range $10^{-11} - 10^{-13}\ m^2 \cdot s^{-1}$. In crystalline materials, the elementary particles are bound in a lattice, thus restricting their mobility.

12.3 Steady-state Diffusion

The treatment is restricted to binary mixtures and one-dimensional diffusion. Two cases are dealt with.

Equimolar Diffusion

See Figure 12.2. A is transported by diffusion from the left to the right while B is transported from the right to the left. Furthermore, $\Phi''_{mole,A} + \Phi''_{mole,B} = 0$. It follows:

$$\frac{c_A - c_{A2}}{c_{A1} - c_{A2}} = \frac{x_2 - x}{x_2 - x_1}. \tag{12.1}$$

$$\Phi''_{mole,A} = -Đ \cdot \frac{c_{A2} - c_{A1}}{x_2 - x_1}. \tag{12.2}$$

There is an analogy with the one-dimensional steady-state heat conduction:

$$\Phi''_h = \frac{\lambda}{d}(T_0 - T_1) = \frac{a\rho \cdot c_p}{d}(T_1 - T_0).$$

See Figure 8.6 of this part.

It is possible to adapt the relationship found for steady-state heat conduction through a cylindrical body having a length of L m to diffusion.

$$\Phi_{mole,A} = -\frac{2\pi Đ L(c_{A1} - c_{A2})}{\ln(R_1/R_2)}\ kmol \cdot s^{-1}. \tag{12.3}$$

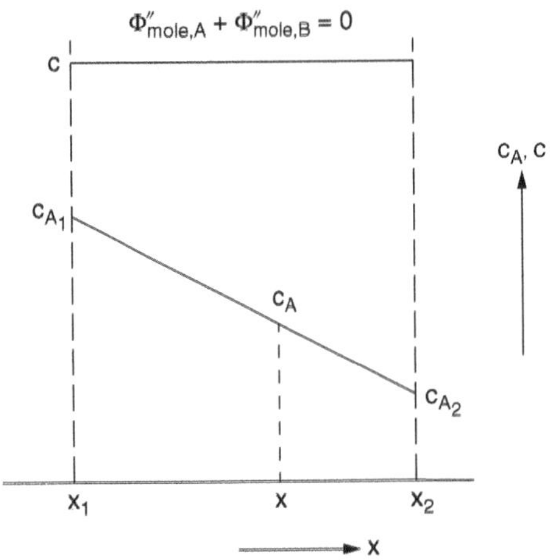

Figure 12.2 Equimolar diffusion. *Source:* Courtesy of Delft Academic Press/VSSD, Delft, The Netherlands.

By the same token, it is possible to adapt the relationship for steady-state heat conduction through a hollow sphere to diffusion:

$$\Phi_{mole,A} = \frac{4\pi Ð(c_{A1} - c_{A2})}{1/R_1 - 1/R_2}. \tag{12.4}$$

Diffusion Through a Stagnant Body

See Figure 12.3. A liquid A is allowed to evaporate in a vertical glass tube over the top of which a stream of vapor-free air is passed, sufficiently rapid for the partial pressure of A in the air flow to be maintained at almost zero. The vapor is transported from the surface of the liquid to the air stream by diffusion. The air is not transported.

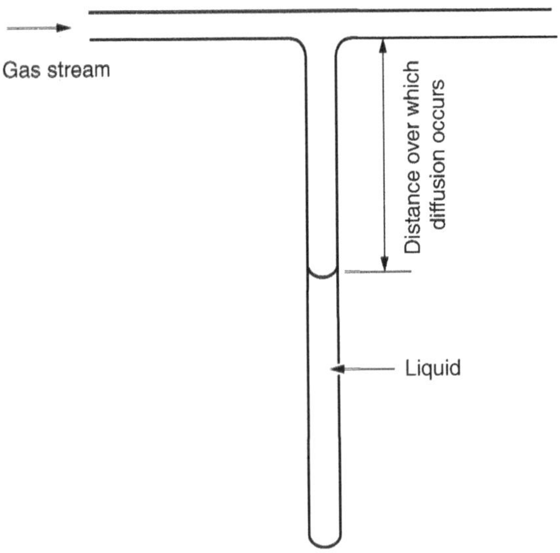

Figure 12.3 Diffusion through a stagnant body.

In order to treat this case, we have to write down a basic equation:

$$\Phi''_{\text{mole,A}} = -Đ \cdot \frac{dc_A}{dx} + \frac{c_A}{c}\left(\Phi''_{\text{mole,A}} + \Phi''_{\text{mole,B}}\right).$$

This equation summarizes that the transport of A in a binary system (B is the other material) consists of two terms. First, there is the diffusion contribution according to Fick's Law. Second, there is the flux resulting from total molar bulk flow. Up till this point, the latter contribution has been assumed to be absent. In this case, $\Phi''_{\text{mole,B}} = 0$, however, $\Phi''_{\text{mole,A}}$ is not equal to zero. Thus, the aforementioned equation becomes:

$$\Phi''_{\text{mole,A}} = -Đ \cdot \frac{dc_A}{dx} + \frac{c_A}{c} \cdot \Phi''_{\text{mole,A}}.$$

The latter equation is rewritten:

$$\Phi''_{\text{mole,A}} = -Đ \cdot \frac{dc_A}{dx} \cdot \frac{c}{c - c_A},$$

$$\Phi''_{\text{mole,A}} \cdot dx = Đc \cdot \frac{d(c - c_A)}{c - c_A}.$$

This is a differential equation which can be integrated. The integration proceeds as follows:

$$\Phi''_{\text{mole,A}} \int_{x_1}^{x_2} dx = Đc \int_{c_{A1}}^{c_{A2}} \frac{d(c - c_A)}{c - c_A}.$$

The result is:

$$\Phi''_{\text{mole,A}} = \frac{Đc}{x_2 - x_1} \cdot \ln \frac{c - c_{A2}}{c - c_{A1}} \ \text{kmol} \cdot \text{m}^{-2} \cdot \text{s}^{-1}. \tag{12.5}$$

It is also possible to integrate between x and x_1 and c_A and c_{A1}. The result is:

$$\Phi''_{\text{mole,A}} = \frac{Đc}{x - x_1} \cdot \ln \frac{c - c_A}{c - c_{A1}} \ \text{kmol} \cdot \text{m}^{-2} \cdot \text{s}^{-1}.$$

On equating the latter equation to Equation 12.5, one obtains:

$$\frac{c - c_A}{c - c_{A1}} = \left(\frac{c - c_{A2}}{c - c_{A1}}\right)^{\frac{x - x_1}{x_2 - x_1}}. \tag{12.6}$$

Equation 12.6 is the distribution of the concentration of A. The concentration of A can be calculated for each distance x.

Example 12.1

A cup of coffee is left on a table and forgotten [2]. See Figure 12.4. The cup is 8 cm high and the original liquid height was 7 cm. The cup is still half full. It is calculated how long ago the cup was poured.

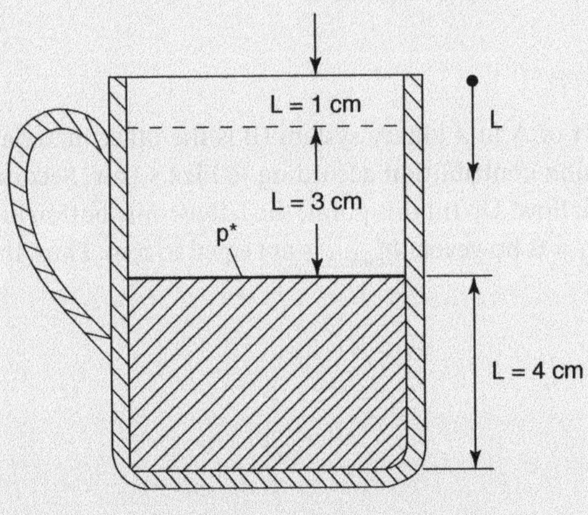

Figure 12.4 Water evaporates from a cup of coffee. *Source:* Transport Phenomena by Beek, W. J. et al. (1999). Reproduced with permission of John Wiley & Sons Inc.

Data

The ambient temperature is 25 °C.

$\rho_l = 997\ \text{kg} \cdot \text{m}^{-3}$.

$Đ = 2.5 \cdot 10^{-6}\ \text{m}^2 \cdot \text{s}^{-1}$.

The relative humidity in the room is 50%.
The saturated water vapor pressure at 25 °C is 3,166 $\text{N} \cdot \text{m}^{-2}$.
The atmospheric pressure is $1.01 \cdot 10^5\ \text{N} \cdot \text{m}^{-2}$.

Calculations

It is assumed that the cup contains water. First, it is realized that we deal with a case of diffusion of water vapor through a stagnant medium, that is, air. Second, the mass flux as a function of the diffusion path is given by Equation 12.5. The calculus is used to be able to arrive at the requested time. The level in the cup decreases with dx in time dt. A mass balance is:

$$\Phi_{mole} \cdot dt = \frac{\rho_l \cdot A}{M} \cdot dx\ \text{kmol} \cdot \text{s}^{-1} \text{ and } \Phi_{mole} = \frac{\rho_l \cdot A}{M} \cdot \frac{dx}{dt}\ \text{kmol} \cdot \text{s}^{-1}.$$

A is the area from which evaporation occurs in m^2 and x is the diffusion path in m. M is the water molecular weight in $\text{kg} \cdot \text{kmol}^{-1}$. Equation 12.5 becomes for this case:

$$\Phi_{mole} = \frac{AĐc}{x} \cdot \ln \frac{c - c_{A2}}{c - c_{A1}}\ \text{kmol} \cdot \text{s}^{-1}.$$

Equating the two expressions for the molar flux:

$$\frac{\rho_l \cdot A}{M} \cdot \frac{dx}{dt} = \frac{AĐc}{x} \cdot \ln \frac{c - c_{A2}}{c - c_{A1}}.$$

c is the concentration in $kmol \cdot m^{-3}$ and follows from the universal gas law $pV^* = RT$:

$$\frac{1}{V^*} = c = \frac{p}{RT} = \frac{1.01 \cdot 10^5}{8{,}314 \cdot 298} = 0.041\ kmol \cdot m^{-3}.$$

The differential equation can be rewritten:

$$dt = \frac{\rho_l}{M Đ c} \cdot \frac{1}{\ln (c - c_{A2})/(c - c_{A1})} \cdot x \cdot dx.$$

The integration of this differential equation proceeds as follows:

$$\int_0^t dt = \frac{\rho_l}{M Đ c} \cdot \frac{1}{\ln (c - c_{A2})/(c - c_{A1})} \cdot \int_{x_1}^{x_2} x \cdot dx,$$

$$t = \frac{\rho_l}{2M Đ c} \cdot \frac{1}{\ln (c - c_{A2})/(c - c_{A1})} \cdot (x_2^2 - x_1^2).$$

Going from symbols to numbers results in:

$$t = \frac{997}{2 \cdot 18 \cdot 2.5 \cdot 10^{-5} \cdot 0.041} \cdot \frac{1}{\ln \dfrac{(1.01 \cdot 10^5 - 3{,}166.4/2)}{(1.01 \cdot 10^5 - 3{,}166.4)}} \cdot (0.04^2 - 0.01^2),$$

$$= 2.52 \cdot 10^6\ s = 4.2\ weeks.$$

On going from symbols to numbers, concentrations were replaced by pressures. That is allowed because pressures are proportional to concentrations.

Concluding Remarks

The transport rate according to Equation 12.5 exceeds the transport rate according to Equation 12.2 by a factor of:

$$\frac{c}{c_{A1} - c_{A2}} \cdot \ln \frac{c - c_{A2}}{c - c_{A1}} = c \cdot \frac{1}{(c - c_{A2}) - (c - c_{A1})} \cdot \ln \frac{c - c_{A2}}{c - c_{A1}} = \frac{c}{(c - c_A)_m} =$$

c divided by the logarithmic mean of $(c - c_A)$.

The results of Equation 12.5 approach the results of Equation 12.2 if $c_A \ll c$.

In practice, situations in between equimolar diffusion and diffusion through a stagnant body will often occur.

Table 12.1 gives physical properties of some materials.

Table 12.1 Physical properties of some materials and constants.

	Air	Water	Bean oil	Stainless steel
	20 °C	20 °C	100 °C	20 °C
ρ, $kg \cdot m^{-3}$	1.20	998	870	7,750
μ, $N \cdot s \cdot m^{-2}$	$1.70 \cdot 10^{-5}$	10^{-3}	$7 \cdot 10^{-3}$	—
$\nu = \mu/\rho$, $m^2 \cdot s$	$1.42 \cdot 10^{-5}$	10^{-6}	$8 \cdot 10^{-6}$	—
c_p, $J \cdot kg^{-1} \cdot K^{-1}$	1,030	4,190	2,120	450
λ, $W \cdot m^{-1} \cdot K^{-1}$	0.025	0.6	0.156	16
σ, $N \cdot m^{-1}$	—	0.07	0.029	—
$Đ_{H2}$, $m \cdot s^{-2}$	$\approx 2 \cdot 10^{-5}$	$5 \cdot 10^{-9}$	$2.4 \cdot 10^{-9}$	—
$a = \lambda/(c_p \cdot \rho)$, $m^2 \cdot s^{-1}$	$2 \cdot 10^{-5}$	$1.43 \cdot 10^{-7}$	$8.5 \cdot 10^{-8}$	$4.6 \cdot 10^{-6}$
$Pr = \nu/a = \mu \cdot c_p/\lambda$, -	0.70	7.0	95	—
$Sc = \nu/Đ$, -	0.71	200	3,333	—
ΔH_v, $J \cdot kg^{-1}$	—	$2.45 \cdot 10^6$	—	—
ΔH_m, $J \cdot kg^{-1}$	—	$3.35 \cdot 10^5$	—	—

Gas law constant $R = 8{,}314\ J \cdot kmol^{-1} \cdot K^{-1}$.
Avogadro constant $N = 6.023 \cdot 10^{26}\ kmol^{-1}$.
Gravitational acceleration constant $g = 9.81\ m \cdot s^{-2}$.
Stefan–Boltzmann constant $\sigma = 5.67 \cdot 10^{-8}\ W \cdot m^{-2} \cdot K^{-4}$.
Volume of ideal gas at Standard Temperature and Pressure $22.41\ m^3 \cdot kmol^{-1}$.
Standard Temperature and Pressure: 273.15 K and $1.013 \cdot 10^5\ N \cdot m^{-2}$.
Source: Transport Phenomena by Beek, W.J. et al. (1999). Reproduced with permission of John Wiley & Sons Inc.

References

1. Bird, R. B., Stewart, W. E., and Lightfoot, E. N. (1960). *Transport Phenomena* (p. 510). New York, NY: Wiley.
2. Beek, W. J., Muttzall, K. M. K., and van Heuven, J. W. (1999). *Transport Phenomena* (p. 252). Chichester, UK: Wiley.

13

Convective Mass Transfer

13.1 Partial and Overall Mass Transfer Coefficients

Figure 13.1 shows an interface between two fluid phases through which a steady-state flux of component A occurs. It is possible to write:

$$\Phi''_{\text{mole,A}} = k''\left(c''_{A,f} - c''_{A,w}\right) \text{ kmol} \cdot \text{m}^{-2} \cdot \text{s}^{-1} \text{ and}$$

$$\Phi''_{\text{mole,A}} = k'\left(c'_{A,w} - c'_{A,f}\right) \text{ kmol} \cdot \text{m}^{-2} \cdot \text{s}^{-1}.$$

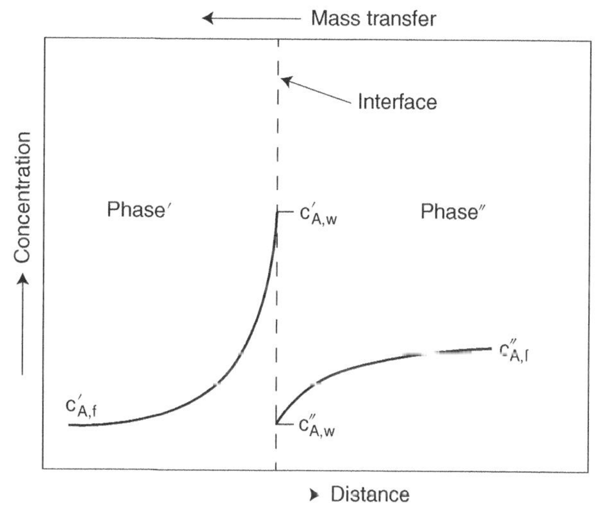

Figure 13.1 Concentration profiles near the interface of two phases. *Source:* Courtesy of Delft Academic Press/VSSD, Delft, The Netherlands.

These expressions state that the flux equals the product of a partial mass transfer coefficient and a driving force. There is an expression for each phase. Equilibrium may be assumed to exist between the two phases at the interface. At low concentrations, the equilibrium can be represented by $c'_{A,w} = m \cdot c''_{A,w}$ in which equation m is the distribution coefficient.

Elimination of $c'_{A,w}$ and $c''_{A,w}$ with this distribution law from the two aforementioned equations results in:

$$\Phi''_{\text{mole,A}} = \left(\frac{1}{k''} + \frac{1}{m \cdot k'}\right)^{-1} \left(c''_{A,f} - \frac{c'_{A,f}}{m}\right) = K''\left(c''_{A,f} - \frac{c'_{A,f}}{m}\right) \text{ kmol} \cdot \text{m}^{-2} \cdot \text{s}^{-1}.$$

Introduction to Chemical Engineering, First Edition. C.M. van 't Land.

The latter equation describes the flux as the product of an overall mass transfer coefficient K″ and a driving force. The equation is relevant for Phase″. However, K″ is still also dependent on k′. A few words on how we can interpret the equation. $c''_{A,f}$ is the actual concentration in Phase″. Considering the fact that the concentration in Phase′ is $c'_{A,f}$, there would be no mass transfer between the two phases if the concentration in Phase″ would be $c'_{A,f}/m$. Thus, in a way, the driving force in Phase″ is $\left(c''_{A,f} - c'_{A,f}/m\right)$.

On multiplying the latter equation with m/m, the following equation is obtained:

$$\Phi''_{mole,A} = \left(\frac{m}{k''} + \frac{1}{k'}\right)^{-1}\left(m \cdot c''_{A,f} - c'_{A,f}\right) = K'\left(m \cdot c''_{A,f} - c'_{A,f}\right)\ \mathrm{kmol \cdot m^{-2} \cdot s^{-1}}.$$

This equation is applicable to Phase′. However, K′ is still also dependent on k″. This equation can be interpreted as follows. The actual concentration in Phase′ is $c'_{A,f}$. Considering the fact that the concentration in Phase″ is $c''_{A,f}$, there would be no mass transfer if the concentration in Phase′ would be $mc''_{A,f}$. So, in a way, the driving force in Phase′ is $\left(m \cdot c''_{A,f} - c'_{A,f}\right)$.

The question may be asked why these models are introduced. The answer is that this approach has been successful for describing mass transfer cases, for example, for liquid/liquid extraction.

When a molar flow Φ_{mole} of a component A must be exchanged between two phases in a mass exchanger, the required area for exchange follows from an equation of the form:

$$\Phi_{mole,A} = A < K\left(c_A - c_A^*\right) > \mathrm{kmol \cdot s^{-1}}.$$

K and c are related to one of the two phases. A is the area available for mass transfer. The logarithmic mean of the driving forces at the inlet and the outlet of the mass exchanger can be calculated for both concurrent and countercurrent flow if K is approximately constant. The approach will be brought one step further when liquid/liquid-extraction is discussed. There is an analogy with the design of heat exchangers for which the design equation is:

$$\Phi_h = UA \cdot \Delta T_m\ \mathrm{W}.$$

Example 13.1

The extraction of benzoic acid from toluene by means of water is tested in a glass pulsed packed column at ambient temperature. See Figure 13.2. The column is filled with glass Raschig rings. The pulsation is carried out to promote mass transfer. Water is the continuous phase and toluene is the dispersed phase: The temperature is ambient.

Figure 13.2 A pulsed packed extraction column.

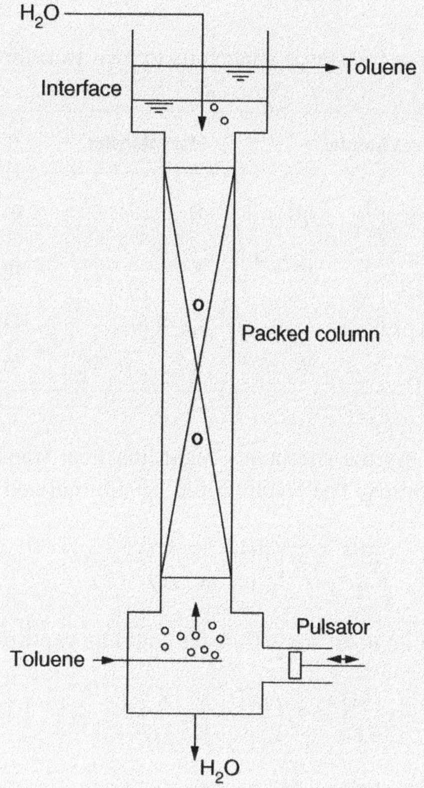

Column data

H = 4 m.

D = 0.05 m.

Raschig ring size 0.01 m.

Pulsator

Stroke: 0.008 m.

Frequency: 2 s^{-1}.

Flows

The concentration of benzoic acid in 1.4 $l \cdot h^{-1}$ of toluene decreases from 0.1 to 0.01 $mol \cdot l^{-1}$. The concentration of benzoic acid in 14 $l \cdot h^{-1}$ of water increases from 0 to 0.01 $mol \cdot l^{-1}$. The concentrations are approximate.

The distribution coefficient of benzoic acid between toluene and water is 8. Benzoic acid is more soluble in toluene than in water. Still, a relatively large water flow can extract benzoic acid from toluene.

It is possible to calculate AK from these test data. K cannot be calculated because A is not known. This is a complication specific for mass transfer in extraction, absorption, distillation, and convective drying equipment. The area of heat exchangers is usually known.

13.2 Mass Transfer Between a Fixed Wall and a Flowing Medium

Mass transfer is analogous to heat transfer. This is summarized in a review.

Heat transfer		Mass transfer	
$a = \frac{\lambda}{c_p \cdot \rho}$	$m^2 \cdot s^{-1}$	$Đ$	$m^2 \cdot s^{-1}$
$\frac{\alpha}{c_p \cdot \rho}$	$m \cdot s^{-1}$	k	$m \cdot s^{-1}$
$c_p \cdot \rho T$	$J \cdot m^{-3}$	c_A or ρ_A	$kmol \cdot m^{-3}$ or $kg \cdot m^{-3}$
Φ''_h	$W \cdot m^{-2}$	$\Phi''_{mole,A}$ or $\Phi''_{m,A}$	$kmol \cdot m^{-2} \cdot s^{-1}$ or $kg \cdot m^{-2} \cdot s^{-1}$

Many measurements regarding heat transfer were carried out in the first half of the previous century. The results could be summarized by equations of the type:

$$\frac{\alpha D}{\lambda} = f_1\left(\frac{\rho v D}{\mu}, \frac{\nu}{a}, \frac{x}{D}\right).$$

In analogy, mass transfer could be captured by:

$$\frac{kD}{Đ} = f_2\left(\frac{\rho v D}{\mu}, \frac{\nu}{Đ}, \frac{x}{D}\right).$$

And $f_1 = f_2$.

$\frac{kD}{Đ}$ is called the Sherwood number (Sh).

$\frac{\nu}{Đ}$ is called the Schmidt number (Sc).

The correlation for mass transfer in a tube is:

$$Sh = 0.027 \cdot Re^{0.80} \cdot Sc^{0.33}. \tag{13.1}$$

The conditions are:

- $\frac{x}{D} > 60$,
- $Re > 10^4$, and
- $Sc > 0.7$.

The relationship is analogous to Equation 9.10 for the transfer of heat in a tube. Sc has replaced Pr. Sc is approximately 1 for gases whereas Sc is in the range 10^2–10^3 for many liquids.

Example 13.2

Air passes through a naphthalene tube having an inside diameter of 0.03 m and a length of 2 m [1]. The air velocity is 15 m · s^{-1} and the air is at 10 °C and atmospheric pressure. The percentage air saturation with naphthalene and the naphthalene sublimation rate is calculated. See Figure 13.3.

Figure 13.3 Concentration profiles in a mass exchanger.

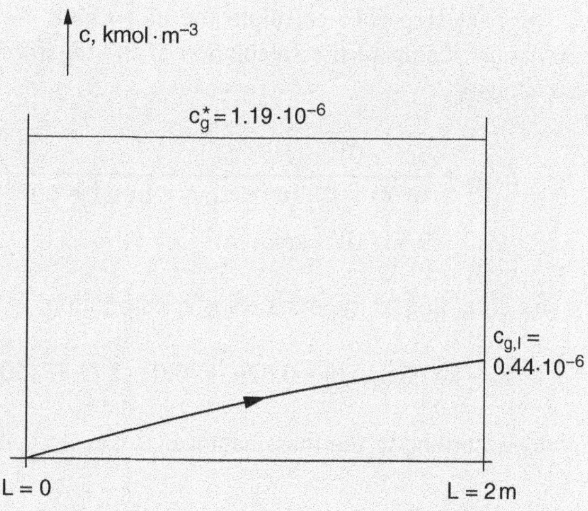

Air properties

$$\rho = 1.24\ \text{kg}\cdot\text{m}^{-3}.$$

$$\mu = 1.8\cdot 10^{-5}\ \text{N}\cdot\text{s}\cdot\text{m}^{-2}.$$

Naphthalene properties

vapor pressure $2.79\ \text{N}\cdot\text{m}^{-2}$.

$$Đ = 5.2\cdot 10^{-6}\ \text{m}^2\cdot\text{s}^{-1}.$$

$$M = 128.2\ \text{kg}\cdot\text{kmol}^{-1}.$$

k is calculated from: $Sh = 0.027\cdot Re^{0.80}\cdot Sc^{0.33}$.

$$Re = \frac{\rho v D}{\mu} = \frac{1.24\cdot 15\cdot 0.03}{1.8\cdot 10^{-5}} = 3.10\cdot 10^4.$$

$$Sc = \frac{\nu}{Đ} = \frac{\mu}{\rho Đ} = \frac{1.8\cdot 10^{-5}}{1.24\cdot 5.2\cdot 10^{-6}} = 2.79.$$

Equation 13.1 can be applied because $Re > 10^4$ and $Sc > 0.7$. The flow is assumed to be adjusted.

$$Sh = 0.027\left(3.10\cdot 10^4\right)^{0.8}\cdot 2.79^{0.33} = 148.4.$$

$$\frac{kD}{Đ} = 148.4 \rightarrow k = \frac{5.2\cdot 10^{-6}\cdot 148.4}{0.03} = 0.026\ \text{m}\cdot\text{s}^{-1}.$$

The naphthalene concentration in the leaving air is assumed. The assumption is: $c_{g,l} = 0.44\cdot 10^{-6}\ \text{kmol}\cdot\text{m}^{-3}$.The saturated naphthalene concentration is:

$$c_g^* = \frac{p}{RT} = \frac{2.79}{8{,}314\cdot 283} = 1.19\cdot 10^{-6}\ \text{kmol}\cdot\text{m}^{-3}.$$

If the assumption is correct, the percentage saturation of the air is 37.0%.

The next step is to calculate the logarithmic mean concentration difference in the mass exchanger. Compare the calculation of the logarithmic mean temperature difference in heat exchangers.

$$\Delta c_{gm} = \frac{(c^* - 0) - (c^* - c_{g,l})}{\ln (c^* - 0)/(c^* - c_{g,l})} = \frac{0.44 \cdot 10^{-6}}{\ln 1.19 \cdot 10^{-6}/(1.19 \cdot 10^{-6} - 0.44 \cdot 10^{-6})} =$$
$$9.53 \cdot 10^{-7}\ \text{kmol} \cdot \text{m}^{-3}.$$

Φ_m according to the mass transfer correlation:

$$\Phi_m = kA \cdot \Delta c_{gm} \cdot M = 0.026 \cdot \pi \cdot 0.03 \cdot 2 \cdot 9.53 \cdot 10^{-7} \cdot 128.2 = 5.97 \cdot 10^{-7}\ \text{kg} \cdot \text{s}^{-1}.$$

Φ_m according to the mass balance:

$$\Phi_m = \frac{\pi}{4} \cdot D^2 v \cdot c_{g,l} \cdot M = \frac{\pi}{4} \cdot 0.03^2 \cdot 15 \cdot 0.44 \cdot 10^{-6} \cdot 128.2 = 5.98 \cdot 10^{-7}\ \text{kg} \cdot \text{s}^{-1}.$$

The assumption of $0.44 \cdot 10^{-6}$ kmol · m^{-3} was correct.

Note 1

Strictly speaking, it is not necessary to assume $c_{g,l} = 0.44 \cdot 10^{-6}$ kmol · m^{-3}. $c_{g,l}$ can be found by equating Φ_m according to the mass transfer correlation and Φ_m according to the mass balance. $c_{g,l}$ is the only unknown variable in the resulting equation. Arithmetically, however, the approach chosen is somewhat simpler.

Note 2

The example treats one of the rare cases where the area for mass transfer is known. Then, the calculation of a mass transfer coefficient is useful. If the area for mass transfer is unknown, we have to fall back on kA or KA. This aspect will be treated for liquid/liquid-extraction.

13.3 Simultaneous Heat and Mass Transfer at Convective Drying

A treatment of the concepts of the wet-bulb temperature and the adiabatic saturation temperature for the system air/water will now be attempted [2]. A start will be made with the wet-bulb temperature. See Figure 13.4. The derivation can be used for warm air temperatures up to 200 °C. Water evaporates from a small wet surface. The evaporation effect does not impact the composition of the air flow. The surface temperature is lower than the gas temperature if the air is unsaturated. This effect is known as "evaporation causes cooling." T_w is called the wet-bulb temperature. T_w is a function of T_g (the dry-bulb temperature) and c_g. T_w is not dependent on the gas flow rate. T_w is the result of the rates of heat and mass transfer. We will now discuss three equations. The first equation states that the heat flux equals the product of the mass flow and the heat of evaporation. The second equation states that the heat flux equals the product of the heat transfer coefficient and the temperature difference. The third equation states that the mass flow equals the product of the mass

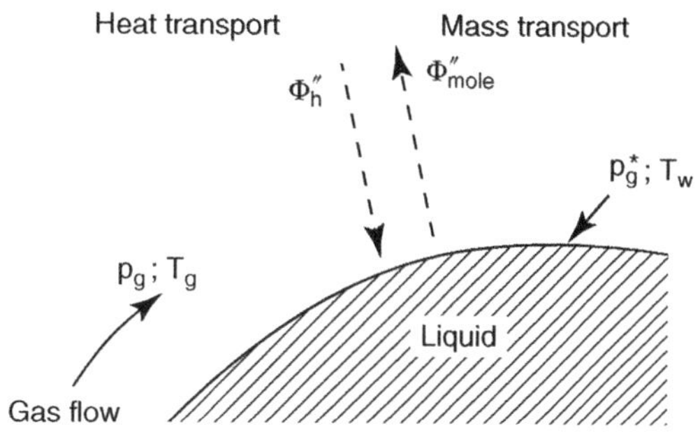

Figure 13.4 Evaporation from a wet surface. *Source:* Transport Phenomena by Beek, W. J. et al. (1999). Reproduced with permission of John Wiley & Sons Inc.

transfer coefficient and the concentration difference. As far as the second and the third equations are concerned, there is an analogy with Ohm's law for the transfer of electricity: $V = iR$ or $i = \frac{1}{R} \cdot V$.

$$\Phi''_h = \Phi''_{mole} \cdot \Delta H \ \mathrm{W \cdot m^{-2}}.$$

$$\Phi''_h = \alpha\left(T_g - T_w\right) \ \mathrm{W \cdot m^{-2}}.$$

$$\Phi''_{mole} = k\left(c_g^* - c_g\right) = \frac{k}{RT}\left(p_g^* - p_g\right) \ \mathrm{kmol \cdot m^{-2} \cdot s^{-1}}.$$

In the third equation, the universal gas law has been applied. Moreover, instead of T_g and T_w, the arithmetical average T of these two variables has been introduced. The reason for this approach is that it enables a relatively simple derivation. However, it limits the validity of the approach as mentioned earlier. Combining the three equations results in:

$$\frac{p_g^* - p_g}{T_g - T_w} = \frac{RT}{\Delta H} \cdot \frac{\alpha}{k}. \tag{13.2}$$

We will now take a look at both the heat transfer coefficient and the mass transfer coefficient. Concerning heat transfer in general, there are correlations to calculate the heat transfer coefficient for geometrically different situations. There is a correlation for the situation depicted in Figure 13.4. The expression has the form:

$$Nu = C \cdot Re^{m} \cdot Pr^{1/3}.$$

Because of the analogy between heat transfer and mass transfer, the correlation for the mass transfer will have the form:

$$Sh = C \cdot Re^{m} \cdot Sc^{1/3}.$$

Dividing the first correlation by the second correlation results in:

$$\frac{\alpha}{k} = c_p \cdot \rho \left(\frac{Sc}{Pr}\right)^{2/3} = c_p \cdot \rho \cdot Le^{2/3}.$$

Le is the Lewis number.

Incorporating the latter result into Equation 13.2 leads to:

$$\frac{p_g^* - p_g}{T_g - T_w} = \frac{RT \cdot c_p \cdot \rho \cdot Le^{2/3}}{\Delta H}. \tag{13.3}$$

Le, c_p, and ρ can be assessed for air at T. p_g^* is the saturated water vapor pressure at T_w. On measuring T_g and T_w, p_g can be calculated.

Example 13.3

The wet-bulb temperature of air at 40 °C is 23 °C. What is the relative humidity of the air at 40 °C?

See Table 13.1 for the saturated water vapor pressure as a function of the temperature between 0 and 40 °C. In Equation 13.3, p_g is the only unknown variable and can thus be calculated.

$p_g^* = 2{,}819\ N \cdot m^{-2}$ (saturated vapor pressure at 23 °C).
$T_g = 40\ °C = 313\ K$.
$T_w = 23\ °C = 296\ K$.
$R = 8{,}314\ J \cdot kmol^{-1} \cdot K^{-1}$.
$T = (23 + 40)/2 = 31.5\ °C = 304.5\ K$.
$c_p = 1{,}010\ J \cdot kg^{-1} \cdot K^{-1}$.
$\rho = 1.17\ kg \cdot m^{-3}$.
$\Delta H = 2.45 \cdot 10^6 \cdot 18\ J \cdot kmol^{-1}$.
$Le^{2/3} = 0.95$.

Table 13.1 Saturated vapor pressure of water.

Temperature, °C	Vapor pressure, $N \cdot m^{-2}$
1	601.4
5	861.5
10	1,215.7
15	1,691.6
20	2,322.9
25	3,150.4
30	4,222.9
35	5,598.3
40	7,344.6

The calculation results in $p_g = 1{,}723\ N \cdot m^{-2}$.

This pressure is 23.4% of the saturated vapor pressure at 40 °C. Thus, the relative humidity of the air at 40 °C is 23.4%.

The discussion will be continued with a treatment of the adiabatic saturation temperature T_S. See Figure 13.5. A more detailed discussion of the adiabatic saturation temperature can be found in Chapter 53 in Part 8 (Crystallization, Solid/liquid Separation, and Drying).

The composition and the temperature of the flowing air are affected when we deal with the adiabatic saturation temperature. Remember that the composition and the temperature of the flowing air are not affected when the wet-bulb temperature is measured. On convective drying, the air flow cools down on performing the drying duty. Cooling continues till the water partial pressure equals the saturated vapor pressure if the solid is not soluble in water. Whereas the wet bulb temperature

Figure 13.5 The adiabatic saturation temperature.

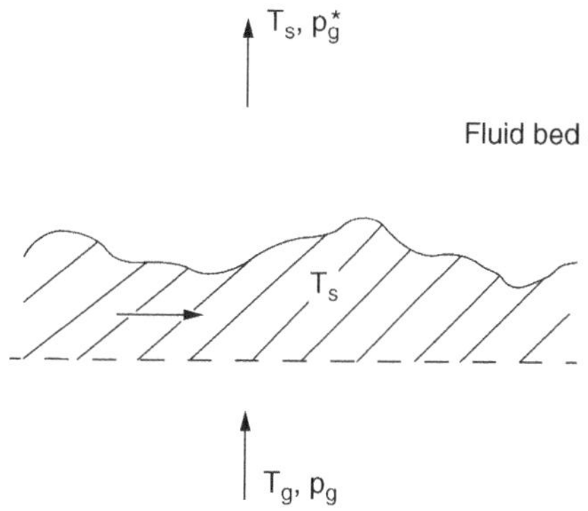

results from the rates of heat and mass transfer, the adiabatic saturation temperature results from heat and mass balances. The amount of water evaporated per kg of dry air:

$$\frac{c_p(T_g - T_s)}{\Delta H}\ \mathrm{kmol\cdot kg^{-1}}.$$

This can also be expressed as:

$$\frac{(p_g^* - p_g)}{\rho RT}\ \mathrm{kmol\cdot kg^{-1}}.$$

T is now the arithmetic average of T_g and T_s. ρ can be approximated by the air specific mass at T. p_g^* is now the saturated vapor pressure at T_s. For the latter expression, an approximation was used. 1 kg of air containing water vapor before drying was compared to 1 kg of air containing water vapor after drying. The approximation is allowed because air of, for example, 200 °C can absorb only 6% water by weight. Thus, by equating the two equations:

$$\frac{p_g^* - p_g}{T_g - T_s} = \frac{RT\cdot c_p\cdot\rho}{\Delta H}. \tag{13.4}$$

This equation resembles Equation 13.3. The latter equation contains $Le^{2/3}$. For the air/water system, $Le^{2/3} = 0.95$. Thus, in practice, $T_w \approx T_s$.

References

1. Griskey, R. G. (1997). *Chemical Engineering for Chemists* (pp. 160–161). Washington, DC: American Chemical Society.
2. Beek, W. J., Muttzall, K. M. K., and van Heuven, J. W. (1999). *Transport Phenomena* (pp. 314–319). Chichester, UK: Wiley.

14

Case Studies Mass Transfer

14.1 Equimolar Diffusion

A wet plate is at a distance of L = 0.1 m from a layer of a drying agent in a desiccator [1]. The drying time of the wet plate is calculated. The following data apply:

$c_g^* = 1.1 \cdot 10^{-3}\ \text{kmol} \cdot \text{m}^{-3}$,

$Đ = 2.5 \cdot 10^{-5}\ \text{m}^2 \cdot \text{s}^{-1}$,

$A = 1.2 \cdot 10^{-3}\ \text{m}^2$, and

$1.2 \cdot 10^{-3}$ kg of water to be evaporated.

Use Equation 12.2. Why is the heat transport not considered?

Solution

$$\Phi''_{\text{mole,A}} = -Đ \cdot \frac{c_{A2} - c_{A1}}{x_2 - x_1}\ \text{kmol} \cdot \text{m}^{-2} \cdot \text{s}^{-1},$$

$$\Phi''_{\text{mole,A}} = -2.5 \cdot 10^{-5} \cdot \frac{1.1 \cdot 10^{-3} - 0}{0.1 - 0} = -2.75 \cdot 10^{-7}\ \text{kmol} \cdot \text{m}^{-2} \cdot \text{s}^{-1}.$$

Number of kmoles of water to be evaporated: $\frac{1.2 \cdot 10^{-3}}{18} = 6.67 \cdot 10^{-5}$.

Time for the evaporation: $\frac{6.67 \cdot 10^{-5}}{2.75 \cdot 10^{-7} \cdot 1.2 \cdot 10^{-3}} = 2.02 \cdot 10^5\ \text{s} = 56.1\ \text{h}$.

The heat transport is not considered because the heat effect is spread out in time. The heat supply for the evaporation occurs by convection and conduction from the surroundings. The evaporation is almost isothermal.

14.2 Diffusion through a Stagnant Body

Through an oversight, instead of Equation 12.5, Equation 12.2 is selected in the example of Chapter 12. Will the time then calculated be shorter or longer than the correct time? What is the ratio between the two times?

Introduction to Chemical Engineering, First Edition. C.M. van 't Land.

Solution

The time calculated with Equation 12.2 is longer than the time calculated with Equation 12.5 because the diffusing molecules are, in the first case, hindered by molecules diffusing into the opposite direction.
The factor is:

$$\frac{101{,}300}{\dfrac{(101{,}300 - 3{,}166.4/2) - (101{,}300 - 3{,}166.4)}{\ln (101{,}300 - 3{,}166.4/2)/(101{,}300 - 3{,}166.4)}} = 1.024.$$

14.3 Sublimation of a Naphthalene Sphere

A moth-ball (naphthalene) having a diameter D_1 of 1 cm is suspended in still air of 10 °C. The initial sublimation rate in $kg \cdot s^{-1}$ is calculated and the sublimation rate in $kg \cdot s^{-1}$ when the diameter D_2 of the sphere is 0.5 cm. The time to have the diameter reduced from 1 to 0.5 cm is also calculated.
The naphthalene properties at 10 °C are:

$$p^* = 2.79\ N \cdot m^{-2},$$

$$Đ = 5.2 \cdot 10^{-6}\ m^2 \cdot s^{-1},$$

$$M = 128.2\ kg \cdot kmol^{-1}, \text{ and}$$

$$\rho = 1{,}150\ kg \cdot m^{-3}.$$

The universal gas constant is $8{,}315\ J \cdot kmol^{-1} \cdot K^{-1}$. In Section 10.2, it was derived that Nu = 2 for heat transfer from a warm sphere in stagnant air. The first calculation is, because of the analogy between heat and mass transfer, to arrive at Sh = 2.

We can assume the partial pressure of naphthalene to be equal to zero at a great distance from the moth-ball and to be equal to the saturation pressure p^* at the surface of the ball (no resistance at the interface). So, the molar flow rate is given by the product of the mass transfer coefficient, area, and driving force.

Initial Rate of Sublimation

$$\Phi_{mole} = kA \cdot c^*\ kmol \cdot s^{-1}.$$

k can be expressed as a function of Đ and D_1 by using Sh = 2. The initial area of the sphere is πD_1^2. c^* can be replaced by p^* by using the universal gas law:

$$p^* V^* = RT \rightarrow \frac{p^*}{c^*} = RT \rightarrow c^* =$$

$$\Phi_{mole} = \qquad (symbols) = \qquad kmol \cdot s^{-1}$$

$$\Phi_m = \qquad kg \cdot s^{-1}$$

Rate of sublimation when the diameter D_2 is 0.5 cm: $\qquad kg \cdot s^{-1}$.

The conclusion is that the rate of sublimation decreases linearly as a function of the diameter. A simple rule is, so to speak, halve the diameter and the rate is halved. That does not immediately make sense as the area for sublimation is divided by four when the diameter is halved. The background is that the area decrease is partly compensated by a greater mass transfer coefficient for smaller diameters.

To find the time to have the diameter halved from 1 to 0.5 cm, we can use the calculus, as we are dealing with a changing phenomenon. The volume decrease of the sphere is dV in a small time dt. A material balance for this small time dt:

$$\Phi_m \cdot dt = -\rho \cdot dV.$$

This is the basic mass balance equation. The minus-sign is used because, as time proceeds, the volume decreases. On substituting:

- $\Phi_m = \Phi_{mole} \cdot M$,
- the expression for Φ_{mole} derived earlier (D_1 is now D), and
- $dV = \pi D^2 \cdot d(D/2)$,

it is possible to obtain a differential equation that can be integrated between t = 0, D = D_1, and t = t, D = D_2.

The result is: $t = \dfrac{\rho RT\left(D_1^2 - D_2^2\right)}{8M\text{Đ}p^*}$ s $\rightarrow t = \text{———} = \quad$ s.

Solution

See Chapter 12. Equation 12.4 reduces to $\Phi_{mole,\,A} = \text{Đ}4\pi \cdot R_1(c_{A1} - c_{A2})$ kmol · s^{-1} when R_2 is very large. On using the concept of the mass transfer coefficient, it is possible to state: $\Phi_{mole,A} = k\pi \cdot D_1^2(c_{A1} - c_{A2})$ kmol · s^{-1}. $2\text{Đ} = k \cdot D_1$ results on dividing the first equation by the second equation and hence $Sh = \dfrac{k \cdot D_1}{\text{Đ}} = 2$. Thus, $k = 2\text{Đ}/D_1$.

The initial rate of sublimation of the naphthalene sphere:

$$\Phi_{mole} = \frac{2\pi\text{Đ} \cdot D_1 \cdot p^*}{RT} = \frac{2\pi \cdot 5.2 \cdot 10^{-6} \cdot 0.01 \cdot 2.79}{8{,}314 \cdot 283} = 3.87 \cdot 10^{-13}\ \text{kmol} \cdot \text{s}^{-1},$$

$$\Phi_m = 128.2 \cdot 3.87 \cdot 10^{-13} = 4.97 \cdot 10^{-11}\ \text{kg} \cdot \text{s}^{-1}.$$

The rate of sublimation when the diameter is 0.5 cm is: $2.48 \cdot 10^{-11}$ kg · s^{-1}.
The derivation to arrive at the equation for the time proceeds as follows:

$$\Phi_m \cdot dt = -\rho \cdot dV,$$

$$\Phi_{mole} \cdot M \cdot dt = -\rho \cdot dV,$$

$$\frac{2\pi D\text{Đ} \cdot p^*}{RT} \cdot M \cdot dt = -\rho\pi \cdot D^2 \cdot d\left(\frac{D}{2}\right),$$

$$dt = -\frac{\rho RT}{4\text{Đ} \cdot p^* \cdot M} \cdot D \cdot dD,$$

$$dt = -\frac{\rho RT}{8\text{Đ} \cdot p^* \cdot M} \cdot dD^2.$$

This is a differential equation which can be integrated. The integration proceeds as follows:

$$\int_{t=0}^{t=t} dt = -\frac{\rho RT}{8\text{Đ} \cdot p^* \cdot M} \cdot \int_{D=D_1}^{D=D_2} dD^2,$$

$$t = -\frac{\rho RT}{8Đ \cdot p^* \cdot M} \cdot \left(D_2^2 - D_1^2\right) = -\frac{1,150 \cdot 8,314 \cdot 283}{8 \cdot 5.2 \cdot 10^{-6} \cdot 2.79 \cdot 128.2} \cdot \left(0.005^2 - 0.01^2\right),$$

$$= 1.36 \cdot 10^7 \text{ s} = 157.9 \text{ days}.$$

Reference

1. Beek, W. J., Muttzall, K. M. K., and van Heuven, J. W. (1999). *Transport Phenomena* (p. 249). Chichester, UK: Wiley.

Notation I

A	Area	m^2
		$m^2 \cdot m^{-3}$
	Constant in the Andrade equation (Chapter 3)	
	Constant in the Orrick and Erbar equation (Chapter 3)	
A_0	Area	m^2
A_1	Area of Body 1	m^2
	Area No. 1	m^2
A_2	Area of Body 2	m^2
	Area No. 2	m^2
a	Head (pressure)	$N \cdot m^{-2}$
	Heat diffusivity ($\lambda/(c_p \cdot \rho)$)	$m^2 \cdot s^{-1}$
	IR absorption coefficient	—
$a_{1,2}$	IR absorption coefficients of Bodies 1 and 2	—
B	Channel height	m
	Constant in the Andrade equation (Chapter 3)	
	Constant in Chapter 8	—
	Constant in the Orrick and Erbar equation (Chapter 3)	
	Plate width	m
	Proportionality factor	—
b	Weir width	m
C	Constant (Chapter 13)	—
	Correction factor (Chapters 2 and 4)	—
	Integration constant (Chapter 4)	$m \cdot s^{-1}$
	Orifice discharge coefficient	—
C_1	Integration constant (Chapter 2)	s
C_w	Drag coefficient	—
c	Concentration	$kmol \cdot m^{-3}$
c^*	Saturated naphthalene concentration	$kmol \cdot m^{-3}$
c_A	Concentration of A	$kmol \cdot m^{-3}$
c_A^*	Saturation concentration of A	$kmol \cdot m^{-3}$
c_B	Concentration of B	$kmol \cdot m^{-3}$
$c_{A1,2}$	Concentration Nos. 1 and 2 of component A	$kmol \cdot m^{-3}$
$c'_{A,f}$	Concentration of component A in the bulk of Liquid′	$kmol \cdot m^{-3}$
$c'_{A,w}$	Concentration of component A at the interface of Liquids′ and ″	$kmol \cdot m^{-3}$

Introduction to Chemical Engineering, First Edition. C.M. van 't Land.

$c''_{A,f}$	Concentration of component A in the bulk of Liquid″	$kmol \cdot m^{-3}$
$c''_{A,w}$	Concentration of component A at the interface of Liquids′ and ″	$kmol \cdot m^{-3}$
c_g	Concentration in the gaseous phase	$kmol \cdot m^{-3}$
c_g^*	Saturation concentration in the gaseous phase	$kmol \cdot m^{-3}$
$c_{g,l}$	Concentration in the gaseous phase at the end of a mass exchanger	$kmol \cdot m^{-3}$
c_p	Specific heat at constant pressure	$J \cdot kg^{-1} \cdot K^{-1}$
c'_p	Specific heat of Liquid′	$J \cdot kg^{-1} \cdot K^{-1}$
c''_p	Specific heat of Liquid″	$J \cdot kg^{-1} \cdot K^{-1}$
c_v	Specific heat at constant volume	$J \cdot kg^{-1} \cdot K^{-1}$
Δc_{gm}	Logarithmic mean concentration difference in the gaseous phase	$kmol \cdot m^{-3}$
D	Diameter of a column, a conduit, a rod, a sphere, a vessel, or a tube	m
	Slab thickness	m
D_1	Initial diameter of a naphthalene sphere	m
D_2	Final diameter of a naphthalene sphere	m
	Inner pipe diameter	m
	Outer pipe diameter	m
Đ	Diffusion coefficient	$m^2 \cdot s^{-1}$
$Đ_{AB}$	Diffusion coefficient of A into B	$m^2 \cdot s^{-1}$
$Đ_{BA}$	Diffusion coefficient of B into A	$m^2 \cdot s^{-1}$
D_h	Hydraulic diameter	m
D_t	Tube diameter	m
d	Diameter of a nozzle or an orifice	m
	Wall thickness	m
d_d	Dirt layer thickness	m
d_p	Particle diameter	m
d_w	Wall thickness	m
E	Amount of methanol	kmol
	Electric field strength	$V \cdot m^{-1}$
e	IR emission coefficient	—
$e_{1,2}$	IR emission coefficients of Bodies 1 and 2	—
F	Area	m^2
	Correction factor (Chapter 9)	—
	Force	N
F_o	Orifice area	m^2
	Rotameter flow channel area	m^2
F_1	Line cross-sectional area	m^2
	Vessel cross-sectional area	m^2
F_2	Contracted area of a jet	m^2
	Hole cross-sectional area	m^2
F_f	Rotameter float area	m^2
F_t	Rotameter tube area	m^2
Fo	Fourier number	—
f	Friction factor	—
	Radiation frequency	s^{-1}
$f_{1,2}$	Function Nos. 1 and 2	—
g	Acceleration due to gravity ($9.81\ m \cdot s^{-2}$)	
H	Channel height	m

	Column height	m
	Enthalpy	$J \cdot kg^{-1}$
$H_{1,2}$	Enthalpies Nos. 1 and 2	$J \cdot kg^{-1}$
ΔH	Liquid evaporation heat	$J \cdot kg^{-1}$
ΔH_m	Fusion heat	$J \cdot kg^{-1}$
ΔH_v	Evaporation heat	$J \cdot kg^{-1}$
h	Height	m
h_o	Liquid height on a weir	m
	Liquid height in a vessel	m
$h_{1,2}$	Height Nos. 1 and 2	m
i	Electrical current	A
K	Consistency index	—
	Overall mass transfer coefficient	$m \cdot s^{-1}$
K′	Overall mass transfer coefficient based on Liquid′	$m \cdot s^{-1}$
K″	Overall mass transfer coefficient based on Liquid″	$m \cdot s^{-1}$
K_w	Friction loss factor	—
K_{w2}	Friction loss factor	—
K_w'	Corrected friction loss factor	—
k	Partial mass transfer coefficient	$m \cdot s^{-1}$
k′	Partial mass transfer coefficient for Liquid′	$m \cdot s^{-1}$
k″	Partial mass transfer coefficient for Liquid″	$m \cdot s^{-1}$
L	Length of an annular channel, a conduit, a cylinder, or a path	m
	Distance	m
$L_{1,2}$	Heat exchanger lengths Nos. 1 and 2	m
Le	Lewis number	—
l	Mean free path length of molecules in the gaseous phase	m
M	Molecular weight	$g \cdot mol^{-1}$ $kg \cdot kmol^{-1}$
m	Distribution coefficient	—
	Flow index	—
	Ratio F_o/F_1	—
m_h	Hydraulic radius	m
N	Avogadro constant ($6.023 \cdot 10^{26}\ kmol^{-1}$)	
Nu	Nusselt number	—
n	Exponent (Chapter 5)	—
P	Amount of purge	kmol
	Power	W
	Specific power input	$W \cdot m^{-3}$
P′	Friction power	W
Pr	Prandtl number	—
p	Pressure	$N \cdot m^{-2}$
p^*	Saturated vapor pressure	$N \cdot m^{-2}$
$p_{1,2}$	Pressure Nos. 1 and 2	$N \cdot m^{-2}$
p_d	Downstream pressure	$N \cdot m^{-2}$
p_g	Water vapor partial pressure	$N \cdot m^{-2}$
p_g^*	Water saturated vapor pressure	$N \cdot m^{-2}$

p_u	Upstream pressure	$N \cdot m^{-2}$
Q	Heat flow	W
R	Amount of recycle	kmol
	Universal gas constant ($8.314\ J \cdot mol^{-1} \cdot K^{-1}$)	
	Electrical resistance	Ohm
	Radius of a tubular flow channel	m
R_1	Electrical resistance No. 1	Ohm
	Inner radius of a cylinder or hollow sphere	m
	Radius of a sphere	m
R_2	Electrical resistance No. 2	Ohm
	Outer radius of a cylinder or hollow sphere	m
R_3	Electrical resistance No. 3	Ohm
Re	Reynolds number	—
$Re_{1,2}$	Reynolds numbers Nos. 1 and 2	—
r	Radius of a cylinder in a tubular flow channel	m
	Radius of a cylinder or sphere	m
S	Wetted conduit perimeter	m
Sc	Schmidt number	—
Sh	Sherwood number	—
T	Momentum	J
	Temperature	°C or K
	Arithmetic average of T_1 and T_2, T_g and T_w, and T_g and T_s	°C or K
T'	Temperature Liquid′	°C or K
T''	Temperature Liquid″	°C or K
T_0	Initial wall or slab temperature	°C or K
	Temperature No. 0	°C or K
T_0'	Temperature Liquid′ at an end of a heat exchanger	°C or K
T_0''	Temperature Liquid″ at an end of a heat exchanger	°C or K
$T_{1,2,3,4}$	Temperature Nos. 1, 2, 3, and 4	°C or K
T_∞	Temperature at a large distance	°C or K
T_{av}	Mean slab temperature	°C or K
T_g	Dry-bulb temperature	°C or K
T_L'	Temperature Liquid′ at the other end of a heat exchanger	°C or K
T_L''	Temperature Liquid″ at the other end of a heat exchanger	°C or K
T_m	Temperature at the center of a slab	°C or K
T_s	Adiabatic saturation temperature	°C or K
T_w	Wet-bulb temperature	°C or K
ΔT_0	Temperature difference at an end of a heat exchanger	K
ΔT_L	Temperature difference at the other end of a heat exchanger	K
ΔT_m	Logarithmic mean temperature difference	K
$\Delta T_{m1,2}$	Logarithmic mean temperature difference Nos. 1 and 2	K
t	Time	s
$t_{1,2,3,4,5}$	Time Nos. 1, 2, 3, 4, 5	s
U	Heat transfer coefficient	$W \cdot m^{-2} \cdot K^{-1}$
	Internal energy	$J \cdot kg^{-1}$
$U_{1,2}$	Heat transfer coefficients Nos. 1 and 2	$W \cdot m^{-2} \cdot K^{-1}$

	Internal energies Nos. 1 and 2	$J \cdot kg^{-1}$
V	Electric tension	V
	Volume	m^3
V^*	Molar volume	$m^3 \cdot kmol^{-1}$
V_f	Rotameter float volume	m^3
v	Velocity	$m \cdot s^{-1}$
$v_{0,1,2}$	Velocity Nos. 0, 1, 2	$m \cdot s^{-1}$
v_{av}	Average velocity	$m \cdot s^{-1}$
v_h	Velocity at height h	$m \cdot s^{-1}$
v_m	Average molecular velocity of gases	$m \cdot s^{-1}$
v_{max}	Maximum velocity	$m \cdot s^{-1}$
v_r	Relative velocity	$m \cdot s^{-1}$
$(v_r)_s$	Swarm relative velocity	$m \cdot s^{-1}$
v_{rx}	Velocity at radius r in x-direction	$m \cdot s^{-1}$
v_x	Velocity at position x	$m \cdot s^{-1}$
	Velocity in x-direction	$m \cdot s^{-1}$
v_y	Velocity in y-direction	$m \cdot s^{-1}$
W	Channel width	m
	Power	W
	Work	W
W_{fr}	Friction work	W
X	Distance in x-direction	m
	Variable for the design of heat exchangers	—
X_a	Distance	m
x	Amount of hydrogen	kmol
	Distance in x-direction	m
	Tube roughness	m
$x_{0,1,2,3,4}$	Distance Nos. 0, 1, 2, 3, 4	m
Y	Parameter for the design of heat exchangers	—
y	Amount of carbon monoxide	kmol
	Distance in y-direction	m
z	Amount of methane	kmol
z'	Empirical orifice factor	—

Greek Symbols

α	Heat transfer coefficient	$W \cdot m^{-2} \cdot K^{-1}$
	Rotameter flow coefficient	—
α'	Heat transfer coefficient for Liquid′	$W \cdot m^{-2} \cdot K^{-1}$
α''	Heat transfer coefficient for Liquid″	$W \cdot m^{-2} \cdot K^{-1}$
α_r	Heat transfer coefficient at IR-radiation	$W \cdot m^{-2} \cdot K^{-1}$
β	Ratio orifice diameter/line diameter	—
δ	Angle	—
	Film thickness	m
	Wall thickness	m

ε	Fraction free space	—
	Specific mass correction factor	—
	Volume fraction continuous phase	—
ε_0	Dielectric permittivity of free space ($8.854 \cdot 10^{-12}$ $F \cdot m^{-1}$)	
ε'_r	Relative dielectric constant of a material	—
ε''_r	Dielectric loss factor of a material	—
Θ	Angle	—
	Correction for thermal conductivity	K
λ	Radiation wavelength	m
	Thermal conductivity	$W \cdot m^{-1} \cdot K^{-1}$
$\lambda_{1,2,3}$	Thermal conductivity Nos. 1, 2, and 3	$W \cdot m^{-1} \cdot K^{-1}$
λ_d	Thermal conductivity of a dirt layer	$W \cdot m^{-1} \cdot K^{-1}$
λ_{max}	IR wavelength of maximum energy	m
λ_w	Thermal conductivity of a wall	$W \cdot m^{-1} \cdot K^{-1}$
μ	Dynamic viscosity	$N \cdot s \cdot m^{-2}$
	Ratio F_2/F_0	—
$\mu_{1,2}$	Dynamic viscosity Nos. 1 and 2	$N \cdot s \cdot m^{-2}$
μ_{eff}	Effective dynamic viscosity	$N \cdot s \cdot m^{-2}$
μ_g	Gas dynamic viscosity	$N \cdot s \cdot m^{-2}$
μ_l	Liquid dynamic viscosity	$N \cdot s \cdot m^{-2}$
μ_w	Dynamic viscosity at the wall	$N \cdot s \cdot m^{-2}$
ν	Kinematic viscosity	$m^2 \cdot s^{-1}$
ρ	Concentration	$kg \cdot m^{-3}$
	Specific mass	$kg \cdot m^{-3}$
ρ_A	Concentration of A	$kg \cdot m^{-3}$
ρ_B	Concentration of B	$kg \cdot m^{-3}$
ρ_b	Bulk density	$kg \cdot m^{-3}$
ρ_f	Fluid specific mass	$kg \cdot m^{-3}$
	Rotameter float material specific mass	$kg \cdot m^{-3}$
ρ_g	Gas specific mass	$kg \cdot m^{-3}$
ρ_l	Liquid specific mass	$kg \cdot m^{-3}$
ρ_s	Solid specific mass	$kg \cdot m^{-3}$
	Bulk density	$kg \cdot m^{-3}$
σ	Electrical conductivity	Ω^{-1}
	Stefan–Boltzmann constant ($5.675 \cdot 10^{-8}$ $W \cdot m^{-2} \cdot K^{-4}$)	
	Surface tension	$N \cdot m^{-1}$
τ	Shear stress	$N \cdot m^{-2}$
τ_0	Shear stress yield value	$N \cdot m^{-2}$
$\tau_{1,2}$	Shear stresse Nos. 1 and 2	$N \cdot m^{-2}$
τ_{rx}	Shear stress at radius r in x-direction	$N \cdot m^{-2}$
τ_w	Shear stress at the wall	$N \cdot m^{-2}$
$\Phi_{1,2}$	Infrared radiation from Bodies 1 and 2	W
Φ''_{12}	Infrared radiation from Body 1 to Body 2	$W \cdot m^{-2}$
Φ''_{21}	Infrared radiation from Body 2 to Body 1	$W \cdot m^{-2}$
Φ_h	Heat flow	W
Φ''_h	Heat flow	$W \cdot m^{-2}$

$\Phi_{h1,2}$	Heat flow Nos. 1 and 2	W
Φ_m	Mass flow	$kg \cdot s^{-1}$
Φ'_m	Mass flow Liquid′	$kg \cdot s^{-1}$
Φ''_m	Mass flow	$kg \cdot m^{-2} \cdot s^{-1}$
	Mass flow Liquid″	$kg \cdot s^{-1}$
$\Phi_{m1,2}$	Mass flow Nos. 1 and 2	$kg \cdot s^{-1}$
$\Phi''_{m,A}$	Mass flow of A	$kg \cdot m^{-2} \cdot s^{-1}$
$\Phi''_{m,B}$	Mass flow of B	$kg \cdot m^{-2} \cdot s^{-1}$
Φ_{mole}	Molar flow	$kmol \cdot s^{-1}$
Φ''_{mole}	Molar flow	$kmol \cdot m^{-2} \cdot s^{-1}$
$\Phi_{mole,A}$	Molar flow of A	$kmol \cdot s^{-1}$
$\Phi''_{mole,A}$	Molar flow of A	$kmol \cdot m^{-2} \cdot s^{-1}$
$\Phi''_{mole,B}$	Molar flow of B	$kmol \cdot m^{-2} \cdot s^{-1}$
Φ_v	Volumetric flow	$m^3 \cdot s^{-1}$
φ	Loss angle	—
Ω	Angular velocity	s^{-1}
ω	Angular velocity	s^{-1}

Part II

Mixing and Stirring

Part II Content

15

Introduction to Mixing and Stirrer Types

A start will be made with an actual case that occurred within Akzo Nobel approximately 50 years ago. A hydrocarbon was oxidized batchwise in a 100-l reactor by passing air through. The temperature was 100 °C and the pressure was atmospheric. Approximately 7.5% by volume of the initial reactor batch was NaOH-50 (a 50% by weight solution of sodium hydroxide in water). Hydrocarbon and NaOH-50 are immiscible. On dispersing the NaOH-50 in the hydrocarbon while passing air through, the oxidation product is an alcohol. However, if the NaOH-50 is not mixed properly, the oxidation product contains the intermediate of the chemical reaction, a hydroperoxide. A turbine impeller was used to disperse the gas and mix the vessel contents. The air oxidation lasted about 8 h. The reaction heat was carried away by means of a jacket and reflux. One day, after closing the air valve, a spouter occurred. What had happened?

Figure 15.1 depicts the air oxidation reactor. The agitator duties are: dispersing NaOH-50 in hydrocarbon, distributing air in the reactor, and heat transfer. Dispersing NaOH-50 is an essential mixing duty because the specific masses are quite different: 1,525 $kg \cdot m^{-3}$ for NaOH-50 and 900 $kg \cdot m^{-3}$ for

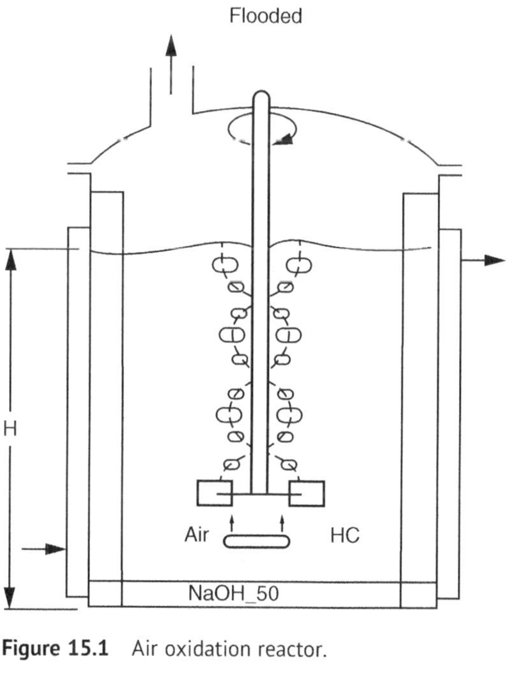

Figure 15.1 Air oxidation reactor.

Introduction to Chemical Engineering, First Edition. C.M. van 't Land.

hydrocarbon. The other two mixing duties are also important. An analysis taught that the impeller had been flooded by too large an air flow. A flooded impeller is, so to speak, drowned in gas and cannot mix the reactor liquid contents properly. On shutting down the air flow, the mixer resumed its mixing action. Now, hydroperoxide and NaOH-50 were suddenly mixed properly. The decomposition of hydroperoxide gave rise to a spouter. The decomposition happened very fast. So, in a short time a lot of heat was generated. The jacket could not carry this heat away in this short time. The consequence was that reaction heat was converted into sensible heat. The reaction mixture was heated up and started to boil.

The remedy was limiting the air flow by an orifice.

In the recent past, micromixing has received special attention; this refers to the production of very small eddies with the purpose of mixing the reacting components as fast as possible. For example, micromixing plays a key role in competitive reactions of the type

$$A + B \rightarrow C. \tag{15.1}$$

$$A + 2B \rightarrow D. \tag{15.2}$$

In this case, the selectivity of the desired Reaction (15.1) can be increased by a fast homogenization of the reaction mixture. If the reaction is, however, of the competitive-consecutive type

$$A + B \rightarrow C. \tag{15.3}$$

$$A + C \rightarrow D. \tag{15.4}$$

then it is important to avoid backmixing of Product C with Reagent A. A plug-flow reactor is used in such a case and mixing is performed with a nozzle or an in-line mixer.

Homogenization refers to the formation of a uniform phase out of several miscible liquids or to the elimination of concentration and temperature gradients during a chemical reaction in the liquid phase.

Three different types of mixers can be distinguished:

- Mechanical mixers,
- Static mixers, and
- Jet mixers.

Mechanical mixers are activated by a separate motor. This category includes mixers in stirred vessels. The mixing energy for static mixers and jet mixers is provided by a pump. A pump passes a liquid through a static mixer causing turbulence. A pump can also pass a liquid through a jet mixer producing a jet in, for example, a storage vessel. The discussion will now be continued with the treatment of agitated vessels. Stirred vessels are important process items in chemical plants. Their main functions are:

- mixing and smoothing out temperature differences in solutions and mixtures,
- suspending solid particles,
- dispersing a liquid into a second liquid,
- distributing gases in liquids and suspensions, for example, at fermentations,
- intensification of mass transfer and the execution of chemical reactions in liquid mixtures and suspensions, and
- heating up and cooling down of liquids.

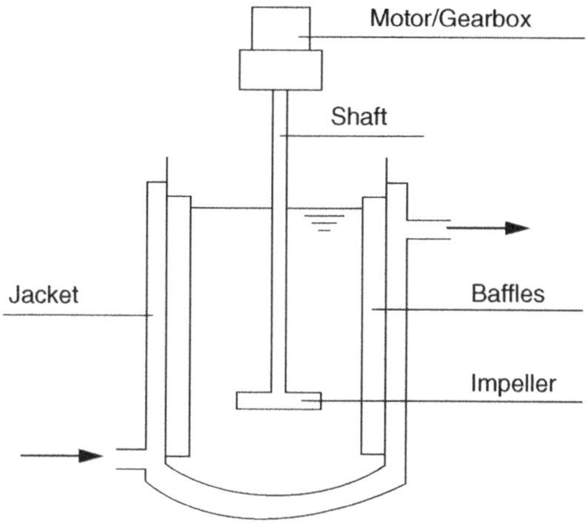

Figure 15.2 Stirred vessel with jacket, baffles, and impeller.

Several of these functions may be carried out simultaneously or successively. For heating or cooling, a vessel is often provided with a jacket or a coil. A jacket can be designed as a half-pipe jacket providing better heat transfer than a conventional jacket. A stirred vessel is depicted in Figure 15.2. The interior auxiliary equipment consists of baffles, feed and drain pipes, coils, and probes. All of these inserts can affect stirring. If an axially positioned stirrer is operated in a vessel without inserts, the liquid is set into rotation, and a vortex is formed. Bulk rotation of the liquid is prevented by the installation of baffles. Fully effective baffling is achieved with four baffles having a width of D/10 (D being the vessel diameter). The baffles are arranged vertically along the entire vessel wall. In order to avoid dead water zones behind the baffles, the baffle width is sometimes reduced to D/12, and they are set at a clearance of D/50 from the wall.

Stirrer types used in the chemical industry are arranged in Figure 15.3 according to the predominant flow pattern they produce, as well as to the range of viscosities over which they can be effectively used [1]. In addition, Figure 15.4 also contains a few stirrers introduced relatively recently [2]. Figure 15.3 will be discussed first.

The discussion will be started with stirrers giving tangential flow. This type of flow is sometimes called radial flow. The turbine stirrer (Rushton turbine, six blades on a disc) is the only high-speed stirrer which imparts a radial motion to the fluid, or at high viscosities, a tangential motion. This stirrer type is effective for low-viscosity liquids and can only be used in baffled vessels. The turbine stirrer causes high levels of shear and is well suited to dispersion processes. The impeller stirrer was developed for use in glass-lined vessels and thus has rounded stirring arms. Usually, one baffle (flow breaker) is used. The cross-beam, grid (also called frame), and blade stirrers belong to the group of low-speed stirrers and are used with $D/d_i = 1.5-2$. They can operate with baffles or, for viscous liquids, without, and are especially well suited to homogenization. As a rule, the low-speed anchor stirrer is operated with a small wall clearance ($D/d_i \approx 1.05$) and is appropriate only for enhancing heat transfer to or from highly viscous liquids.

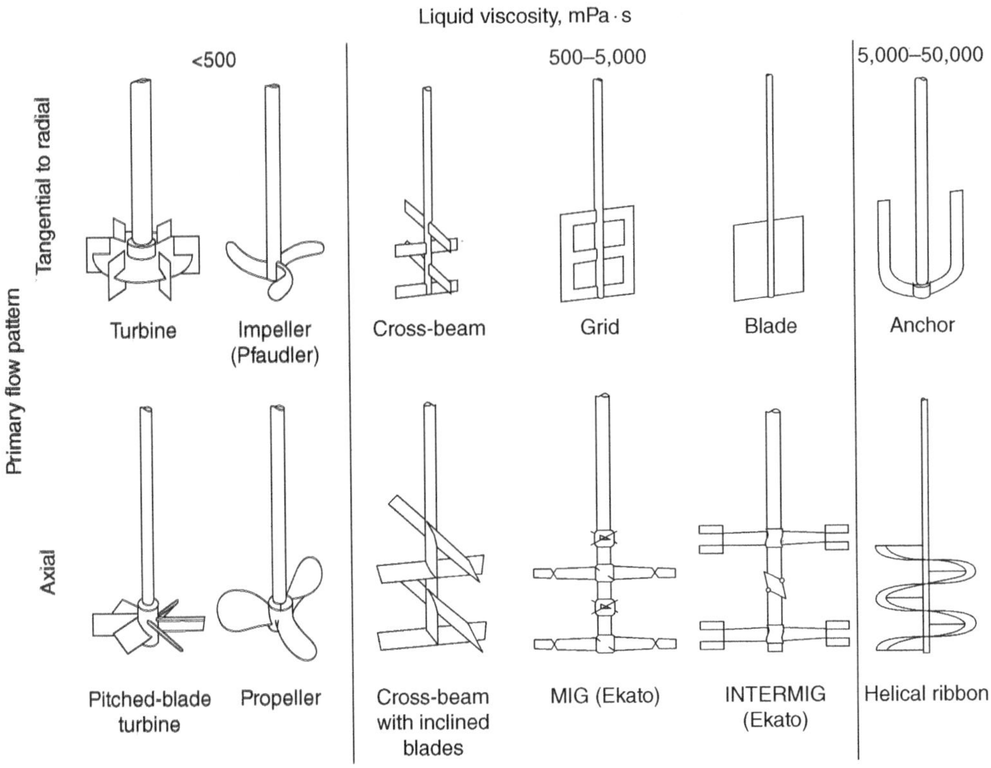

Figure 15.3 Common stirrer types. *Source:* Ullmann's [1], Wiley Online Library.

Next, the stirrers giving axial flow patterns will receive attention. The pitched-blade turbine and the propeller stirrer belong to the group of high-speed mixers. As a rule, both are used for low-viscosity liquids in baffled vessels. They are well suited to homogenization and suspension of solids. The next category includes the cross-beam stirrer with pitched blades and the MIG and INTERMIG stirrers of Ekato in Germany. These stirrers are operated at low speeds, with $D/d_i \approx 1.5$ (with baffles) or $D/d_i = 1.1$ (without baffles). A characteristic of the MIG and INTERMIG stirrers is that, at the periphery, the process fluid is pumped upward, and, centrally, downward. Alternatively, the other way round. The low-speed helical ribbon stirrer with small wall clearance ($D/d_i \approx 1.05$) is operated in such a way that it drives the liquid downward along the wall. Under these conditions, it is the best suited of all stirrer types for the homogenization of highly viscous liquids.

Now, the stirrers in Figure 15.4 will be discussed shortly [2]. The first two stirrers are so-called hydrofoils. They are in-between propellers and pitched-blade turbines. Generally, they are easier to manufacture than propellers. The Lightnin A310 is often used for suspending particulate materials,

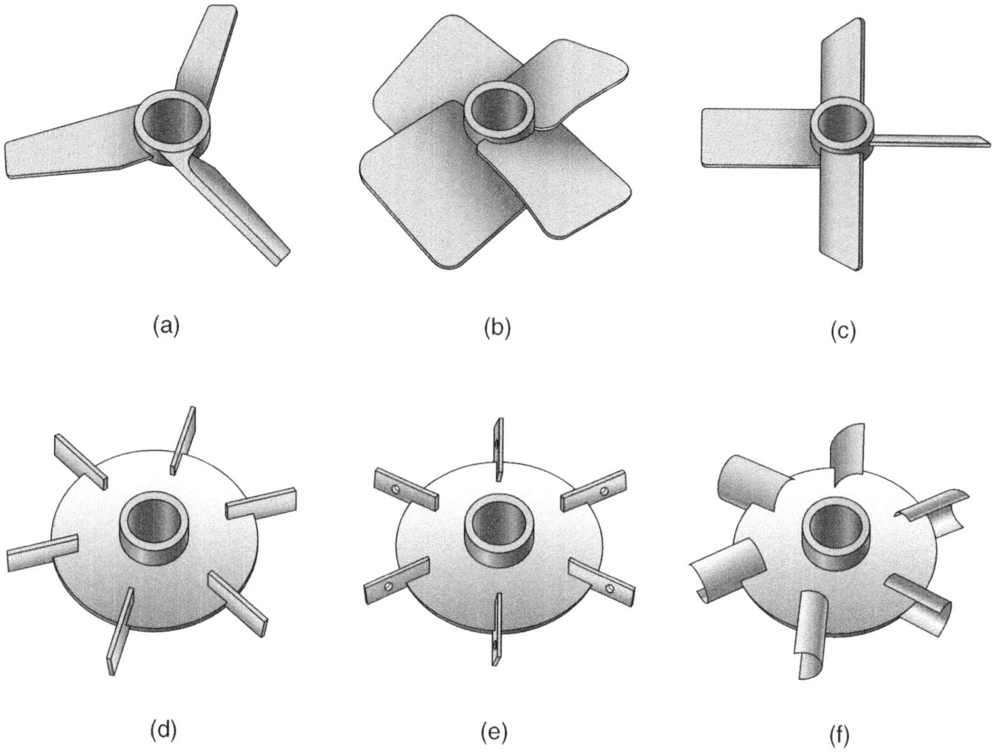

Figure 15.4 Six relatively modern stirrers. (a) Lightnin A310, (b) Lightnin A315, (c) 45° pitched-blade turbine (PBT45), (d) standard Rushton turbine with six blades (DT6), (e) six-blade Rushton turbine with one hole on each blade (DTH6), and (f) concave six-blade impeller (CD6). *Source:* Courtesy of the Journal of Chemical Engineering of Japan, Tokyo, Japan.

whereas the A315 is used for the distribution of gases. The sixth stirrer is a concave Rushton turbine. It is the CD6 of Chemineer. During rotation, the hollow side is the front side. This stirrer is also used for the distribution of gases. The bonus with regard to conventional Rushton turbines is that it is possible to distribute a certain gas flow with less power.

A rotor-stator stirrer is depicted in Figure 15.5. The rotor consists of a blade or paddle stirrer enclosed by a ring of baffles (stator). As a result, high levels of shear are exerted on a small volume. A sawtooth disk is also shown in Figure 15.5. With this stirrer, the liquid is accelerated radially in a thin ring away from the center and then quickly decelerated. High levels of shear can thus be achieved, even without a stator ring or baffles. Both stirrer types are well suited to emulsification and dispersion over a wide range of viscosities, for example, in the production of paint pigments.

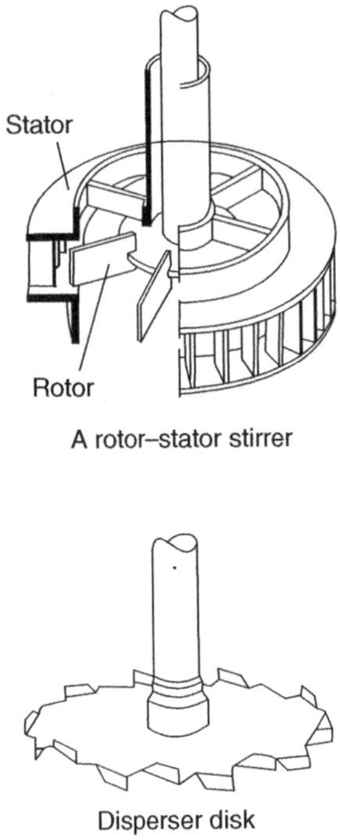

Figure 15.5 The rotor-stator stirrer and the sawtooth disk. *Source:* Ullmann's [1], Wiley Online Library.

References

1. Ullmann (2002). *Processes & Process Engineering—Stirring. Ullmann's Encyclopedia of Industrial Chemistry*. Wiley Online Library.
2. Zhu, Y., Bandopadhayay, P. C., and Wu, J. (2001). Measurement of gas–liquid mass transfer in an agitated vessel—A comparison between different impellers. *Journal of Chemical Engineering of Japan*, 34 (5), 579–584.

16

Mixing Time

16.1 Introduction

The mixing time θ is the time required for a mixer to achieve the desired degree of homogeneity. The most common methods for determining the mixing time are the Schlieren method and chemical methods.

In the Schlieren method, two miscible liquids having different refractive indices are mixed; the complete homogenization of the liquids is indicated by the disappearance of streaks (Schlieren in German).

Chemical measurement methods rely on the changing of a color. The color change is due to an indicator. The addition of a pink aqueous solution of sodium hydroxide to an aqueous solution of sulfuric acid is an example. The color changes to blue. Phenolphthalein is the indicator in this case. The addition of an aqueous solution of sodium thiosulfate to an aqueous solution of iodine to which starch has been added is a second example. The latter solution is blue, and the color disappears on adding the former solution.

Furthermore, it is possible to use a tracer. In a mixing-time experiment represented in Figure 16.1, a tracer material is mixed with a liquid in a vessel [1]. The tracer solution has the same viscosity and specific mass as the liquid in the vessel. By means of a suitable detector, the tracer concentration is measured as a function of time at a particular location in the vessel. When the amount of tracer is known, it is a simple matter to calculate the equilibrium concentration C_∞. The tracer concentration as a function of time at Point A will approach the equilibrium concentration as shown in Figure 16.1. It should be noted, however, that the mixing time depends on the way in which the tracer is added and the location of Point A [1].

A consecutive dye reaction (diazotization) is used for very short mixing times. The degree of mixing can be determined from the selectivity.

Introduction to Chemical Engineering, First Edition. C.M. van 't Land.

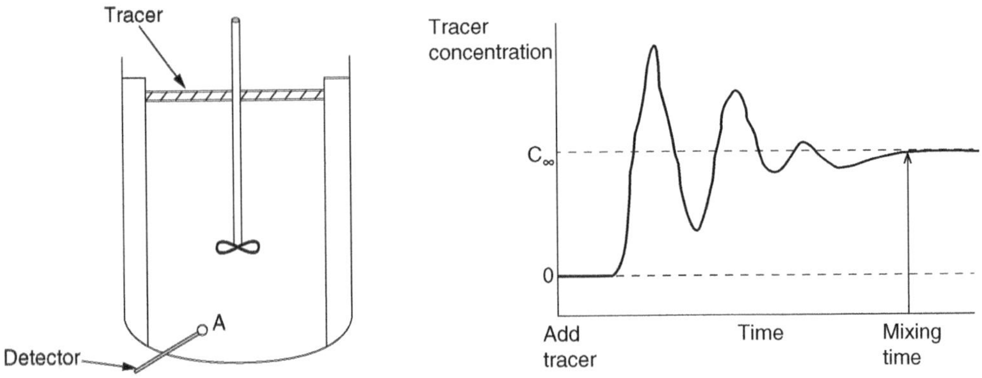

Figure 16.1 Mixing-time measurement. *Source:* Courtesy of Elsevier, Amsterdam, The Netherlands.

16.2 Approach of Beek et al.

The discussion will be restricted to propellers and turbines in fully baffled vessels [2]. The pumping capacity of a propeller is:

$$\Phi_V = v_{ax} \cdot \frac{\pi}{4} \cdot d_i^2 \ m^3 \cdot s^{-1}.$$

The pumping capacity of a turbine is

$$\Phi_V = v_{rad} \cdot \pi \cdot d_i \cdot h \ m^3 \cdot s^{-1}.$$

Both v_{ax} and v_{rad} are proportional to the tip speed of the blades ($\pi n \cdot d_i$), whereas h is proportional to d_i. Thus, it is found for both types of agitators:

$$\Phi_V = K_V \cdot n \cdot d_i^3 \ m^3 \cdot s^{-1}.$$

If $Re > 10^4$, $K_V = 1.3$ for a turbine ($b/d_i = 1/5$) and $K_V = 0.4 - 0.5$ for a propeller (pitch one to two times the propeller diameter). This is the first time that the Reynolds number is mentioned in the context of mixing. In the field of stirring and mixing, it is customary to define the Reynolds number as follows:

$$Re = \frac{\rho n \cdot d_i^2}{\mu} = \frac{n \cdot d_i^2}{\nu}.$$

Thus, the linear velocity in Re is replaced by nd_i. The linear tip velocity of a stirrer is $\pi n \cdot d_i$. On introducing the linear tip velocity into Re, π is omitted.

As a rule-of-thumb, the mixing time is approximately four times the circulation time of the vessel contents:

$$\theta \approx 4 \cdot \frac{V}{\Phi_V} \text{ s.} \tag{16.1}$$

Substitution of the expression for Φ_V in Equation 16.1 leads to:

$$n\theta \approx \frac{4V}{K_V \cdot d_i^3}.$$

This means that $n\theta$ is constant in the turbulent region. $n\theta$ is the number of agitator rotations required to accomplish mixing.

Example 16.1

What value of θ is obtained when a liquid having a kinematic viscosity of 10^{-5} $m^2 \cdot s^{-1}$ is mixed in a baffled vessel having a diameter of 1 m and a flat bottom with a turbine rotating at 5 s^{-1}? The liquid height is also 1 m. d_i = 0.3 m according to Figure 16.2 and $h_i = d_i$.

$$Re = \frac{n \cdot d_i^2}{\nu} = \frac{5 \cdot 0.3^2}{10^{-5}} = 4.5 \cdot 10^4.$$

$$\Phi_V = 1.3 \cdot 5 \cdot 0.3^3 = 0.176 \text{ m}^3 \cdot \text{s}^{-1}.$$

$$V = \frac{\pi}{4} \cdot 1^2 \cdot 1 = 0.785 \text{ m}^3.$$

$$\theta \approx 4 \cdot \frac{0.785}{0.176} = 18 \text{ s.}$$

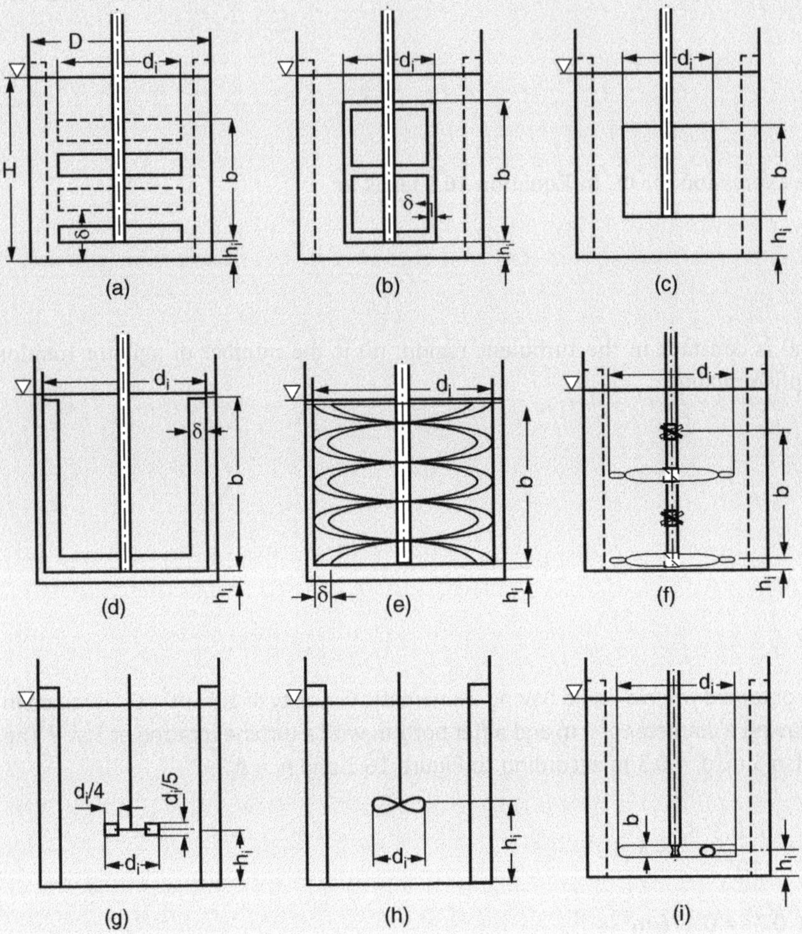

	Type of stirrer	D/d_i	h_i/d_i	b_i/d_i	δ/d_i	
a	Cross-beam	1.5	0.15	1.0	0.15	
b	Frame	2	0.2	1.5	0.1	
c	Blade	2	0.4	1.0		
d	Anchor	1.02	0.01	1.0	0.1	
e	Helical ribbon	1.02	0.01	1.0	0.1	Double helix, pitch $0.5 \cdot d_i$
f	MIG®	1.43	0.15	1.0		4 Beams
g	Turbine	3.33	1.0			6 Paddles
h	Propeller	3.33	1.5			3 Vanes, $\alpha = 25°$
i	Impeller	1.5	0.25	0.15		

Figure 16.2 Dimensions and installation conditions of stirrer types for vessels with H/D = 1. Baffles indicated by dashed lines mean that the stirrer concerned is utilized in both baffled and unbaffled tanks. Baffle width 0.1 · D. *Source:* Ullmann's [3], Wiley Online Library.

16.3 Approach of Zlokarnik

First, the mixing-time characteristic for miscible liquids having approximately the same specific mass and the same viscosity is considered. In this type of homogenization, the mixing time θ for a given stirrer type, vessel size, and vessel inserts depends on the rotational speed n, the stirrer diameter d_i, and the kinematic viscosity. In terms of dimensional analysis, this can be expressed as $n\theta = f(Re)$. In this equation, known as the mixing-time characteristic, $n\theta$ represents the mixing number and $Re = n \cdot d_i^2/\nu$. A number of geometries are shown in Figure 16.2 [3]. The corresponding mixing-time characteristics are in Figure 16.3 [3]. Stirrers a, b, c, d, e, and h were tested in transparent vessels having diameters of 190, 200, 290, 400, and 600 mm [4]. The test temperature was 20 °C. The test liquids were Newtonian, and the viscosity range was $1 - 4 \cdot 10^4$ cP. The baffle width was 10% of the vessel diameter. The mixing times were determined by a chemical method, that is, the neutralization of alkali by the addition of acid. Use was made of an indicator. Depending on the range of Reynolds numbers, the following stirrers can produce homogenization with the minimum number of rotations: helical ribbon mixer, blade stirrer, and the MIG stirrer, the last two in baffled vessels.

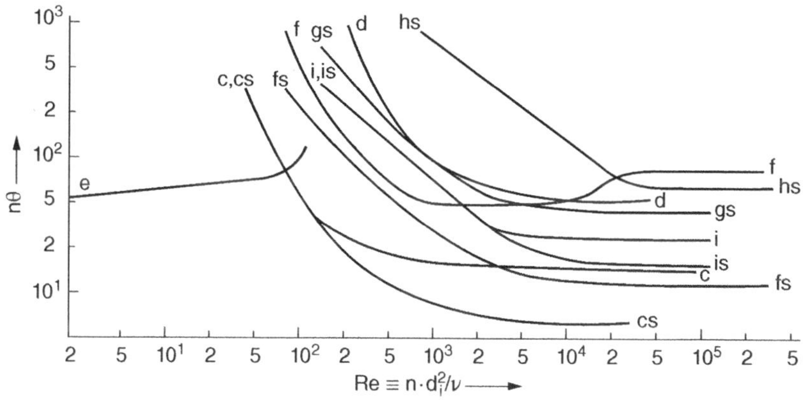

Mixing-time characteristics of the stirrer types in Figure 16.2
Cross-beam stirrer a = 1.8 · c (i.e., the mixing time is 1.8 times that of Curve c); as = 1.8 · cs; grid stirrer; b = 1.25 · c; bs = 1.25 · cs; blade stirrer c, cs; anchor stirrer d; helical ribbon stirrer e; MIG-stirrer f, fs; turbine stirrer gs; propeller stirrer hs; impeller stirrer i, is. In each case s indicates the presence of baffles.
Example: What value of θ is obtained when a liquid with $v = 10^{-4}\ m^2\ s^{-1}$ (= 1 St) is mixed in a baffled vessel using a cross-beam stirrer (as = 1.8 · cs) with d = 1 m and $n = 1\ s^{-1}$?
$Re = n \cdot d_i^2/\nu = 10^4$.
$n\theta = 1.26 \times 10$.
$\theta \approx 13$ s.

Figure 16.3 Mixing-time characteristics of the stirrers in Figure 16.2. (s, baffled tank). *Source:* Ullmann's [3], Wiley Online Library.

Example 16.2

What value of θ is obtained when a liquid having a kinematic viscosity of $10^{-5}\ m^2 \cdot s^{-1}$ is mixed in a baffled vessel having a diameter of 1 m with a turbine rotating at $5\ s^{-1}$? The geometry is according to Figure 16.2.

The method of Beek et al. was applied to this problem in Section 16.1 and 18 s were found. $Re = 4.5 \cdot 10^4$ and Line gs in Figure 16.3 gives $n\theta = 45$. The mixing time θ is 9 s. Probably, the method of Beek et al. has a safety margin. Possibly, the vessel contents are already mixed after two to three times the circulation time. Zlokarnik carried out specific experiments. Furthermore, Zlokarnik worked with $b/d_i = 1/4$, whereas Beek et al. indicate $b/d_i = 1/5$. That means that Zlokarnik's turbine blade width is larger than the blade width of Beek et al. In addition, Zlokarnik specifies the turbine blade height as $d_i/5$, whereas Beek et al. do not specify the blade height.

Note 16.1

High-speed stirrers, like the Rushton turbine and the propeller, are used widely in the chemical industry. In Figure 16.2, $D/d_i = 3.33$ and $d_i/D = 0.3$. In the Standard Tank Configuration, often used for high-speed stirrers, $d_i/D = 1/3 = 0.333$ and $D/d_i = 3$. Thus, the aforementioned data are not applicable to the Standard Tank Configuration. Still, the data are presented because the set is comprehensive. The mixing times obtained can be multiplied with $(3/3.33)^3 = 0.73$ to obtain mixing times for the Standard Tank Configuration. See Equation 16.1.

The Standard Tank Configuration comprises:

- $D/d_i = 3$,
- $D/H = 1$,
- D/h_i (h_i is called the clearance), and
- 4 baffles.

Usually, a vessel is equipped with a dished bottom (the radius is $0.8 \cdot D$) or a "Klöpperboden" (a German standard, the radius equals D).

Note 16.2

According to Zlokarnik, $n\theta$ is constant for $Re > 10^4$. This is confirmed by Harnby et al. [5]. According to these authors, at both low and high Reynolds numbers, regions exist where the total number of impeller revolutions, $n\theta$, required to achieve mixing, is constant. Kraume and Zehner [6] also find, for both the Rushton turbine and the pitched-blade turbine, that $n\theta$ is constant for $Re > 10^4$. Their constant value for the Rushton turbine in the turbulent region is 35, whereas Zlokarnik's value is 45. Part of this difference can be explained by the fact that Kraume and Zehner's stirrer had a diameter of 125 mm in a vessel having a diameter of 400 mm. Zlokarnik's geometry would imply a vessel diameter of 417 mm for a stirrer diameter of 125 mm. Thus, Kraume and Zehner would, for $n\theta$, probably have obtained $(417/400)^3 \cdot 35 = 39.5$ in that case. See Equation 16.1.

Note 16.3

The applicability of Zlokarnik's method to larger vessels is not certain.

Second, the mixing-time characteristic for miscible liquids not having approximately the same specific mass and the same viscosity is discussed. In that homogenization, coarse mixing is effected by the specific mass difference. However, the elimination of the final concentration gradients takes place under conditions characterized by the properties of the homogeneous mixture, ρ_{av} and ν_{av}.

$$\theta = f(n, d_i, \rho_{av}, \nu_{av}, g \cdot \Delta\rho).$$

$$n\theta = f(\mathrm{Re}, \mathrm{Gr}), \text{ in which } \mathrm{Re} = \frac{n \cdot d_i^2}{\nu} \text{ and } \mathrm{Gr} = \frac{d_i^3 \cdot g \cdot \Delta\rho}{\nu_{av}^2 \cdot \rho_{av}}.$$

Gr is the Grashof number. It is often used to describe transport phenomena influenced by specific mass differences. It is also called the Archimedes number (Ar).

The mixing-time characteristic of a cross-beam stirrer in a vessel according to Figure 16.4 is [7, 8]:

$$\sqrt{n\theta} = 51.6 \cdot \mathrm{Re}^{-1}\left(\mathrm{Gr}^{1/3} + 3\right). \tag{16.2}$$

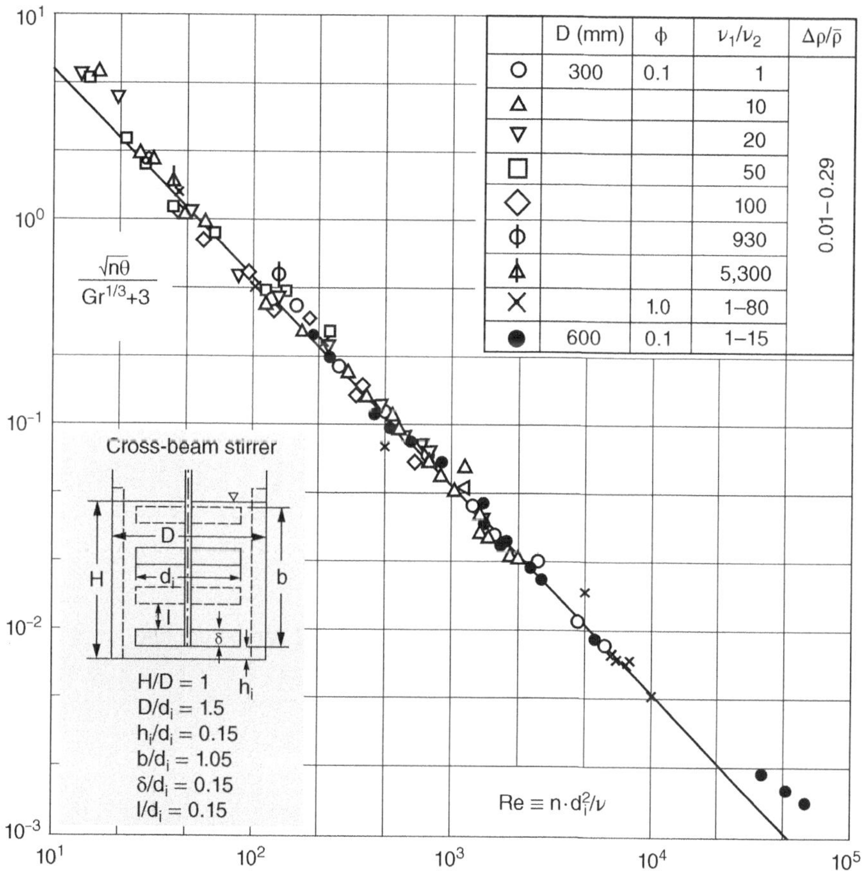

Figure 16.4 Mixing-time characteristic for cross-beam stirrers at H/D = 1 in material systems with specific mass and viscosity differences, $\Phi = \nu_2/\nu_1$. *Source:* Stirring – Theory and Practice by Zlokarnik, M. (2001). Reproduced with permission of John Wiley & Sons Inc.

The conditions for the applicability of this equation are

$10 < Re < 10^5$, and

$100 < Gr < 10^{11}$.

The results were obtained in baffled tanks having diameters of 300 and 600 mm. The test liquids were aqueous sugar solutions, glycerol/water-mixtures, and brine.

Example 16.3

A baffled vessel is equipped with a cross-beam stirrer. The geometry is according to Figure 16.4. The stirrer diameter is 1 m and the rotational speed is 0.1 s^{-1}. Monochlorobenzene (MCB) and sun flower oil (SFO) are mixed. Each component occupies 50% of the stirred volume. The temperature is 25 °C. The mixing time is calculated by using Equation 16.2. The relevant physical properties are [9]:

	MCB	SFO
ρ, $kg \cdot m^{-3}$	1,106	919
μ, $N \cdot s \cdot m^{-2}$	$7.5 \cdot 10^{-4}$	$5.5 \cdot 10^{-2}$

$$\rho_{av} = 0.5 \cdot 1{,}106 + 0.5 \cdot 919 = 1{,}012.5 \text{ kg} \cdot m^{-3}.$$

The viscosity of the mixture is calculated by means of the method of Nissan and Grunberg [10]. The result is $2.55 \cdot 10^{-3}$ $N \cdot s \cdot m^{-2}$.

$$Gr = \frac{d_i^3 \cdot g \cdot \Delta\rho}{\nu_{av}^2 \cdot \rho_{av}} = \frac{1^3 \cdot 9.81(1{,}106 - 919)}{(2.55 \cdot 10^{-3}/1{,}012.5)^2(1{,}012.5)} = 2.86 \cdot 10^{11}.$$

$$Re = \frac{n \cdot d_i^2}{\nu_{av}} = \frac{0.1 \cdot 1^2}{(2.55 \cdot 10^{-3}/1{,}012.5)} = 4.0 \cdot 10^4.$$

Equation 16.2: $\sqrt{n\theta} = 51.6 \cdot Re^{-1}\left(Gr^{1/3} + 3\right)$.

$$\sqrt{n\theta} = 51.6 \cdot \frac{1}{4.0 \cdot 10^4}\left(\left(2.86 \cdot 10^{11}\right)^{1/3} + 3\right) = 6.0.$$

$n\theta = 36$,

$\theta = 360$ s,

Gr is greater than 10^{11}. However, it is thought that Equation 16.2 can be applied. Figure 16.3 predicts the mixing time for $\Delta\rho = 0$ and $\Delta\nu = 0$.

Taking $\nu = \dfrac{2.55 \cdot 10^{-3}}{1{,}012.5} = 2.49 \cdot 10^{-6}$ $m^2 \cdot s^{-1}$.

$Re = 4.0 \cdot 10^4$.

$n\theta = 1.8 \cdot 7 = 12.6$.

$\theta = 126$ s.

References

1. Harnby, N., Edwards, M. F., and Nienow, A. W. (2000). *Mixing in the Process Industries* (pp. 150–151). Oxford, UK: Butterworth-Heinemann.
2. Beek, W. J., Muttzall, K. M. K., and van Heuven, J. W. (1999). *Transport Phenomena* (pp. 128–133). Chichester, UK: Wiley.
3. Ullmann (2002). *Processes & Process Engineering—Stirring. Ullmann's Encyclopedia of Industrial Chemistry*. Wiley Online Library.
4. Zlokarnik, M. (1967). Stirrer suitability for the homogenization of liquid mixtures. *Chemie-Ingenieur-Technik*, 39, 539–548. (in German).
5. Harnby, N., Edwards, M. F., and Nienow, A. W. (2000). *Mixing in the Process Industries* (p. 152). Oxford, UK: Butterworth-Heinemann.
6. Kraume, M., and Zehner, P. (2001). Experience with experimental standards for measurements of various parameters in stirred tanks: A comparative test. *Transactions of the Institution of Chemical Engineers*, 79 (Part A), 811–818.
7. Zlokarnik, M. (1970). Influence of differences of specific mass and of viscosity on the mixing time on homogenizing liquid mixtures. *Chemie-Ingenieur-Technik*, 42, 1009–1011. (in German).
8. Zlokarnik, M. (2001). *Stirring—Theory and Practice* (pp. 110–112). Weinheim, Germany: Wiley-VCH.
9. Pacek, A. W., Chamsart, S., Nienow, A. W., and Bakker, A. (1999). The influence of impeller type on mean drop size distribution in an agitated vessel. *Chemical Engineering Science*, 54, 4211–4222.
10. Poling, B. E., Prausnitz, J. M., and O'Connell, J. P. (2001). *The Properties of Liquids and Gases* (pp. 9.70–9.80). New York, NY: McGraw-Hill.

17

Power Consumption

A typical power curve for a stirrer is shown in Figure 17.1. The power consumption in a baffled vessel is, in the turbulent region, higher than the power consumption in a vessel without baffles. This is caused by the obstruction to the flow in the vessel.

The shape of the power curve of baffled systems can be physically interpreted as follows [1]. Generally, the force on an obstacle in a flow is described with a drag coefficient C_W:

$$F = C_W \cdot A \cdot 0.5 \cdot \rho \cdot v_r^2. \tag{17.1}$$

A is the largest cross-sectional area of the obstacle perpendicular to v_r, the relative undisturbed velocity. Let us consider, for instance, one of the six blades of a turbine stirrer. Equation 17.1 can be applied. It can be stated:

- $A \propto d_i^2$,
- $v_r^2 \propto n^2 \cdot d_i^2$, and
- C_W is a constant for high Re numbers (turbulent region).

Thus, in the turbulent region, $F \propto \rho \cdot n^2 \cdot d_i^4$. The power needed to overcome this force is equal to force $\times$ $2\pi \cdot$ arm $\times$ number of revolutions per unit of time. In other words, $P \propto \rho \cdot n^3 \cdot d_i^5$. The proportionality constant is called P_o, the Power number.

$$P = P_o \cdot \rho \cdot n^3 \cdot d_i^5 \ \mathrm{W} \quad \left(\mathrm{Re} > 10^4\right). \tag{17.2}$$

In Equation 17.1, $C_W \propto 1/\mathrm{Re}$ in the laminar region (Stokes' law). It follows that $P_o \propto 1/\mathrm{Re}$ in the laminar region. Thus, in that region:

$$P = P_o \cdot \mu \cdot n^3 \cdot d_i^5 \ \mathrm{W} \ (\mathrm{Re} < 10). \tag{17.3}$$

The shape of the power curve in Figure 17.1 resembles the curve relating the drag coefficient for a sphere to Re. Generally, Equation 17.2 is used for design purposes.

For Re > 10^4, the Power number P_o is a constant. In that region, P_o is a function of the design of the stirring system only. For Re < 10^4, P_o is a function of the design of the stirring system and of Re.

Figure 17.2 gives the power characteristics of the stirrer types in Figure 15.3 when the geometries of Figure 16.2 are used [2]. For other geometries, see the literature. The characters used in Figures 16.2 and 16.3 are also used in Figure 17.2. For instance, Line c is applicable to the blade stirrer depicted in Figure 16.2. Line cs is applicable to the same blade stirrer in a baffled vessel. The character s indicates the presence of baffles. It is derived from the German word "Stromstörer," meaning an obstruction to flow. A table in Figure 17.2 gives the Power number P_o both for Re = 1 and in the turbulent region. In the turbulent region, the Power numbers are

Introduction to Chemical Engineering, First Edition. C.M. van 't Land.

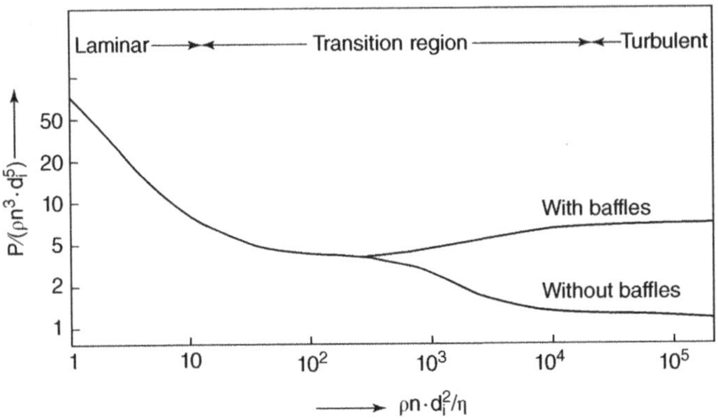

Figure 17.1 A typical power consumption curve for a turbine stirrer. *Source:* Transport Phenomena by Beek W. J. et al. (1999). Reproduced with permission of John Wiley & Sons, Inc.

practically constant, for example, the number of a Rushton turbine in a baffled vessel in the turbulent region is 5.0. The stirrers of Figure 15.4 are not in Figure 17.2. For some of these stirrers, the Power numbers in baffled vessels in the turbulent region are:

- Lightnin A310: 0.25,
- Lightnin A315: 0.75,

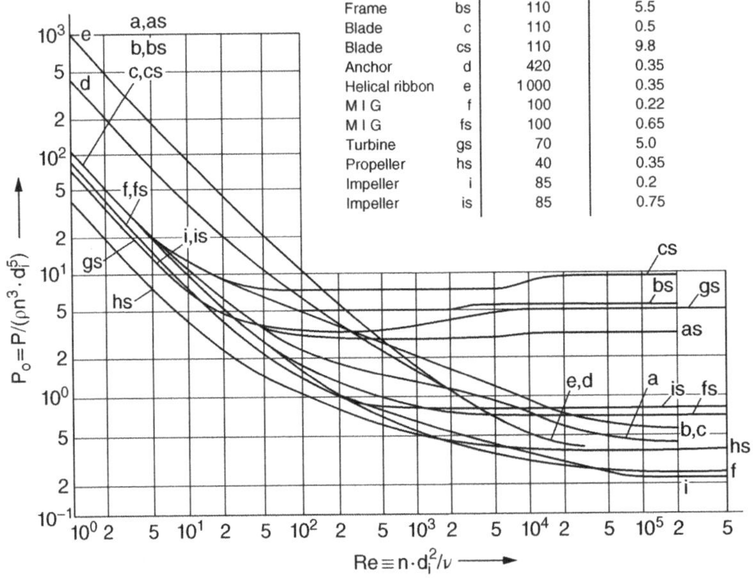

Type of stirrer		$P_o(Re = 1)$	$P_o(Re = 10^5)$
Cross-beam	a	110	0.4
Cross-beam	as	110	3.2
Frame	b	110	0.5
Frame	bs	110	5.5
Blade	c	110	0.5
Blade	cs	110	9.8
Anchor	d	420	0.35
Helical ribbon	e	1 000	0.35
M I G	f	100	0.22
M I G	fs	100	0.65
Turbine	gs	70	5.0
Propeller	hs	40	0.35
Impeller	i	85	0.2
Impeller	is	85	0.75

Figure 17.2 Power characteristics of the stirrer types in Figure 16.2. (s, baffled tank). *Source:* Ullmann's [2], Wiley Online Library.

- 45° pitched-blade turbine (four blades): 1.4, and
- Chemineer CD6: 3.5.

Note that these Power numbers are approximate as the geometries are not defined. The Power numbers are (not too strongly) dependent on the geometry of the stirring system. See also Table 18.5.

Example 17.1

A liquid having a specific mass of 1,000 kg · m^{-3} and a viscosity of 1 N · s · m^{-2} is stirred with a blade stirrer in a vessel having a diameter of 1 m. The geometry is according to Figure 16.2. Baffles are not present. The rotational speed and the mixing power for a mixing time of 50 s are calculated.

Figure 16.3 is consulted. For $n = 40/60\ s^{-1}$, $Re = \dfrac{1{,}000 \cdot 40 \cdot 0.5^2}{60 \cdot 1} = 166.7$.

Then, $n\theta = 33$ and $\theta = 50$ s.

Figure 17.2 gives $P_o = 4$ for Re = 166.7.

$$P = 4 \cdot 1{,}000(40/60)^3 \cdot 0.5^5 = 37\ W.$$

Example 17.2

A liquid having a specific mass of 1,000 kg · m^{-3} and a viscosity of 10^{-2} N · s · m^{-2} is stirred with a turbine stirrer in a baffled vessel having a diameter of 1 m. The geometry is according to Figure 16.2. The rotational speed for the dissipation of 1 kW · m^{-3} is calculated.

Vessel volume: $\frac{\pi}{4} \cdot 1^2 \cdot 1 = 0.785\ m^3$.

$$P = 1{,}000 \cdot 0.785 = 785\ W.$$

The flow regime is assumed to be turbulent. According to Figure 17.2, $P_o = 5.0$.

$$785 = 5.0 \cdot 1{,}000 \cdot n^3(0.3)^5\ W \rightarrow n = 4.0\ s^{-1}\ (240\ rpm).$$

$$Re = \frac{1{,}000 \cdot 4.0 \cdot 0.3^2}{10^{-2}} = 36{,}000\ (turbulent).$$

Note 17.1

The vessels in Figure 16.2 have flat bottoms. Process vessels often have a dished bottom or a "Klöpperboden." Strictly speaking, Figure 17.2 cannot be applied to those vessels. However, the differences are minor. Moreover, there is much literature concerning power curves.

Note 17.2

Figure 17.2 can also be applied to industrial scales.

References

1. Beek, W. J., Muttzall, K. M. K., and van Heuven, J. W. (1999). *Transport Phenomena* (pp. 130–133). Chichester, UK: Wiley.
2. Ullmann (2002). *Processes & Process Engineering—Stirring. Ullmann's Encyclopedia of Industrial Chemistry*. Wiley Online Library.

18

Suspensions

18.1 Introduction

Agitators are used in stirred tanks containing particulate solids and liquids in many applications. The widest range of applications is covered by suspensions in which the particles are present in sufficiently low concentrations. Thus, they can be considered to have a small effect on fluid viscosity. Furthermore, the stability of the particle is generally unaffected by the agitation. Within this class of applications would fall dissolution, chemical reaction including catalysis, ion-exchange and adsorption, crystallization, and precipitation. This chapter is concerned specifically with the suspension of solids, present in relatively low concentration, under turbulent flow, that is, impeller $Re > 10^4$.

The discussion will focus on the rotational speed at which all particles are just suspended. In many cases, it will be sufficient that the liquid phase is well mixed, and all particles of the solid phase are in suspension, so that the whole solid surface is well exposed to the liquid. There are many empirical correlations in the literature. Generally, they are different. However, the correlation of Zwietering [1] is broadly similar to a large number of them. Therefore, Zwietering's work will be discussed in some detail.

The effect of rotating stirrers on the suspension of solids was studied by Zwietering in transparent vessels of glass and Perspex, which were fitted with four baffles along the wall. Visual observations were made. The suspension was considered complete when particles did not reside longer than one second on the bottom. The cylindrical vessels were filled in general to a height equal to the vessel diameter. Flat bottoms were used in the majority of the experiments. However, according to Zwietering, the influence of the shape of the bottom is small. Figure 18.1 shows the types of stirrers used in Zwietering's experiments, whereas Tables 18.1 and 18.2 contain, respectively, the dimensions of the vessels and the stirrers. Table 18.3 contains the specific mass and the size of the sand and salt particles used, whereas Table 18.4 contains the specific mass and the viscosity of the five liquids used.

The amount of solids in the suspension was varied from 0.5 to 20% by weight. The experiments were carried out at ambient temperature. Experiments with sand and water were carried out in all six vessels. More than 1,000 experiments were carried out with this system. The experiments with other liquids were carried out in the three smallest vessels only. All propellers pumped downward.

The results are presented in Figures 18.2–18.6.

D: vessel diameter in m.

d_i: agitator diameter in m.

h_i: height of the agitator above the bottom in m.

Introduction to Chemical Engineering, First Edition. C.M. van 't Land.

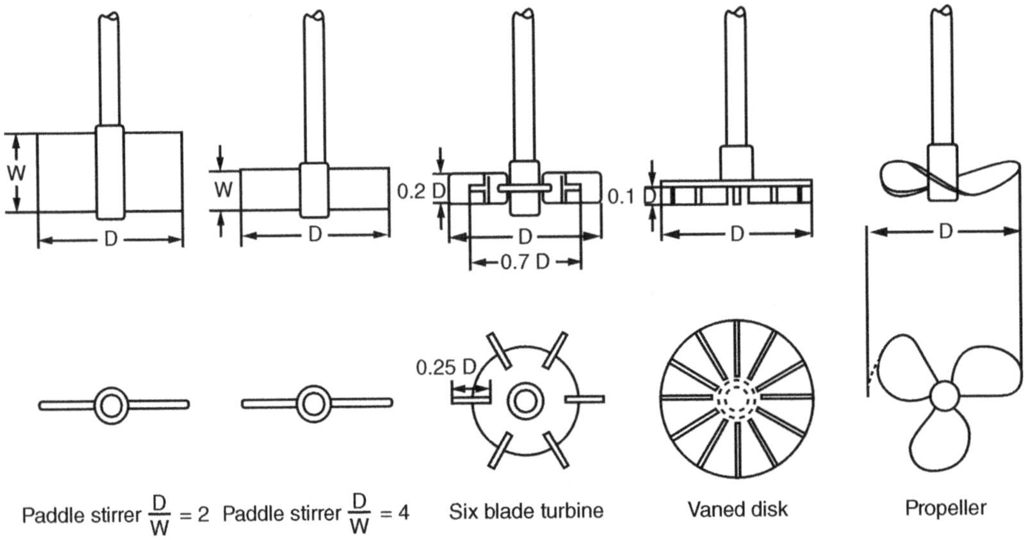

Figure 18.1 Five types of stirrers used in the experiments. *Source:* Courtesy of Elsevier, Amsterdam, The Netherlands.

Table 18.1 Dimensions of the vessels.

Vessel diameter, m	0.154	0.192	0.24	0.29	0.45	0.60
Vessel volume, l	2.9	5.5	11	19	70	170

Source: Zwietering [1].

Table 18.2 Types and dimensions of stirrers.

Type of stirrer	**Diameter, m**					
Paddles, $h_i/d_i = 0.5$	0.06	0.08	0.112	0.16		
Paddles, $h_i/d_i = 0.25$	0.06	0.08	0.16	0.224		
Turbines	0.06	0.08	0.12	0.16	0.20	
Vaned disks	0.06	0.08	0.10	0.12	0.16	0.20
Propellers	0.05	0.07	0.115			

Source: Zwietering [1].

Table 18.3 Specific mass and size of the solid particles.

Solid	ρ_s, **kg · m^{-3}**	**Particle size, μm**		
Sand	2,600	125–150	250–350	710–850
Sodium chloride	2,160	125–150	150–250	250–350

Source: Zwietering [1].

Table 18.4 Specific mass and viscosity of the liquids.

Liquid	ρ_L, kg · m^{-3}	μ, cP	ν, m^2 · s^{-1}
Water	1,000	1.0	10^{-6}
Acetone	790	0.31	$3.9 \cdot 10^{-7}$
Carbon tetrachloride	1,600	1.0	$6.5 \cdot 10^{-7}$
Potassium carbonate solution	1,440	5.0	$3.5 \cdot 10^{-6}$
Oil	840	9.3	$1.11 \cdot 10^{-5}$

Source: Zwietering [1].

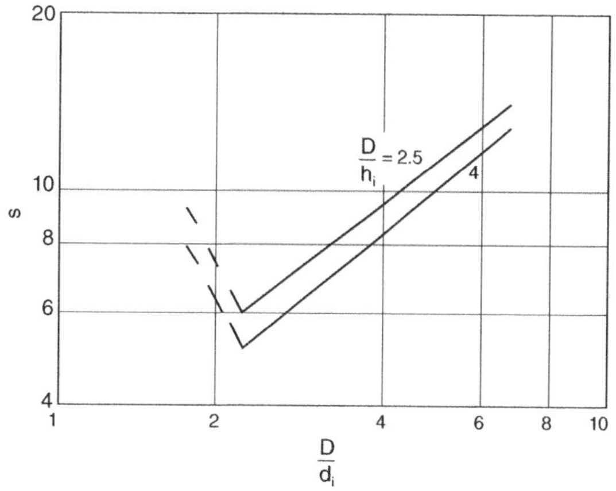

Figure 18.2 s-Values for propeller. *Source:* Courtesy of Elsevier, Amsterdam, The Netherlands.

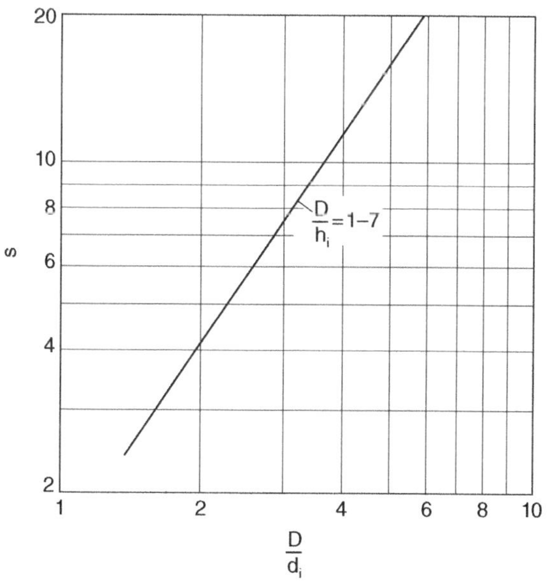

Figure 18.3 s-Values for disk turbine. *Source:* Courtesy of Elsevier, Amsterdam, The Netherlands.

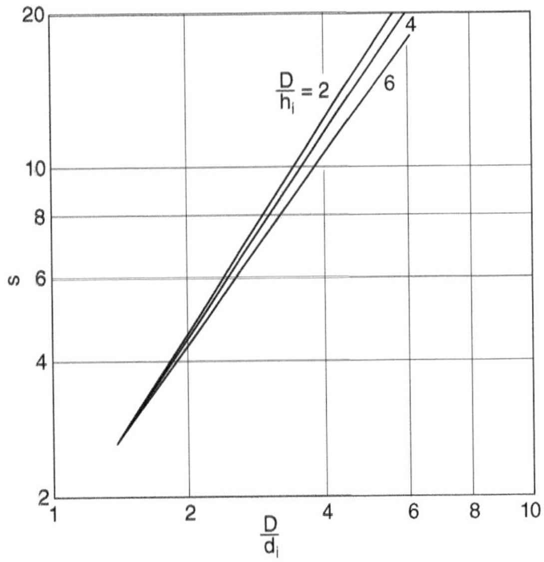

Figure 18.4 s-Values for vaned disk. *Source:* Courtesy of Elsevier, Amsterdam, The Netherlands.

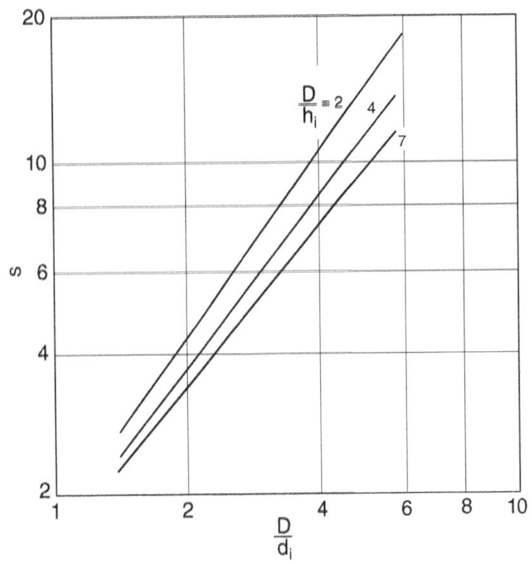

Figure 18.5 s-Values for flat paddle with height $0.5 \cdot d_i$. *Source:* Courtesy of Elsevier, Amsterdam, The Netherlands.

Zwietering's criterion is:

$$n^* = \frac{s \cdot \nu^{0.1} \cdot d_p^{0.2}(g \cdot \Delta\rho/\rho_l)^{0.45}(100 \cdot a/(1-a))^{0.13}}{d_i^{0.85}}. \tag{18.1}$$

This empirical expression is not dimensionally homogeneous. Zwietering used metric units.

n^*: the stirrer speed at which all particles are just suspended.

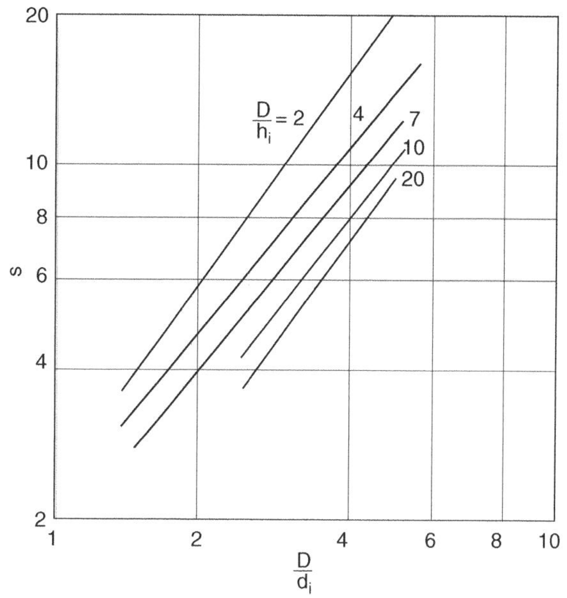

Figure 18.6 s-Values for flat paddle with height $0.25 \cdot d_i$. *Source:* Courtesy of Elsevier, Amsterdam, The Netherlands.

a: mass fraction solids.

Note that the viscosity is the viscosity of the suspending liquid.

s can be selected for a given geometry from Figures 18.2–18.6. Next, n^* can be calculated. The lines for the propeller show a break, which corresponds with a change in the flow pattern. It is found that for $D/d_i > 2$, the solid particles move radially outward along the bottom, and the last piles of particles occur at the periphery. However, if $D/d_i < 2$, the solids move inward and centrally up into the propeller.

n^* is approximately proportional to the square root of the specific mass difference. Furthermore, the dependency of n^* on the other independent variables is as follows:

- ν 1,000 times greater: n^* 2 times greater,
- d_p 10 times larger: n^* 1.6 times greater, and
- a two times greater: n^* 1.11 times greater.

Example 18.1

A process vessel has a diameter of 0.5 m. It is fitted with a dished bottom, four baffles, and a flat blade turbine agitator having a diameter of 0.15 m. The stirrer is located 0.15 m above the bottom. Sand having an average particle size of 300 μm is suspended in water. The weight fraction solids is 0.1. See Figure 18.1 concerning physical properties. n^* is calculated.

Figure 18.3: s = 9.

$$n^* = \frac{9(10^{-6})^{0.1}(0.3 \cdot 10^{-3})^{0.2}(9.81 \cdot 1{,}600/1{,}000)^{0.45}(100 \cdot 0.1/(1-0.1))^{0.13}}{0.15^{0.85}} = 10.6\ \mathrm{s}^{-1}.$$

18.2 Power Consumption

The suspension viscosity is needed for the calculation of the power consumption. It is recommended to use the formula [2]:

$$\mu_{susp} = \frac{\mu_l}{(1 - X/X')^{1.8}}. \quad (18.2)$$

X: volume fraction suspended material,
X′: volume fraction solid material in a settled bed.

Example 18.2

The power consumption of the stirrer of Example 18.1 is calculated.

Equation 18.2 is used to calculate the suspension's viscosity. X is calculated as follows.

	kg	$kg \cdot m^{-3}$	m^3	Volume fraction
Sand	0.1	2,600	$3.85 \cdot 10^{-5}$	0.041
Water	0.9 +	1,000	$90.00 \cdot 10^{-5}$ +	0.959 +
	1.0		$93.85 \cdot 10^{-5}$	1.000

X = 0.041.

As a rule-of-thumb, the bulk density of a crystallized particulate material is approximately 55% of the specific mass. Thus, the volume fraction solid material in a settled bed X′ is about 0.55.

$$\mu_{susp} = \frac{10^{-3}}{(1 - 0.041/0.55)^{1.8}} = 1.15 \cdot 10^{-3}\ N \cdot s \cdot m^{-2}.$$

The suspension's specific mass is calculated as follows.

$$\rho_{susp} = \frac{1.0}{93.85 \cdot 10^{-5}} = 1{,}066\ kg \cdot m^{-3}.$$

$$Re = \frac{1{,}066 \cdot 10.6 \cdot 0.15^2}{1.15 \cdot 10^{-3}} = 2.21 \cdot 10^5 \text{ (turbulent)}.$$

Figure 17.2: $P_o = 5.0$.

$$P = 5.0 \cdot 1{,}066 \cdot 10.6^3 \cdot 0.15^5 = 482\ W.$$

The vessel volume is 0.1 m^3 and $P/V = 482/0.1 = 4{,}820\ W \cdot m^{-3}$.

Example 18.3

The process vessel of Example 18.1 is equipped with a propeller having the same diameter and location as the turbine of the previous two examples. n* and the power consumption of the propeller are calculated.

Figure 18.2: s = 7.5. (Note that Zwietering did not define the propeller's slope!)

$$n^* = 10.6 \cdot \frac{7.5}{9} = 8.83\ \mathrm{s}^{-1}$$ (see Example 18.1).

Figure 17.2: $P_o = 0.35$ (turbulent region, $\alpha = 25° \rightarrow$ pitch is 1.47 times the diameter, see also Figure 16.2).

$$P = 0.35 \cdot 1{,}066 \cdot 8.83^3 \cdot 0.15^5 = 19.5\ \mathrm{W}.$$

$$\frac{P}{V} = \frac{19.5}{0.1} = 195\ \mathrm{W \cdot m^{-3}}.$$

A propeller consumes less power than a turbine agitator.

Power consumption on scaling-up:

$$P = P_o \cdot \rho \cdot n^3 \cdot d_i^5\ \mathrm{W}.$$

$$P \propto n^3 \cdot d_i^5.$$

$$\frac{P}{V} \propto n^3 \cdot d_i^2.$$

Zwietering: $n \cdot d_i^{0.85}$ is constant, thus, $n \propto \frac{1}{d_i^{0.85}}$.

Substitution of the latter expression into the first expression leads to

$$\frac{P}{V} \propto \frac{d_i^2}{d_i^{2.55}} = \frac{1}{d_i^{0.55}}.$$

Thus, P/V would decrease with a factor 1.46 when the volume increases with a factor 8. However, Zwietering recommends to keep P/V constant on scaling-up his results to larger equipment. On extrapolating Zwietering's results, the situation in, for example, a 0.1-m^3 vessel is considered first. Next, the specific power input in $\mathrm{W \cdot m^{-3}}$ obtained on small scale is kept constant. The geometry is also kept constant.

18.3 Further Work

Harnby et al. [3] mention work of Chapman et al., who worked on scales ranging from 0.3 to 1.8 m and found an equation similar to Zwietering's equation. They also mention further work confirming Zwietering's correlation. Table 18.5 is from their book and contains P_o- and s-values for a range of impellers and geometries.

Table 18.5 P_o- and s-values for a range of impellers and geometries (flat-bottomed vessels, H = D, four baffles, baffle width 0.1 · D).

Impeller type	h_i/D	d_i/D	P_o	s
4 MFD (four-bladed 45° pitch turbine pumping down)[a]	1/5	1/3	1.4	5.7
	1/4	1/3		6.2
	1/4	1/2		5.8
	1/4	1/4		7.1
6 MFD (six-bladed 45° pitch turbine pumping down)[a]	1/4	1/2	1.6	5.7
A310 (trade name Lightning)	1/5	1/3	0.25	7.6
	1/4	1/3		7.9
	1/4	1/2		6.0
HE3 (trade name Chemineer)	1/4	1/2	0.25	6.2
	1/4	1/3		7.2
Intermig (2 off) (trade name Ekato)	1/6, 2/3[b]	7/10	0.7	7.4
6 MFU	1/4	1/2	1.6	6.9

[a]MFD: Mixed Flow Design.
[b]Lower and upper arm, respectively.

References

1. Zwietering, T. N. (1958). Suspending of solid particles in liquid by agitators. *Chemical Engineering Science*, 8, 244–253.
2. Orr, C., and DallaValle, J. M. (1954). Heat-transfer properties of liquid-solid suspensions. *Chemical Engineering Progress Symposium Series*, 50, 29–45.
3. Harnby, N., Edwards, M. F., and Nienow, A. W. (2000). *Mixing in the Process Industries* (pp. 370–376). Oxford, UK: Butterworth-Heinemann.

19

Liquid/Liquid Dispersions

Van Heuven and Beek [1] have studied turbulently stirred liquid/liquid dispersions at ambient temperature. These dispersions were obtained in seven cylindrical vessels with a turbine impeller and four baffles (width $0.1 \cdot D$). The volumes of the relatively small vessels were in the range 1.56–71 l. One large vessel had a volume of $1.36\ m^3$. In the largest vessel, the impeller diameter was one third of the vessel diameter. For the other vessels, $d_i = 0.3 \cdot D$. The vessels had flat bottoms. The liquid height was equal to the vessel diameter. The stirrer location was halfway the liquid height. The stirrer blade height was $0.2 \cdot d_i$ and the stirrer blade length was $0.25 \cdot d_i$. Water was the continuous phase and benzene, octanol, or styrene were dispersed. These systems occur at extractions.

The volume fraction dispersed phase was varied from 0 to 0.4. The average droplet sizes were in the range 20–150 μm. The time required to reach stationary conditions varied and was checked. At all experiments, the flow was turbulent, that is, $Re > 10^4$. The authors give for the average droplet size:

$$\frac{d_p}{d_i} = 0.047(1 + 2.5 \cdot \Phi)\ We^{-0.6}. \tag{19.1}$$

In this equation, d_p is the surface-to-volume average droplet diameter in m $\left(\Sigma n \cdot d_p^3 / \Sigma n \cdot d_p^2\right)$. This average diameter is also called the Sauter diameter. Φ is the volume fraction dispersed phase. We is the Weber number:

$$We = \frac{\rho_{av} \cdot n^2 \cdot d_i^3}{\sigma_m}.$$

The Weber number appears when dispersions of immiscible liquids are made and the interfacial tension plays a role.

ρ_{av} is the specific mass of the mixture in $kg \cdot m^{-3}$, and σ_m is the interfacial tension of the dispersion in $N \cdot m^{-1}$.

The factor $(1 + 2.5 \cdot \Phi)$ indicates that, for a given rotational speed, the droplet size doubles when Φ increases from 0 to 0.4. The factor 2.5 is a measure of the tendency to coalesce. For a given Φ, d_p is constant when $d_i \cdot We^{-0.6}$ is constant.

$$d_i \cdot We^{-0.6} = d_i \cdot \frac{\sigma_m^{0.6}}{\rho_{av}^{0.6} \cdot n^{1.2} \cdot d_i^{1.8}} = \frac{\sigma_m^{0.6}}{\rho_{av}^{0.6}} \cdot \frac{1}{n^{1.2} \cdot d_i^{1.8}} \cdot$$

$n^{1.2} \cdot d_i^{0.8}$ constant is equivalent to $n^3 \cdot d_i^2$ constant and that means constant power input in $W \cdot m^{-3}$. Thus, the scale-up criterion is to keep the power input in $W \cdot m^{-3}$ constant.

Introduction to Chemical Engineering, First Edition. C.M. van 't Land.

Example 19.1

A vessel has a diameter of 1 m and is equipped with a six-bladed turbine agitator. The vessel is fully baffled. The geometry is according to Figure 16.2, however, the stirrer height equals 0.5 · D. The vessel is filled with a mixture of 20 volume % of benzene in water. The liquid height equals the vessel diameter. The rotational speed of the impeller to give a d_p of 50 μm is calculated. The power input per unit volume is also calculated. The physical data of benzene and water at 20 °C are

	ρ, $kg \cdot m^{-3}$	μ, $N \cdot s \cdot m^{-2}$
Benzene	880	$6.49 \cdot 10^{-4}$
Water	998	10^{-3}

The interfacial tension σ_m is 0.035 $N \cdot m^{-1}$.
Equation 19.1 is used.

$$We^{-0.6} = \frac{50 \cdot 10^{-6}}{0.3 \cdot 0.047(1 + 2.5 \cdot 0.2)} = 2.36 \cdot 10^{-3}.$$

$$We = \frac{\rho_{av} \cdot n^2 \cdot d_i^3}{\sigma_m} = 24{,}000.$$

$$\rho_{av} = 0.8 \cdot 998 + 0.2 \cdot 880 = 974\ kg \cdot m^{-3}.$$

$$n = 5.7\ s^{-1}.$$

It will be checked whether the flow is turbulent. The criterion is: $Re > 10^4$.
The viscosity of the dispersion can be calculated by means of Einstein's equation:

$$\mu_m = \mu_c(1 + a\Phi + b \cdot \Phi^2 + \cdots), \text{ with} \tag{19.2}$$

a = 2.5 and b = 7.2 for spheres.

$$\mu_m = 10^{-3}(1 + 2.5 \cdot 0.2 + 7.2 \cdot 0.04) = 1.8 \cdot 10^{-3}\ N \cdot s \cdot m^{-2}.$$

$$Re = \frac{\rho_{av} \cdot n \cdot d_i^2}{\mu_m} = \frac{974 \cdot 5.7 \cdot 0.3^2}{1.8 \cdot 10^{-3}} = 2.78 \cdot 10^5 \text{ (turbulent)}.$$

$P_0 = 5.0$ according to Figure 17.2.

$$P = 5.0 \cdot 974 \cdot 5.7^3 \cdot 0.3^5 = 2{,}192\ W.$$

The vessel volume is 0.785 m^3, and the power input is 2.8 $kW \cdot m^{-3}$.

The authors also give a criterion to calculate the minimum impeller speed. This is the impeller speed at which "lakes" of the dispersed phase occur neither at the surface nor at the bottom. Generally, if the flow is turbulent, "lakes" will not be found. Note that this contribution does not refer to a homogeneous dispersion. It also concerns pure liquids. Surface active agents are not present. On substituting Equation 19.1 into the criterion for the minimum impeller speed and rewriting it, the following equation is obtained:

$$n_{min} = \frac{g \cdot \Delta\rho \cdot \sigma_m^{0.2} \cdot \mu_m^{0.2} \cdot 36(1 + 2.5 \cdot \Phi)^{2.33}}{\left(d_i^2 \cdot \rho_{av}^{1.4}\right)^{0.385}}. \tag{19.3}$$

The criterion for the minimum impeller speed was obtained by interpreting experimental results in the aforementioned stirred vessels. Benzene, hexane, octanol, and styrene were dispersed in water. The volume fraction dispersed phase was varied from 0 to 0.35. The average droplet sizes were in the range 30–240 μm. The minimum stirrer speed varied from 1 (in the largest vessel) to 10 (in the smallest vessel).

Equation 19.3 leads to the criterion that to avoid "lakes" at the surface or at the bottom, $n_{min} \propto d_i^{-0.77}$. Substitution of this result into $P/V \propto n^3 \cdot d_i^2$ gives $P/V \propto d_i^{-0.31}$. Thus, P/V can decrease on scaling-up. The consequences of this finding will be discussed shortly. Let us assume a desired droplet size was obtained on a small scale. It was also observed that phase segregation did not occur on this small scale. If scaling-up is carried out by keeping the specific power input constant, we can be relatively confident that the same average droplet size will be obtained on a large scale. It then also stands to reason that phase segregation will not occur on a large scale.

Example 19.2

The minimum stirrer speed for the mixing duty of Example 19.1 is calculated by using Equation 19.3.

$$n_{min} = \frac{9.81 \cdot 118 \cdot 0.035^{0.2} \cdot \left(1.8 \cdot 10^{-3}\right)^{0.2} \cdot 36(1 + 2.5 \cdot 0.2)^{2.33}}{\left(0.3^2 \cdot 974^{1.4}\right)^{0.385}},$$

$$= 2.5\ s^{-1}.$$

In this chapter, the work of Van Heuven and Beek was discussed. It should be mentioned that there is also literature giving different results. This aspect will be discussed shortly for Equation 19.1. The factor 2.5 is rather low and is applicable to systems that do not coalesce easily. Higher values can be found in the literature, for example, 3.5 or even higher. The factor 0.047 is low as well. Values in the range 0.05–0.08 can be found in the literature. The results of the aforementioned two authors are applicable to low-viscosity liquids that do not coalesce readily. There is, however, a general understanding that scaling-up should be carried out by keeping the specific power input in $kW \cdot m^{3}$ constant.

Reference

1. van Heuven, J. W., and Beek, W. J. (1970). Scaling-up rules for turbulent liquid/liquid dispersions in stirred vessels. *De Ingenieur*, 82 (44), Ch51–Ch60. (in Dutch).

20

Gas Distribution

20.1 Introduction

This chapter treats gas dispersion by a single impeller in a vessel, fitted with four baffles, in which the liquid depth is roughly equal to the vessel diameter. See Figure 20.1. Gas is introduced into the vessel at a point directly below the agitator. The gas rises up to the impeller and is then dispersed throughout the vessel by the pumping action of the rotating impeller blades. Air oxidation of a hydrocarbon is a typical example.

The six-blade disc impeller, also known as the Rushton turbine, is the most commonly used impeller for gas dispersion. As a result, this impeller has been intensively studied by researchers and a number of design rules were developed. The Rushton turbine will be discussed in Section 20.2. Chemineer developed concave impellers for gas dispersion. See Figure 20.2. These impellers can, at equal rotational speeds, handle larger gas quantities than the Rushton turbine. For three-phase processes, a four-blade or a six-blade 45° pitched-blade turbine is recommended. See Figure 20.6. For processes in which solid suspension is the most difficult requirement (i.e., low gas flow rate and dense solids), the impeller should pump downward. Vice versa for a high gas flow rate and light solids. The four-blade pitched-blade 45° turbine pumping downward will be discussed in Section 20.3.

The material in this chapter can be used for the distribution of gas in low-viscosity liquids. Moreover, it is limited to pressures < 10 bar.

20.2 Turbine

Various gas/liquid flow patterns for a radial flow impeller can be seen in Figure 20.3.

Flooding

Flooding is defined as the condition under which the impeller is overwhelmed by gas. This occurs at low impeller speeds or high gas flow rates. The gas is not distributed properly. Flooding must be avoided.

Loading

On increasing the impeller speed or decreasing the gas flow rate, the impeller begins to distribute the gas radially. However, the gas is not yet swept into the region below the turbine.

Introduction to Chemical Engineering, First Edition. C.M. van 't Land.

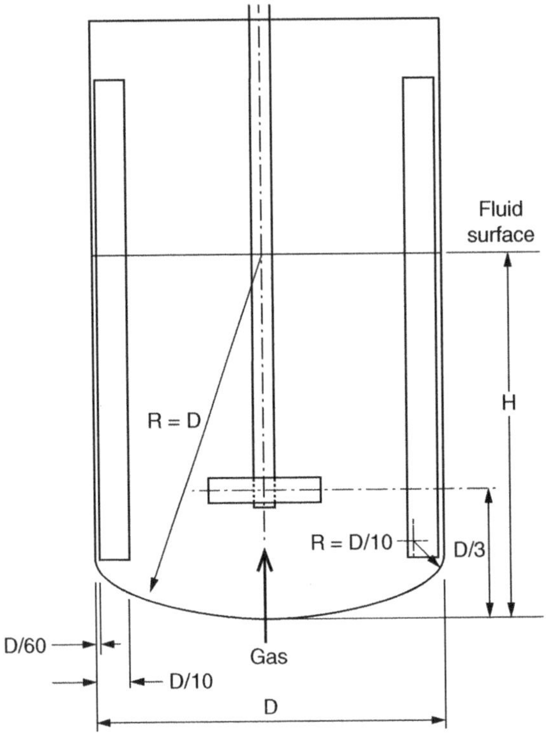

Figure 20.1 Vessel fitted with a single impeller.

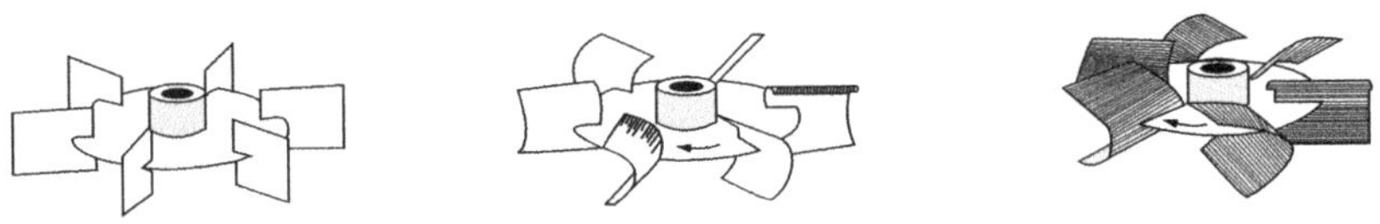

Figure 20.2 Impellers for gas dispersion.

Complete Dispersion

This condition is depicted in Figure 20.3 (d). Bubbles are swept into the region below the agitator. There is little recirculation of gas bubbles back into the impeller region.

Recirculation

A further increase in impeller speed, or a further decrease in gas flow rate, results in the recirculation of gas bubbles back into the impeller region. The transition from complete dispersion to recirculation is a gradual process, not as abrupt as the flooding/loading and loading/complete dispersion transitions.

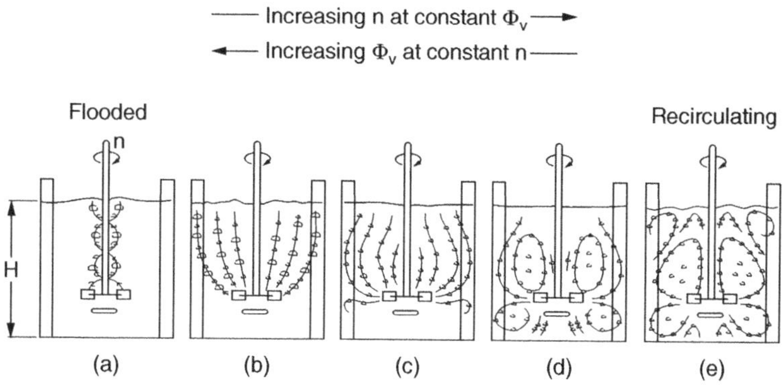

Figure 20.3 Gas/liquid flow patterns for a radial flow impeller.

For processes that are not mass-transfer limited, for example, slow chemical reactions in the liquid phase, the degree of gas dispersion is not critical and loading will suffice. For gas/liquid reactions where rapid mass transfer is required, for example, fast chemical reactions in the liquid phase, complete dispersion is necessary.

Typical superficial gas velocities are in the range 0.01–0.02 $m \cdot s^{-1}$. Typical impeller tip speeds are in the range $1 - 10\ m \cdot s^{-1}$.

Transition Speeds

In the three following equations, metric units are to be used.

The minimum impeller speed required to obtain loading and avoid flooding for a given impeller diameter and gas flow rate is given by [1]:

$$n_L = \frac{(0.33 \cdot \Phi_V \cdot D^{3.5})^{0.33}}{d_i^{2.5}}\ s^{-1}. \tag{20.1}$$

Φ_V is the gas flow rate in $m^3 \cdot s^{-1}$, D is the vessel diameter in m, and d_i is the impeller diameter in m.

The minimum agitator speed required to completely disperse a given gas flow rate for a given impeller diameter is given by [1]:

$$n_{CD} = \frac{\sqrt{(15.7 \cdot \Phi_V \sqrt{D})}}{d_i^2}\ s^{-1}. \tag{20.2}$$

The minimum impeller speed required to ensure a high degree of gas recirculation for a given gas flow rate and for a given impeller diameter is given by [2]:

$$n_R = \frac{(7.4 \cdot \Phi_V)^{0.2} D}{d_i^2}\ s^{-1}. \tag{20.3}$$

These equations can be applied for the distribution of air into water, for d_i/D-ratios from 0.25 to 0.5, impeller clearance ratios h_i/D from 0.25 to 0.4, and vessels up to 1.2 m in diameter. The point or small ring sparger (diameter $< d_i$) should be mounted directly under the impeller disc at a clearance $< 0.2 \cdot d_i$.

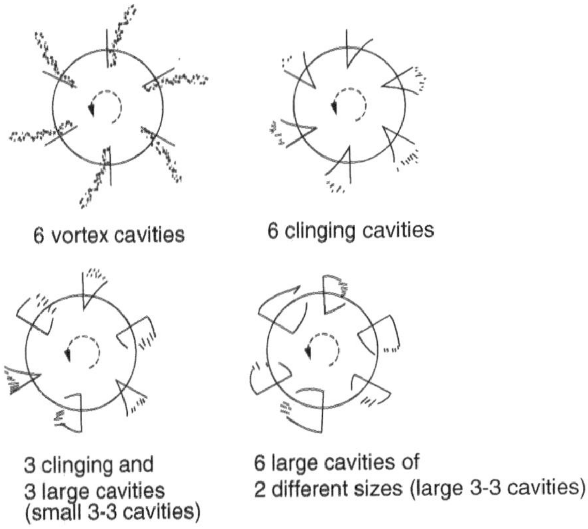

Figure 20.4 Cavity regimes for a turbine impeller.

Note

For the CD-6 impeller of Chemineer (see Figure 20.2), the factors 0.33 and 15.7 in Equations 20.1 and 20.2 are replaced by 0.14 and 7.85, respectively.

Power Consumption

On gassing a turbine, gas cavities develop behind the blades. Gas cavities reduce the drag of the impeller and hence have a significant effect on the power consumption of the turbine. The shape and size of the gas cavities depend on impeller speed and gas flow rate. See Figure 20.4. This figure can be explained as follows. If, at a certain rotational speed, a small amount of gas is administered, one sees six vortex cavities. On increasing the gas flow, six clinging cavities develop. At still larger gas flows, six small 3-3 cavities and eventually six large 3-3 cavities can be found. The next step is that ragged cavities (flooding) are obtained. The transition from ragged cavities to six large 3-3 cavities is defined by Equation 20.1. The transition from six large 3-3 cavities to six small 3-3 cavities occurs when Q, the Flow number, is approximately 0.1. The definition of the Flow number is $\Phi_V/(nd_i^3)$.

Zlokarnik [3] found that the Power number of a gassed turbine can, for the air/water-system, be given as a function of Q. See Figure 20.5. The equation is:

$$P_o = 1.5 + \left(0.5 \cdot Q^{0.075} + 1{,}600 \cdot Q^{2.6}\right)^{-1}. \tag{20.4}$$

Validity ranges: $Q \leq 0.15$, $Fr \geq 0.65$, $Re \geq 10^4$, and $D/d_i \geq 3.3$. To apply this equation to an extended D/d_i-range ($D/d_i \geq 2.2$), Q must be replaced by $A = Q(1 + 38(D/d_i)^{-5})$. Fr, the Froude number, is $n^2 \cdot d_i/g$. It is proportional to the ratio of the radial acceleration of the periphery of the stirrer and the acceleration due to gravity. Furthermore, $H/D = 1$ and $h_i/d_i = 1$. The author worked with baffled vessels (baffle width $0.1 \cdot D$) made of acryl glass and having diameters of 200, 300, and 450 mm. He used, at all the test work, a standard Rushton turbine having a diameter of 90 mm. The vessels had Klöpper bottoms. The temperature was ambient. The air was passed on to the stirrer by means of a pipe having a diameter of 10 mm and ending centrally 10 mm under the stirrer. The air left the pipe through a 2-mm hole.

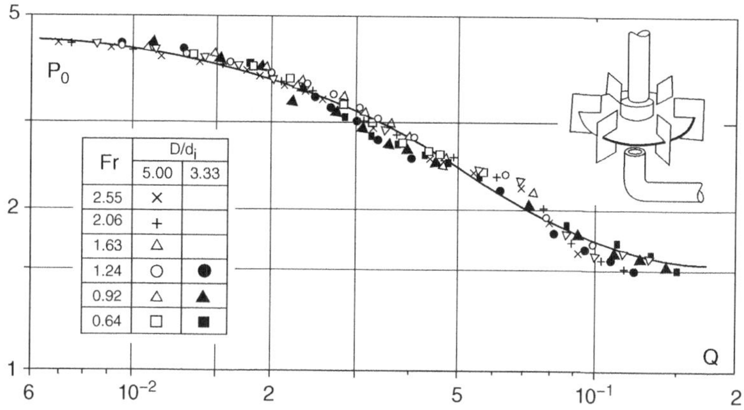

Figure 20.5 Power characteristic $P_0(Q)$ of a turbine stirrer in the air/water system; $H/D = 1$, $h_i/d_i = 1$. *Source:* Stirring by Zlokarnik, M. (2001). Reproduced with permission of John Wiley & Sons Inc.

Zlokarnik [4] also describes test work with air and more viscous liquids. The liquid having the highest viscosity was pentane diol ($0.123\ N \cdot s \cdot m^{-2}$). His results show that Equation 20.4 is also approximately correct for air and these more viscous liquids.

Gas Hold-up

Van 't Riet [5] correlated data concerning the distribution of air in water at atmospheric pressure and ambient temperature by means of Rushton turbines and impellers resembling Rushton turbines. Most measurements were carried out in a cylindrical baffled vessel having a diameter of 44 cm. His vessels had flat bottoms. In the 44-cm vessel, impellers having a diameter of 17.6 cm were used, $h_i = 1/3 \cdot D$. His results can be expressed as:

$$\varepsilon = 0.13 \cdot \varepsilon_V^{1/3} \cdot v_S^{2/3}.$$

ε is the fraction gas volume of the vessel contents. ε_V is the gassed power input in W divided by $(\pi/4)\ D^3$. The unit of D is m. v_S is the superficial gas velocity if the vessel would be empty.

Thus, the gas hold-up is moderately dependent on the specific power input and strongly dependent on the superficial gas velocity. Van 't Riet filled the 44-cm vessel with a liquid height of 75 cm to avoid air entrainment from the atmosphere at high stirrer speeds. However, ε is the hold-up in the stirred region only. Thus, in practice, the hold-up in the Standard Tank Configuration will be greater due to air entrainment. It is recommended to replace the coefficient 0.13 by 0.17:

$$\varepsilon = 0.17 \cdot \varepsilon_V^{1/3} \cdot v_S^{2/3}. \tag{20.5}$$

Van 't Riet found that the addition of salts leads to an increase of ε due to a reduced bubble size.

Example 20.1

A fully baffled process vessel with a dished bottom has a diameter of 1 m. The vessel is equipped with a Rushton turbine and $d_i = 0.4\ m \cdot h_i$ is also 0.4 m, the sparger ring has a diameter of $0.8 \cdot d_i$ and the sparger ring clearance is $0.1 \cdot d_i$. H/D = 1. $0.01\ m^3 \cdot s^{-1}$ should be completely dispersed in water. The temperature is ambient. A rotational impeller speed is selected, the power consumption and the gas residence time are calculated.

Equation 20.1: $n_L = \dfrac{(0.33 \cdot 0.01 \cdot 1^{3.5})^{0.33}}{0.4^{2.5}} = 0.7\ s^{-1}.$

Equation 20.2: $n_{CD} = \frac{\sqrt{(15.7 \cdot 0.01\sqrt{1})}}{0.4^2} = 2.5\,s^{-1}.$

Equation 20.3: $n_R = \frac{(7.4 \cdot 0.01)^{0.2} \cdot 1}{0.4^2} = 3.7\,s^{-1}.$

An impeller speed of 3 s^{-1} is selected.

Ungassed Power Consumption

$$P = 4.5 \cdot 1{,}000 \cdot 3^3 \cdot 0.4^5 = 1{,}244\,W.$$

4.5 is taken as the ungassed Power number instead of the more usual 5.0. The reason is that 4.5 is the maximum ordinate in Figure 20.5.

Gassed Power Consumption

The applicability of Equation 20.4 is checked.

$$Q = \frac{0.01}{3 \cdot 0.4^3} = 0.052.$$

$$Fr = \frac{3^2 \cdot 0.40}{9.81} = 0.37.$$

$$Re = \frac{1{,}000 \cdot 3 \cdot 0.4^2}{10^{-3}} = 4.8 \cdot 10^5.$$

$D/d_i = 2.5.$

The conclusion of the check is that Equation 20.4 can be used if Q is replaced by A. The fact that Fr is smaller than 0.65 is accepted.

$$A = 0.052\left(1 + 38\left(\frac{1}{0.4}\right)^{-5}\right) = 0.072.$$

$$P_o = 1.5 + (0.5 \cdot 0.072^{0.075} + 1{,}600 \cdot 0.072^{2.6})^{-1} = 2.0.$$

$$P = \left(\frac{2.0}{4.5}\right)1{,}244 = 553\,W.$$

Gas Residence Time

The volume of a dished head for a 3-mm wall thickness follows from [6]:

$$0.050 \cdot D^3 + 1.65 \cdot 0.003 \cdot D^2 = 0.055\,m^3.$$

The cylindrical volume is $0.9(\pi/4)1^2 = 0.7\,m^3$.
Total liquid volume: $0.055 + 0.7 = 0.76\,m^3$.

$$\varepsilon_V = 553/0.76 = 727.6\ \mathrm{W \cdot m^{-3}}.$$

$$v_S = \frac{0.01}{(\pi/4)1^2} = 0.013\ \mathrm{m \cdot s^{-1}}.$$

$$\varepsilon = 0.17 \cdot 727.6^{1/3} \cdot 0.013^{2/3} = 0.085.$$

$$0.085 = \frac{V_G}{V_G + V_L} = \frac{V_G}{V_G + 0.76} \rightarrow V_G = 0.071\ \mathrm{m^3}.$$

$$\tau = 0.071/0.01 = 7.1\ \mathrm{s}.$$

Note 20.1

The calculated P_o-value can be found in Figure 20.5 if A is taken as the abcissa.

Note 20.2

Here too, 4.5 is taken as the ungassed Power number.

20.3 Pitched-Blade Turbine Pumping Downward

See Figure 20.6. Three transitions were discussed for the turbine in Section 20.2. For the pitched-blade turbine pumping downward, there is only one transition: from direct to indirect gas loading. With direct gas loading, gas bubbles rise directly into the impeller blades. With indirect gas loading, gas bubbles are swept away from the impeller to be recirculated into the impeller region by the liquid flow. In the direct loading regime, large pockets of gas escape from the dispersion at irregular intervals showing the instability of the dispersion. This behavior is analogous to the flooded condition of a disc turbine and should be avoided.

The minimum impeller speed required to avoid direct loading for a given impeller diameter and gas flow rate is given by the following equation [7, 8]. Again, metric units should be used.

$$n_L = \frac{11.7\sqrt{\Phi_V}}{d_i^{1.63}}. \tag{20.6}$$

This correlation was derived for a four-blade, 45°-pitch blade turbine with $D/d_i = 2$, a small ring sparger of $0.7 \cdot d_i$ mounted directly under the impeller, and a vessel diameter of 0.56 m. Air was dispersed in water. It is recommended to select a rotational speed at least 20% higher than calculated from Equation 20.6.

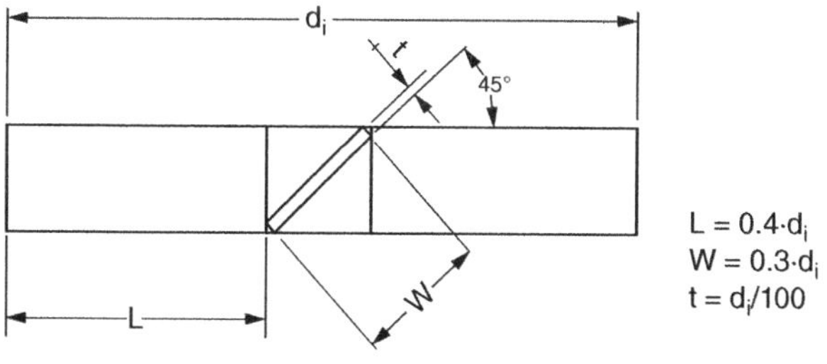

Figure 20.6 Four-blade, 45° pitched-blade turbine.

Power Consumption

For design purposes, assume that in the indirect loading regime, the gassed power consumption is 90% of the ungassed power consumption. The Power number of the four-blade, 45°-pitch blade turbine is 1.4.

Gas Hold-up

Equation 20.5 can be applied.

Example 20.2

See Example 20.1. The dispersion duty of that example is now taken care of by a four-blade, 45°-pitch-blade turbine having a diameter of 0.5 m and pumping downward. A ring sparger with a diameter of 0.35 m is used.

Equation 20.6: $n_L = \dfrac{11.7\sqrt{0.01}}{0.5^{1.63}} = 3.6\ s^{-1}$.

Select $1.2 \cdot 3.6 = 4.3\ s^{-1}$.

Ungassed Power Consumption

$$P = 1.4 \cdot 1{,}000 \cdot 4.3^3 \cdot 0.5^5 = 3{,}478\ W.$$

Gassed Power Consumption

$$0.9 \cdot 3{,}478 = 3{,}130\ W.$$

Gas Residence Time

See Example 20.1.

$$\varepsilon_V = 3{,}130/0.76 = 4{,}118\ W \cdot m^{-3}.$$

Equation 20.5: $\varepsilon = 0.17 \cdot 4{,}118^{1/3} \cdot 0.013^{2/3} = 0.15$.

$$0.15 = \frac{V_G}{V_G + V_L} = \frac{V_G}{V_G + 0.76} \rightarrow V_G = 0.135\ m^3.$$

$$\tau = 0.135/0.01 = 13.5\ s.$$

20.4 Turbine Scale Up

First, the turbine rotational speed at which the gas is just completely dispersed is considered. On keeping the gas flow in VVM (vessel volumes per minute) constant, it can be derived from Equation 20.2 that the specific power input is proportional to $d_i^{1.25}$. See Appendix 20.1. For example, on doubling the agitator diameter, the specific power increases with a factor 2.4. Doubling the agitator diameter means a capacity increase with a factor 8. Thus, on keeping the specific power input constant on scaling-up, the quality of the dispersion decreases. However, the gas residence time increases on scaling-up if VVM is kept constant. This is because the superficial gas velocity increases on keeping VVM constant and hence the gas hold-up increases, see Equation 20.5. It is possible that these two effects compensate each other to some extent.

For many gas dispersion processes, the liquid phase chemical reaction does not proceed at the optimum rate because the transfer from the gas phase to the liquid phase is limiting. However, dispersing the gas completely is the best one we can do. It is common practice to scale-up by keeping VVM and specific power input in $W \cdot m^{-3}$ or $W \cdot kg^{-1}$ constant. On doing so, it should be checked whether the impeller is at least loaded. This will be elucidated in the next section.

20.5 Batch Air Oxidation of a Hydrocarbon

The air is distributed by means of a Rushton turbine. Small-scale experiments and pilot-plant oxidations have taught that the desired process result can be obtained by keeping $W \cdot kg^{-1}$ and VVM constant. The specific power input is $2.5\ W \cdot kg^{-1}$. The air is dispersed at 0.84 VVM. The air flow is expressed in $nm^3 \cdot min^{-1}$.

The stirrer rotational speed in an existing modified reactor is calculated. Furthermore, the gas distribution and the gas hold-up are checked.

The initial charge of the industrial reactor is $15.8\ m^3$. This corresponds to $1.48 \cdot 10^4$ kg ($\rho = 937\ kg \cdot m^{-3}$). It is estimated that, on small scale, the gassed power input was 35% of the ungassed power input. It is also estimated that the gassed power input on large scale is 35% of the ungassed power input on large scale. The latter estimation will be checked. The air oxidation proceeds at 112 °C. The gas leaves the liquid at atmospheric pressure.

The gassed power input is $2.5 \cdot 1.48 \cdot 10^4 = 3.7 \cdot 10^4$ W.

Ungassed power consumption $3.7 \cdot 10^4/0.35 = 1.06 \cdot 10^5$.

The diameter of the baffled reactor is 2.75 m and both d_i and h_i are one third of the reactor diameter, that is, 0.92 m.

The ungassed power input on a large scale is used to calculate the large-scale stirrer rotational speed.

$$1.06 \cdot 10^5 = 4.5 \cdot 937 \cdot n^3 \cdot 0.92^5\ W \rightarrow n = 3.4\ s^{-1}.$$

4.5 is taken as the Power number instead of the more common 5.0; 4.5 is the Power number in Figure 20.5.

Gas Distribution

$$0.84 \cdot 15.8 = 13.3\ nm^3 \cdot min^{-1} = 0.22\ nm^3 \cdot s^{-1}.$$

At 112 °C and 0.3 barg: $0.22 \cdot \dfrac{(273+112)}{273} \cdot \dfrac{1}{1.3} = 0.24\ m^3 \cdot s^{-1}$.

Equation 20.1: $n_L = \dfrac{(0.33 \cdot 0.24 \cdot 2.75^{3.5})^{0.33}}{0.92^{2.5}} = 1.7\ s^{-1}$.

Equation 20.2: $n_{CD} = \dfrac{\left(15.7 \cdot 0.24\sqrt{2.75}\right)^{0.5}}{0.92^2} = 3.0\ s^{-1}$.

Equation 20.3: $n_R = \dfrac{(7.4 \cdot 0.24)^{0.2} \cdot 2.75}{0.92^2} = 3.6\ s^{-1}$.

The gas flow is completely dispersed.

Check the power input on industrial scale.

First, it is assumed that the system behaves like the air/water-system. The applicability of Equation 20.4 is checked.

$$Q = \frac{0.24}{3.4 \cdot 0.92^3} = 0.09.$$

$$Fr = \frac{3.4^2 \cdot 0.92}{9.81} = 1.08.$$

$$\text{Re} = \frac{937 \cdot 3.4 \cdot 0.92^2}{10^{-3}} = 2.7 \cdot 10^6.$$

The water viscosity was taken.

$$\frac{D}{d_i} = \frac{2.75}{0.92} = 3.0.$$

The conclusion is that Equation 20.4 can be used if Q is replaced by A.

$$A = Q\left(1 + 38\left(\frac{D}{d_i}\right)^{-5}\right) = 0.09\left(1 + 38 \cdot 3.0^{-5}\right) = 0.104.$$

$$P_o = 1.5 + \left(0.5 \cdot 0.104^{0.075} + 1{,}600 \cdot 0.104^{2.6}\right) = 1.7.$$

$$P = \left(\frac{1.7}{4.5}\right) 1.06 \cdot 10^5 = 4.00 \cdot 10^4 \text{ W}.$$

$$\frac{4.00 \cdot 10^4}{1.48 \cdot 10^4} = 2.7 \text{ W} \cdot \text{kg}^{-1}.$$

This specific power input is approximately equal to the specific power input on small scale.
Check the gas hold-up.
Equation 20.5: $\varepsilon = 0.17 \cdot \varepsilon_V^{1/3} \cdot v_S^{2/3}$.

$$\varepsilon_V = \frac{4.00 \cdot 10^4}{15.8} = 2.53 \cdot 10^3 \text{ W} \cdot \text{m}^{-3}.$$

$$v_S = \frac{0.24}{(\pi/4)2.75^2} = 0.04 \text{ m} \cdot \text{s}^{-1}.$$

$$\varepsilon = 0.17\left(2.53 \cdot 10^3\right)^{1/3} \cdot 0.04^{2/3} = 0.27.$$

$$0.27 = \frac{V_G}{V_G + 15.8} \rightarrow V_G = 5.8 \text{ m}^3.$$

Thus, the liquid level will rise considerably in the industrial reactor. On a small scale, because of the lower superficial gas velocity, this effect will be less pronounced. The large-scale gas residence time will also be longer than the small-scale gas residence time.

20.6 Remark

The material of this lecture serves to ascertain that the impeller is not flooded and that the vessel does not contain too much gas. The mass transfer between gas and liquid is a matter of $k_l a$-values and that subject is treated in Chapter 21.

Appendix 20.1

$$P/V \propto n^3 \cdot d_i^2. \quad (1)$$

$$\Phi_v \propto d_i^3 \text{ (VVM constant)}. \quad (2)$$

$$\text{Equation (20.2)}: n \propto \frac{\Phi_V^{1/2} \cdot d_i^{1/4}}{d_i^2} = \frac{\Phi_V^{1/2}}{d_i^{1\,3/4}}. \quad (3)$$

$$\text{Using (2) in (3)}: n \propto \frac{d_i^{1\,1/2}}{d_i^{1\,3/4}} = \frac{1}{d_i^{1/4}}. \tag{4}$$

$$\text{Using (4) in (1)}: P/V \propto d_i^2 = d_i^{1\,1/4} = d_i^{1.25}. \tag{5}$$

References

1. Nienow, A. W., Warmoeskerken, M. M. C. G., Smith, J. M., and Konno, M. (1985). *On the flooding/loading transition and the complete dispersal condition in aerated vessels agitated by a Rushton-turbine.* 5th European Conference on Mixing. Germany: Würzburg.
2. Nienow, A. W., and Wisdom, D. J. (1977). *The effect of scale and geometry on flooding, recirculation, and power in gassed stirred vessels.* Second European Conference on Mixing. Cambridge, UK.
3. Zlokarnik, M. (2001). *Stirring* (p. 85). Weinheim, Germany: Wiley-VCH.
4. Zlokarnik, M. (1973). Power consumption of gassed liquids. *Chemie-Ingenieur-Technik*, 45 (10a), 689–692. (in German).
5. van't Riet, K. (1975). *Turbine Agitator Hydrodynamics and Dispersion Performance* (p. 91). Meppel, The Netherlands: Krips Repro. (in Dutch).
6. Perry, R. H., and Green, D. W. (1997). *Perry's Chemical Engineers' Handbook* (pp. 10–140). New York, NY: McGraw-Hill.
7. Chapman, C. M., Nienow, A. W., Cooke, M., and Middleton, J. C. (1983). Particle–gas–liquid mixing in stirred vessels. Part I: Particle–liquid mixing. *Chemical Engineering Research and Design*, 61 (3), 71–81.
8. Chapman, C. M., Nienow, A. W., Cooke, M., and Middleton, J. C. (1983). Particle–gas–liquid mixing in stirred vessels. Part II: Gas–liquid mixing. *Chemical Engineering Research and Design*, 61 (3), 82–95.

21

Physical Gas Absorption

21.1 Introduction

Gas distribution in a stirred vessel is discussed in Chapter 6. Gas is distributed to effect mass transfer, either from the gas to the vessel contents or from the vessel contents to the gas phase. This chapter treats the efficiency of mass transfer from the gaseous phase through a liquid boundary layer. Generally, the study of this aspect is indicated as measuring $k_l \cdot a$-values. $k_l \cdot a$ is the product of k_l, the liquid mass transfer coefficient, and a, the interfacial area in m^2 per m^3 of liquid. It is customary to consider the product $k_l \cdot a$ because the interfacial area is not well known. Compare this to a heat exchanger with a known area for heat transfer. It is then possible to separate the area for heat transfer and the heat transfer coefficient. Since mixing has little effect on k_l, it is the interfacial area, which is a function of the mixing. Hence, $k_l \cdot a$ is a function of the superficial velocity v_s and ε_v, the power input per unit volume of liquid. Note that the superficial velocity is calculated for an empty vessel.

21.2 $k_l \cdot a$ Measurements

In industrial gas/liquid contacting in stirred vessels, the following starting points are applicable in many cases:

- the depletion of the gaseous phase is small, meaning that the resistance to mass transfer is in the liquid phase ($k_g \gg k_l$), and
- the liquid phase is well-mixed.

For a semi-batch process (for example, passing air through an amount of water absorbing oxygen), the following equation is applicable:

$$\Phi_{mole} = k_l \cdot a(c_{sat} - c) \text{ kmol} \cdot m^{-3} \cdot s^{-1}.$$

This equation can also be written as:

$$\frac{dc}{dt} = k_l \cdot a(c_{sat} - c) \text{ kmol} \cdot m^{-3} \cdot s^{-1}.$$

In words: The speed with which the oxygen concentration in the liquid phase increases is proportional to the driving force. The driving force is the difference between the saturation

Introduction to Chemical Engineering, First Edition. C.M. van 't Land.

concentration and the actual concentration. The actual concentration is a function of time. It is possible to rewrite the differential equation and to integrate it:

$$\int_{c_0}^{c} \frac{dc}{(c_{sat} - c)} = k_l \cdot a \int_{0}^{t} dt,$$

$$-\int_{c_0}^{c} \frac{d(c_{sat} - c)}{(c_{sat} - c)} = k_l \cdot at,$$

$$-(\ln (c_{sat} - c) - \ln (c_{sat} - c_0)) = k_l \cdot at,$$

$$\ln \frac{(c_{sat} - c_0)}{(c_{sat} - c)} = k_l \cdot at,$$

$$\frac{c - c_0}{c_{sat} - c_0} = 1 - e^{-k_l \cdot at}.$$

If the initial concentration is 0, the expression becomes

$$\frac{c}{c_{sat}} = 1 - e^{-k_l \cdot at}. \quad (21.1)$$

The absorption of oxygen in water will be discussed. In that case, $k_l a$-measurements can be carried out as follows. Nitrogen is passed through a stirred vessel containing demineralized water to remove dissolved oxygen. At a certain point in time, the nitrogen flow is replaced by an air flow, and the concentration of dissolved oxygen is measured as a function of time by means of polarographic oxygen electrodes. Figure 21.1 is a typical plot of this equation [1]. The diameter of the stirred, fully baffled vessel is 0.39 m, and the unaerated liquid height is also 0.39 m. The vessel bottom is flat. Both the diameter and the height of the impeller are 0.13 m. The impeller is a 45° pitched-blade turbine having four blades. The Power number of this impeller is 1.4. Its rotational speed is 700 rpm. The superficial air velocity is $5 \cdot 10^{-3}$ m $\cdot$ s^{-1}.

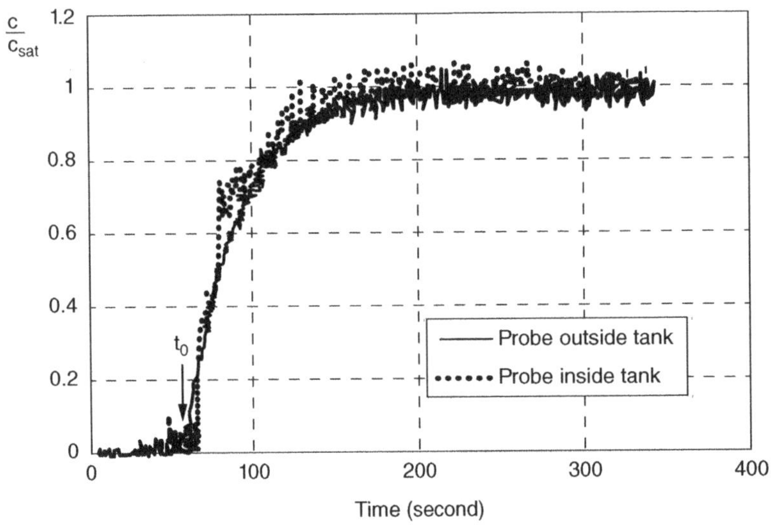

Figure 21.1 Plot for the calculation of $k_l a$. *Source:* Courtesy of The Journal of Chemical Engineering of Japan, Tokyo, Japan.

It is useful to discuss the reliability of this method shortly. Reference is made to the experiment of Figure 21.1. First, the depletion of the gaseous phase. The depletion of the gaseous phase must be minimal, otherwise c_{sat} changes. The vessel liquid volume is 46.6 l and the air flow is $35.8\ \text{l} \cdot \text{min}^{-1}$. Thus, the oxygen flow is $7.5\ \text{l} \cdot \text{min}^{-1}$, that is, $10.0\ \text{g} \cdot \text{min}^{-1}$. In 2 min, 20 g of oxygen is fed to the stirred vessel. The solubility of oxygen in pure water at ambient temperature is $42.9\ \text{mg} \cdot \text{l}^{-1} \cdot \text{bar}^{-1}$[2]. Thus, the vessel absorption capacity is $0.21 \cdot 46.6 \cdot 0.0429 = 0.42$ g (the oxygen partial pressure is 0.21 bar). The actual absorption lasted about 2 min and only 0.42 g of oxygen were absorbed. So, the amount of oxygen fed to the vessel is approximately 50 times larger than the amount actually absorbed. So, initially the oxygen concentration in the exiting air was approximately 20% by volume, gradually recovering to 21% by volume. The resistance to mass transfer is in the liquid phase mainly.

Typically, diffusion coefficients for gases are approximately a factor of 10^4–10^5 greater than for liquids and thus also $k_g \gg k_l$, that is, the resistance by gaseous transport $1/k_g$ is negligible compared with $1/k_l$. The only important exception in which $k_g \ll k_l$ has to be assumed is the chemisorption of a component from a gas mixture in, for example, flue gas scrubbing, since here depletion of the gas to be absorbed is involved and the concentration of the gaseous component to be absorbed is already zero at the interface due to the extremely fast chemical reaction. Since gas scrubbing is not normally carried out in stirred tanks, this exception will not be considered here.

Second, the mixing of the liquid phase. The assumption that this phase is perfectly mixed is not too bold.

Example 21.1

$k_l \cdot a$ is calculated from the data in Figure 21.1.

The experiment starts when 61 s have elapsed. When 100 s have elapsed, $c/c_{sat} = 0.72$. The experimental time is 39 s. The relevant expression is

$$0.72 = 1 - e^{-k_l \cdot a \cdot 39},$$

$$k_l \cdot a = 0.0326\ \text{s}^{-1}.$$

The following equation correlates many literature data for oxygen absorption from air by water without electrolytes and is accurate within approximately 20–40% [3]:

$$k_l \cdot a = 2.6 \cdot 10^{-2} \cdot \varepsilon_v^{0.4} \cdot v_s^{0.5}\ \text{s}^{-1}. \tag{21.2}$$

ε_v is the power input in $\text{W} \cdot \text{m}^{-3}$, and v_s is the superficial gas velocity in $\text{m} \cdot \text{s}^{-1}$ if the vessel would be empty. Note that both a and ε_v are related to 1 m^3 of liquid and not dispersion. The scale of the work is up to 2.6 m^3, whereas ε_v varied from 0.5 to 10 $\text{kW} \cdot \text{m}^{-3}$.

The following equation correlates many literature data for oxygen absorption from air by water containing electrolytes and is accurate within approximately 20–40% accuracy [3]:

$$k_l \cdot a = 2.0 \cdot 10^{-3} \cdot \varepsilon_v^{0.7} \cdot v_s^{0.2}\ \text{s}^{-1}. \tag{21.3}$$

The remarks concerning the units made for Equation 21.2 also apply for Equation 21.3. The scale of the work is between 2 l and 4.4 m^3, whereas ε_v varied from 0.5 to 10 $\text{kW} \cdot \text{m}^{-3}$.

21.3 Power Consumption on Scaling Up

The absorption of oxygen from air in water in the absence of electrolytes is considered first. VVM (Vessel Volume per Minute) is assumed to be constant.

VVM constant means $v_s \infty d_i$. Equation 21.2: $k_l \cdot a \infty \varepsilon_v^{0.4} \cdot v_s^{0.5} \infty$.

$\left(n^3 \cdot d_i^2\right)^{0.4} \cdot d_i^{0.5} = n^{1.2} \cdot d_i^{1.3} \rightarrow$ for $k_l \cdot a$ constant: $n^{1.2} \infty 1/d_i^{1.3} \rightarrow n \infty 1/d_i^{1.08}$.

Substitution of this result in $P/V \infty n^3 \cdot d_i^2$ gives $P/V \infty 1/d_i^{1.24}$. The meaning thereof is that the specific power input can decrease on scaling-up; the factor is 0.42 for a capacity increase with a factor 8.

Next, the absorption of oxygen from air in water in the presence of electrolytes is considered. Now, VVM constant and Equation 21.3 lead to $P/V \infty 1/d_i^{0.28}$. This means that, also in this case, the specific power input can decrease on scaling-up, and the factor is 0.82 for a capacity increase with a factor 8.

21.4 Remarks

1) Impeller type and number have little influence on $k_l \cdot a$ as long as ε_v and v_s are kept constant [3].
2) The gas sparger should be placed directly under the impeller. In case of a ring sparger, its diameter should be slightly smaller than the impeller diameter [3].
3) To estimate $k_l \cdot a$ for gas/liquid pairs other than oxygen/water, use either the film or penetration (or surface renewal) theory, whichever gives the more conservative estimate.

   Film theory: $(k_l \cdot a)_1/(k_l \cdot a)_2 = Đ_1/Đ_2$.
   Penetration theory: $(k_l \cdot a)_1/(k_l \cdot a)_2 = \sqrt{Đ_1}/\sqrt{Đ_2}$.

4) In general, $k_l \cdot a$ increases with increasing temperature, decreasing gas/liquid interfacial tension, and decreasing liquid viscosity. Typical $k_l \cdot a$-values for systems not containing electrolytes are in the range 0.01–0.1 s^{-1}.

References

1. Zhu, Y., Bandopadhayay, P. C., and Wu, J. (2001). Measurement of gas–liquid mass transfer in an agitated vessel—A comparison between different impellers. *Journal of Chemical Engineering of Japan*, 34 (5), 579–584.
2. Zlokarnik, M. (2001). *Stirring. Theory and Practice* (p. 135). Weinheim, Germany: Wiley-VCH Verlag GmbH.
3. van 't Riet, K. (1979). Review of measuring methods and results in nonviscous gas–liquid mass transfer in stirred vessels. *Industrial and Engineering Chemistry Process Design and Development*, 18 (3), 357–364.

22

Heat Transfer in Stirred Vessels

22.1 Introduction

Stirred vessels are important process items in chemical plants. Their main functions are summarized in Chapter 15. Heating up or cooling down of liquids in stirred vessels occurs frequently. It is also possible to supply or to extract heat under isothermal conditions.

For heating or cooling, a vessel is often provided with a jacket or a half-pipe jacket. See Figure 15.2. A half-pipe jacket provides better heat transfer than a conventional jacket. Coils can be used to provide additional heat exchanging area; however, they impact the flow in the vessel.

The stirred liquids are assumed to be Newtonian. The VDI Heat Atlas was consulted for the heat transfer correlations in Sections 22.2, 22.3, and 22.4 [1]. The correlation in Section 22.5 can also be found in the Heat Atlas [2].

22.2 Heat Transfer Jacket Wall/Process Liquid

Nu-correlations often have the following form:

$$\mathrm{Nu} = \mathrm{C} \cdot \mathrm{Re}^{\mathrm{a}} \cdot \mathrm{Pr}^{\mathrm{b}} \cdot \left(\frac{\mu}{\mu_{\mathrm{w}}}\right)^{\mathrm{c}}. \tag{22.1}$$

with

$a \approx 2/3$,

$b \approx 1/3$,

$c \approx 1/7\ (0.14)$, and

C capturing geometrical aspects.

Turbines and Pitched-Blade Turbines

The following correlation is valid for the Standard Tank Configuration (see Chapter 16):

$$\mathrm{Nu}_{\mathrm{v}} = 0.66(\sin\gamma)^{0.5} \cdot \mathrm{Re}^{2/3} \cdot \mathrm{Pr}^{1/3} \cdot \left(\frac{\mu}{\mu_{\mathrm{w}}}\right)^{0.14}. \tag{22.2}$$

The conditions for the applicability of this equation are:

$10^2 \leq \mathrm{Re} \leq 4 \cdot 10^5$,

$2 \leq \mathrm{Pr} \leq 2 \cdot 10^3$,

$h/d_i = 1/5$ (for pitched-blade turbines the sloping height),

$b/d_i = 1/4$ (for Rushton turbines),

$Z = 6$, and

$45° \leq \gamma \leq 90°$.

Z is the number of blades, whereas γ is the angle between the stirrer blades and a horizontal plane. The subscript v in the Nusselt number means that the linear dimension in Nu is the vessel diameter. For other geometrical ratios, see the VDI Heat Atlas [1].

Propellers

The following correlation is valid for the Standard Tank Configuration (see Chapter 16):

$$Nu_v = 0.505 \cdot \Psi_p \cdot Re^{2/3} \cdot Pr^{1/3} \cdot \left(\frac{\mu}{\mu_w}\right)^{0.14}. \quad (22.3)$$

The conditions for the applicability of this equation are:

$1.7 \cdot 10^4 \leq Re \leq 9 \cdot 10^5$,

$1.9 \leq Pr \leq 2.4$,

$0.4 \leq S/d_i \leq \infty$ (normal range $0.5 \leq S/d_i \leq 1.5$), and

$1 \leq Z \leq 45$ (normal range $2 \leq Z \leq 5$).

Ψ_p is a factor specific for a number of blades and a slope of the blades. For a marine-type propeller with three blades and a pitch S equal to the impeller diameter, $\Psi_p = 1$. $\Psi_p = 1.1$ when the pitch of the same type of propeller equals 1.5 times the impeller diameter. The subscript v of Nu in Equation 22.3 means that the linear dimension in Nu is the vessel diameter.

To obtain Equation 22.3, water was taken as a test liquid. The liquid temperature was not varied over a large range. That is the reason that the range for Pr is rather small. For design purposes, Equation 22.3 can also be used for organic liquids having higher Prandtl numbers. Usually, a propeller pumps the liquid to the bottom of the vessel.

On comparing Equations 22.2 and 22.3, it strikes that the heat transfer performance is similar for equal speeds and diameters. However, the turbine power consumption is much greater. Propellers generate high liquid velocities, which is favorable for heat transfer.

Pfaudler impellers

Pfaudler impellers are glass-lined impellers made by Messrs. Pfaudler in Germany. They are used when corrosive liquids are processed, for example, acid chlorides. See Figure 22.1.

The following equation is valid for a Pfaudler impeller:

$$Nu_v = 0.349 \cdot Re^{0.719} \cdot Pr^{0.264} \cdot \left(\frac{\mu}{\mu_w}\right)^{0.14}. \quad (22.4)$$

The conditions and data for the applicability of this equation are:

$4.5 \cdot 10^3 \leq Re \leq 5.7 \cdot 10^4$,

$840 \leq Pr \leq 6.3 \cdot 10^3$,

$d_i/D = 0.575$,

$h/D = 0.15$,

$h_i/D = 0.092$,

$R/D = 0.4$, and

$\beta = 15°$.

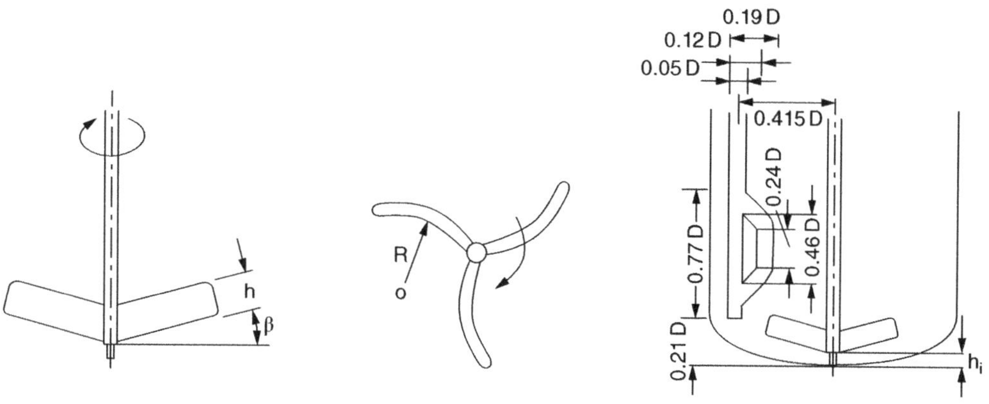

Figure 22.1 Pfaudler impeller. *Source:* Courtesy of Springer-Verlag, Berlin, Germany.

There is one baffle, and the liquid height equals the vessel diameter D.
The subscript v in Equation 22.4 means that the linear dimension in Nu is the vessel diameter.

So far, the turbine, propeller, and Pfaudler impeller were treated. Correlations for paddle stirrers and anchor stirrers can be found in the Heat Atlas [1].

Example 22.1

A process vessel is filled with water. The vessel contents are stirred by means of a propeller and heated by means of a jacket from 20 to 80 °C (the average temperature is 50 °C). The wall temperature is kept at 95 °C. The coefficient for the heat transfer from the vessel wall to the vessel contents is calculated.

Process Vessel Data

Standard Tank Configuration

$D = 1\ \text{m}.$

$n = 2.4\ \text{s}^{-1}.$

$Z = 3.$

$S = 0.35\ \text{m}.$

Water data at 50 °C

$\rho = 988\ \text{kg} \cdot \text{m}^{-3}.$

$c_p = 4{,}185\ \text{J} \cdot \text{kg}^{-1} \cdot \text{K}^{-1}.$

$\lambda = 0.641\ \text{W} \cdot \text{m}^{-1} \cdot \text{K}^{-1}.$

$\mu = 0.546 \cdot 10^{-3}\ \text{N} \cdot \text{s} \cdot \text{m}^{-2}.$

Water viscosity at 95 °C

$\mu_w = 0.299 \cdot 10^{-3}\ \text{N} \cdot \text{s} \cdot \text{m}^{-2}.$

$$Pr = \frac{\mu \cdot c_p}{\lambda} = \frac{0.546 \cdot 10^{-3} \cdot 4{,}185}{0.641} = 3.56,$$

$$Re = \frac{\rho n \cdot d_i^2}{\mu} = \frac{988 \cdot 2.4(1/3)^2}{0.546 \cdot 10^{-3}} = 4.83 \cdot 10^5,$$

$$S/d_i = 1.05.$$

Equation 22.3 can be used. $\Psi_p = 1.02$

$$Nu_v = 0.505 \cdot 1.02(4.83 \cdot 10^5)^{2/3}(3.56)^{1/3}\left(\frac{0.546}{0.299}\right)^{0.14} = 5{,}268,$$

$$\alpha = \frac{Nu_v \cdot \lambda}{D} = \frac{5{,}268 \cdot 0.641}{1} = 3{,}377\ W \cdot m^{-2} \cdot K^{-1}.$$

22.3 Heat Transfer Coil Wall/Process Liquid

Helical coils in baffled vessels will be discussed. See Figure 22.2.

The next equation is valid for turbines:

$$Nu_c = 0.225\left(\frac{d_i}{D}\right)^{0.18}\left(\frac{d_c}{D}\right)^{0.52}\left(\frac{D_c}{D}\right)^{-0.27}\left(\frac{h_i}{D}\right)^{0.14}\left(\frac{Z}{4}\right)^{0.28} \cdot Re^{2/3} \cdot Pr^{1/3} \cdot \left(\frac{\mu}{\mu_w}\right)^{0.14}. \quad (22.5)$$

The part preceding Re to the power 2/3 concerns the geometry. The conditions for the applicability of this equation are:

$1{,}500 \leq Re \leq 7.7 \cdot 10^5$,

$2 \leq Pr \leq 1{,}190$,

$0.2 \leq d_i/D \leq 0.4$,

$0.023 \leq d_c/D \leq 0.054$,

$0.5 \leq D_c/D \leq 0.7$,

$0.27 \leq h_i/D \leq 0.63$, and

$4 \leq Z \leq 6$.

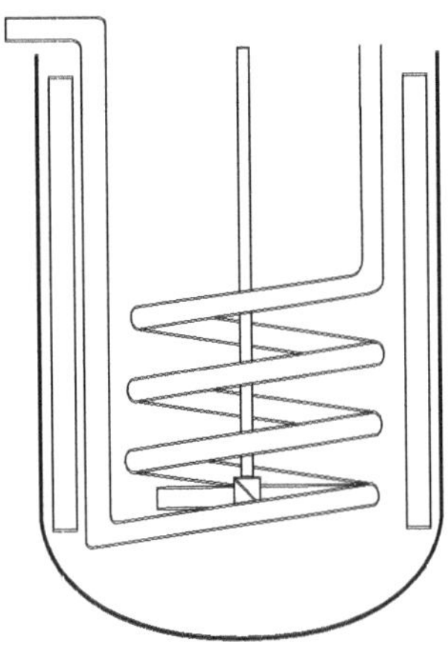

Figure 22.2 Vessel with stirrer, helical coil, and baffles. *Source:* Courtesy of Elsevier, Amsterdam, The Netherlands.

The linear dimension appearing in Nu is the outside coil tube diameter d_c. The diameter of the coil itself is D_c. d_i and h_i are, respectively, the impeller diameter and height.

The equation for propellers is:

$$Nu_c = 0.0573\left(\frac{h_i}{D}\right)^{-0.254}\left(\frac{S}{d_i}\right)^{2.33}\left(\frac{d_c}{D}\right)^{0.572}\left(\frac{S_c}{d_c}\right)^{-0.018}(n_b)^{-0.07}\left(\frac{w_b}{D}\right)^{-0.058}$$

$$\times Re^{0.67}\cdot Pr^{0.41}\cdot\left(\frac{\mu}{\mu_w}\right)^{0.034}. \qquad (22.6)$$

The part preceding Re to the power 2/3 concerns the geometry. The conditions for the applicability of this equation are:

$1.3\cdot 10^4 \le Re \le 1.1\cdot 10^6,$

$3 \le Pr \le 300,$

$0.16 \le h_i/D \le 0.82,$

$1.0 \le S/d_i \le 1.33,$

$0.014 \le d_c/D \le 0.042,$

$1.2 \le S_c/d_c \le 8.0,$

$2 \le n_b \le 6,$

$0.06 \le w_b/D \le 0.17,$

$0.16 \le d_i/D \le 0.5,$

$1.4 \le D_c/d_i \le 4.1,$ and

$0.5 \le D_c/D \le 0.7.$

S is the propeller pitch, whereas S_c is the coil pitch. n_b and w_b are, respectively, the number and the width of the baffles in the vessel. The linear dimension appearing in Nu is the outside coil tube diameter.

Turbines and propellers were treated in this section. Correlations for paddle stirrers and anchor stirrers can be found in the Heat Atlas [1].

Example 22.2

A process vessel is filled with water. The vessel contents are stirred by means of a turbine and heated by means of a coil. The coefficient for the heat transfer from the coil to the vessel contents is calculated. The coil wall temperature is 95 °C and water is heated from 20 to 80 °C (average temperature 50 °C).

Process Vessel Data

$D = 1$ m.

$n = 1.5\ s^{-1}$.

$d_i/D = 1/3$.

$d_c = 0.022$ m.

$D_c/D = 0.7$.

$h_i/D = 1/3$.

$Z = 6$.

Water Data

See Example 22.1 of this chapter.

$$Pr = 3.56.$$

$$Re = \frac{\rho n \cdot d_i^2}{\mu} = \frac{988 \cdot 1.5(1/3)^2}{0.546 \cdot 10^{-3}} = 3.0 \cdot 10^5.$$

Equation 22.5 can be used.

$$Nu_c = 0.225\left(\frac{1}{3}\right)^{0.18}\left(\frac{0.022}{1}\right)^{0.52}\left(\frac{0.7}{1}\right)^{-0.27}\left(\frac{1}{3}\right)^{0.14}\left(\frac{6}{4}\right)^{0.28}$$

$$\times (3.0 \cdot 10^5)^{2/3}(3.56)^{1/3}\left(\frac{0.546}{0.299}\right)^{0.14} = 198.8.$$

$$\alpha = \frac{Nu_c \cdot \lambda}{d_c} = \frac{198.8 \cdot 0.641}{0.022} = 5{,}792\ W \cdot m^{-2} \cdot K^{-1}.$$

22.4 Heat Transfer Jacket Medium/Vessel Wall

The approach outlined in this section was developed by Lehrer from experiments with a stirred vessel having a diameter of 610 mm and a jacket gap of 25.4 mm (1 in) [1]. The liquid height was equal to the vessel diameter. Lehrer worked with two different inlet nozzles. A radial nozzle and a tangential nozzle had diameters of 19.1 and 12.7 mm, respectively. The vessel was equipped with four baffles. Several paddle stirrers were used. The vessel contents were cooled. Water was the cooling medium ($5 < Pr < 7$). The recommended correlation is:

$$Nu_j = \frac{0.03 \cdot Re_j^{0.75} \cdot Pr(\mu/\mu_w)^{0.14}}{1 + \left(1.74(Pr - 1)/Re_j^{0.125}\right)}, \tag{22.7}$$

δ is the jacket gap in m.

$$d_j = \left(\frac{8}{3}\right)^{0.5} \cdot \delta\ m,$$

d_j is to be used in both Re_j and Nu_j.

$$v_j = \sqrt{v_o \cdot v_s} + v_a,$$

v_j is to be used in Re_j,
v_o is the linear flow velocity in the inlet nozzle,
v_s and v_a are defined differently for radial and tangential nozzles.

Radial Nozzles

$$v_s = \frac{\Phi_m}{\rho(\pi/4)\left(D_j^2 - D_o^2\right)}. \tag{22.8}$$

$$v_a = 0.5\sqrt{2g \cdot h_j \cdot \beta \cdot \Delta T},$$

v_a is the velocity due to free convection.

β is the coefficient of thermal expansion in K^{-1}, and ΔT is the temperature increase of the cooling medium in K. v_a is added to the geometric mean velocity when the free convection velocity and the geometric mean have the same direction. v_a is subtracted from the geometric mean velocity when v_a opposes the geometric mean velocity.

Tangential Nozzles

$$v_s = \frac{\Phi_m}{\rho \cdot h_j \cdot \delta}. \tag{22.9}$$

$$v_a = 0.$$

The ranges of Re_j were for the radial nozzle and the tangential nozzle, respectively: $9{,}000 \leq Re_j \leq 4 \cdot 10^4$ and $2.2 \cdot 10^4 \leq Re_j \leq 8 \cdot 10^4$. The accuracy of Equation 22.7 is ±30%. The approach is relatively conservative. According to the Heat Atlas, Lehrer's approach can be used for both cooling and heating the vessel contents. The Heat Atlas also discusses a second method.

Example 22.3

The coefficient for the heat transfer from a vessel wall to cooling water in a jacket is calculated. The vessel contents are at 80 °C.

Vessel Data

$D_o = 0.6$ m.

$\delta = 0.025$ m.

$h_j = 0.6$ m.

$d_o = 0.025$ m (tangential nozzle).

Process Data

$\Phi_m = 2.5\ kg \cdot s^{-1}$.

$T_i = 20$ °C.

$T_0 = 24$ °C.

$T_w = 38$ °C (estimated).

Water Data

$\rho = 998.2\ kg \cdot m^{-3}$ (20 °C).

$\rho = 997.8\ kg \cdot m^{-3}$ (22 °C).

$c_p = 4{,}180\ J \cdot kg^{-1} \cdot K^{-1}$.

$\lambda = 0.603\ W \cdot m^{-1} \cdot K^{-1}$.

$\mu = 0.961 \cdot 10^{-3}\ N \cdot s \cdot m^{-2}$ (22 °C).

$\mu = 0.682 \cdot 10^{-3}\ N \cdot s \cdot m^{-2}$ (38 °C).

$d_j = (8/3)^{0.5} \cdot 0.025 = 0.0408$ m.

$$v_o = \frac{2.5}{998.2(\pi/4)0.025^2} = 5.10\,m\cdot s^{-1}.$$

$$v_s = \frac{2.5}{997.8\cdot 0.6\cdot 0.025^2} = 0.167\,m\cdot s^{-1}.$$

$$v_a = 0\,m\cdot s^{-1}.$$

$$v_j = \sqrt{5.10\cdot 0.167} + 0 = 0.923\,m\cdot s^{-1}.$$

$$Re_j = \frac{\rho\cdot v_j\cdot d_j}{\mu} = \frac{997.8\cdot 0.923\cdot 0.0408}{0.961\cdot 10^{-3}} = 3.9\cdot 10^4.$$

$$Pr = \frac{\mu\cdot c_p}{\lambda} = \frac{0.961\cdot 10^{-3}\cdot 4{,}180}{0.603} = 6.66.$$

$$Nu_j = \frac{0.03(3.9\cdot 10^4)^{0.75}\cdot 6.66(0.961/0.682)^{0.14}}{1+\frac{1.74(6.66-1)}{(3.9\cdot 10^4)^{0.125}}} = 160.4.$$

$$\alpha = \frac{Nu_j\cdot\lambda}{d_j} = \frac{160.4\cdot 0.603}{0.0408} = 2{,}371\,W\cdot m^{-2}\cdot K^{-1}.$$

22.5 Heat Transfer Coil Medium/Coil Wall

The following correlation for turbulent flow has an accuracy of ±15% [2].

$$Nu = \frac{(\xi/8)\,Re\cdot Pr(Pr/Pr_w)^{0.14}}{1+12.7\sqrt{\xi/8}\cdot\left(Pr^{2/3}-1\right)}, \tag{22.10}$$

$$\xi = \left(\frac{0.3164}{Re^{0.25}} + 0.03\left(\frac{d_c}{D_c}\right)^{0.5}\right)\left(\frac{\mu_w}{\mu}\right)^{0.27}.$$

The condition for the applicability of the equation is $Re > 2.2\cdot 10^4$.

The correlation is based on tests with air ($Pr = 0.7$) and water ($2 < Pr < 5$). The viscosity correction is based on the experimental results of heating.

22.6 Batch Heating and Cooling

In batch operations, it is often necessary to calculate the time, t, needed to heat or cool the contents of a process vessel from temperature T_1 to T_2. The temperature of the medium in the jacket or coil is often constant or approximately constant. That is, for example, the case when steam condenses in a jacket or coil. Let T_w be the wall temperature and A the area of a heat transfer surface. When T is the temperature of the liquid at any time t, the heat balance in a small time dt is:

$$M\cdot c_p\cdot dT = UA(T_w - T)dt.$$

This expression can be re-written:

$$\frac{dT}{dt} = \frac{UA}{M\cdot c_p}(T_w - T).$$

In words, the rate of temperature rise is proportional to the temperature difference (the driving force).

Integration of the equation is possible when the differential equation is re-arranged.

$$\int_{T_1}^{T_2} \frac{dT}{T_W - T} = \frac{UA}{M \cdot c_p} \int_0^t dt.$$

And the result of the integration is:

$$t = \frac{M \cdot c_p}{UA} \ln \frac{T_W - T_1}{T_W - T_2}. \tag{22.11}$$

The derivation in this section is applicable if:

- the heat capacity of the vessel is negligible,
- there is no heat exchange with the surroundings,
- U is constant,
- there is no reaction heat, and
- the energy input by the stirrer is negligible.

References

1. VDI (2010). *VDI Heat Atlas—N3: Heat Transfer and Power Consumption in Stirred Vessels*. Düsseldorf, Germany: VDI.
2. VDI (2010). *VDI Heat Atlas—G3: Heat Transfer in Helically Coiled Tubes*. Düsseldorf, Germany: VDI.

23

Scale Up of Mixing

23.1 Introduction

First, the product physical properties on small scale and on large scale should be identical. Second, the geometry is kept constant. On testing on small scale, the desired process result was found on adjusting a stirrer rotational speed n_1. Scale-up comprises the prediction of the large-scale stirrer speed n_2.

Example 23.1

A process is carried out in a 60-l baffled vessel having a diameter of 0.4 m, a cylindrical liquid height of 0.4 m, and a dished bottom. The total liquid volume is 54 l. A catalyst is suspended and the reaction heat is transferred to a jacket. The vessel is equipped with a propeller pumping downward and having a diameter of $0.3 \cdot D = 0.12$ m. A rotational speed of 10 s^{-1} is selected. The viscosity of the suspension is 0.05 $N \cdot s \cdot m^{-2}$ and its specific mass is 1,200 $kg \cdot m^{-3}$. It is desired to scale-up to 60 m^3 (factor 1,000).

$$Re_1 = \frac{\rho_{susp} \cdot n_1 \cdot d_{i1}^2}{\mu_{susp}} = \frac{1{,}200 \cdot 10 \cdot 0.12^2}{0.05} = 3{,}456.$$

$P_o = 0.35$ according to Figure 17.2.

$$P = 0.35 \cdot \rho \cdot n_1^3 \cdot d_{i1}^5 = 0.35 \cdot 1{,}200 \cdot 10^3 \cdot 0.12^5 = 10.5 \text{ W}.$$

$$\varepsilon = \frac{10.5}{1.2 \cdot 54} = 0.16 \text{ W} \cdot \text{kg}^{-1}.$$

The large-scale power consumption is: $1.2 \cdot 5.4 \cdot 10^4 \cdot 0.16 = 1.04 \cdot 10^4$ W = 10.4 kW.

$$1.04 \cdot 10^4 = 0.35 \cdot 1{,}200 \cdot n_2^3 \cdot 1.2^5 \text{ W} \rightarrow n_2 = 2.1 \text{ s}^{-1}.$$

$$Re_2 = \frac{\rho_{susp} \cdot n_2 \cdot d_{i2}^2}{\mu_{susp}} = \frac{1{,}200 \cdot 2.1 \cdot 1.2^2}{0.05} = 7.3 \cdot 10^4.$$

Introduction to Chemical Engineering, First Edition. C.M. van 't Land.

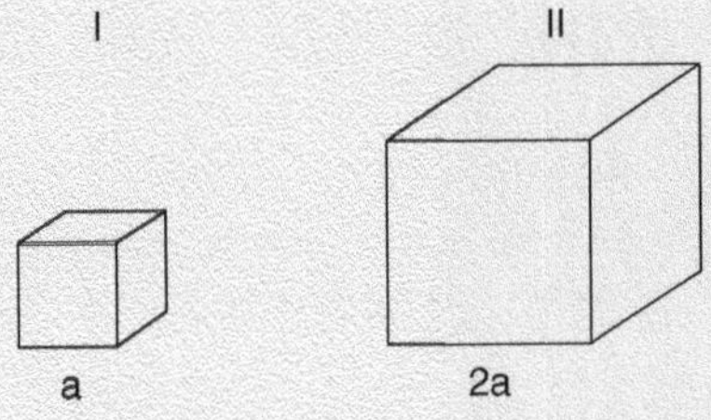

			Factor
Edge	a	2a	2
Area	$6a^2$	$24a^2$	4
Volume	a^3	$8a^3$	8
m^2/m^3	$\frac{6}{a}$	$\frac{3}{a}$	$\frac{1}{2}$

$\sqrt[3]{8} = 2$

Figure 23.1 Decrease of $m^2 \cdot m^{-3}$ on scaling-up.

According to Equation 22.3, $Nu_v \propto Re^{2/3}$. Thus, $(Nu_{v2}/Nu_{v1}) = (7.3 \cdot 10^4/3{,}456)^{2/3} = 7.64$. That means that $(\alpha_2 \cdot D_2/\alpha_1 \cdot D_1) = 7.64 \rightarrow \alpha_2 = 0.764 \cdot \alpha_1$. On keeping the power input in $W \cdot kg^{-1}$ constant, the heat transfer coefficient on a large scale will be smaller than the heat transfer coefficient on a small scale. The physical reason for this decrease is that the distance between the stirrer and the wall increases on scaling-up. That will cause a decrease of the velocity at which the vessel contents flow past the heat exchanging surface.

On scaling-up, the heat generation increases with a factor of 1,000 ($\propto D^3$). The jacket heat transfer area increases with a factor of 100 ($\propto D^2$). Thus, per kg of reacting mass, the heat transfer area decreases with a factor of 10. See Figure 23.1. To cater for this effect, it will be necessary to slow down the chemical reaction or to install additional cooling possibilities.

However, on keeping the specific power input constant, both the suspension characteristics and the heat transfer coefficient can be kept approximately constant.

23.2 Homogenization

The mixing time θ is the most important parameter at homogenization. Mixing time is the subject of Chapter 16. Figure 16.3 gives the mixing-time characteristics of a number of agitators. $n\theta$ is plotted as a function of Re in this figure. θ is the mixing time in s, and n is the rotational speed in s^{-1}. So, the product $n\theta$ is the number of agitator rotations required to accomplish mixing. It makes sense that this number is high at low rotational speeds and low at high rotational speeds (turbulence promotes mixing). In Figure 16.3, $n\theta$ decreases when Re increases for $10^2 < Re < 10^4$. $n\theta$ is constant for $Re > 10^4$ [1, 2]. At both high and low Reynolds numbers, regions exist where the total number of impeller revolutions, $n\theta$, required to achieve mixing, is constant.

A further reference gives that for both the Rushton turbine and the pitched-blade turbine, nθ is 35 when Re > 10^4 [3]. Reference [1] states that nθ is 45 for these two impellers. One reason for the difference is that the two geometries are slightly different.

The consequence of nθ is constant when Re > 10^4 is that n is constant on scaling-up if we want θ to be constant on scaling-up. On realizing that $P/V \propto n^3 \cdot d_i^2$, keeping n constant means that $P/V \propto d_i^2$. See Figure 23.2. The implication is that the specific power input in, for example, $W \cdot kg^{-1}$, for a scale-up factor of, for example, 100, would have to increase by a factor of 10 if it is desired to keep the mixing time constant. The temptation is to consider such an increase as exaggerated and to compromise with a smaller factor. However, there are cases that rapid attainment of uniformity is critical for a desired process result. An example of such a case was found [4]. A plant for the production of a red pigment had to be torn down immediately after start-up because it produced a brown pigment. A red pigment had been obtained in a laboratory by fast mixing of two reactants. The latter mixing caused a high supersaturation and the red polymorph of the chemical reaction precipitated. The mixing in the industrial reactor was less intense and under these conditions the brown polymorph precipitated. The production in several small reactors was successful.

The importance of the mixing time can be investigated on small scale by carrying out experiments at which the mixing time is varied by, for example, variation of the rotational speed of the impeller. To carry out miniplant studies in a reactor having a small d_i/D-ratio is a further possibility.

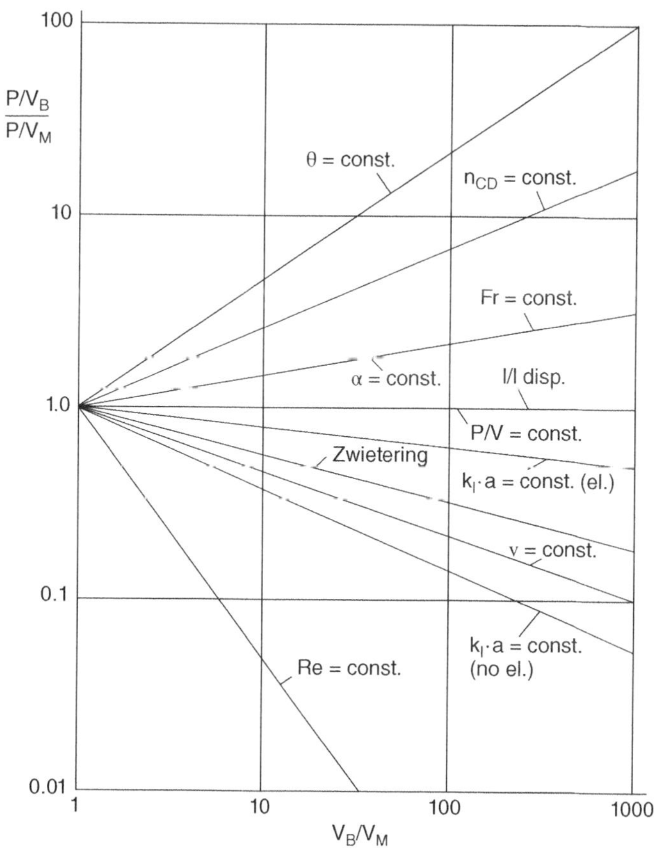

Figure 23.2 Specific power input ratios as a function of scale on keeping various aspects constant.

23.3 Suspensions

Stirring suspensions is discussed in Chapter 18. The application of Zwietering's rule would imply that the specific energy input in, for example, $W \cdot kg^{-1}$, could decrease on scaling-up [5]. See Figure 23.2. The author, however, recommends to keep the specific power input constant on scaling-up. Zwietering's results were obtained for low-viscosity liquids, particle sizes between 125 and 850 μm, and slurry densities in the range 0–20% by weight.

23.4 Liquid/Liquid Dispersions

Liquid/liquid dispersions are the subject of Chapter 19. One type of these mixtures are emulsions. Liquid/liquid dispersions are also met at extractions. On making these mixtures, it is important to keep the specific power input in, for example, $W \cdot kg^{-1}$ constant. See the relevant line in Figure 23.2.

23.5 Gas Distribution

The subject of gas distribution in stirred vessels is treated in Chapter 20. The turbine rotational speed at which the gas is just completely dispersed is considered in that chapter. It follows from Equation 20.2 at constant VVM (vessel volume per minute) that the specific power input in, for example, $W \cdot kg^{-1}$, is proportional to $d_i^{1.25}$. See Section 20.4 of Chapter 20 and Figure 23.2. Doubling the stirrer diameter corresponds with an eightfold capacity increase. The quality of the dispersion decreases if, on scaling-up, the specific power input and VVM are kept constant. On the other hand, the gas residence time will then increase. This is because VVM is kept constant. Then, $\Phi_v \propto D^3$, whereas the cross-sectional area $\propto D^2$. Hence, $v_s \propto D$ and a higher v_s leads to a greater ε according to Equation 20.5. This means a longer gas residence time. To some extent, the poorer dispersion will be compensated by a greater gas hold-up.

23.6 $k_l \cdot a$

In Chapter 21, work concerning the distribution of air in water and the absorption of oxygen therefrom was described. First, the case without electrolytes is treated. It is shown that then, if VVM (vessel volume per minute) is kept constant, the specific energy input in, for example, $W \cdot kg^{-1}$, should be proportional to $1/d_i^{1.24}$ to keep $k_l \cdot a$ constant. This means that the specific power input in, for example, $W \cdot kg^{-1}$, can decrease on scaling-up, and the factor is 0.42 for a capacity increase of a factor of 8. Second, the case of the presence of electrolytes is discussed. Now, the specific energy input in, for example, $W \cdot kg^{-1}$, should be proportional to $1/d_i^{0.28}$ to keep $k_l \cdot a$ constant. This means that, also in this case, the specific power input can decrease on scaling-up. The factor is 0.82 for a capacity increase of a factor of 8.

23.7 Heat Transfer

Chapter 22 contains correlations concerning heat transfer in stirred vessels. Generally, Nu is proportional to $Re^{2/3}$. Thus, it follows:

$$\alpha \cdot d_i \propto Re^{2/3} \rightarrow \alpha \propto \frac{n^{2/3} \cdot d_i^{4/3}}{d_i} = n^{2/3} \cdot d_i^{1/3},$$

α is constant if $n^2 \cdot d_i$. is constant, that is, $n \propto 1/\sqrt{d_i}$.

$$\frac{P}{V} \propto n^3 \cdot d_i^2.$$

Substitution of $n \propto 1/\sqrt{d_i}$ into this expression leads to:

$$\frac{P}{V} \propto \sqrt{d_i}.$$

This means that an eightfold capacity increase, implying a doubling of the agitator diameter, is accompanied by a 40% increase (factor $\sqrt{2}$) in specific power input in, for example, $W \cdot kg^{-1}$. See Figure 23.2.

The Froude number gives the ratio between the radial acceleration of the periphery of the agitator and the acceleration due to gravity:

$$Fr = \frac{2\pi^2 \cdot n^2 \cdot d_i}{g}.$$

In the field of mixing and stirring, it is customary to use a modified number:

$$Fr = \frac{n^2 \cdot d_i}{g}.$$

It follows: $Fr \propto n^2 \cdot d_i$

If Fr is kept constant, α is kept constant at the same time. The heat transfer coefficient will decrease on scaling-up if the specific power input in, for example, $W \cdot kg^{-1}$, is kept constant. See Section 23.1.

References

1. Ullmann (2002). *Processes & Process Engineering—Stirring. Ullmann's Encyclopedia of Industrial Chemistry*. Wiley Online Library.
2. Harnby, N., Edwards, M. F., and Nienow, A. W. (2000). *Mixing in the Process Industries* (p. 152). Oxford, UK: Butterworth-Heinemann.
3. Kraume, M., and Zehner, P. (2001). Experience with experimental standards for measurements of various parameters in stirred tanks: A comparative test. *Transactions of the Institution of Chemical Engineers*, 79 (Part A), 811–818.
4. Witkamp, G. J. (2000). A lot of confidence. *NPT Procestechnologie*, 6 (2), 8–9. (in Dutch).
5. Zwietering, T. N. (1958). Suspending of solid particles in liquid by agitators. *Chemical Engineering Science*, 8, 244–253.

24

Case Studies Mixing and Stirring

24.1 Mixing Time—Comparison of Stirrers

A baffled process vessel has a diameter of 1 m. The process liquid has a specific mass of 1,000 $kg \cdot m^{-3}$ and a viscosity of 10^{-1} $N \cdot s \cdot m^{-2}$. It is requested to calculate the mixing time and the power input for these impellers:

- propeller $d_i/D = 0.3$ and $n = 10\ s^{-1}$,
- turbine $d_i/D = 0.3$ and $n = 5\ s^{-1}$,
- cross-beam stirrer $d_i/D = 0.67$ and $n = 1\ s^{-1}$, and
- blade stirrer $d_i/D = 0.5$ and $n = 1\ s^{-1}$.

The geometries are according to Figure 16.2. Figures 16.3 and 17.2 can be consulted. The calculations can be summarized in a table. $Re = \rho n \cdot d_i^2/\mu$

	n, s	Re	nθ, -	θ, s	P_o, -	P, W
Propeller						
Turbine						
Cross-beam						
Blade						

Solution

	n, s^{-1}	Re, -	nθ, -	θ, s	P_o, -	P, W
Propeller	10	9,000	170	17	0.35	850
Turbine	5	4,500	50	10	5.0	1,519
Cross-beam	1	4,489	12.6	12.6	3.0	405
Blade	1	2,500	7.5	7.5	8.0	250

A blade stirrer is an optimum agitator for homogenization.

Introduction to Chemical Engineering, First Edition. C.M. van 't Land.

24.2 Mixing Time–Scale Up of Process

A process is scaled up from 1 to 1,000-l scale. The diameter of the small vessel is 0.1 m and the diameter of the large vessel is 1.0 m. The vessel impellers are Rushton turbines having a diameter of $0.3 \cdot D$. The process liquid has a specific mass of $1{,}000\ kg \cdot m^{-3}$ and a viscosity of $10^{-1}\ N \cdot s \cdot m^{-2}$. It is requested to calculate the mixing time and the specific power input on the two scales. Figures 16.3 and 17.2 can-be consulted. The calculations can be summarized in a table. $Re = \rho n \cdot d_i^2/\mu$.

D, m	d_i, m	n, s^{-1}	Re, -	nθ, -	θ, s	P_o, -	P, W
0.1		50					
1.0		10					

Can the mixing time in this case be kept constant on scaling-up?

Solution

D, m	d_i, m	n, s^{-1}	Re, -	nθ, -	θ, s	P_o, -	P, W	$W \cdot kg^{-1}$
0.1	0.03	50	450	200	4	3.5	10.6	10.6
1.0	0.3	10	9,000	50	5	5.0	12,150	12.2

The mixing time can be kept constant on scaling-up. This a special case as the small-scale flow is laminar.

24.3 Suspensions

Experiments concerning the suspension of 1-mm glass beads in water in a vessel having a diameter of 0.4 m were reported [1]. Figure 24.1 summarizes results for a 45° pitched-blade turbine having six blades. It is assumed that the impeller pumps downward. Both the impeller diameter and the impeller height are 0.125 m. What would be the minimum rotational speed of this agitator according to Zwietering's criterion for a solids volume of 0.05? Table 18.5 can be consulted to estimate an s-value. The kinematic viscosity of the suspending liquid is $10^{-6}\ m^2 \cdot s^{-1}$ and the specific mass of the solids is $2{,}600\ kg \cdot m^{-3}$. Compare the result with the results of Figure 24.1.

Solution

$$D = 0.4\ m.$$

$$d_i = 0.125\ m.$$

$$h_i = 0.125\ m.$$

$$d_i/D = 0.3,\ h_i/D = 0.3 \rightarrow s = 6.$$

$$d_p = 0.001\ m.$$

$$\rho_s = 2{,}600\ kg \cdot m^{-3}.$$

$$\rho_l = 1{,}000\ kg \cdot m^{-3}.$$

$$\Delta\rho = 2{,}600 - 1{,}000 = 1{,}600\ kg \cdot m^{-3}.$$

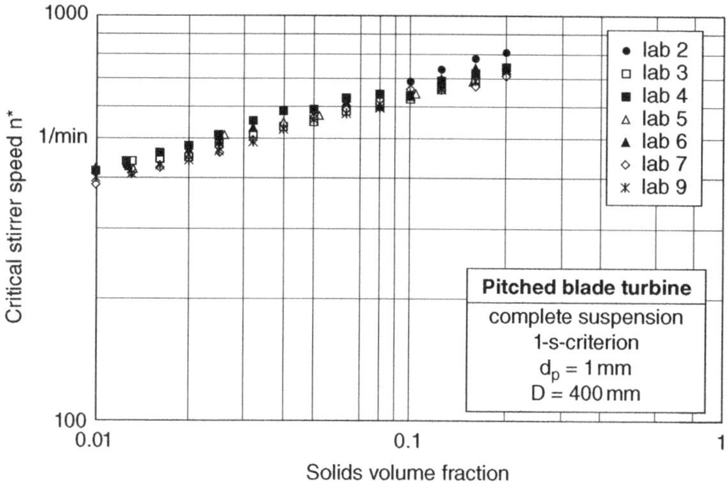

Figure 24.1 Critical stirrer speed for complete suspension of 1 mm glass beads with 45° pitched-blade turbine having six blades. *Source:* Courtesy of The Institution of Chemical Engineers, Rugby, UK.

$$g = 9.81\ \mathrm{m \cdot s^{-2}}.$$

$$a = \frac{0.05 \cdot 2{,}600}{0.05 \cdot 2{,}600 + 0.95 \cdot 1{,}000} = 0.120.$$

Zwietering's criterion yields $n^* = 10.7\ \mathrm{s^{-1}}$, that is, 642 $\mathrm{min^{-1}}$. The seven laboratories find 575 $\mathrm{min^{-1}}$.

24.4 Air Oxidation Optimization

The air oxidation discussed in Section 20.5 of Chapter 20 lasts 8 h. Initially, the oxygen concentration in the leaving air is 2–3%. At the end, the concentration rises to 9% and the oxidation is stopped for process safety reasons. Specifically, at that point an area is entered in which it is possible to ignite an explosion in the gaseous phase. The mass transfer rate of oxygen determines the reaction rate in the first phase of the process. It is requested to consider the possibility to halve the reaction time by doubling the air flow rate. The specific power input on passing gas through the vessel contents is kept constant at 2.5 $\mathrm{W \cdot kg^{-1}}$. This can be assumed to be the case if the ungassed power input mentioned in Section 20.5 is kept constant. The actual large-scale power input should be checked. A turbine stirrer having a diameter of 1.2 m is in focus. D = 2.75 m.

Ungassed power consumption is $1.06 \cdot 10^5$ W.

$$1.06 \cdot 10^5 = \quad \cdot \quad \cdot n^3 \cdot \quad \mathrm{W} \rightarrow n = \quad \mathrm{s^{-1}}.$$

The gas flow is twice the gas flow in Section 20.5, that is, $\quad \mathrm{m^3 \cdot s^{-1}}$.

$$\text{Equation 20.1: } n_L = \frac{(0.33 \cdot \Phi_V \cdot D^{3.5})^{0.33}}{d_i^{2.5}} = \quad \mathrm{s^{-1}}.$$

$$\text{Equation 20.2: } n_{CD} = \frac{(15.7 \cdot \Phi_V \cdot D^{0.5})^{0.5}}{d_i^2} = \quad \mathrm{s^{-1}}.$$

$$\text{Equation 20.3: } n_R = \frac{(7.4 \cdot \Phi_V)^{0.2} D}{d_i^2} = \quad \mathrm{s^{-1}}.$$

Establish whether the impeller is flooded, loaded, disperses the air completely, or recycles the gas. The applicability of Equation 20.4 is checked.

$$\mathrm{Re} = \frac{\rho n \cdot d_i^2}{\mu} =$$

The viscosity is 10^{-3} N · s · m^{-2}.

$$\mathrm{Fr} = \frac{n^2 \cdot d_i}{g} =$$

$$Q = \frac{\Phi_V}{n \cdot d_i^3} =$$

$$\frac{D}{d_i} =$$

$$A = Q\left(1 + 38\left(\frac{D}{d_i}\right)^{-5}\right) =$$

$$P_o = 1.5 + \left(0.5 \cdot A^{0.075} + 1{,}600 \cdot A^{2.6}\right)^{-1} =$$

$$P = \frac{}{4.5} \cdot 1.06 \cdot 10^5 = \qquad \mathrm{W}.$$

$$\frac{}{1.48 \cdot 10^4} = \qquad \mathrm{W \cdot kg^{-1}}.$$

Is this specific power input accepted?

Equation 20.4: $\varepsilon = 0.17 \cdot \varepsilon_V^{1/3} \cdot v_s^{2/3} =$

$$\varepsilon = \frac{V_G}{V_G + 15.8} \rightarrow V_G = \qquad \mathrm{m^3}.$$

The liquid volume would expand from 15.8 m^3 to m^3. Is that practical? How would that affect the mixing pattern? Is the idea to double the air flow rate accepted or rejected?

Solution

Stirrer Rotational Speed

$$1.06 \cdot 10^5 = 4.5 \cdot 937 \cdot n^3 \cdot 1.2^5 \rightarrow n = 2.2\,\mathrm{s^{-1}}.$$

Gas Distribution

$$2 \cdot 0.24 = 0.48\,\mathrm{m^3 \cdot s^{-1}}.$$

$$n_L = 1.1\,\mathrm{s^{-1}}.$$

$$n_{CD} = 2.45\,\mathrm{s^{-1}}.$$

$$n_R = 2.46\,\mathrm{s^{-1}}.$$

The impeller is loaded.
Check the applicability of Equation 20.4

$$\mathrm{Re} = 2.97 \cdot 10^6.$$

$$\mathrm{Fr} = 0.6.$$

$$Q = 0.13.$$

$$\frac{D}{d_i} = 2.3.$$

Equation 20.4 can be applied.

$A = 0.21.$

$P_o = 1.54.$

$P = 3.6 \cdot 10^4$ W.

The specific power input of 2.4 $W \cdot kg^{-1}$ is accepted.

$\varepsilon_V = 2.4 \cdot 937 = 2{,}249\ W \cdot m^{-3}.$

$v_S = 0.08\ m \cdot s^{-1}.$

$\varepsilon = 0.41.$

$V_G = 11.0\ m^3.$

Thus, the volume of the dispersion would be almost twice the liquid volume. That cannot be accommodated in the reactor. The idea is rejected.

24.5 Calculating $k_l \cdot a$

$k_l \cdot a$ was calculated from data in Figure 21.1. The result was $k_l \cdot a = 0.0326\ s^{-1}$. It is requested to calculate $k_l \cdot a$ for this situation using Equation 21.1. The diameter of the fully baffled vessel containing demineralized water is 0.39 m and the unaerated liquid height is also 0.39 m. The vessel bottom is flat. Both the diameter and the height of the impeller are 0.13 m. The impeller is a 45° pitched-blade turbine having four blades. Its rotational speed is 700 rpm. The Power number of this impeller is 1.4. The linear air velocity v_s is $5 \cdot 10^{-3}\ m \cdot s^{-1}$. First, check whether the impeller is directly or indirectly loaded.

$$\Phi_V = \frac{\pi}{4} \cdot D^2 \cdot v_S = \qquad m^3 \cdot s^{-1}.$$

$$n_L = \frac{11.7\sqrt{\Phi_V}}{d_i^{1.63}} = \frac{\qquad}{\qquad} = \qquad s^{-1}.$$

The impeller power consumption is given by the following equation:

$$P = P_o \cdot \rho_l \cdot n^3 \cdot d_i^5 = \qquad W.$$

$$V = \frac{\pi}{4} \cdot D^2 \cdot H = \qquad m^3.$$

$$\frac{P}{V} = \varepsilon_V = \qquad W \cdot m^{-3}.$$

Equation 21.1: $k_l \cdot a = 2.6 \cdot 10^{-2} \cdot \varepsilon_V^{0.40} \cdot v_S^{0.5} = \qquad s^{-1}.$
Compare the measurement to the calculation.

Solution

A calculation based on Figure 21.1 leads to $k_l \cdot a = 0.0326\ s^{-1}$.

$\Phi_V = 5.97 \cdot 10^{-4}\ m^3 \cdot s^{-1}.$

$n_L = 7.95\ s^{-1},$

$7.95\ s^{-1}$ is 477 rpm. The actual rotational speed is 700 rpm, so the impeller is indirectly loaded.

$$P = 0.9 \cdot 1.4 \cdot 1{,}000 \cdot 11.67^3 \cdot 0.13^5 = 74.35\,\text{W}.$$

$$\frac{P}{V} = \varepsilon_V = 1{,}596\,\text{W} \cdot \text{m}^{-3}.$$

Equation 21.1: $k_l \cdot a = 2.6 \cdot 10^{-2} \cdot 1{,}596^{0.4}(5 \cdot 10^{-3})^{0.5} = 0.035\ \text{s}^{-1}$.
The result of this calculation compares well to the calculation based on Figure 21.1.

24.6 Heating Toluene in a Stirred Vessel

Toluene is heated by means of warm water of 80 °C from 10 to 30 °C in a jacketed baffled stainless steel process vessel equipped with a propeller stirrer. The coefficient for the heat transfer from the warm water to the vessel wall is 2,500 $\text{W} \cdot \text{m}^{-2} \cdot \text{K}^{-1}$. It is requested to calculate the process time.

Vessel Data

Standard Tank Configuration with dished bottom.

$$D = 1\,\text{m}.$$

Wall thickness 3 mm,
λ stainless steel 18 $\text{W} \cdot \text{m}^{-1} \cdot \text{K}^{-1}$.

Stirrer Data

$$S = d_i.$$

Three blades

$$n = 5\,\text{s}^{-1}.$$

Toluene Data

Boiling point 110.7 °C

$$\rho = 868\,\text{kg} \cdot \text{m}^{-3}.$$

$$\mu = 0.586 \cdot 10^{-3}\,\text{N} \cdot \text{s} \cdot \text{m}^{-2}.$$

$$c_p = 1{,}720\,\text{J} \cdot \text{kg}^{-1} \cdot \text{K}^{-1}.$$

$$\lambda = 0.140\,\text{W} \cdot \text{m}^{-1} \cdot \text{K}^{-1}.$$

Process Heat Transfer Coefficient

$$Re = \frac{\rho n \cdot d_i^2}{\mu} =$$

$$Pr = \frac{\mu \cdot c_p}{\lambda} =$$

μ/μ_w is approximately 1.
Check the applicability of Equation 22.3.

$$Nu_v =$$

$$\alpha = \qquad \text{W} \cdot \text{m}^{-2} \cdot \text{K}^{-1}.$$

Overall heat transfer coefficient:

$$\frac{1}{U} = \frac{1}{2{,}500} + \frac{0.003}{18} + \frac{1}{\alpha} =$$

$$U = \qquad W \cdot m^{-2} \cdot K^{-1}.$$

Heating Time

Toluene volume 0.76 m^3 (see Section 20.2).

$$M = \qquad kg.$$

$$A = \qquad m^2.$$

Use Equation 22.11:

$$t = \frac{M \cdot c_p}{UA} \ln \frac{80 - 10}{80 - 30} = \qquad s.$$

Solution

Process Heat Transfer Coefficient

$$Re = 8.1 \cdot 10^5.$$

$$Pr = 7.2.$$

$$Nu_v = 0.505 \cdot 1\left(8.1 \cdot 10^5\right)^{2/3} (7.2)^{1/3} = 8{,}473.$$

$$\alpha = 1{,}186\ W \cdot m^{-2} \cdot K^{-1}.$$

Overall Heat Transfer Coefficient

$$U = 709\ W \cdot m^{-2} \cdot K^{-1}.$$

Heating Time

$$t = \frac{660 \cdot 1{,}720}{709 \cdot 3.9} \ln \frac{80 - 10}{80 - 30} = 138\ s.$$

24.7 Overall Heat Transfer Coefficient of a Jacketed Reactor

A jacketed, agitated reactor consists of a vertical cylinder having a diameter of 1.5 m, a hemispherical base, and a flat flanged top. The jacket is fitted to the cylindrical section only and extends to a height of 1 m. The spacing between the jacket and the vessel wall is 75 mm. The jacket is fitted with a spiral baffle. The pitch of the spiral is 200 mm. The jacket is used to cool the reactor contents; $3.2 \cdot 10^4\ kg \cdot h^{-1}$ of chilled water enters the jacket at 10 °C and leaves at 20 °C.

The reactor is fitted with a flat blade disc turbine agitator having a diameter of 0.5 m and running at 210 rpm ($Z = 6$, $h/d_i = 1/5$, $b/d_i = 1/4$). The vessel is baffled and constructed of stainless steel plate 10 mm thick ($\lambda = 16\ W \cdot m^{-1} \cdot K^{-1}$).

It is requested to estimate the heat transfer coefficient at the outside wall of the reactor, at the inside wall of the reactor, and the overall coefficient in the clean condition.

The correlation for the heat transfer in the jacket is $Nu = 0.027 \cdot Re^{0.8} \cdot Pr^{0.33}$. The linear dimension in Nu is the hydraulic diameter D_h. The conditions for the applicability of the correlation are $Re > 10^4$, $Pr > 0.7$, $x/D_h > 60$. The correlation for the heat transfer in the reactor can be found in Chapter 22. Viscosity corrections can be omitted. It is requested to check the applicability of the correlations.

Physical Properties

	Water (15 °C)	Reactor contents
ρ, $kg \cdot m^{-3}$	1,000	850
μ, $N \cdot s \cdot m^{-2}$	0.001	0.08
λ, $W \cdot m^{-1} \cdot K^{-1}$	0.6	0.4
c_p, $J \cdot kg^{-1} \cdot K^{-1}$	4,200	2,650

Jacket

$$Re = \frac{\rho v \cdot D_h}{\mu} =$$

Use the hydraulic diameter D_h, that is, four times the cross-sectional area of the flow channel divided by the periphery.

$$Pr = \frac{\mu \cdot c_p}{\lambda} =$$

$$Nu =$$

Check the applicability of the correlation.
Use D_h also in Nu.

$\alpha_1 =$ $W \cdot m^{-2} \cdot K^{-1}$.

Reactor

$$Re = \frac{\rho n \cdot d_i^2}{\mu} =$$

$$Pr = \frac{\mu \cdot c_p}{\lambda} =$$

Check the applicability of the chosen correlation.

$$Nu_v =$$

$\alpha_2 =$ $W \cdot m^{-2} \cdot K^{-1}$.

Overall Heat Transfer Coefficient

$$\frac{1}{U} = \frac{1}{\alpha_1} + \frac{\delta}{\lambda} + \frac{1}{\alpha_2}$$

$U =$ $W \cdot m^{-2} \cdot K^{-1}$.

Calculate the temperature difference between the jacket and the reactor contents by dividing the heat load by the area and U. The reactor temperature is arrived at by adding the average jacket temperature to this temperature difference.

Solution

Jacket

$$D_h = \frac{4 \cdot 0.2 \cdot 0.075}{2(0.2 + 0.075)} = 0.109\ \text{m.}$$

$$v = \frac{3.2 \cdot 10^4}{1{,}000 \cdot 3{,}600 \cdot 0.2 \cdot 0.075} = 0.59\ \text{m} \cdot \text{s}^{-1}.$$

$$\text{Re} = \frac{1{,}000 \cdot 0.59 \cdot 0.109}{0.001} = 6.4 \cdot 10^4.$$

$$\text{Pr} = \frac{0.001 \cdot 4{,}200}{0.6} = 7.0.$$

The correlation for the jacket can be used.

$$\text{Nu} = 0.027 \cdot \left(6.4 \cdot 10^4\right)^{0.8} (7.0)^{0.33} = 359.$$

$$\alpha_1 = 1{,}976\ \text{W} \cdot \text{m}^{-2} \cdot \text{K}^{-1}.$$

Reactor

$$\text{Re} = \frac{850 \cdot 3.5 \cdot 0.5^2}{0.08} = 9{,}297.$$

$$\text{Pr} = \frac{0.08 \cdot 2{,}650}{0.4} = 530.$$

Equation 22.2 can be used.

$$\text{Nu}_v = 0.66 \cdot (9{,}297)^{2/3} (530)^{1/3} (1)^{0.14} = 2{,}361.$$

$$\alpha_2 = 630\ \text{W} \cdot \text{m}^{-2} \cdot \text{K}^{-1}.$$

Overall Heat Transfer Coefficient

$$\frac{1}{U} = \frac{1}{1{,}976} + \frac{0.01}{16} + \frac{1}{630} = 2.72 \cdot 10^{-3},$$

$$U = 368\ \text{W} \cdot \text{m}^{-2} \cdot \text{K}^{-1}.$$

The heat load is: $\dfrac{32.000 \cdot 10 \cdot 4{,}200}{3{,}600} = 3.7 \cdot 10^5\ \text{W}.$

$$\Delta T = \frac{3.7 \cdot 10^5}{368 \cdot \pi \cdot 1.5 \cdot 1} = 213\ \text{K.}$$

The temperature of the reactor contents is $213 + 15 = 228\ °\text{C}$.

24.8 Scale Up of Mixing

A scale-up from 20 U.S. gallon to 2,500 U.S. gallon is mentioned [2]. It is stated that, if the power input per unit volume (P/V) is held constant, circulation time increases threefold, tip speed by 70%, and the impeller Reynolds number by a factor of 8.5.

The first statement of these three statements is checked. Turbulent flow is assumed. P/V constant means $n^3 \cdot d_i^2$ is constant. Hence, $n \propto 1/d_i^{2/3}$. $n\theta$ is constant. So, $\theta \propto 1/n$. Note: θ is the mixing time and not the circulation time. Substitution of the first equation into the second equation results in $\theta \propto d_i^{2/3}$. $d_{i2}/d_{i1} = (2{,}500/20)^{1/3} = 5.0$ and $5.0^{2/3} = 2.9$.

It is requested to check the other two statements by adopting reasonings analogous to this reasoning.

Solution

Tip Speed

$v \propto n \cdot d_i$ and (P/V) constant implies $n \propto 1/d_i^{2/3}$. Substitution of the second equation into the first equation results in $v \propto d_i/d_i^{2/3} = d_i^{1/3}$. So, $v_2/v_1 = (d_{i2}/d_{i1})^{1/3} = ((2{,}500/20)^{1/3})^{1/3} = 1.7$. The tip velocity on large scale increases by 70%.

Reynolds Number

$Re \propto n \cdot d_i^2$ and P/V constant implies $n \propto 1/d_i^{2/3}$. Substitution of the second equation into the first equation results in $Re \propto d_i^2/d_i^{2/3} = d_i^{4/3}$. So, $Re_2/Re_1 = ((2{,}500/20)^{1/3})^{4/3} = 8.5$.

References

1. Kraume, M., and Zehner, P. (2001). Experience with experimental standards for measurements of various parameters in stirred tanks. *Transactions of the Institution of Chemical Engineers*, 79 (Part A), 811–818.
2. Leng, D. E. (1991). Succeed at scale up. *Chemical Engineering Progress*, 87, 23–31.

Notation II

A	Area	m^2
	Modified Flow number	—
Ar	Archimedes number	—
a	Exponent in Equation (22.1)	—
	Factor in Equation (19.2)	—
	Interfacial area per m^3 of liquid	m^{-1}
	Mass fraction solids	—
b	Exponent in Equation (22.1)	—
	Factor in Equation (19.2)	—
	Turbine stirrer blade width	m
C	Constant in Equation (22.1)	—
C_w	Drag coefficient	—
c	Oxygen concentration	$kmol \cdot m^{-3}$
	Exponent in Equation (22.1)	—
c_0	Initial oxygen concentration	$kmol \cdot m^{-3}$
c_∞	Equilibrium concentration	$kmol \cdot m^{-3}$
c_p	Specific heat	$J \cdot kg^{-1} \cdot K^{-1}$
c_{sat}	Oxygen saturation concentration	$kmol \cdot m^{-3}$
D	Vessel diameter	m
D_0	Inner jacket diameter	m
$D_{1,2}$	Vessel diameters Nos. 1 and 2	m
$Đ_{1,2}$	Diffusion coefficients Nos. 1 and 2	$m^2 \cdot s^{-1}$
D_c	Diameter at which a coil is wound	m
D_h	Hydraulic diameter	m
D_j	Outer jacket diameter	m
d_0	Nozzle diameter	m
d_c	Coil tube diameter	m
d_j	Jacket parameter	m
d_i	Impeller diameter	m
$d_{i1,2}$	Impeller diameters Nos. 1 and 2	m
d_p	Particle size	m
	Droplet diameter	m
F	Force	N
Fr	Froude number	—

Gr	Grashof number	—
g	Acceleration due to gravity	$m \cdot s^{-2}$
H	Liquid height in a vessel	m
h	Turbine impeller blade height	m
	Pfaudler impeller stirring element height	m
h_i	Height of the impeller above the vessel bottom	m
h_j	Jacket height	m
K_v	Proportionality constant in Section 16.2	—
k_g	Mass transfer coefficient in the gas phase	$m \cdot s^{-1}$
k_l	Mass transfer coefficient in the liquid phase	$m \cdot s^{-1}$
M	Mass	kg
Nu	Nusselt number	—
Nu_c	Nusselt number based on d_c	—
Nu_j	Nusselt number based on d_j	—
Nu_v	Nusselt number based on D	—
$Nu_{v1,2}$	Nusselt numbers based on D_1 and D_2	—
n	Impeller rotational speed	s^{-1}
n^*	Impeller speed at which all particles are just suspended	s^{-1}
$n_{1,2}$	Impeller rotational speeds Nos. 1 and 2	s^{-1}
n_b	Number of baffles	—
n_{CD}	Impeller speed for complete gas dispersion	s^{-1}
n_L	Minimum impeller speed to avoid flooding	s^{-1}
n_{min}	Minimum impeller speed to avoid segregation of two liquid phases	s^{-1}
n_R	Impeller speed for gas recirculation	s^{-1}
P	Power	W
P_0	Power number	—
Pr	Prandtl number	—
Pr_w	Prandtl number at wall conditions	—
Q	Flow number	—
R	Radius of a Pfaudler impeller stirring element	m
Re	Reynolds number	—
$Re_{1,2}$	Reynolds numbers Nos. 1 and 2	—
Re_j	Reynolds number based on d_j	—
S	Propeller pitch	m
S_c	Coil pitch	m
s	Zwietering parameter	—
T	Temperature	°C or K
T_0	Temperature leaving flow	°C
$T_{1,2}$	Temperatures Nos. 1 and 2	°C or K
T_i	Temperature entering flow	°C
T_w	Temperature at the wall	°C or K
t	Time	s
U	Overall heat transfer coefficient	$W \cdot m^{-2} \cdot K^{-1}$
V	Liquid volume in a vessel	m^3

V_G	Gas volume in a gassed vessel	m^3
V_L	Liquid volume in a gassed vessel	m^3
v_0	Velocity in a jacket inlet nozzle	$m \cdot s^{-1}$
$v_{1,2}$	Circumferential impeller velocities Nos. 1 and 2	$m \cdot s^{-1}$
v_a	Free convection velocity in a jacket	$m \cdot s^{-1}$
v_{ax}	Axial velocity in a stirred vessel	$m \cdot s^{-1}$
v_j	Velocity to be used in a Reynolds jacket number	$m \cdot s^{-1}$
v_r	Relative undisturbed velocity	$m \cdot s^{-1}$
v_{rad}	Radial velocity in a stirred vessel	$m \cdot s^{-1}$
v_s	Velocity in a jacket	$m \cdot s^{-1}$
	Superficial gas velocity in an empty vessel	$m \cdot s^{-1}$
We	Weber number	—
w_b	Baffle width	m
X	Volume fraction suspended material	—
X'	Volume fraction solids in a settled bed	—
x	Distance	m
Z	Number of blades of a turbine impeller	—

Greek Symbols

α	Heat transfer coefficient	$W \cdot m^{-2} \cdot K^{-1}$
	Propeller blade angle	—
$\alpha_{1,2}$	Heat transfer coefficients Nos. 1 and 2	$W \cdot m^{-2} \cdot K^{-1}$
β	Angle of a part of a Pfaudler impeller	—
	Thermal expansion coefficient	K^{-1}
γ	Turbine impeller blade angle	—
δ	Jacket gap	m
ε	Fraction gas volume of vessel contents	—
	Power input per kg	$W \cdot kg^{-1}$
ε_v	Power input per m^3 of liquid	$W \cdot m^{-3}$
θ	Mixing time	s
	Volume fraction dispersed phase	—
λ	Thermal conductivity	$W \cdot m^{-1} \cdot K^{-1}$
μ	Dynamic viscosity	$N \cdot s \cdot m^{-2}$
μ_c	Dynamic viscosity of a continuous phase	$N \cdot s \cdot m^{-2}$
μ_l	Dynamic viscosity of a liquid	$N \cdot s \cdot m^{-2}$
μ_m	Dynamic viscosity of a liquid/liquid-dispersion	$N \cdot s \cdot m^{-2}$
μ_{susp}	Dynamic viscosity of a suspension	$N \cdot s \cdot m^{-2}$
μ_w	Dynamic viscosity at the wall	$N \cdot s \cdot m^{-2}$
ν	Kinematic viscosity	$m^2 \cdot s^{-1}$
ν_{av}	Average kinematic viscosity	$m^2 \cdot s^{-1}$
ξ	Coil heat transfer parameter	—
ρ	Specific mass	$kg \cdot m^{-3}$
ρ_{av}	Average specific mass	$kg \cdot m^{-3}$
ρ_l	Liquid specific mass	$kg \cdot m^{-3}$

ρ_s	Solid specific mass	$kg \cdot m^{-3}$
ρ_{susp}	Suspension specific mass	$kg \cdot m^{-3}$
σ_m	Interfacial tension of a liquid/liquid-dispersion	$N \cdot m^{-1}$
τ	Residence time	s
Φ	Volume fraction dispersed phase	—
Φ_m	Mass flow rate	$kg \cdot s^{-1}$
Φ_{mole}	Absorbed oxygen flow	$kmole \cdot m^{-3} \cdot s^{-1}$
Φ_v	Impeller pumping capacity	$m^3 \cdot s^{-1}$
	Gas flow	$m^3 \cdot s^{-1}$
Ψ_p	Propeller parameter	—

Part III

Chemical Reactors

Part III Content

25

Chemical Reaction Engineering—An Introduction

The chemical reactor is the heart of any chemical plant. Raw materials from a different chemical process or purchased externally must usually be purified to a suitable composition for the reactor to handle. After leaving the reactor, the unconverted reactants, any solvents, and all byproducts must be separated from the desired product before it is sold or used as a reactant in a different chemical process.

25.1 Fluidized Catalytic Cracking (FCC)

The FCC reactor is the most important piece of equipment in the chemical industry [1]. Catalytic cracking is the primary chemical process in a petroleum refinery. The incoming crude is first desalted and passed through an atmospheric-pressure distillation column, which separates it into fractions. The heavy gas oil stream is cracked into smaller hydrocarbons suitable for gasoline. In plain words: Diesel fuel is converted into gasoline because the market demands much gasoline and less diesel fuel.

Reactions in hydrocarbon cracking may be represented as:

$$\text{hydrocarbons} \rightarrow \text{smaller hydrocarbons} + \text{coke} + \text{hydrogen}.$$

A representative reaction is the cracking of hexadecane into octane and octene:

$$\text{n-C}_{16}\text{H}_{34} \rightarrow \text{n-C}_8\text{H}_{18} + \text{n-C}_8\text{H}_{16},$$

with the side-reaction:

$$\text{n-C}_{16}\text{H}_{34} \rightarrow 16\text{C} + 17\text{H}_2.$$

Initially, cracking was carried out in fixed beds. Complications in fixed beds were:

- heat had to be supplied to heat the reactants to the desired temperature and overcome the endothermicities of the cracking reaction, and
- the reactor had to be shut down periodically for coke removal.

Both problems were overcome by the development in the 1940s of a fluidized bed system, which replaces the need to shut down to remove coke from the catalyst.

Introduction to Chemical Engineering, First Edition. C.M. van 't Land.

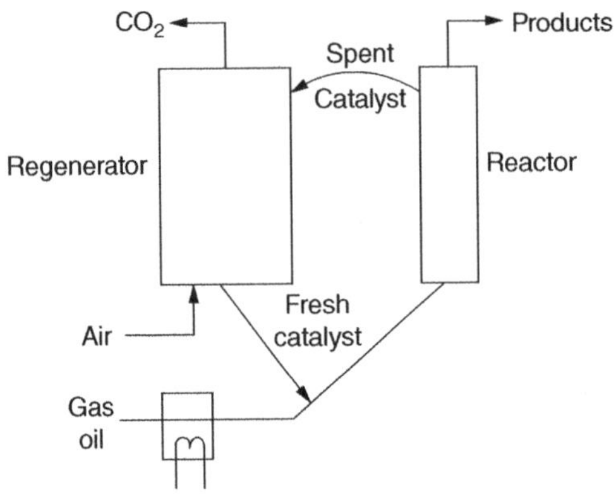

	Reactor	Regenerator
Feed	Gas oil	Air
Products	Alkanes, olefins, H_2	CO_2
Temperature	550 °C	650 °C
Pressure	2 atm	2 atm
Gas residence time	<1 s	Several seconds
Catalyst residence time	<1 s	Several seconds
Flow pattern	PFR (riser)	CSTR
Heat	Absorbed	Generated

Figure 25.1 Block scheme and operating conditions of a Fluidized Catalytic Cracking Reactor. *Source:* Courtesy of Oxford University Press, Oxford, UK.

The FCC reactor is really two reactors with catalyst particles cycled between them. The evaporated gas oil is fed along with fresh catalyst to the first, called the reactor, and the spent catalyst is recycled to the reactor.

The strength of FCC is that the reactor–regenerator combination solves both the heat management and coking problems simultaneously. Burning off the coke is strongly exothermic, and this reaction in the regenerator supplies the heat (carried with the regenerated catalyst particles) for the endothermic cracking reactions in the reactor.

Typical operating conditions of the reactor and the regenerator are in Figure 25.1.

25.2 Kinetic Rate Data and Transport Phenomena

Compounds participating in a chemical reaction usually occur in the liquid or gaseous phase. The mobility of the elementary particles is often too restricted in the solid phase.

Only one phase is present when a reaction is homogeneous. At a heterogeneous reaction, more than one phase is present. In the latter case, physical factors related to mass transfer can affect the progress of a reaction even more than chemical affinity. The combustion of coal in air illustrates the importance of mass transfer between a solid and a gas phase. The reaction of coal combustion in air is very rapid, but conventional industrial power plants attain only 1% of the rate possible because oxygen cannot be supplied rapidly enough to the coal surface [2]. See Figure 25.2. The data

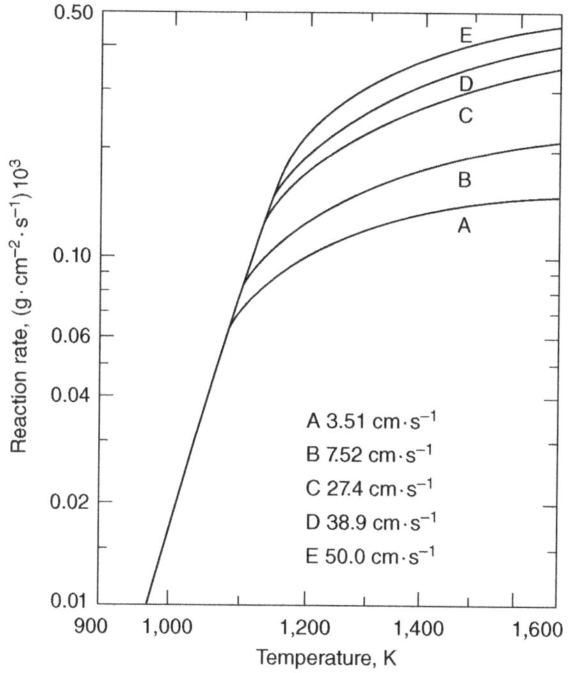

Figure 25.2 The reaction rate of coal combustion in air above 1,100 K depends strongly on the air velocity. *Source:* Courtesy of the American Chemical Society, Washington DC.

show that air velocity is the controlling factor in the chemical reaction. Up to approximately 1,100 K (827 °C), the combustion reaction is solely a function of the temperature. The air velocity becomes a defining variable at higher temperatures.

The nitration of liquid benzene by a mixture of nitric acid and sulfuric acid dissolved in water (the combination is called mixed acid) is a further example. The chemical reaction is rapid, and the rate of nitration is determined by the transport of the reactants through the interface between the two phases. The latter transport, in turn, is a function of the agitation intensity.

25.3 Reactor Types

See Figure 25.3.

Batch and Semi-Batch Reactor

To a batch reactor, chemicals are added and allowed to react for a given time period. Conversion and other dependent variables are a function of time. Then, the products and unreacted starting materials are removed. Batch reactors are sometimes equipped with a distillation column. To manage the heat transfer, a batch reactor is often designed with a jacket or a coil. To promote heat transfer and efficiency, a batch reactor has an agitator and baffles.

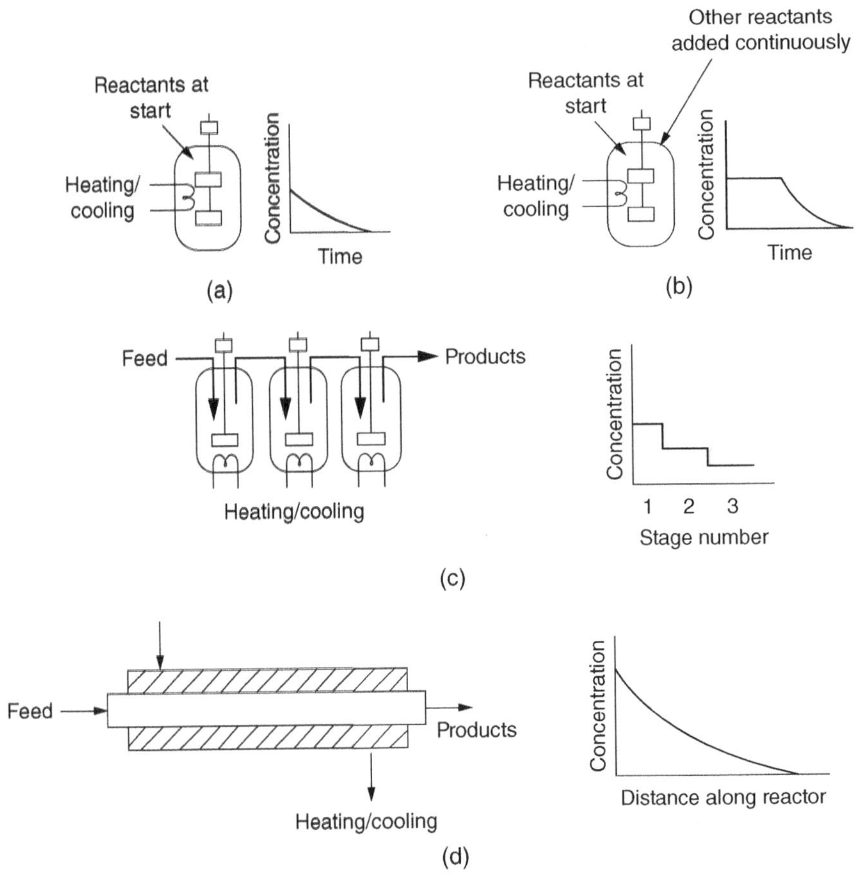

Figure 25.3 Reactor types: (a) batch, (b) semi-batch, (c) continuous stirred tank reactor, and (d) plug flow reactor.

Batch reactors are used for small-scale production of expensive products. They are also justified for multiple, low-volume products or when continuous flow is difficult as in the manufacture of, for example, polymer resins. Batch reactors are used often to develop continuous processes. A batch reactor can be used as a semi-batch reactor. In that case, one or more components are added or removed continuously. For example, gradual addition of gaseous chlorine to a stirred vessel containing benzene and a catalyst results in higher yields of di- and trichlorobenzene than the inclusion of liquid chlorine in the original batch. In addition to better yields and selectivity, continuous addition or removal assists in controlling temperature, particularly when the reaction is strongly exothermic. Thus, use of a semi-batch reactor intrinsically permits more stable and safer operation than in a batch reactor.

Plug Flow Reactor (PFR)

The plug flow reactor is a vessel through which flow is continuous, usually at steady state and configured so that conversion and other dependent variables are functions of position and therewith of time. In the ideal plug flow reactor, the fluids flow as if they were solid plugs or pistons. That statement is more or less valid for turbulent flow. Reaction time is the same for all flowing material at any given tube cross-section; hence, there is a relationship between time in the batch reactor and position in the PFR. Plug flow reactors resemble batch reactors in providing high initial driving forces, which diminish as the reactions progress down the tube.

In a few instances, the reactor may simply consist of an empty tube (homogeneous reaction), but it will usually contain a bed of catalyst (heterogeneous reaction) or occasionally a bed of inert material intended to improve heat transfer. The PFR can be found in petrochemical plants and in bulk chemicals production plants generally. Well-known examples are:

- the homogeneous reaction $2NO + O_2 \rightarrow 2NO_2$, a step in the synthesis of HNO_3, nitric acid,
- the heterogeneous reaction $CO + 2H_2 \leftrightarrow CH_3OH$, the reaction to produce methanol from synthesis gas, and
- the heterogeneous reaction $N_2 + 3H_2 \leftrightarrow 2NH_3$, the Haber-Bosch reaction to produce ammonia.

The first reaction is an irreversible one, whereas the other two reactions are reversible.

Continuous Stirred Tank Reactor (CSTR)

Reactants and products are continuously added and withdrawn respectively to and from a continuous stirred tank reactor. Because the composition of the mixture leaving a CSTR is the same as the composition of the reactor contents, the driving forces, that is, the reactants concentrations, are necessarily low. For positive-order reactions, a CSTR requires the largest volume of the reactor types to obtain desired conversions. When high conversions of reactants are needed, several CSTRs in series can be used. The larger the number of reactors, the closer performance approaches that of a tubular reactor. Continuous stirred tank reactors in series are simpler and easier to design for isothermal operation than plug flow reactors.

25.4 Batch Reactions Versus Continuous Reactions

- Large productions, that is, larger than 100,000 tpa are preferably made continuously as this leads to relatively low investment figures.
- Having stationary conditions is an advantage of continuous manufacture. The status of the plant at, for example, 3 o'clock in the morning equals the status at, for example, 6 o'clock in the evening. The conditions are always the same. An easy check is always possible.
- Continuous productions should take place 7 days a week. Three-shift operations are unfavorable.
- Continuous plants are dedicated plants. They cannot be used for multipurpose productions.
- Some reactions last very long, for example, the production of pulp from wood chips. In the acid bisulfite process, the chips are treated for 3–7 h at 125–143 °C, while the pH is 1–2. Until recently, batch digesters were used exclusively. However, modern pulp technology is dominated by continuous digesters. The background is the desire to obtain uniform pulping results. The investment figures for continuous pulp plants are higher than for batch pulp plants.
- Reactions having small margins should be carried out batchwise.

Example 25.1

It is possible to convert 1,12-dodecane dioic acid (DA) into 1,12-dodecane-peroxi-dioic acid (DPDA). The reaction equation is:

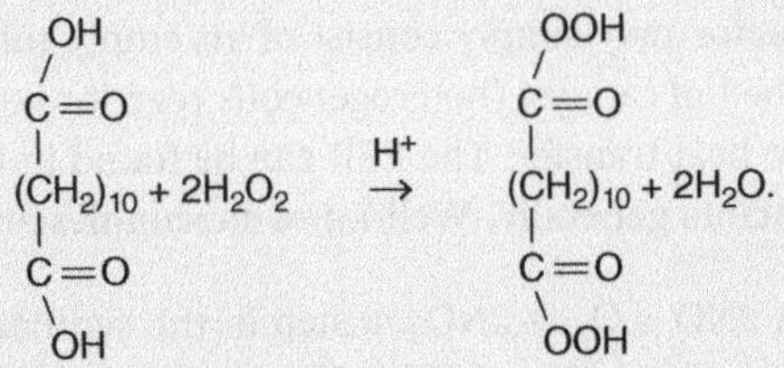

This reaction is carried out batchwise. The pressure of the reaction is atmospheric, whereas the temperature is ambient. First, an aqueous solution of hydrogen peroxide (H_2O_2) and sulfuric acid is prepared, and the heat of dilution is carried away. The composition of this mixture is checked. Next, DA powder is dosed. The powder dissolves and is peroxidized. Peroxidation is short for superoxidation and that means that extra oxygen is introduced into a molecule. The reaction is catalyzed by H^+-ions from sulfuric acid. The peroxidized powder precipitates. A postreaction starts when all powder has been dosed. The reaction is brought to completion. Next, water is added to the reactor contents, which stops all chemical activities. Finally, the precipitate is separated from the reactor contents by filtration and washed.

Attempts have been made to carry out this reaction continuously. These attempts failed because of the process safety. At too high an acid concentration, the di-peroxide decomposes which may give rise to a runaway reaction. The margins are too small for continuous production. On producing batchwise, the initial acid concentration is high. As the reaction proceeds, the acid is diluted. One moves away from the dangerous area.

- On manufacturing batchwise, corrections can be made when the batch is ready. It is also possible to mix batches. On producing continuously, any off-spec material has already left the reactor. To put it differently, on manufacturing batchwise, product certification is possible.
- In batch reactions, materials having very high viscosities can be handled. Batch reactors can be equipped with sturdy mixers.
- Starting-up and shutting down a plant can be critical stages. A continuous reactor is normally at stationary conditions. A batch reactor is started up and shut down intermittently.
- Equipment can be used more efficiently in continuous processes than in batch processes. For example, in a batch process, a reactant must be added to a reactor in 5 min and during the transfer be heated from ambient to 100 °C. In a continuous process, the mass flow in $kg \cdot h^{-1}$ would be much smaller and the heat exchanger would be smaller than the heat exchanger of a batch process.

25.5 Adiabatic Temperature Rise

The concept of the adiabatic temperature rise is explained as follows. For example, reactants A and B react with each other in a closed batch reactor. The reaction is exothermic and the heat of the reaction cannot be carried away. The heat development is absorbed by the reactor contents as sensible heat, and this causes the temperature of the reactor contents to rise.

Example 25.2

A room contains 9.1% methane (CH_4) by volume [3]. By turning on the light switch, a spark is created. The spark causes a vapor/air explosion. It is assumed that the room can withstand the pressure due to the explosion. The temperature and pressure in the room due to the explosion are calculated. The methane combustion equation is:

$$CH_4 + 2O_2 \rightarrow CO_2 + 2H_2O,$$

and this reaction has a reaction heat of $8.04 \cdot 10^5\ J \cdot mol^{-1}$ of CH_4. On taking the oxygen content of air as 20% by volume, the methane concentration is adequate to bind all oxygen. For a basis of 1 mol of CH_4, there are 2 mol of O_2, and 8 mol of N_2 before reaction: a total of 11 mol. The number of moles after the reaction is also 11. The assumption of a specific heat of 3.5·R is reasonable for small molecules such as N_2 at moderate temperatures. It is now possible to calculate the adiabatic temperature rise:

$$\Delta T_{ad} = \frac{8.04 \cdot 10^5}{11 \cdot 3.5 \cdot 8{,}314} = 2{,}512\ K.$$

In °C, the maximum room temperature will become 20 + 2,512 = 2,532 °C.

If the initial pressure is 1 bar, the maximum pressure after the explosion will be: $\frac{2{,}532 + 273}{20 + 273} = 9.6$ bar.

Even mildly exothermic reactions have adiabatic temperature rises much above 100 K, and total combustion processes for stoichiometric mixtures in air have adiabatic temperature rises above 2,000 K.

Example 25.3

A reaction with a reaction heat of $10^5\ J \cdot mol^{-1}$ takes place in an aqueous solution with a reactant concentration of $2\ mol \cdot l^{-1}$ and an initial temperature of 50 °C. What is the temperature if this reaction goes to completion in an adiabatic reactor?

$$T = 50 + \frac{2 \cdot 10^5}{4{,}190} = 97.7\ °C.$$

It is important to be aware of the heat absorbed or liberated in a chemical reactor and of the adiabatic temperature rise.

25.6 Recycle

Figure 25.4 illustrates the different product distributions effected by changing individual recycle rates for a methane chlorination process [4].

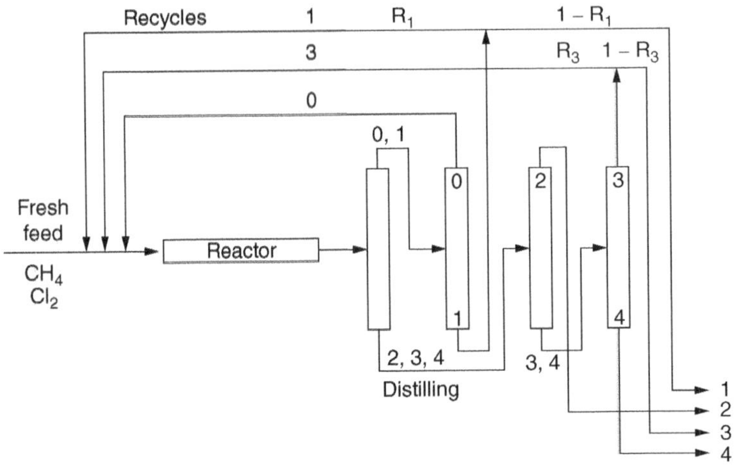

Recycle ratios		Fraction of CH_4 converted into			
R_1	R_3	CH_3Cl	CH_2Cl_2	$CHCl_3$	CCl_4
0	0	0.29	0.42	0.26	0.03
1	0	0	0.56	0.39	0.05
0	1	0.15	0.36	0	0.49
1	1	0	0.38	0	0.62

Figure 25.4 Effects of product recycle on methane chlorination product distributions, where 0 = methane, 1 = methyl chloride, 2 = methylene chloride, 3 = chloroform, and 4 = carbon tetrachloride. *Source:* Reactor Design and Operation by Westerterp, K. R. et al. (1984). Reproduced with permission of John Wiley & Sons, Inc.

25.7 Process Intensification

The savings obtained by process intensification (PI) can be considerable.

Reactive distillation (RD) is an important category of PI. RD is the combination of a reaction step and a distillation step. The methyl acetate production by Eastman Chemical Co. is an example of RD [5]. This process basically consists of one reactive distillation column instead of, conventionally, a reactor and nine distillation columns. See Figure 25.5. Because of the presence of three azeotropes in the reaction mixture, distillation as a subsequent step becomes complicated.

A further example of PI concerns the synthesis of an additive [6]. A cooperation between Forschungs Zentrum Kartsruhe (FZK) in Germany and Dutch State Mines (DSM) in The Netherlands resulted in the replacement of a 10-m^3 batch reactor by a 3-l continuous reactor. See Figure 25.6. The production capacity of 1,700 $kg \cdot h^{-1}$ was maintained. The location is DSM at Linz in Austria. The additive is probably made via the maleic acid anhydride route. It concerns two strongly exothermic reactions in series, which are catalyzed by H_2SO_4. The following description concerns the continuous reactor. It has flow channels of approximately 100 μm. The cooling medium is passed through the reactor in crossflow and the heat load of the reactor is 100 kW. The reactions proceed at a high rate, which is important, because the intermediate is very unstable. The continuous reactor is a PFR and, hence, the intermediate does not "see" the final product. That is important because the two compounds can react with each other. The yield of the continuous reaction is 15% higher than the yield of the batch reaction because the side reactions

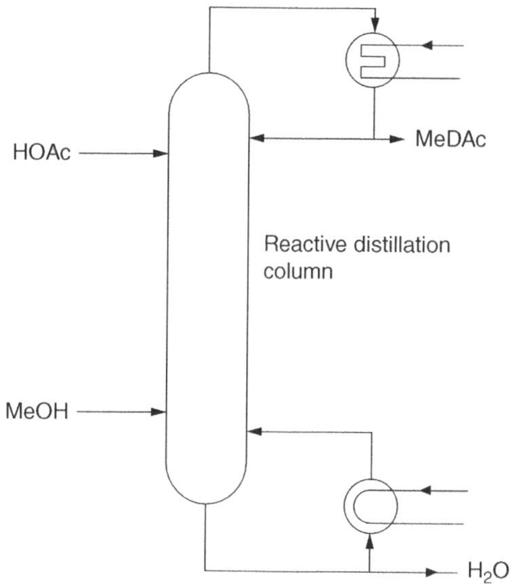

Figure 25.5 Reactive distillation to produce methyl acetate as an example of Process Intensification.

Figure 25.6 Replacing a 10-m^3 batch reactor by a continuous 3-l reactor. *Source:* Courtesy of Koninklijke DSM N. V., Heerlen, The Netherlands.

are avoided. An aspect is that Re in the flow channels predicts laminar flow. This would not be a bonus for the reaction as the viscosities of the reactants are strongly different. However, turbulent mixing appears to occur on microscale. The concept has been proven by three campaigns at which, per campaign, 300 t were produced.

A further account states that the product is diacetone acrylamide and that the reaction conditions are 180 °C and 20 bar [7]. A Ritter reaction is mentioned. The production started in 2005.

PI is summarized as follows. In the first step, the tasks to be performed by the process are defined. The technologies to accomplish the tasks are selected in the second step. For example, a separation step might be accomplished by membranes. In the last step, the technologies are translated into equipment. If there is synergy between technologies, they can be integrated into one piece of equipment [6].

References

1. Schmidt, L. D. (1998). *The Engineering of Chemical Reactions* (pp. 63–65). Oxford, UK: Oxford University Press.
2. Griskey, R. G. (1996). *Chemical Engineering for Chemists* (p. 17). Washington, DC: American Chemical Society.
3. Schmidt, L. D. (1998). *The Engineering of Chemical Reactions* (pp. 55–56). Oxford, UK: Oxford University Press.
4. Westerterp, K. R., van Swaay, W. P. M., and Beenackers, A. A. C. M. (1984). *Chemical Reactor Design and Operation* (pp. 122–123). New York, NY: Wiley.
5. Agreda, V. H., and Heise, W. H. (1990). High-purity methyl acetate via reactive distillation. *Chemical Engineering Progress*, 36 (2), 40–46.
6. van Ede, J. (2007). Microreactors. *Chemisch Weekblad*, 103 (7), 20–21. (in Dutch).
7. Dijkgraaf, A. (2008). Intensify or drown. *Chemisch Weekblad*, 104 (24), 16–18. (in Dutch).

26

A Few Typical Chemical Reactors

26.1 The Carbo-V-Process of Choren

See Figure 26.1. The first three characters of Choren stand for carbon, hydrogen, and oxygen, whereas the last three characters stand for "renewable." Making diesel oil out of wood is depicted schematically. The first step is the gasification of dry wood at 400–500 °C. The gases are burnt in a combustion chamber with a short measure of oxygen at 1,300 °C. The reaction product is synthesis gas, a mixture of carbon monoxide and hydrogen. When the composition of the entering gases is simplified to methane, CH_4, the reaction equation is:

$$CH_4 + \frac{1}{2}O_2 \rightarrow CO + 2H_2.$$

The reaction equation of the complete combustion of methane is:

$$CH_4 + 2O_2 \rightarrow CO_2 + 2H_2O.$$

Synthesis gas was given that name because it can be used for many purposes, for example, the production of methanol, solvents, and diesel oil.

The next steps are gas cooling, filtration, and purification. CO and H_2 are converted into diesel oil in a second reactor. The reaction equation for that step is:

$$nCO + (2n + 1)H_2 \rightarrow C_nH_{2n+2} + nH_2O \; (\Delta H_R = -167\,\text{kJ per mol CO}).$$

A typical component of diesel oil is $C_{16}H_{34}$, hexadecane. This type of process is termed oligomerization. It is the combination of a number of molecules to a large molecule. If the number is very large, the process is termed polymerization. The reaction in the second reactor is a Fischer–Tropsch synthesis, named after two German chemists who invented the process in 1925. The reaction typically proceeds at 250 °C and 20–30 bar with iron (Fe) as a catalyst. CO and H_2 for the Fischer–Tropsch synthesis were initially made by the gasification of coal.

26.2 Coal Gasification

See Figure 26.2. Making synthesis gas out of methane yields a mixture of CO and $2H_2$. See Section 26.1. The C/H ratio of methane 1/4. The C/H ratio of coal is smaller, for example, 1/2. Thus, the molecular ratio of hydrogen and carbon monoxide in the gaseous product of

Introduction to Chemical Engineering, First Edition. C.M. van 't Land.

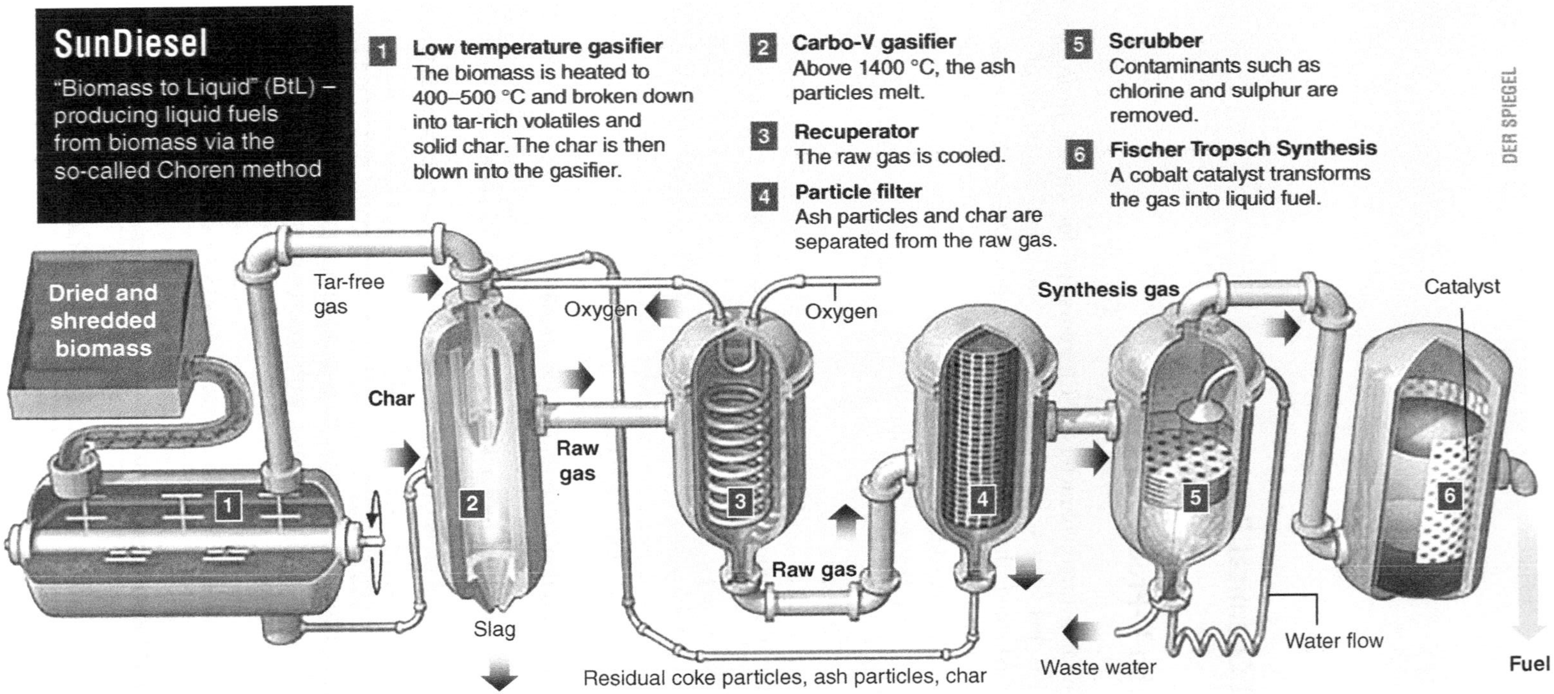

Figure 26.1 The Carbo-V process of Choren. *Source:* Courtesy of Der Spiegel, Issue 40/2004, Hamburg, Germany.

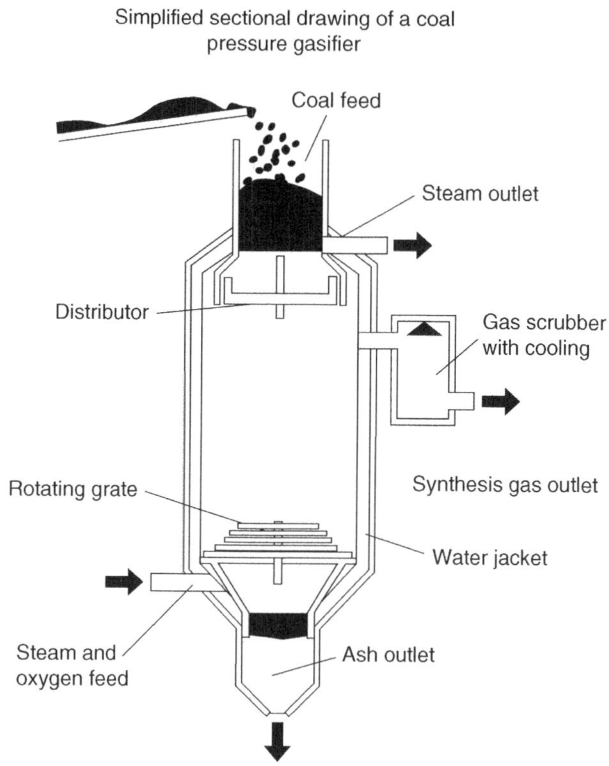

Figure 26.2 Coal gasification. *Source:* Courtesy of KSB SE & Co. KGaA, Frankenthal, Germany.

coal gasification to produce synthesis gas is smaller than two. To increase the calorific value of the gas, steam is also fed to the reactor. When the composition of coal is simplified to CH_2, the reaction equations are:

$$CH_2 + \frac{1}{2}O_2 \rightarrow CO + H_2, \text{and}$$

$$CO + H_2O \rightarrow CO_2 + H_2.$$

The latter reaction is called the water gas shift reaction. The gas produced by the reactor consists of CO, H_2, and small amounts of CO_2 and CH_4.

26.3 Biofuels

First, ethanol manufacture is discussed. Starchy grains like corn are ground up and mixed with water. Other grains must first soak up to 48 h. In a fermenting tank, enzymes convert the starches into a simple sugar named dextrose. Yeast is added and fermentation starts. The simple sugar is converted into ethanol and carbon dioxide. The pressure is atmospheric and the temperature is 36–37 °C. After 40–50 h, the tank contents contain 6% by volume of ethanol.

Suspended solid material is separated from the liquid tank contents by means of filtration. The next step is distillation.

Second, the biodiesel production is treated. Oils pressed from plants like rapeseed and soybean contain fats called triglycerides and are heated. After heating, methanol and a catalyst such as potassium hydroxide (KOH) are added. This produces methyl esters of carboxylic acids. The latter compounds can be used as biodiesel. Glycerol is an important byproduct.

26.4 Pyrogenic Silica

The chemical formula of pyrogenic silica is SiO_2. The indication "pyrogenic" means that the material is produced by means of a flame. It is also possible to make silica by means of a wet chemical process. The production of the latter material by far exceeds the production of the former material. Pyrogenic silica is applied to, for example, improve the flowing characteristics of many powders. Figure 26.3 is a block scheme of the production of the material. The liquid raw material silicon tetrachloride is evaporated and the gas is mixed with air. The boiling point of the tetrachloride is 57.6 °C. The gaseous flow then enters a combustion chamber where the tetrachloride reacts with hydrogen and oxygen. The equation of the overall chemical reaction is:

$$2H_2 + O_2 + SiCl_4 \rightarrow SiO_2 + 4HCl.$$

The process is called a flame hydrolysis because the hydrogen and oxygen react to water, which subsequently reacts with silicon tetrachloride. The flow leaving the continuous combustion chamber consists of gaseous hydrochloric acid and silica particles. It is cooled, and the two phases are separated.

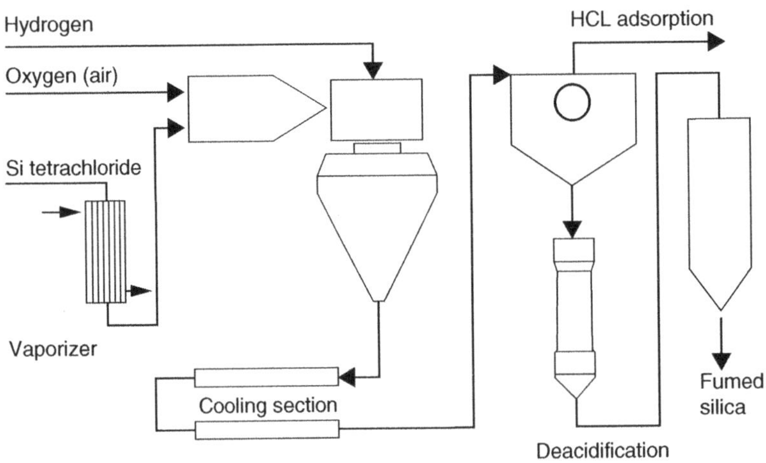

Figure 26.3 The production of pyrogenic silica. *Source:* Courtesy of Evonik, Germany.

26.5 Microwaves

See Figure 26.4. It depicts the polymerization of 2-ethyl-2-oxazoline in both a plug flow reactor and a continuous stirred tank reactor. An initiator is used and energy is supplied by microwaves.

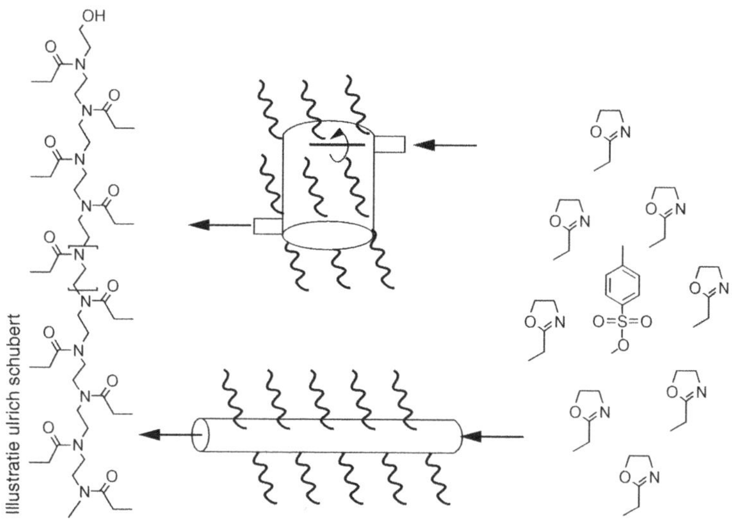

Figure 26.4 Microwaves. *Source:* Courtesy of Schubert, U.S., Jena, Germany.

27

The Order of a Reaction

27.1 The Rate of a Reaction

The rate of a chemical reaction is a function of:

- the temperature,
- the concentration, and
- the presence of catalysts.

The temperature dependency of the chemical reaction rate will be discussed in Chapter 28. The reaction rate almost always increases when the concentrations of the reactants increase. For gas-phase reactions, concentrations are a function of pressure. Catalysts increase the reaction rate. They play a role in the chemical reaction, however, chemically and masswise they are not affected by the chemical reaction. Physically, they may have undergone a change at the chemical reaction. As an example, the role of manganese dioxide, MnO_2, at the decomposition of potassium chlorate, $KClO_3$, is mentioned. The latter compound is decomposed into potassium chloride, KCl, and oxygen, O_2. The catalyst is crystalline before the reaction and amorphous after the reaction. Chemically, it is the same material and its mass has not changed either. Promoters are substances enhancing the activity of catalysts. The same goes for co-catalysts. Inhibitors are substances preventing a chemical reaction to occur or slowing down its reaction rate.

27.2 Introductory Remarks on the Order of a Reaction

Experimental data of chemical kinetics can be obtained by measuring concentrations of reactants and products during batch reactions. Usually, the temperature is kept constant at these experiments. On plotting the concentration of a reactant or reactants as a function of time, it can in many instances be found that the results can be described by means of differential equations of the general form $dC_A/dt = f(C_A, C_B, ..., C_N)$. It is also possible that a concentration can be described by an integrated form of a differential equation. C_A is the concentration of a particular reactant followed as a function of time.

In many instances, the rate, which will be written as the rate of decrease of the concentration of reactant A, $-dC_A/dt$, is found to depend on the product of concentration terms. For example:

$$-\frac{dC_A}{dt} = k \cdot C_A^a \cdot C_B^b ... C_N^n.$$

Introduction to Chemical Engineering, First Edition. C.M. van 't Land.

The order of a reaction is defined as the sum of the exponents of the concentration terms in this rate law. The rate law reflects that the possibility for molecules to react with each other becomes favorable when they are close to each other and vice versa.

The concept of molecularity of a reaction is used to indicate the molecular mechanism by which it proceeds. For example, the decomposition of hydrogen iodide into hydrogen and iodine involves two molecules, ($2HI \rightarrow H_2 + I_2$), and is therefore called a bimolecular reaction. All bimolecular reactions are second order, but the converse is not true. The concept of molecularity should be applied only to individual elementary reactions. When a reaction proceeds through several stages, we do not talk about its molecularity, since one step may involve two molecules, a further one three, and so on.

The considerations in this chapter apply only when the reacting molecules are free to meet each other in the reaction system and do not experience physical limitations.

27.3 First-Order Reaction

$$A \rightarrow B + C.$$

This reaction is first order if the progress of the isothermal batch reaction can be described by

$$-\frac{dC_A}{dt} = k \cdot C_A,$$

$$\frac{dC_A}{C_A} = -k \cdot dt.$$

Integration occurs from $C_A = C_{Ao}$ at $t = 0$ to $C_A = C_A$ and $t = t$. k is the reaction rate constant.

$$\int_{C_{Ao}}^{C_A} dC_A = -k \int_0^t dt.$$

$$\text{The result is: } t = -\frac{1}{k} \ln \frac{C_A}{C_{Ao}} \text{ s.} \quad (27.1)$$

The conversion of A is $x_A = \frac{C_{Ao} - C_A}{C_{Ao}} = 1 - \frac{C_A}{C_{Ao}}$.

Conversion of A is, in concentration terms, what was converted divided by what was initially present.

The decomposition of nitrogen pentoxide is an example of a first-order reaction [1].

$$2N_2O_5 \rightarrow 4NO_2 + O_2.$$

This is an overall equation; the reaction that determines the rate is probably a unimolecular reaction (e.g., $N_2O_5 \rightarrow N_2O_4 + O$). Table 27.1 contains data for the gas phase reaction at 45 °C.

Consider, for example, the situation after 6,000 s. It then follows from Equation 27.1:

$$6{,}000 = -\frac{1}{k} \ln \frac{18}{348.4} \rightarrow k = 4.94 \cdot 10^{-4}\,s^{-1}.$$

The series of measurements confirms the first order of the reaction. Plotting $\ln C_A/C_{Ao}$ as a function of time results in a straight line.

Table 27.1 Decomposition of nitrogen pentoxide at 45 °C.

Time t (s)	p_{N2O5} (mm Hg)
0	348.4
600	247
1,200	185
1,800	140
2,400	105
3,000	78
3,600	58
4,200	44
4,800	33
5,400	24
6,000	18
7,200	10
8,400	5
9,600	3
∞	0

A further example of a reaction that can be described by the first-order law as expressed by Equation 27.1 is the inversion of cane sugar in diluted hydrochloric acid (HCl). The conversion involves two different molecules, that is, cane sugar and water. However, the excess of water is such that the water concentration is not changed appreciably. This type of reaction is called pseudo-unimolecular. The reaction equation is:

$$\underset{\text{cane-sugar}}{C_{12}H_{22}O_{11}} + H_2O \xrightarrow{H^+} \underset{\text{glucose}}{C_6H_{12}O_5} + \underset{\text{fructose}}{C_6H_{12}O_5}.$$

Cane sugar is optically active (dextrorotatory), as are glucose and dextrose (both laevorotatory), and hence the name inversion. The conversion can be followed by means of a polarimeter.

The equation for the inversion is:

$$k = \frac{1}{t} \ln \frac{r_0 - r_\infty}{r_t - r_\infty} \quad s^{-1}.$$

See Equation 27.1. r_0 is the initial rotation, r_∞ is the rotation after a long while, and r_t is the rotation after a time t. The amount of unchanged cane sugar is proportional to $(r_t - r_\infty)$.

Example 27.1

Cane-sugar is hydrolyzed in diluted hydrochloric acid. In 0.9 N HCl and at 25 °C, with $r_0 = +24.09°$ and $r_\infty = -10.74°$, $r_t = = 10.02°$ after 46 min [2].

$$k = \frac{1}{46 \cdot 60} \ln \frac{34.83}{20.76} = 1.87 \cdot 10^{-4}\, s^{-1}.$$

A further example of a first-order reaction is radioactive decay, for example, $Ra \rightarrow Rn + \alpha$. This is a unimolecular reaction following a first-order law.

27.4 Second-Order Reactions

$$A + B \rightarrow C + D.$$

Case 1

The initial concentrations of A and B are equal. The reaction is second order when the progress of the batch reaction can be described at a certain temperature by:

$$-\frac{dC_A}{dt} = k \cdot C_A \cdot C_B = k \cdot C_A^2,$$

$$\frac{dC_A}{C_A} = -k \cdot dt.$$

Integration occurs from $C_A = C_{Ao}$ at $t = 0$ to $C_A = C_A$ at $t = t$.

$$\int_{C_{A0}}^{C_A} \frac{dC_A}{C_A} = -k \int_0^t dt,$$

$$t = \frac{1}{k}\left(\frac{1}{C_A} - \frac{1}{C_{A0}}\right) \text{ s.} \tag{27.2}$$

$$x_A = \frac{C_{A0} - C_A}{C_{A0}} \quad (x_A \text{ is the conversion}).$$

Combining the latter two equations results in:

$$t = \frac{x_A}{k \cdot C_A} \text{ s.}$$

Example 27.2

Ethyl acetate is hydrolyzed by sodium hydroxide in an aqueous solution [3]. The temperature is 25 °C and the initial concentration of ethyl acetate and sodium hydroxide is 16.00 $mmol \cdot l^{-1}$.

$$CH_3COOC_2H_5 + NaOH \rightarrow CH_3COONa + C_2H_5OH.$$

See Table 27.2. Consider the situation after 1,500 s.

$$1{,}500 = \frac{1}{k}\left(\frac{1}{4.32} - \frac{1}{16.00}\right) \rightarrow k = 1.13 \cdot 10^{-4}\, l \cdot mmol^{-1} \cdot s^{-1}.$$

See Equation 27.2. The series of measurements in Table 27.2 confirms the second order. Plotting $(1/C_A - 1/C_{Ao})$ as a function of time results in a straight line.

Table 27.2 Hydrolysis of ethyl acetate.

Time t (s)	C_A (mmol · l^{-1})
300	10.24
900	6.13
1,500	4.32
2,100	3.41
3,300	2.31

Case 2

The initial concentrations of A and B are not equal. The reaction is second order when the progress of the batch reaction at a certain temperature can be described by:

$$-\frac{dC_A}{dt} = k \cdot C_A \cdot C_B.$$

The initial concentrations of A and B are C_{A0} and C_{B0}.

$$C_B = C_{B0} - (C_{A0} - C_A) = C_A + C_{B0} - C_{A0}.$$

$$\frac{dC_A}{C_A(C_A + C_{B0} - C_{A0})} = -k \cdot dt.$$

To be able to integrate the left-hand side, it is necessary to separate the integrand into two partial fractions:

$$\frac{1}{C_A(C_A + C_{B0} - C_{A0})} = \frac{D}{C_A} + \frac{E}{C_A + C_{B0} - C_{A0}}.$$

D and E can be calculated as follows. Multiply both sides of the equation with the denominator of the left-hand side:

$$1 = D(C_A + C_{B0} - C_{A0}) + E \cdot C_A.$$

This equation is valid for all values of C_A. When $C_A = C_{A0} - C_{B0}$, $E = 1/C_A = 1/(C_{A0} - C_{B0})$. When $C_A = 0$, $D = 1/(C_{B0} - C_{A0})$. It follows:

$$\frac{1}{C_A(C_A + C_{B0} - C_{A0})} = \frac{1}{C_A(C_{B0} - C_{A0})} + \frac{1}{(C_A + C_{B0} - C_{A0})(C_{A0} - C_{B0})}.$$

The integration of the first term of the right-hand side is attempted first.

$$\int_{C_{A0}}^{C_A} \frac{dC_A}{C_A(C_{B0} - C_{A0})} = \frac{1}{C_{B0} - C_{A0}} \ln \frac{C_A}{C_{A0}}.$$

Next, the integration of the second term is attempted.

$$\int_{C_{A0}}^{C_A} \frac{dC_A}{(C_A + C_{B0} - C_{A0})(C_{A0} - C_{B0})} = \frac{1}{C_{A0} - C_{B0}} \ln \frac{C_A + C_{B0} - C_{A0}}{C_{B0}}.$$

Adding the two results: $\frac{1}{C_{B0} - C_{A0}} \ln \frac{C_A \cdot C_{B0}}{C_{A0}(C_A + C_{B0} - C_{A0})}$.

The final result is:

$$t = \frac{1}{k(C_{A0} - C_{B0})} \ln \frac{C_A \cdot C_{B0}}{C_{A0}(C_A + C_{B0} - C_{A0})} \text{ s.} \quad (27.3)$$

Example 27.3

The example concerns a case in which the initial concentrations of A and B are unequal [4]. However, the concentrations are such that the stoichiometry is met. The treatment is similar to the treatment leading to Equation 27.3. The reaction between ethylene dibromide and potassium iodide in 99% methanol is second order.

$$C_2H_4Br_2 + 3KI \rightarrow C_2H_4 + 2KBr + KI_3.$$

Sealed bulbs containing the reaction mixture were kept in a thermostat at 59.7 °C. At intervals of 2 or 3 min, a bulb was withdrawn, and its contents were analyzed for I_2 (KI_3) by means of a thiosulfate titration.

The equation becomes for this case:

$$t = \frac{1}{k(3 \cdot C_{A0} - C_{B0})} \ln \frac{C_{B0} \cdot C_A}{C_{A0}(3 \cdot C_A + C_{B0} - 3 \cdot C_{A0})} \text{ min.}$$

The plot of the right-hand side of this equation as a function of time is a straight line with $k = 0.299\ l \cdot mol^{-1} \cdot min^{-1}$. See Figure 27.1. In this figure, $x = C_{A0} - C_A$, $a = C_{A0}$, and $b = C_{B0}$. A and a refer to ethylene dibromide, whereas B and b refer to potassium iodide.

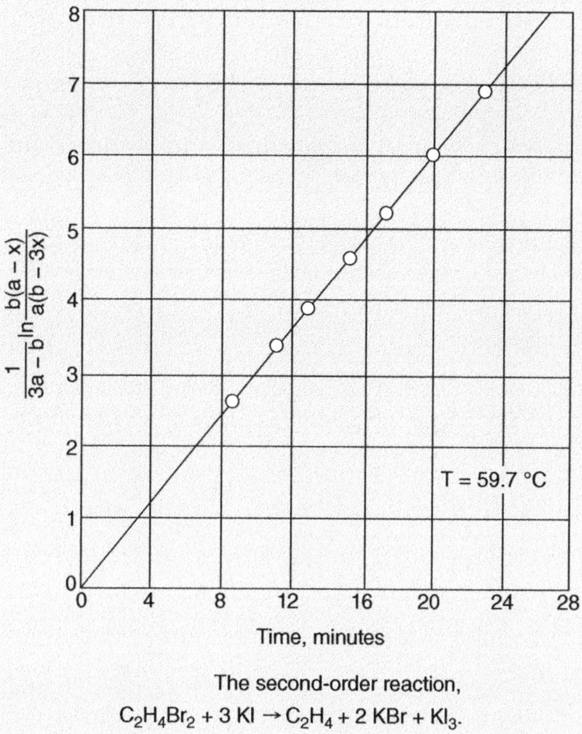

Figure 27.1 Plotting the kinetics of a second-order reaction with unequal initial concentrations of A and B. *Source*: Courtesy of the American Chemical Society, Washington DC.

References

1. Moore, W. J. (1963). *Physical Chemistry* (p. 261). London, UK: Longmans Green and Co Ltd.
2. Mee, A. J. (1956). *Physical Chemistry* (pp. 363–365). London, UK: Heinemann.
3. Mee, A. J. (1956). *Physical Chemistry* (pp. 366–368). London, UK: Heinemann.
4. Moore, W. J. (1963). *Physical Chemistry* (p. 262). London, UK: Longmans Green and Co Ltd.

28

The Rate of Chemical Reactions as a Function of Temperature

28.1 Arrhenius' Law

For many reactions, and particularly elementary reactions, the rate expression can be written as a product of a temperature-dependent term and a composition-dependent term, or:

$$r_i = f_1(\text{temperature}) \cdot f_2(\text{composition}) = k \cdot f_2(\text{composition}).$$

For such reactions, the temperature-dependent term, the reaction rate constant, has been found in practically all cases to be well represented by Arrhenius' law:

$$k = k_0 \cdot e^{-E/RT}. \tag{28.1}$$

In this equation, k_0 is called the frequency or pre-exponential factor, and E is called the activation energy of the reaction. This expression fits experiment well over wide temperature ranges and is strongly suggested from various standpoints as being a very good approximation to the true temperature dependency. At the same concentration, but at two different temperatures, Arrhenius' law indicates:

$$\ln \frac{r_2}{r_1} = \ln \frac{k_2}{k_1} = \frac{E}{R}\left(\frac{1}{T_1} - \frac{1}{T_2}\right). \tag{28.2}$$

The temperature dependency of reactions is determined by the activation energy and temperature level of the reaction, as indicated in Table 28.1. These findings are summarized as follows:

1) From Arrhenius' law, a plot of ln k versus 1/T gives a straight line, with large slope for large E and small slope for small E.

Introduction to Chemical Engineering, First Edition. C.M. van 't Land.

Table 28.1 Temperature rise (K) needed to double the rate of reaction for activation energies and average temperatures shown.

	Activation energy, kJ · mol^{-1}			
Average temperature, °C	**40**	**160**	**280**	**400**
0	11	2.7	1.5	1.1
400	65	16	9.3	6.5
1,000	233	58	33	23
2,000	744	185	106	74

2) Reactions with high activation energies are very temperature-sensitive, reactions with low-activation energies are relatively temperature-insensitive.
3) Any given reaction is much more temperature-sensitive at low temperature than at high temperature.
4) From Arrhenius' law, the value of the frequency factor k_0 does not affect the temperature sensitivity.

28.2 How to Influence Chemical Reaction Rates

A different way of writing Equation 28.1 is:

$$\ln k = \ln k_0 - \frac{E}{RT}.$$

Thus, a high reaction rate constant is obtained when k_0 is high. k_0, the frequency factor, is a measure of the efficiency of the collisions between reacting molecules. Chemists are studying the possibility to increase this efficiency by, for example, laser radiation [1].

A high reaction rate can also be obtained by a low activation energy. That can be accomplished by a catalyst. There are two broad classes of catalysts: those that operate at close to ambient temperature with biochemical systems, and the man-made catalysts that operate at high temperature. The biochemical catalysts are called enzymes. The man-made catalysts, mostly solids, usually aim to cause the high-temperature rupture or synthesis of materials. These reactions play an important role in many industrial processes, such as the production of methanol, sulfuric acid, and various petrochemicals, polymers, paints, and plastics. It is estimated that well over 50% of all chemical products produced today are made by using catalysts.

It is also possible to increase the reaction rate by increasing the temperature. However, if k_0 is large, or E is small, the temperature-dependency is modest.

Reference

1. Dijkgraaf, A. (2011). Playing billiard on an industrial scale. *Chemisch Weekblad*, 108 (5), 13. (in Dutch).

29

Chemical Reaction Engineering—A Quantitative Approach

29.1 Introduction

The three main reactor types are the batch reactor, the plug flow reactor (PFR), and the continuous stirred tank reactor (CSTR), and they will be discussed quantitatively in this chapter. For Sections 29.2–29.4, three basic assumptions will be made:

- irreversible reactions,
- isothermal operation, and
- liquid-phase reactions without change of specific mass or gas-phase reactions at which the number of moles reacting equals the number of moles produced.

Section 29.5 concerns reactor choice. Staging is discussed in Section 29.6. It implies giving a CSTR some characteristics of a PFR. Reversible reactions will be dealt with in Section 29.7. In that section, the reactions are also carried out isothermally. Furthermore, for liquid-phase reactions, the specific mass is constant, and for gas-phase reactions, the number of moles is constant.

29.2 Batch Reactor

First-Order Reaction

$$A \rightarrow B + C.$$

$$\frac{dC_A}{dt} = -k \cdot C_A.$$

From Section 27.2:

$$t = -\frac{1}{k} \ln \frac{C_A}{C_{A0}}.$$

The conversion is: $x_A = \dfrac{C_{A0} - C_A}{C_{A0}} = 1 - \dfrac{C_A}{C_{A0}}.$

Combining the latter two equations leads to:

$$t = -\frac{1}{k} \ln \left(1 - x_A\right) \text{s}.$$

Introduction to Chemical Engineering, First Edition. C.M. van 't Land.

Example 29.1

The reaction $A \rightarrow B + C$ has $k = 0.02\ s^{-1}$. The times required for 10, 50, 90, 99, and 99.9% conversion in a batch reactor are compared.

$$10\% : t = -\frac{1}{0.02} \ln (1-0.1) = 5.3\ s.$$

$$50\% : t = -\frac{1}{0.02} \ln (1-0.5) = 34.7\ s.$$

$$90\% : t = -\frac{1}{0.02} \ln (1-0.9) = 115\ s.$$

$$99\% : t = -\frac{1}{0.02} \ln (1-0.99) = 230\ s.$$

$$99.9\% : t = -\frac{1}{0.02} \ln (1-0.990) = 345\ s.$$

The reaction time increases as the required conversion increases.

Second-order reaction with equal initial concentrations of the reactants

$$A + B \rightarrow C + D.$$

$$\frac{dC_A}{dt} = -k \cdot C_A^2.$$

From Section 27.3:

$$t = \frac{1}{k}\left(\frac{1}{C_A} - \frac{1}{C_{A0}}\right) s.$$

The conversion is: $x_A = \dfrac{C_{A0} - C_A}{C_{A0}}$.

Combining the latter two equations results in:

$$t = \frac{x_A}{k \cdot C_A} s.$$

Example 29.2

The reaction $A + B \rightarrow C + D$ obeys second order kinetics with $k = 0.029\ l \cdot mol^{-1} \cdot s^{-1}$. The initial concentrations are $C_{A0} = C_{B0} = 1\ mol \cdot l^{-1}$. The times required for 10, 50, 90, 99, and 99.9% conversion in a batch reactor are compared.

$$10\% : t = \frac{0.1}{0.029 \cdot 0.9} = 3.8\ s.$$

$$50\% : t = \frac{0.5}{0.029 \cdot 0.5} = 34.5\ s.$$

$$90\% : t = \frac{0.9}{0.029 \cdot 0.1} = 310\ s.$$

$$99\% : t = \frac{0.99}{0.029 \cdot 0.01} = 3{,}414\ s.$$

$$99.9\% : t = \frac{0.999}{0.029 \cdot 10^{-3}} = 34{,}448\ s.$$

The time for 50% conversion is equal to the time for 50% conversion of Example 29.1. For conversions smaller than 50%, the second-order reaction is faster than the first-order reaction. The opposite is true for conversions greater than 50%. The background is that, for the second-order reaction, two different molecules have to come together: "It takes two to tango." In a diluted system, that is difficult. For a first-order reaction, a contact with a different entity is not required.

Note that, for a first-order reaction, it is, for the calculation of the time for a conversion, not necessary to know concentrations. C_{A0} and C_A are required for the calculation of the time for a conversion for a second-order reaction with equal initial concentrations of two reactants.

29.3 Plug Flow Reactor

A PFR has a certain length L. The residence time is the length of the reactor divided by the linear velocity of the fluid flowing through the PFR. This time is comparable to the batch time t of a batch reactor. The process in a PFR equals the process in a batch reactor, however, the time coordinate of a batch reactor is replaced by the length coordinate of a PFR. Having traveled a length L in a PFR means having resided for $L/v = t$ s in the PFR. Having resided for t s in a PFR is equivalent to having resided for t s in a batch reactor.

Example 29.3

The first-order reaction $A \rightarrow B + C$ occurs in a PFR with 90% conversion. If $k = 1\ min^{-1}$ and $\Phi_v = 8\ l \cdot min^{-1}$, what residence time and reactor volume will be required?

$$t = -\frac{1}{1} \ln(1 - x_A) = -\ln(1 - 0.9) = 2.3\ min.$$

$$V = \Phi_v \cdot t = 8 \cdot 2.3 = 18.4\ l.$$

How long would a 2-cm tube be and what would be the fluid velocity?

$$L = \frac{18.4 \cdot 10^{-3}}{(\pi/4)0.02^2} = 58.6\ m.$$

$$v = \frac{58.6}{2.3 \cdot 60} = 0.42\ m \cdot s^{-1}.$$

The flow of the fluid through the tube reactor should be turbulent.

Example 29.4

The second-order reaction $A + B \rightarrow C + D$ occurs in a PFR with 90% conversion. If $k = 1\ l \cdot mol^{-1} \cdot min^{-1}$, $C_{A0} = C_{B0} = 2\ mol \cdot l^{-1}$, and $\Phi_v = 8\ l \cdot min^{-1}$, what residence time and reactor volume will be required?

$$t = \frac{x_A}{k \cdot C_A} = \frac{0.9}{1 \cdot 0.2} = 4.5\ \text{min and } V = 4.5 \cdot 8 = 36\ l.$$

29.4 Continuous Stirred Tank Reactor (CSTR)

First-Order Reaction

$$A \rightarrow B + C.$$

On carrying out the reaction in a batch reactor or a plug flow reactor, the following law is applicable:

$$-\frac{dc_A}{dt} = k \cdot C_A.$$

This law describes the change of C_A as a function of time. However, in a CSTR, C_A is constant. It equals C_A in the flow leaving the reactor because the reactor is well mixed. C_A in the flow entering the reactor, C_{A0}, is greater than C_A in the reactor. Therefore, the aforementioned law is replaced by the following expression:

$$\frac{\Delta C_A}{\Delta t} = k \cdot C_A, \text{ and}$$

$$\Delta C_A = k \cdot C_A \cdot \Delta t.$$

In this expression, ΔC_A is the difference between C_{A0} and C_A and Δt is the residence time. On replacing Δt by t and on introducing x_A, the conversion, the following expression is obtained:

$$t = \frac{x_A}{k(1 - x_A)}\ s. \qquad (29.1)$$

It is recalled that $x_A = \frac{C_{A0} - C_A}{C_{A0}}$.

Example 29.5

The reaction $A \rightarrow B + C$ occurs in a CSTR with 90% conversion. If $k = 1\ min^{-1}$ and $\Phi_v = 8\ l \cdot min^{-1}$, what residence time and reactor volume would be required?

$$t = \frac{0.9}{1(1 - 0.9)} = 9\ \text{min.}$$

$$V = 9 \cdot 8 = 72\ l.$$

The same reaction was discussed in Examples 29.3 and 29.5. In Example 29.3, the reaction was carried out in a plug flow reactor. In Example 29.5, the reaction was carried out in a CSTR. The calculated reactor volume was 18.4 l in Example 29.3, and it was 72 l in Example 29.5. This is because when we introduce fresh feed into the CSTR, we immediately dilute it to the exit concentration, and we therefore see lower rates and hence longer residence times are required. In a PFR, the concentration of reactants gradually reduces as we travel along the length of the reactor, so, although the rate at the exit of the PFR is similar to that in the CSTR, the upstream rates are higher. In a batch reactor, at short times, we have higher rates and the rate gradually decreases as time passes.

Imagine having a reaction taking place in the liquid phase where the reactant is green and the product is colorless. The feed to the reactor will be green. In the case of a CSTR, the exit stream will be a light shade of green. This will be the same color as the color of the liquid within the reactor. In the case of the PFR, the exit stream will also be a light shade of green; however, the color of the stream will gradually change from inlet to outlet. For reactions of positive order, lighter shades of green would of course correspond to higher reaction rates. See Figure 29.1.

Second-order reaction A + B → C + D with equal initial concentrations of A and B

On carrying out the reaction in a PFR or in a batch reactor, the following equation applies:

$$-\frac{dC_A}{dt} = k \cdot C_A^2.$$

This equation can for a CSTR be replaced by:

$$t = \frac{C_{A0} - C_A}{k \cdot C_A^2} = \frac{C_{A0} \cdot x_A}{k \cdot C_A^2}. \tag{29.2}$$

The reasoning is similar to the reasoning for a first-order reaction in this section.

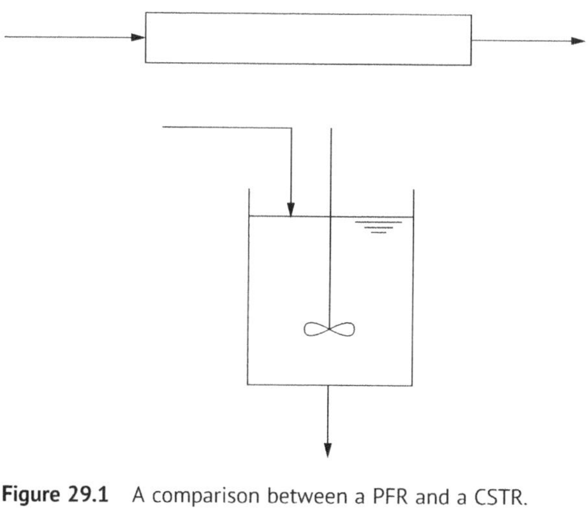

Figure 29.1 A comparison between a PFR and a CSTR.

Example 29.6

The second-order reaction $A + B \rightarrow C + D$ occurs in a CSTR with 90% conversion. If $k = 1\ l \cdot mol^{-1} \cdot min^{-1}$, $C_{A0} = C_{B0} = 2\ mol \cdot l^{-1}$, and $\Phi_v = 8\ l \cdot min^{-1}$, what residence time and reactor volume will be required?

$$t = \frac{2 \cdot 0.9}{1 \cdot 0.2^2} = 45\ min.$$

$$V = 8 \cdot 45 = 360\ l.$$

Note that this reactor is 10 times larger than the PFR of Example 29.4. The same reaction was discussed in Examples 29.4 and 29.6.

Example 29.7

The reaction $A \rightarrow B + C$ occurs in two CSTRs (both of the same size) in series with 90% conversion. If $k = 1\ min^{-1}$ and $\Phi_v = 8\ l \cdot min^{-1}$, what residence times and reactor volumes will be required? See Figure 29.2.

For the first reactor: $t = \frac{C_{A0} - C_{Ai}}{k \cdot C_{Ai}} = \frac{1}{k}\left(\frac{C_{A0}}{C_{Ai}} - 1\right).$

For the second reactor: $t = \frac{C_{Ai} - C_{Ae}}{k \cdot C_{Ae}} = \frac{1}{k}\left(\frac{C_{Ai}}{C_{Ae}} - 1\right).$

$\frac{C_{A0}}{C_{Ai}} = \frac{C_{Ai}}{C_{Ae}}$ because the two reactors have the same size.

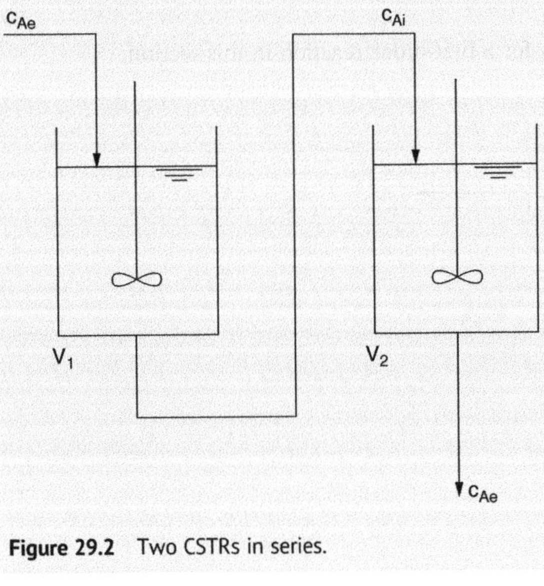

Figure 29.2 Two CSTRs in series.

Multiply both sides with $\frac{C_{A0}}{C_{Ai}}$ to obtain $\left(\frac{C_{A0}}{C_{Ai}}\right)^2 = \frac{C_{A0}}{C_{Ae}}$,

$\frac{C_{A0}}{C_{Ae}} = 10$, so $\frac{C_{A0}}{C_{Ai}} = \sqrt{10}$.

Going back to the equation for the first reactor, we now have:

$t = \frac{1}{1}\left(\sqrt{10} - 1\right) = 2.16$ min and $V_1 = 8 \cdot 2.16 = 17.3$ l.

$V_1 = V_2$ and the total reactor volume is $2 \cdot 17.3 = 34.6$ l.

In Examples 29.3, 29.5, and 29.7, the same chemical reaction was treated. The reaction systems were, respectively, a PFR, a CSTR, and two CSTRs in series. A survey of the required volumes:

PFR	Example 29.3	18.4 l
Two CSTRs	Example 29.7	34.6 l
CSTR	Example 29.5	72 l

For three CSTRs, the total volume will be between 34.6 and 18.4 l. For the same conversion, as the number of CSTRs increases, the total volume approaches that of a PFR (this true for all types of reactions of positive order, not just first order).

29.5 Reactor Choice

Batch reactors were compared to continuous reactors in Chapter 25. Plug flow reactors and continuous stirred tank reactors were treated quantitatively in this chapter. For reactions of positive order, a CSTR is always larger than a PFR. This would suggest that a PFR is always a better choice than a CSTR. However, that is not the case, and in Chapter 32, this aspect will be discussed.

29.6 Staging

Reactors with more than one impeller on a shaft exist. Sometimes, the reactor is even divided into compartments. That is called staging. It is possible to obtain higher conversions in such a reactor than in a CSTR.

Consider the reaction $A + B \rightarrow C$. The excess of B is considerable; thus, the reaction obeys first-order kinetics in A. It could concern the removal of an undesired impurity from a waste stream.

CSTR:

$$\frac{C_{Ae}}{C_{A0}} = \frac{1}{kt + 1}.$$

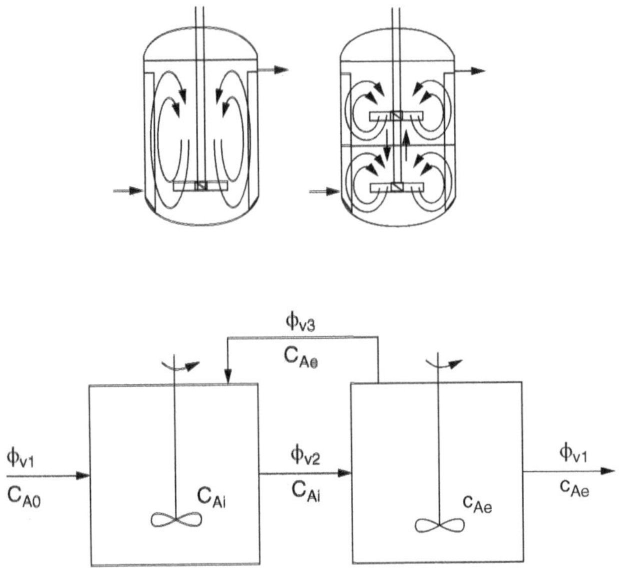

Figure 29.3 Two CSTRs in series with backmixing.

Two CSTRs in series:

$$\frac{C_{Ae}}{C_{A0}} = \frac{1}{(1 + kt)^2}.$$

Two CSTRs in series with backmixing:

$$\frac{C_{Ae}}{C_{A0}} = \frac{1}{1 + 2kt + (kt)^2/(1 + f)}. \tag{29.3}$$

f is the ratio between the backmixing from the second reactor to the first reactor and the feed: $f = \Phi_{v3}/\Phi_{v1}$. See Figure 29.3. The definition of t is $t = V/\Phi_{v1}$. V is the volume of any single reactor.

The expression for the two CSTRs with backmixing is derived as follows. The mass balance equation for the second reactor is:

$$\Phi_{v2} \cdot C_{Ai} - \Phi_{v3} \cdot C_{Ae} - \Phi_{v1} \cdot C_{Ae} = kV \cdot C_{Ae}.$$

According to the definition of f:

$$\Phi_{v3} = f \cdot \Phi_{v1} \text{ and } \Phi_{v2} = \Phi_{v1}(1 + f).$$

Substitutions lead to:

$$\frac{C_{Ae}}{C_{Ai}} = \frac{f + 1}{kt + f + 1}.$$

The mass balance equation for the first reactor is:

$$\Phi_{v1} \cdot C_{A0} + \Phi_{v3} \cdot C_{Ae} - \Phi_{v2} \cdot C_{Ai} = kV \cdot C_{Ai}.$$

Using the former expressions leads to:

$$\frac{C_{Ai}}{C_{A0}} = \frac{kt + 1 + f}{(kt + 1 + f)^2 - f(1 + f)}.$$

Equation 29.3 is obtained by a multiplication: $\frac{C_{Ae}}{C_{Ai}} \cdot \frac{C_{Ai}}{C_{A0}} = \frac{C_{Ae}}{C_{A0}}$.

Example 29.8

The removal of an undesired impurity in a waste stream obeys first-order kinetics and the rate constant is $k = 0.1\ \text{min}^{-1}$. $1\ \text{m}^3$ of effective reactor volume is available and the feed flow is $4\ \text{l} \cdot \text{min}^{-1}$.

CSTR ($1\ \text{m}^3$):

$$\frac{C_{Ae}}{C_{A0}} = \frac{1}{0.1 \cdot 250 + 1} = 0.038 \rightarrow x_A = 1 - 0.038 = 0.962.$$

Two CSTRs in series (each having a volume of $0.5\ \text{m}^3$):

$$\frac{C_{Ae}}{C_{A0}} = \frac{1}{(0.1 \cdot 125 + 1)^2} = 0.0055 \rightarrow x_A = 1 - 0.0055 = 0.9945.$$

Two CSTRs in series, f = 1 (each CSTR has a volume of $0.5\ \text{m}^3$):

$$\frac{C_{Ae}}{C_{A0}} = \frac{1}{1 + 2 \cdot 0.1 \cdot 125 + \frac{(0.1 \cdot 125)^2}{1 + 1}} = 0.0096 \rightarrow x_A = 1 - 0.0096 = 0.9904.$$

The conclusion is that staging is effective, even when there is backmixing.

29.7 Reversible Reactions

First Case

$$A \leftrightarrow B.$$

Both reactions are first order.

PFR

$$-\frac{dC_A}{dt} = k_f \cdot C_A - k_b \cdot C_B \text{ and } C_B = C_{Ao} - C_A.$$

If $C_A = C_{A0}$ and $C_B = 0$ at $t = 0$, we have

$$\int_{C_{A0}}^{C_A} \frac{dC_A}{k_f \cdot C_A - k_b(C_{A0} - C_A)} = -\int_0^t dt.$$

Integration leads to:

$$t = \frac{1}{k_f + k_b} \ln \frac{k_f \cdot C_{A0}}{k_f \cdot C_A - k_b(C_{A0} - C_A)}. \tag{29.4}$$

CSTR

$$\Phi_v(C_{A0} - C_A) = (k_f \cdot C_A - k_b \cdot C_B)V \rightarrow$$
$$t = \frac{V}{\Phi_v} = \frac{C_{A0} - C_A}{k_f \cdot C_A - k_b \cdot C_B}. \tag{29.5}$$

Example 29.9

For the reactions $A \leftrightarrow B, r = k_f \cdot C_A - k_b \cdot C_B$, the residence times for 50% conversion in a CSTR and in a PFR are calculated. $k_f = 0.6\ \text{min}^{-1}$, $k_b = 0.2\ \text{min}^{-1}$, $C_{A0} = 1\ \text{mol} \cdot \text{l}^{-1}$, and $C_{B0} = 0\ \text{mol} \cdot \text{l}^{-1}$.

$$\text{PFR} : t = \frac{1}{0.6 + 0.2} \ln \frac{0.6 \cdot 1}{0.6 \cdot 0.5 - 0.1 \cdot 0.5} = 1.09\ \text{min}.$$

$$\text{CSTR} : t = \frac{1 - 0.5}{0.6 \cdot 0.5 - 0.2 \cdot 0.5} = 2.5\ \text{min}.$$

Trying to calculate in Example 29.9, the required times in a PFR and in a CSTR for 90% conversion would fail. The reason is that we cannot go beyond the concentration of A at which the rate of the forward reaction equals the rate of the backward reaction. That concentration of A is called the equilibrium concentration. Although the rate constant of the backward reaction is smaller than the rate constant of the forward reaction, that effect is, at the equilibrium concentration, compensated by the fact that the concentration of B exceeds the concentration of A. The following expression is valid at the equilibrium concentration of A: $k_f \cdot C_A = k_b \cdot C_B$. Furthermore, $C_A + C_B = 1$. It follows that, at equilibrium, $C_A = 0.23\ \text{mol} \cdot \text{l}^{-1}$ and $C_B = 0.77\ \text{mol} \cdot \text{l}^{-1}$. The equilibrium conversion is 77%. This is the maximum conversion obtainable for these kinetics in any single reactor.

Second Case

$$A \leftrightarrow 2B.$$

The forward reaction is first order, the backward reaction is second order.

PFR

$$-\frac{dC_A}{dt} = k_f \cdot C_A - k_b \cdot C_B^2 \text{ and } C_B = 2(C_{A0} - C_A).$$

If $C_A = C_{A0}$ and $C_B = 0$ at $t = 0$, we have

$$\int_{C_{A0}}^{C_A} \frac{dC_A}{k_f \cdot C_A - 4k_b(C_{A0} - C_A)^2} = -\int_0^t dt.$$

The left-hand side of this equation can be integrated by splitting the integrand, a fraction, up into partial fractions. It can be written as:

$$\frac{1}{k_f \cdot C_A - 4k_b(C_{A0} - C_A)^2} = -\frac{1}{4k_b} \cdot \frac{1}{(C_A - C_{A1})(C_A - C_{A2})},$$
$$= -\frac{1}{4k_b} \cdot \left(\frac{D}{C_A - C_{A1}} + \frac{E}{C_A - C_{A2}}\right).$$

The technique of splitting up a fraction into partial fractions was discussed in Chapter 27. C_{A1} and C_{A2} are the roots of the quadratic equation:

$$\frac{k_f \cdot C_A - 4 \cdot k_b (C_{A0} - C_A)^2}{-4 \cdot k_b}.$$

$$D = \frac{1}{C_{A1} - C_{A2}} \text{ and } E = -D.$$

The result is $t = \left(\ln \frac{C_{A0} - C_{A1}}{C_A - C_{A1}} - \ln \frac{C_{A0} - C_{A2}}{C_A - C_{A2}} \right) \frac{D}{4 \cdot k_b}$ s.

This case illustrates that the solution of mass balance equations, when the kinetics are high order and reversible, is relatively complicated for a PFR.

CSTR

$$\Phi_v (C_{A0} - C_A) = V \left(k_f \cdot C_A - k_b \cdot C_B^2 \right) \rightarrow$$

$$t = \frac{V}{\Phi_v} = \frac{C_{A0} - C_A}{k_f \cdot C_A - k_b \cdot C_B^2} = \frac{C_{A0} - C_A}{k_f \cdot C_A - 4k_b (C_{A0} - C_A)^2} \text{ s.}$$

This case also illustrates that the solution of mass balance equations, when the kinetics are high order and reversible, is not complicated for a CSTR.

30

A Plant Modification: From Batchwise to Continuous Manufacture

30.1 Introduction

The modification concerns the manufacture of an intermediate for the production of polyurethane by Huntsman at Rotterdam in The Netherlands [1]. The incentive to study the change from batch to continuous was the need to extend the plant capacity. Two liquid raw materials, A and B, react with each other. Next, a catalyst is added and a second reaction takes place. The first reaction is more critical than the second one. The first reaction produces oligomeric molecules. The smallest molecule can schematically be indicated as ABA. That molecule is called di. A molecule, which can schematically be indicated as ABABA, is called tri. Further molecules are formed, like a molecule called tet. A competes with molecules already formed for the reaction with B. A great excess of A promotes the formation of di. The addition of a catalyst causes additional rearrangements.

30.2 Batchwise Production

The semi-batch reaction is preceded by heating the full amount of A to the reaction temperature. Then, B is gradually added while the heat of the reaction is carried away by circulating the reactor contents through an external loop with a heat exchanger. Next, a catalyst is added, and the mixture is stirred to accomplish the rearrangements of the molecules.

30.3 Continuous Manufacture

First, the post-stir was made continuous. After the dosing of the full amount of B, the reactor contents were passed through a PFR with the catalyst. The modification appeared successful and only half the amount of catalyst was required. Second, the main reaction was made continuous. A was pumped through a second PFR and B was dosed at various points to the PFR: multistage addition. See Figure 30.1. This design was chosen on the basis of laboratory work in a glass PFR. By this procedure, the same percentages by weight of di, tri, tet, and further molecules as at the semi-batch reaction could be achieved. The former batch reactor was used as a buffer between the two PFRs.

Introduction to Chemical Engineering, First Edition. C.M. van 't Land.

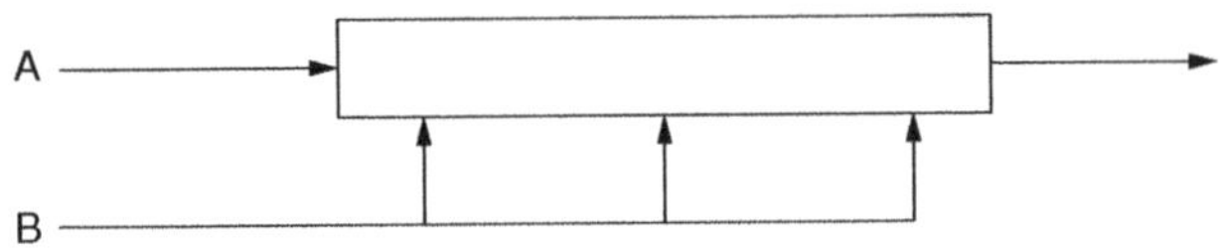

Figure 30.1 Multistage addition.

Reference

1. Dijkgraaf, A. (2004). Seamless from batch to continuous. *Chemisch Weekblad*, 100 (5), 20–21. (in Dutch).

31

Intrinsic Continuous Process Safeguarding

31.1 Summary

Two principle techniques exist for process safety in the chemical industry; these methods are based on the maintenance of either reaction conditions or chemical/physical stability in order to provide a safe working environment. From experience gained in the manufacture of organic peroxides, it is concluded that the latter technique (intrinsic continuous process safeguarding), aiming at prevention, is to be preferred. Intrinsic continuous process safeguarding is based on the stability of the reaction system and reaction conditions and minimizes the possibility of dangerous occurrences resulting from human error or equipment failure. Qualitative and quantitative hazard analysis is used to determine conditions for chemical/physical stability of a system. A flow calorimeter suitable for evaluation of thermochemical characteristics of the reaction system is mentioned. Process research, process development, and safety research should cooperate closely on process safety aspects from the start of the R&D synthetic work.

31.2 Introduction

Intrinsic continuous process safeguarding is part of the manufacturing technology for organic peroxides. It can also be applied to other areas. Organic peroxides are compounds containing a relatively weak –O–O– bond. This is the cause of their thermal instability. Organic peroxides can decompose at relatively low temperatures to form free radicals. The free radicals can act as initiators for polymerization reactions. A free radical combines with, for example, an alkene molecule. The combination reacts with a further alkene molecule, and so on: a polymerization reaction.

When organic peroxides degrade heat is developed. A fire or a thermal explosion results if this heat cannot be carried away. Incidents are known of peroxides decomposing violently during production, storage, or transport, and this has resulted in personnel and material damage. In retrospect, the most serious accidents occurred between 1935 and 1975. In this period, the production increased from 10-folds to hundreds of metric tpa. The majority of serious incidents occurred during the reactions to make the organic peroxides. Much data have been accumulated from these incidents, and by careful analysis, it has been possible to extract the important safety factors.

Introduction to Chemical Engineering, First Edition. C.M. van 't Land.

In the past, process safety has often been formulated after a study of product safety. However, direct process safety data are required.

31.3 The Production of Organic Peroxides

A typical process flow diagram for the production of a typical organic peroxide, a di-acyl peroxide, is shown in Figure 31.1. The reaction occurs batchwise. An acid chloride is added to an aqueous solution of NaOH and H_2O_2.

$$2\text{-}R\text{-}C(=O)\text{-}Cl + Na_2O_2 \rightarrow R\text{-}C(=O)\text{-}O\text{-}O\text{-}C(=O)\text{-}R + 2\ NaCl.$$

The reaction proceeds at ambient temperature in an open, stirred stainless steel vessel. The organic peroxide is obtained as a fine precipitate. The solid/liquid system is converted into a liquid/liquid system by the injection of live steam. The liquid organic peroxide is separated from the aqueous phase by means of a disk centrifuge. Finally, the product is solidified on a belt flaker.

The production of other organic peroxides proceeds similarly. The temperatures and pressures are low. Enameled reactors are also used. Catalysts are applied at a number of reactions. The example discussed is an alkaline process. There are also acid processes.

31.4 Intrinsically Safe Processes

Intrinsically safe processes are characterized by high chemical and physical stability, that is, stable raw materials/products and weakly exothermic reactions whose rates are only slightly affected by temperature variation [1]. While it is always desirable to use an intrinsically safe process, for chemical or economic reasons, it is not always possible for the manufacture of certain chemicals. This results in the use of a safeguarded process. Processes can be either intrinsically or extrinsically safeguarded.

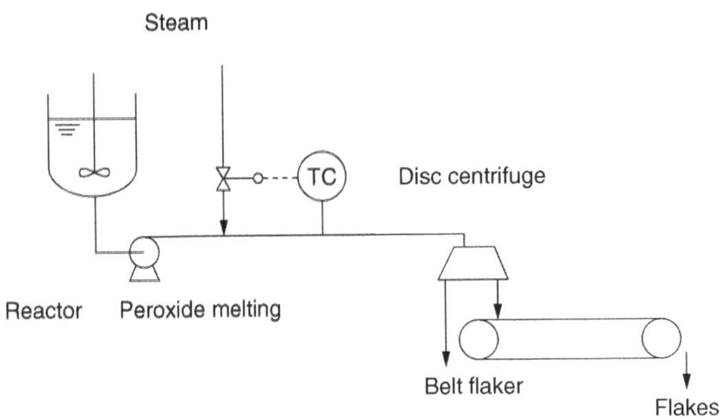

Figure 31.1 The manufacture of a di-acyl peroxide.

31.5 Intrinsic Process Safeguarding

This is a safeguarding originating from the core of the process and is consequently directly and completely based on the reaction system and the reaction conditions [2]. The safeguarding is continuously present or active; an activation is not required. Continuously functioning systems like pumps and stirrers can be part of the safeguarding. Two types of protection can be distinguished:

- the reaction conditions control the progress of the decomposition when the critical area has been entered, or
- the reaction conditions are sufficiently far from the critical area.

Sometimes, dangerous processes can be intrinsically protected in a simple way. An example is the use of strongly diluted systems; the temperature rise caused by the undesired reaction's heat development is fixed by the heat capacity of the system. See the description of the manufacture of a di-acyl peroxide in Section 31.3.

The air oxidation of hydrocarbons is a further example. The oxygen level of the off-gas controls the air supply in order to stay out of the vapor explosion area.

31.6 Extrinsic Process Safeguarding

Extrinsic process safeguarding operates by maintenance of safe process conditions by means of intervention by instrumentation or personnel. Hence, extrinsic process safeguarding is indirectly related to the reaction system. The safeguarding starts working upon a signal. Thus, the method comes down to symptom fighting. Compared to intrinsic process protection, significant drawbacks exist:

- Extrinsic safeguarding requires more than one stage to achieve safety—either human or instrumental, for example, the "sounding" of a signal and the subsequent reaction. Its functioning is dependent on the maintenance and checking of the equipment. Proper functioning is not continuously apparent.
- Extrinsic process protection often requires the provision of complicated process instrumentation and control. Redundancy is often practiced.
- Extrinsic process protection cannot protect if the reaction gets out of control rapidly.

For these reasons, extrinsic safeguarding is appropriate only as complementary safeguarding by providing protection in places through which entering the hazardous area is improbable and as secondary protection by drawing up a second line of defense behind the intrinsic protection line.

31.7 Additional Remarks

Research departments and process development departments operate along different lines. Much effort is required to find an integrated approach to process safeguarding.

Some resistances to be overcome are reviewed.

- There is too much confidence in extrinsic safeguarding—often advocated by chemical and mechanical engineers—and this attitude does not motivate chemists to design processes with built-in safety.

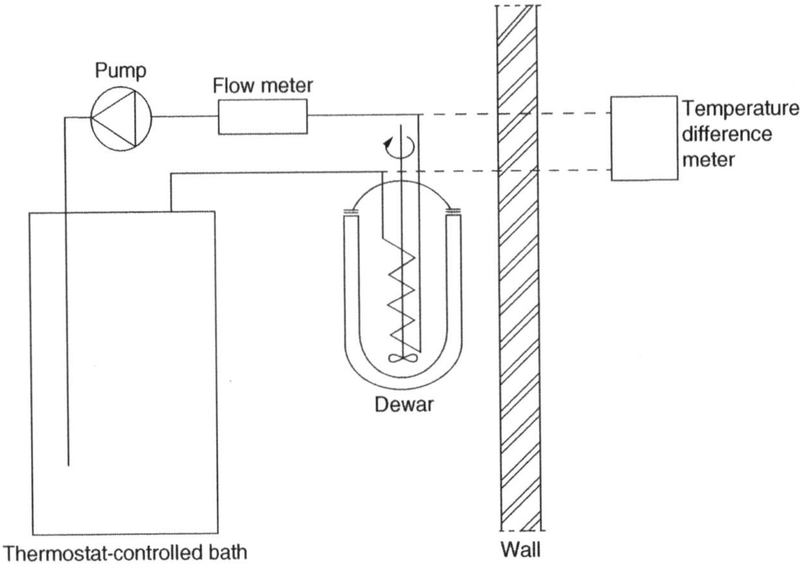

Figure 31.2 Flow calorimeter. *Source:* Courtesy of the American Chemical Society, Washington DC, USA.

- The view is sometimes held that a chemically unattainable ideal is being sought; however, even violent decompositions that occur during undesired reactions of peroxides can be controlled simply.
- Serious damage can occur to measuring equipment when the critical area is examined. This, unfortunately, can inhibit the use of the equipment. A suitable flow calorimeter system includes a simple replaceable measuring cell and is placed behind a barrier wall. See Figure 31.2.

In practice, the application of intrinsic continuous process safeguarding often appears to render economic advantages.

Figure 31.3 illustrates the line of thought.

31.8 Practical Approach

Qualitative Analysis of Synthetic Routes

Product and process research, process development, and safety research departments study the various possibilities to synthesize the compound. The isothermal and adiabatic heat development are considered. A provisional choice is made.

Quantitative Analysis of Synthetic Routes

The provisional choice is now being investigated quantitatively to verify its correctness. Flow calorimeter experiments are carried out in the critical area.

Realization of the Chemical/Physical Basis

Design is based on the previous point. Extrinsic safeguarding completes the stable chemical/physical basis.

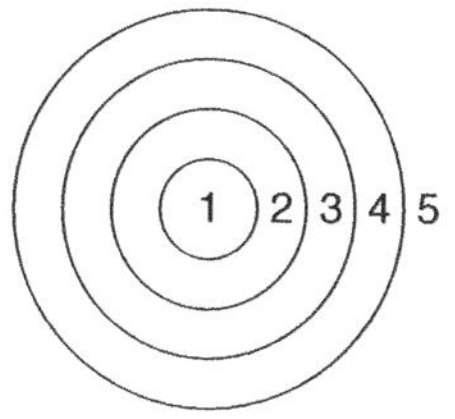

1. Process
2. Intrinsic process safeguarding
3. Extrinsic process protection and procedures
4. Barrier wall
5. Emergency procedures

Figure 31.3 Process protection.

31.9 Examples

The Synthesis of a Dialkyl Peroxide

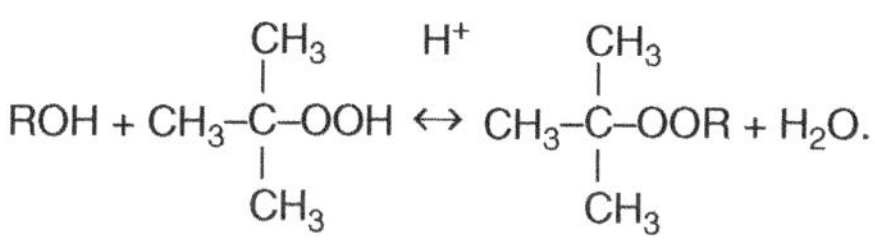

This equation shows the reaction between an alcohol and tertiary butyl hydroperoxide, catalyzed by a strong acid. The reaction is reversible, so the synthesis proceeds when water is removed. The possibilities are:

- water binding to a large quantity of moderately concentrated strong acid (for example, sulfuric acid),
- water binding to a drying salt (for example, calcium chloride or magnesium sulfate),
- water stripping with an inert gas, and
- steam distillation of toluene.

A high acid concentration favors the synthesis; however, too high an acid concentration can result in the initiation of a violent peroxide decomposition. Such a decomposition can occur either at the start or at the end of the reaction.

The decomposition reactions are:

$$R_1OOR_2 + H^+ \rightarrow R_1OH + R_2O^+.$$

$$R_2O^+ + H_2O \rightarrow H^+ + \text{decomposition products} + \text{heat}.$$

Process research was carried out concerning a synthesis in a closed reactor. The reaction water was removed by means of steam distillation. Such a process was preferred from an environmental point of view. Toluene was used for this distillation. The water production stops at the end. If the distillation is continued, the acid is concentrated and the decomposition starts. Three to four instrumental methods to detect the endpoint were tested. However, it still remained a possibility to pass the endpoint if the instruments would fail to take action.

Further process research revealed that the decomposition could be controlled by the steam distillation of toluene at 20–40 Torr and 25–30 °C. The temperature could be controlled by recycling the condensed toluene. In the decomposition stage, the toluene flow is larger than in the reaction stage. However, a large condenser can handle this toluene flow.

The Synthesis of a Peroxiketal

This example concerns the method of binding the reaction water by means of a large quantity of acid (for example, sulfuric acid 70% by weight). For chemical reasons, it is sometimes necessary to select a process in which the sulfuric acid and a reaction component are added, and the other reactant is subsequently dosed. Since the whole quantity of sulfuric acid is available from the start (and the acid concentration is high), the critical stage of this process is at the start of the reaction. In a process for the manufacture of a peroxiketal according to this method, it appeared from flow calorimeter experiments that decomposition started at a reaction temperature of 40 °C. Work at 50 °C and calculations based thereon showed that the heat development in the large-scale reactor could not be controlled.

To achieve a safe margin for the reaction temperature of the synthesis of this peroxide, whose decomposition can progress very exothermally in an acid medium, a temperature of 25 °C was prescribed. At this temperature, the influence of the sulfuric acid concentration was subsequently investigated, using the flow calorimeter. It was found that if double the amount of sulfuric acid was used, a spontaneous explosion occurred immediately at the start of the reaction without any warning. The reason is that the water also produced by the reaction could not dilute the acid to a safe level. On using less acid, the water produced first dilutes the acid satisfactorily. Extrinsic protection would not get the time to come into action. This dangerous situation could be avoided by reducing the concentration of the sulfuric acid in the process; the flow calorimeter indicated that under these circumstances, the reaction is kept out of the critical area at the overdose mentioned. The yield is not affected by this process modification.

Reactions with Epoxides

The large-scale manufacture of these compounds has in practice given rise to incidents, for example, at the reaction between epichlorhydrin and an N-substituted aniline [3].

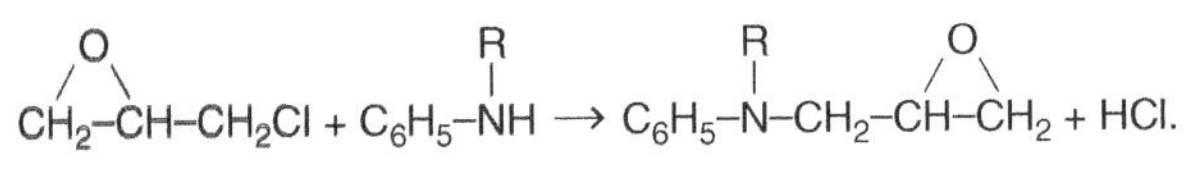

This reaction was carried out batchwise by adding the two reactants together and heating to 60 °C. At 60 °C, the reaction started and manually a switch from heating to cooling was made. One day, an operator made a mistake and the temperature rose to 70 °C. The reaction could not be kept under control at 70 °C. The temperature rose to 120 °C in 10 min and an explosion followed. By that time, the area had been evacuated. Personnel damage did not occur.

A different reaction system had to be designed. The outcome of process research was to first add the full amount of the N-substituted aniline to the reactor and to add the epichlorhydrin in portions next. Furthermore, the reaction temperature was raised. Each portion reacted immediately, the reactor temperature rose, and it was reduced by subsequent cooling. This proved to be a safe procedure. In addition, the reactor capacity increased because of the higher reaction rate.

References

1. Kletz, T. A. (1978). Inherently safer design: The growth of an idea. *Process Safety Progress*, 15 (1), 5–8.
2. Gerritsen, H. G., and van 't Land, C. M. (1985). Intrinsic continuous process safeguarding. *Industrial & Engineering Chemistry Process Design and Development*, 24, 893–896.
3. Schierwater, F. -W. (1971). The safe operation of exothermic reactions, especially in the liquid phase. *Industrial Chemical Engineering Symposium Series*, 34, 47–49.

32

Reactor Choice and Scale Up

32.1 Introduction

First, the contents of Chapter 25, Chemical Reactors, are reviewed shortly. Kinetic rate data and transport phenomena are compared. Physical factors related to mass transfer can affect the progress of a reaction even more than chemical affinity. The main reactor types are the batch reactor, the PFR, and the CSTR. Batch reactions are compared to continuous reactions.

In Chapter 27, the concept of the order of a reaction was introduced.

Methods to calculate reactor sizes are developed in Chapter 29. For irreversible reactions of positive order, for a given conversion, a PFR is always smaller than a CSTR. It is explained in Section 32.2 that sometimes a CSTR is to be preferred because of yield reasons. That section treats parallel reactions.

Section 32.3 reviews physical effects playing a role at chemical reactions.

32.2 Parallel Reactions

Let us first review the situation with different kinetics for two irreversible parallel reactions such as:

$$A \rightarrow B, r_1 = k_1 \cdot C_A^{m1}.$$

$$A \rightarrow C, r_2 = k_2 \cdot C_A^{m2}.$$

For the PFR or batch reactor the equation for A is:

$$-\frac{dC_A}{dt} = k_1 \cdot C_A^{m1} + k_2 \cdot C_A^{m2}.$$

A simple example with $m_1 = 1$ and $m_2 = 2$ is considered.

For these reactions in a PFR, the mass balance equation becomes:

$$t = -\int_{C_{A0}}^{C_A} \frac{dC_A}{k_1 \cdot C_A + k_2 \cdot C_A^2},$$

which equation must be solved by partial fractions or numerically.

For the CSTR, the equation is:

$C_{A0} - C_A = t\left(k_1 \cdot C_A^{m1} + k_2 \cdot C_A^{m2}\right)$, which is a polynomial in C_A.

Introduction to Chemical Engineering, First Edition. C.M. van 't Land.

When, also in this case, $m_1 = 1$ and $m_2 = 2$, the solution for the CSTR is a quadratic equation:

$$C_{A0} - C_A = t(k_1 \cdot C_A + k_2 \cdot C_A^2) \rightarrow t = \frac{C_{A0} - C_A}{k_1 \cdot C_A + k_2 \cdot C_A^2}.$$

The following expression can also be derived directly:

$$C_A = \frac{-(1 + k_1 \cdot t) \pm \sqrt{(1 + k_1 \cdot t)^2 + 4 \cdot k_2 \cdot t \cdot C_{A0}}}{2 \cdot k_2 \cdot t}.$$

Example 32.1

The selectivities of forming B in a CSTR and in a PFR at 90% conversion of A are calculated. The initial concentration of A is 8 mol · l^{-1}. There are two reactions by which A is converted:

$$A \rightarrow B, r_1 = 4.$$
$$A \rightarrow C, r_2 = k_2 \cdot C_A = C_A.$$

The rates are in mol · l^{-1} · min^{-1}. The first reaction that produces B is zero-th order in C_A and is the desired one, whereas the reaction which produces C is first order in C_A and is undesired.

Carrying out the two reactions in a PFR is discussed first. The mass balance equation for A is:

$$-\frac{dC_A}{dt} = r_1 + k_2 \cdot C_A = 4 + C_A,$$

which equation can be integrated to yield:

$$t = -\int_{C_{A0}}^{C_A} \frac{dC_A}{4 + C_A} = \ln\frac{4 + C_{A0}}{4 + C_A} = \ln\frac{4 + 8}{4 + 0.8} = 0.92 \text{ min}.$$

The mass balance equation for B is:

$$\frac{dC_B}{dt} = 4 \rightarrow C_B = 4t = 3.68 \text{ mol} \cdot l^{-1}.$$

whereas 90% of A has been converted, that is, 7.2 mol · l^{-1}, a concentration of B of 3.68 mol · l^{-1} has been obtained. Thus, the selectivity is 3.68/7.2 = 0.51. By definition, the yield is the product of conversion and selectivity: 0.9 · 0.51 = 0.46.

Next, carrying out the reaction in a CSTR is treated. The mass balance equation is:

$$C_{A0} - C_A = t(4 + C_A),$$
$$t = \frac{C_{A0} - C_A}{4 + C_A} = \frac{8 - 0.8}{4 + 0.8} = 1.5 \text{ min}.$$
$$C_B = 4t = 6 \text{ mol} \cdot l^{-1}.$$

Now, when 90% of A has been converted, that is, 7.2 mol · l^{-1}, a concentration of B of 6 mol · l^{-1} has been obtained. So, the selectivity is 6/7.2 = 0.83, and the yield is 0.9 · 0.83 = 0.75.

The CSTR requires in this example a longer residence time than the PFR. That makes sense as the kinetics are positive order. Furthermore, the CSTR gives a higher selectivity to B. It is possible to work in practice with a relatively large reactor to avoid continual losses of raw material as one would have in a PFR. The heart of the matter in this example is that in a CSTR, a low-order reaction is favored rather than a high-order reaction.

32.3 Physical Effects

In many instances, physical effects play a role on carrying out chemical reactions. If that is the case and the reaction is carried out successfully on a small scale, a large reactor having the same geometrical proportions as the small reactor can be designed. Often, it is satisfactory to maintain on stirring the power input in, for example, $kW \cdot m^{-3}$. This aspect is discussed in Part II, Mixing and Stirring. For the suspension of solid particles and to attain a liquid/liquid dispersion that usually appears to be good practice. To distribute gases, it is common practice to also use this criterion. However, the quality of the gas distribution decreases. Concerning heat transfer, an important aspect is that the area of a jacket in square meter per cubic meter reactor content is proportional to 1/D. That means that this quotient decreases by a factor of 10 when the vessel volume increases with a factor of 1,000. When the mixing time is important, as can be the case when the reaction is or the reactions are fast, it is not enough to maintain the power input.

33

Case Studies Chemical Reaction Engineering

33.1 Order of a Reaction

First Application

The following data have been obtained for the reaction between acetic acid anhydride and water [1].

$C_{A0} = 0.1160$

t, s	87	417	544	852
$C_{A0} - C_A$	0.02378	0.07748	0.08848	0.10360

The reaction equation is:

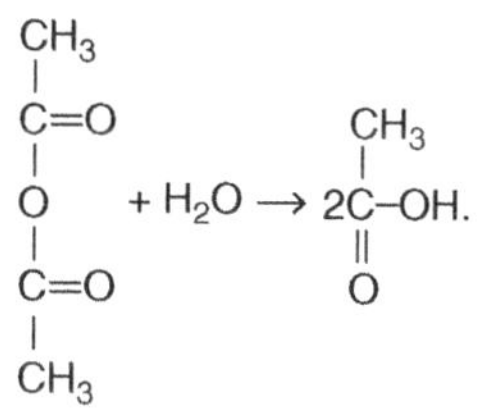

C_{A0} is the concentration of acetic acid anhydride present initially in $mol \cdot l^{-1}$. Find the order of the reaction, and whether it agrees with the chemical equation. If not, attempt an explanation.

Second Application

The order of the reaction between methyl iodide and pyridine was investigated. The reaction equation is:

$$CH_3I + C_5H_5N \rightarrow C_5H_5NCH_3.$$

C_{A0} and C_{B0} are the initial concentrations in $mol \cdot l^{-1}$ of methyl iodide and pyridine respectively.

$C_{A0} = 0.03.$

$C_{B0} = 0.06.$

x is the concentration of the quaternary iodide $C_5H_5NCH_3I$ formed in $mol \cdot l^{-1}$. The reaction was carried out 60 °C. The results are:

t, min	17.3	31.4	53.7	97.1
x	0.01	0.015	0.02	0.025

Introduction to Chemical Engineering, First Edition. C.M. van 't Land.

Check whether the reaction follows the bimolecular law and calculate the velocity constant.

Complete the following listing for the case that both C_{A0} and C_{B0} are 0.03 mol · l^{-1}. The temperature remains 60 °C.

t, min	17.3	31.4	53.7	97.1
X				

Explain the difference between the two listings.

Solution

First Application

The measured results obey the unimolecular law:

$$t = -\frac{1}{k}\ln\frac{C_A}{C_{A0}}\ s.$$

This is Equation 27.1 and $k = 2.6 \cdot 10^{-3}\ s^{-1}$.

The active mass of the water does not change appreciably in dilute solutions, and so the unimolecular law is obeyed.

Second Application

The reaction is bimolecular and the following expression can be used:

$$k = \frac{1}{t(C_{A0} - C_{B0})}\ln\frac{C_{B0} \cdot C_A}{C_{A0}(C_A + C_{B0} - C_{A0})}\ l \cdot mol^{-1} \cdot min^{-1}.$$

This is Equation 27.3.

$$k = 0.43\ l \cdot mol^{-1} \cdot min^{-1}.$$

The listing for the case that both C_{A0} and C_{B0} are 0.03 mol · l^{-1} is as follows:

t, min	17.3	31.4	53.7	97.1
x	$5.5 \cdot 10^{-3}$	$8.6 \cdot 10^{-3}$	0.0123	0.0167

These results can be obtained by applying the following expression:

$$t = \frac{1}{k}\left(\frac{1}{C_A} - \frac{1}{C_{A0}}\right)\ min.$$

This is Equation 27.2. On applying the latter equation, the rate constant k obtained from the first series of results can be used. That follows from the fact that the temperature remains 60 °C. At each time, the numbers are smaller than the numbers in the first listing. The reason is that the number of collisions per unit of time between the reacting molecules is, in the second case, smaller than in the first case. Therefore, the reaction makes more progress in the first case than in the second case. Note that not all collisions lead to a reaction.

33.2 Chemical Reaction Rate as a Function of Temperature

First Application

The pyrolysis (thermal decomposition in the absence of oxygen) of ethane proceeds with an activation energy of about 300 kJ · mol^{-1}. How much faster is the decomposition at 650 °C than at 500 °C [2]?

Second Application

At 1,100 K, n-nonane cracks (breaks down into smaller molecules) 20 times faster than at 1,000 K [2]. Calculate the activation energy for this decomposition.

Solution

First Application

Equation 28.2 can be used.

$$\ln \frac{r_1}{r_2} = \frac{3 \cdot 10^5}{8,314}\left(\frac{1}{773} - \frac{1}{923}\right) = 7.586,$$

$$\frac{r_1}{r_2} = 1{,}970.$$

Second Application

Equation 28.2 can also be used in this case.

$$\ln 20 = \frac{E}{R}\left(\frac{1}{1{,}000} - \frac{1}{1{,}100}\right),$$

$$E = 274\ \text{kJ} \cdot \text{mol}^{-1}.$$

33.3 Reactor Size

A second-order liquid-phase reaction is to be carried out in a PFR. The reaction equation is: A + B ⟶ Products, with

$$r_A = k \cdot C_A \cdot C_B \text{ and } k = 10^{-6}\ \text{m}^3 \cdot \text{s}^{-1} \cdot \text{mol}^{-1}.$$

The inlet concentration of both A and B is 10 mol · m^{-3} and the flow rate is 10^{-5} m^3 · s^{-1}. Calculate the reactor volume required for 50% conversion.

Solution

The information for this case can be found in Section 29.2.

$$t = \frac{x_A}{k \cdot C_A}, \Phi_v \cdot t = V, V = \Phi_v \cdot \frac{x_A}{k \cdot C_A} = 10^{-5} \cdot \frac{0.5}{10^{-6} \cdot 0.5 \cdot 10} = 1\ \text{m}^3.$$

33.4 Reversible Reactions

The first-order reversible reaction $A \leftrightarrow B$ is conducted in an isothermal PFR. The rate constants are 0.06 and 0.04 s^{-1} for the forward and backward reaction, respectively. The feed contains A, does not contain B, and the initial concentration of A in the feed, C_{A0}, is 1 $mol \cdot m^{-3}$. The feed flow is 0.01 $m^3 \cdot s^{-1}$. See Figure 33.1.

a) Calculate the reactor volume for 50% conversion.
b) Calculate the reactor volume for 50% conversion when half of the flow leaving the reactor is recycled and mixed with the feed. This case simulates the case of a PFR with backmixing, an undesired phenomenon in PFRs. Note: the flow through the reactor is now 0.02 $m^3 \cdot s^{-1}$. The equation for a PFR with the aforementioned recycle is:

$$t = \frac{1}{k_f + k_b} \ln \frac{(C_A + C_{A0})(k_f + k_b) - 2 \cdot k_b \cdot C_{A0}}{2 \cdot C_A(k_f + k_b) - 2 \cdot k_b \cdot C_{A0}} \ s.$$

This formula can be obtained by integrating an expression for a PFR in Section 29.7 of Chapter 29, however, with different limits for the left-hand side. The limits for this case are $(C_{A0} + C_A)/2$ and C_A, respectively.

c) If the recycle is increased substantially, the reactor system becomes a CSTR. Calculate the reactor volume for 50% conversion in a CSTR.
d) Why, in going from (a) to (c) through (b), does the reactor volume increase?
e) What is the limiting conversion in any reactor type? See the text in Section 29.7 of Chapter 29.

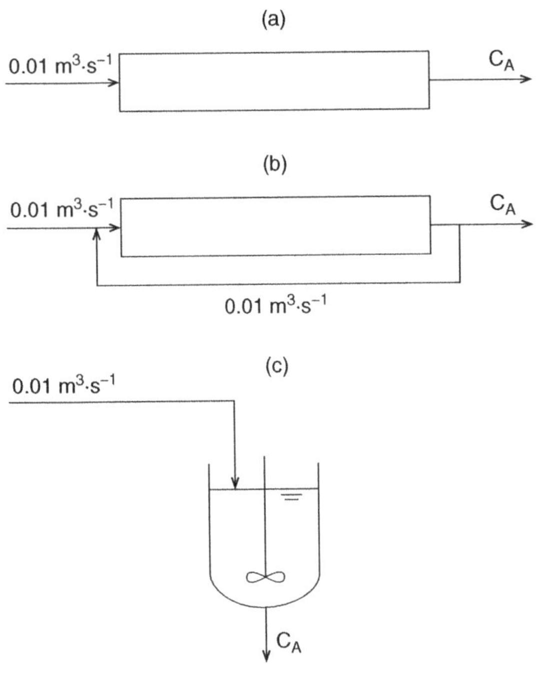

Figure 33.1 Various schemes for a reversible reaction.

Solution

a) The following equation can be applied:

$$t = \frac{1}{k_f + k_b} \ln \frac{k_f \cdot C_{A0}}{k_f \cdot C_A - k_b(C_{A0} - C_A)} \text{ s},$$

This is Equation 29.4.

$$t = \frac{1}{0.06 + 0.04} \ln \frac{0.06 \cdot 1}{0.06 \cdot 0.5 - 0.04(1 - 0.5)} = 17.9 \text{ s},$$

$$t = \frac{V}{\Phi_v}, V = t \cdot \Phi_v = 17.9 \cdot 0.01 = 0.179 \text{ m}^3.$$

b) The equation is:

$$t = \frac{1}{k_f + k_b} \ln \frac{(C_A + C_{A0})(k_f + k_b) - 2 \cdot k_b \cdot C_{A0}}{2 \cdot C_A(k_f + k_b) - 2 \cdot k_b \cdot C_{A0}} \text{ s},$$

$$t = \frac{1}{0.06 + 0.04} \ln \frac{(0.5 + 1)(0.06 + 0.04) - 2 \cdot 0.04 \cdot 1}{2 \cdot 0.5(0.06 + 0.04) - 2 \cdot 0.04 \cdot 1} = 12.5 \text{ s},$$

$$t = \frac{V}{\Phi_v}, V = t \cdot \Phi_v = 12.5 \cdot 0.02 = 0.250 \text{ m}^3.$$

c) The equation is:

$$t = \frac{C_{A0} - C_A}{k_f \cdot C_A - k_b(C_{A0} - C_A)} \text{ s}.$$

This is Equation 29.5.

$$t = \frac{1 - 0.5}{0.06 \cdot 0.5 - 0.04(1 - 0.5)} = 50 \text{ s},$$

$$t = \frac{V}{\Phi_v}, V = t \cdot \Phi_v = 50 \cdot 0.01 = 0.5 \text{ m}^3.$$

d) For chemical reactions of positive order, a CSTR is, for a certain conversion, always larger than a PFR. Case (a) concerns a PFR, case (c) concerns a CSTR, and case (b) is in between.

e) $k_f \cdot C_{A,eq} = k_b \cdot C_{B,eq} = k_b(1 - C_{A,eq})$,
$k_f \cdot C_{A,eq} = k_b - k_b \cdot C_{A,eq}$,

$$C_{A,eq} = \frac{k_b}{k_f + k_b} = \frac{0.04}{0.06 + 0.04} = 0.4.$$

Thus, the equilibrium conversion is 60%. This is as far as we can go in any reactor type and size.

33.5 Competing Reactions

Consider a reaction forming P, which competes with a wasteful one:

$$A + B \rightarrow P.$$

$$B + B \rightarrow X.$$

The selectivity calls for a high concentration of A simultaneously with a low level of B, that is, plug flow for A and perfect mixing for B. Three different schemes are depicted in Figure 33.2. Which is the best scheme to handle the combination of reactions?

Solution

The best scheme to handle the combination is a series of CSTRs, with A flowing from tank to tank and B dosed to each tank, that is, B is distributed to achieve its rapid dilution, which results in less dimerization.

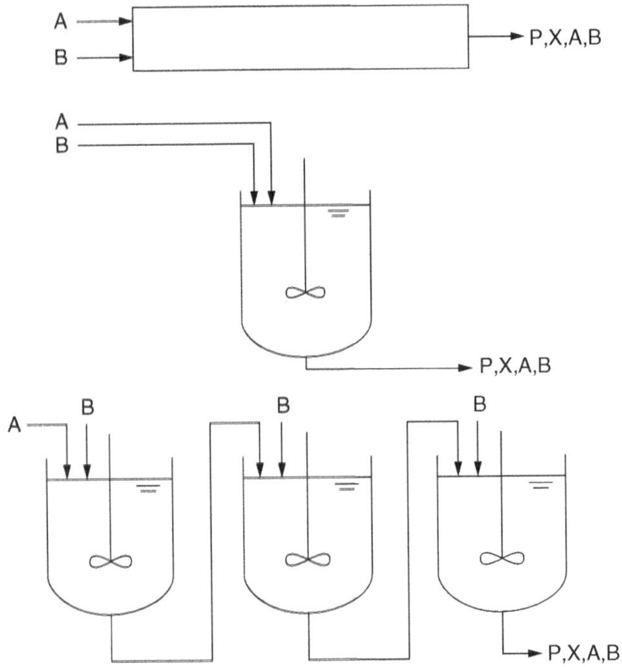

Figure 33.2 Process schemes for competing reactions.

33.6 The Hydrolysis of Acetic Acid Anhydride

An Akzo Nobel plant wants to remove an underground storage vessel, which was used for the storage of acetic acid anhydride in the past (A. M. van den Braken, private communication, 2000). Inspection of the vessel reveals that 2 t of the chemical is still present. The plant has permission to drain acetic acid via a biological treatment unit. Acetic acid anhydride can be converted into acetic acid by hydrolysis. The reaction heat is known: $-57.8\ \text{kJ} \cdot \text{mol}^{-1}$. The equation for the chemical reaction can be found in Section 33.1.

Acetic acid anhydride hydrolysis is known as a dangerous reaction [3, 4]. Initially, the reaction is slow because water and acetic acid anhydride are immiscible liquids. In the next stage, the reaction is accelerated because water, acetic acid anhydride, and acetic acid form one phase.

What is the maximum adiabatic temperature rise on hydrolyzing acetic acid anhydride to acetic acid 70% by weight? How can 2 t of acetic acid anhydride be hydrolyzed safely in a reactor to acetic acid anhydride 70% by weight? Table 33.1 contains physical properties.

Solution

The first objective is to calculate ΔT, the adiabatic temperature rise in K.
ΔT appears in the expression: $Q = M \cdot c_p \cdot \Delta T$ J. Q is the total amount of heat developed by the conversion in J, M is the mass involved in kg, and c_p is the specific heat of the final material in $J \cdot kg^{-1} \cdot K^{-1}$. If Q, M, and c_p are known, ΔT can be calculated.

$$Q = \frac{2{,}000}{102} \cdot 10^3 \cdot 57.8 \cdot 10^3 = 1.133 \cdot 10^9\ \text{J}.$$

$$M = \frac{120}{102} \cdot 2{,}000 \cdot \frac{1}{0.7} = 3{,}361\ \text{kg}.$$

$$c_p = 0.7 \cdot 2{,}060 + 0.3 \cdot 4{,}190 = 2{,}699\ \text{J} \cdot \text{kg}^{-1} \cdot \text{K}^{-1}.$$

$$\Delta T = \frac{Q}{M \cdot c_p} = \frac{1.133 \cdot 10^9}{3{,}361 \cdot 2{,}699} = 124.9\ \text{K}.$$

The second objective is to design a method to carry out the hydrolysis safely. The hydrolysis is carried out batchwise; 1,361 kg of water is added to a 5-m^3 reactor. This amount of water is meant for the hydrolysis and the dilution to a 70% by weight aqueous solution of acetic acid. The water temperature is raised to 40 °C. A catalyst is added next. Then, the anhydride is added in 100-kg amounts. Per 100 kg of anhydride, the temperature of the reactor contents rises 5–10 K. After each amount of 100 kg, the temperature is lowered to 40 °C.

Table 33.1 Physical properties for the hydrolysis of acetic acid anhydride.

Property	Acetic acid	Acetic acid anhydride	Water
Molecular weight, $g \cdot mol^{-1}$	60	102	18
c_p, $J \cdot kg^{-1} \cdot K^{-1}$	2,060	1,850	4,190

33.7 Cumene Air Oxidation

Cumene is oxidized batchwise in a stirred reactor at atmospheric pressure and 112 °C by means of air in the presence of NaOH 50% by weight to yield phenyl-i-propanol. Cumene, NaOH-50, and a catalyst are mixed and heated to 112 °C. Then, air distribution is started. The reaction time is 8 h. The question is raised whether it is possible to shorten the reaction time by reacting at 130 °C. Table 33.2 contains vapor pressure data of cumene and NaOH-50 at a few relevant temperatures. It is requested to answer the question.

Table 33.2 Vapor pressures in mm Hg for cumene air oxidation.

Material	110 °C	120 °C	130 °C
Cumene	200	300	400
NaOH-50	180	290	400

Solution

Saturated vapor pressures in mm Hg at 130 °C are given.

Cumene	400
NaOH-50	400+
	800

The sum of these saturated vapor pressures exceeds 760 mm Hg, the atmospheric pressure. So, the mixture boils. Raising the temperature to 130 °C is not a good idea.

References

1. Mee, A. J. (1960). *Physical Chemistry* (p. 698). London, UK: Heinemann.
2. Levenspiel, O. (1999). *Chemical Reaction Engineering* (p. 34). Boston, MA: Wiley.
3. Benson, G. (1947). Perchloric acid cleanings. *Chemical and Engineering News*, 25, 3458.
4. Leigh, W. R. D., and Krzeminski, Z. S. (1962). The uncatalysed reaction of acetic acid anhydride with water. *Chemistry and Industry*, 26, 778–779.

Notation III

Concentrations are in mole · l^{-1}, mol · m^{-3}, or kmol · m^{-3}.

$C_{A,B,N}$	Concentrations of components A, B, and N	
C_{A0}	Initial concentration of A	
$C_{A1,2}$	Roots of a quadratic equation in which C_A appears	
C_{Ae}	Final concentration of A	
$C_{A,eq}$	Equilibrium concentration of A	
C_{Ai}	Intermediate concentration of A	
C_{B0}	Initial concentration of B	
$C_{B,eq}$	Equilibrium concentration of B	
c_p	Specific heat	$J \cdot kg^{-1} \cdot K^{-1}$
E	Activation energy	$kJ \cdot mol^{-1}$
f	Backmixing ratio	—
ΔH_R	Reaction heat	$kJ \cdot mol^{-1}$

Reaction rate constants are in s^{-1} or min^{-1} for first-order reactions. They are in $l \cdot mol^{-1} \cdot s^{-1}$, $l \cdot mmol^{-1} \cdot s^{-1}$, or $l \cdot mol^{-1} \cdot min^{-1}$ for second-order reactions.

k	Reaction rate constant	
k_0	Frequency or pre-exponential factor	
$k_{1,2}$	Reaction rate constants Nos. 1 and 2	
k_b	Reaction rate constant of a backward reaction	
k_f	Reaction rate constant of a forward reaction	
L	Plug flow reactor length	m
M	Mass	kg
n	Number of moles	—
Q	Amount of heat	J
R	Universal gas constant (8.314 $J \cdot mol^{-1} \cdot K^{-1}$)	
r_0	Initial rotation of polarized light	—
$r_{1,2}$	Reaction rates Nos. 1 and 2	$mol \cdot s^{-1}$
r_A	Reaction rate of component A	$mol \cdot m^{-3} \cdot s^{-1}$
r_t	Rotation of polarized light at time t	—
r_∞	Rotation of polarized light at $t = \infty$	—
T	Temperature	°C or K

Introduction to Chemical Engineering, First Edition. C.M. van 't Land.

$T_{1,2}$	Temperatures Nos. 1 and 2	°C or K
ΔT_{ad}	Adiabatic temperature rise	K
t	Time	s or min
V	Reactor volume	l
$V_{1,2}$	Reactor volumes Nos. 1 and 2	l
v	Velocity	$m \cdot s^{-1}$
x	Concentration	$mol \cdot l^{-1}$
x_A	Conversion of A	—

Greek Symbols

Φ_v	Flow	$l \cdot min^{-1}$
$\Phi_{v1,2,3}$	Flows Nos. 1, 2, and 3	$l \cdot min^{-1}$

Part IV

Distillation

Part IV Content

34
Continuous Distillation

34.1 Introduction

Distillation is the most widely used separation process. Liquid mixtures can be separated into their several components on the basis of their different vapor pressures. Distillation is the key operation of the oil refinery. The columns for this separation process have to be designed with a larger range in capacity than any other types of process equipment, with single columns from 0.3 to 10 m in diameter and from 3 m upward to 75 m in height.

34.2 Vapor–Liquid Equilibrium

A temperature-composition diagram for a binary system shows the composition of the vapor in equilibrium with a liquid of given composition. For example, such a diagram for the system benzene–toluene at atmospheric pressure can be seen in the upper part of Figure 34.1. The boiling point of benzene is 80 °C whereas the boiling point of toluene is 110 °C.

Generally, temperature-composition diagrams are for constant pressure conditions. The majority of industrial distillations take place at substantially constant pressure.

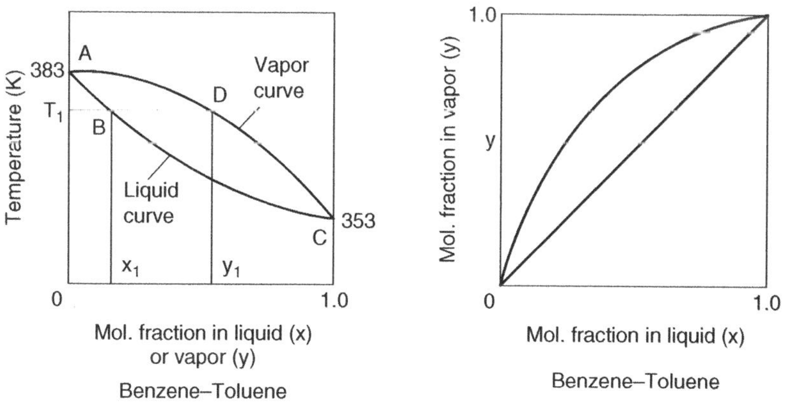

Figure 34.1 Temperature-composition diagram and x/y-diagram for the system benzene–toluene. *Source:* Courtesy of Elsevier, Amsterdam, The Netherlands.

Introduction to Chemical Engineering, First Edition. C.M. van 't Land.

In Figure 34.1, the curve ABC in the temperature-composition diagram shows the composition of the liquid that boils at any given temperature. It is called the boiling point curve. The curve ADC shows the corresponding composition of the vapor at that temperature. It is called the dew point curve. Thus, a liquid of composition x_1 will boil at temperature T_1, and the vapor in equilibrium is indicated by point D of composition y_1. It is seen that for any liquid composition x, the vapor formed will be richer in the more volatile component. x is the mole fraction of the more volatile component in the liquid, and y in the vapor. Examples of mixtures giving these types of curves are benzene–toluene, n-heptane–toluene, and carbon disulfide–carbon tetrachloride.

Figure 34.1 also shows the x/y-diagram for the system benzene–toluene at atmospheric pressure. It can be obtained directly from the temperature-composition diagram. It shows the mole fraction of benzene in the vapor phase which is in equilibrium with the mole fraction of benzene in the liquid phase. Note that it does not contain temperature data. The diagram is useful for distillation purposes as will be discussed shortly.

Figure 34.2 shows temperature-composition diagrams and x/y-diagrams that are different from those shown in Figure 34.1. They are for the systems acetone–carbon disulphide and

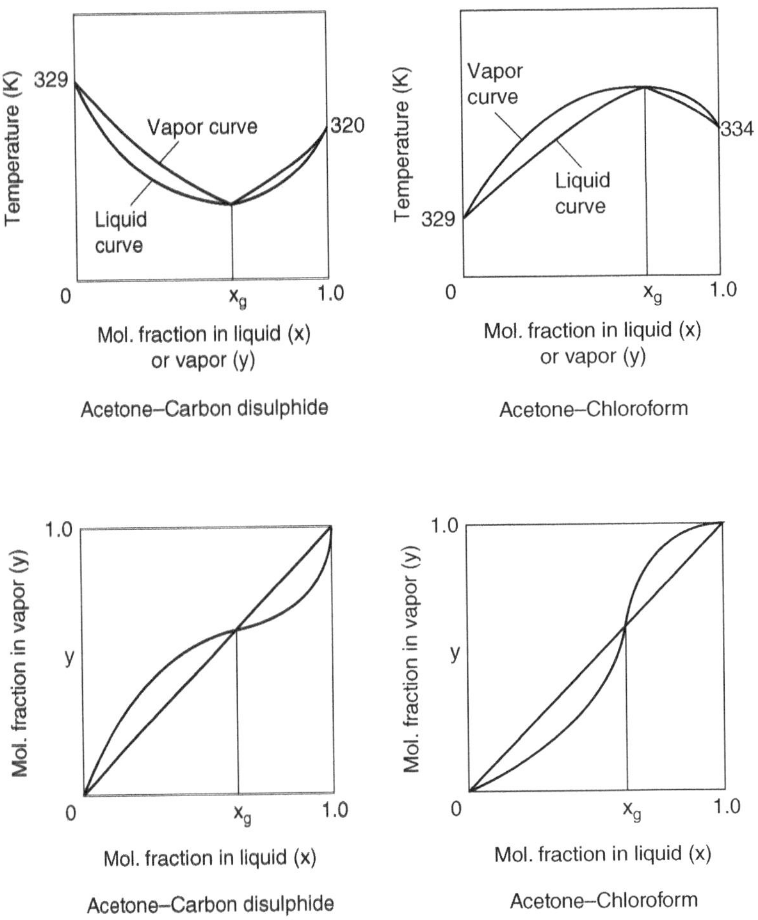

Figure 34.2 Temperature-composition diagrams and x/y-diagrams for two further systems. *Source:* Courtesy of Elsevier, Amsterdam, The Netherlands.

acetone–chloroform at atmospheric pressure. In the upper two diagrams, $x = 0$ means that we have pure acetone having a boiling point of 56 °C. First, the system acetone–carbon disulphide. From pure acetone up till an acetone content of x_g, the system behaves like the system benzene–toluene. The vapor contains less acetone than the liquid. Carbon disulphide is, with its boiling point of 47 °C, the more volatile component. However, beyond x_g, the situation is reversed and the vapor contains more acetone than the liquid. The mixture having a composition of x_g is called an azeotropic mixture. According to the x/y-diagram, it would not be possible to further separate acetone–carbon disulphide mixtures. In such a case, special methods are necessary to effect separation and these are discussed in Section 36.2 of Chapter 36. More information concerning azeotropy is provided in that chapter as well. Second, the acetone–chloroform system. Also in this case, we see an azeotropic mixture. x is now the mole fraction of the less volatile component.

Figure 34.3 depicts a typical temperature-composition diagram. ABJ is the boiling curve and ADJ the dew point curve. If a liquid mixture of composition x_2 is at a temperature T_3 below its boiling point T_2, as shown by point G on the diagram, then on heating at constant pressure, the following changes will occur:

(a) When the temperature reaches T_2 in point B, the liquid will start to boil, and some vapor of composition y_2 is formed, as shown by point E.
(b) On further heating, the composition of the liquid will change because of the loss of the more volatile component to the vapor and the boiling point will therefore rise to, for example, T'. At this temperature, the liquid will have a composition represented by point L, and the vapor a

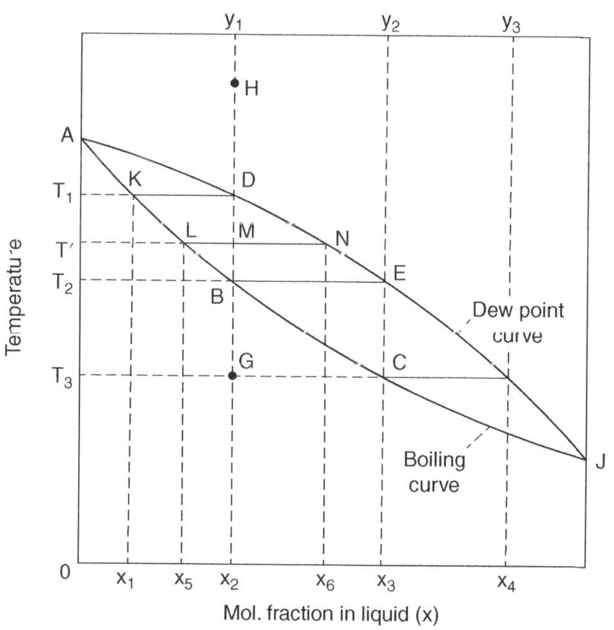

Figure 34.3 A typical temperature-composition diagram. *Source:* Courtesy of Elsevier, Amsterdam, The Netherlands.

composition represented by point N. Since material is not lost from the system, there will be a change in the proportion of liquid to vapor, the molar ratio of the liquid and the vapor being MN/ML. This is called the Lever Rule. The proof proceeds as follows:

Point L: m_l moles of liquid,
Point N: m_v moles of vapor,
Amount of moles of A in M: $x_2(m_l + m_v)$,
Amount of moles of A in M is also: $x_5 \cdot m_l + x_6 \cdot m_v$,

$$x_2(m_l + m_v) = x_5 \cdot m_l + x_6 \cdot m_v,$$

$$\frac{m_l}{m_v} = \frac{x_6 - x_2}{x_2 - x_5} = \frac{MN}{ML}.$$

(c) On further heating to T_1, all of the liquid is vaporized to give vapor D of the same composition $y_1 = x_2$ as the original liquid.

Partial vaporization of the liquid gives a vapor richer in the more volatile component than the liquid. If the vapor initially formed, as for instance at point E, is at once removed by condensation, then a liquid of composition x_3 is obtained, represented by point C. The step BEC may be regarded as an ideal stage, since the liquid passes from composition x_2 to a liquid of composition x_3.

Cooling superheated vapor represented by point H to D will lead to condensation and the first drop of liquid has a composition K. Further cooling to T' will give liquid L and vapor N. Thus, partial condensation brings about enrichment of the vapor in the more volatile component in the same way as partial vaporization. The industrial distillation column is, in essence, a series of units in which these two processes of partial vaporization and partial condensation are effected simultaneously.

34.3 The Fractionating Column

Figure 34.4 depicts a typical fractionating column built with perforated trays. The column is a cylindrical structure divided into sections by trays which permit the upward flow of vapor and the downward flow of liquid. The liquid reflux flows across each tray, over a weir, down a downcomer to the tray below. See Figure 34.5. The vapor rising to the top tray passes to a condenser and then through an accumulator or reflux drum and a reflux divider, where part is withdrawn as the overhead product D, and the remainder is returned to the top tray as reflux R.

The liquid in the base of the column is frequently heated, either by condensing steam or by a hot thermal oil stream, and vapor rises to the bottom tray. In Figure 34.4, heating is achieved in an external reboiler.

The vapor generated in the reboiler is returned to the bottom of the column with a composition y_s, and enters the bottom tray where it is partially condensed. Revaporization also occurs and the two processes result in vapor of composition y_1. This operation of partial condensation of the rising vapor and partial vaporization of the reflux is repeated on each tray. Vapor of composition y_t from the top tray is condensed to give the top product D and the reflux R, both of the same composition y_t. The feed stream is introduced on some intermediate tray where the liquid has approximately the same composition as the feed. The part of the column above the feed point is called the rectifying section and the part below the feed point is called the stripping section. The vapor rising from an

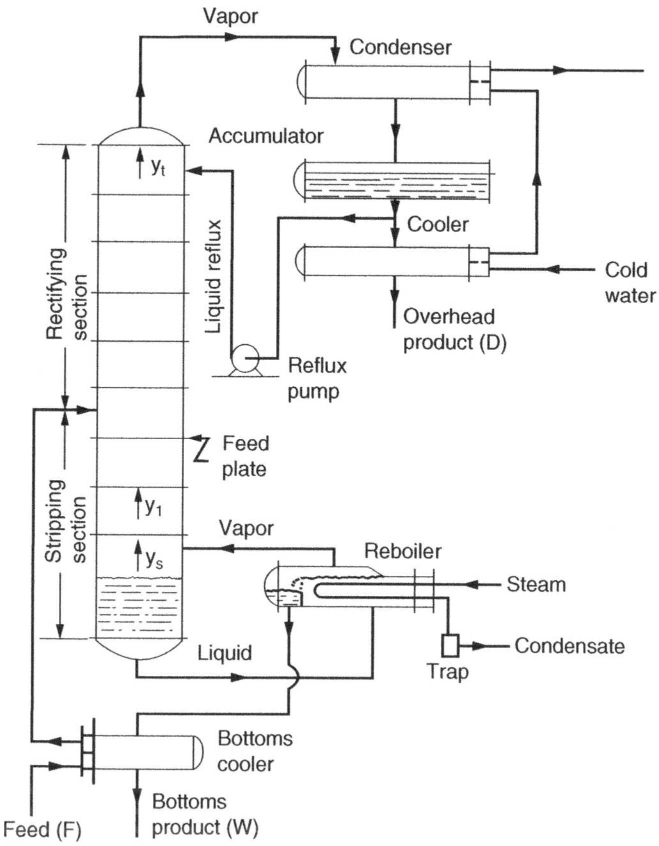

Figure 34.4 Continuous fractionating column with rectifying and stripping sections. *Source:* Courtesy of Elsevier, Amsterdam, The Netherlands.

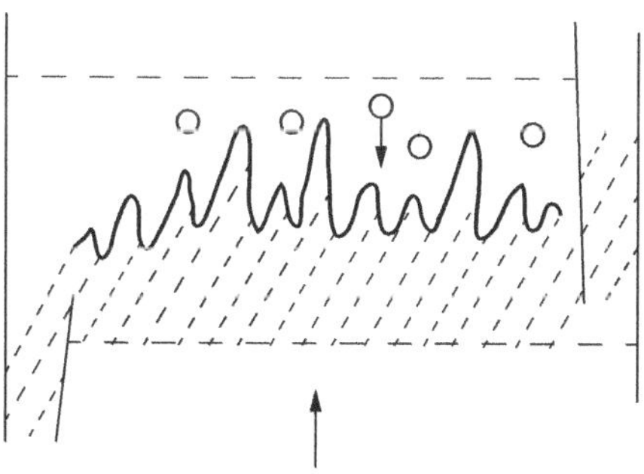

Figure 34.5 Mass transfer on a tray. *Source:* Courtesy of Deft Academic Press/VSSD, Delft, The Netherlands.

ideal tray is in equilibrium with the liquid leaving the tray. A smaller degree of enrichment will occur in practice.

The perforated tray is known as the sieve tray and has perforations of up to approximately 12 mm. Valve units and bubble caps are alternative arrangements for promoting heat and mass transfer on a tray. Besides tray columns, packed columns may be distinguished. Packed columns are used for relatively small capacities. The aim is, in all cases, to promote good mixing of vapor and liquid with a low pressure drop across the tray or packing.

The number of moles passing in each direction from vapor to liquid and in reverse will be approximately the same since the heat given out by 1 mol of the vapor on condensing is approximately equal to the heat required to vaporize 1 mol of the liquid. This is the concept of constant molar overflow. The background is that the molar enthalpies of evaporation of many materials are approximately 40 $kJ \cdot mol^{-1}$. This is one of the reasons why distillation is usually described in molar terms, so using molar flows and mole fractions for composition. In the arrangement discussed above, distillation is carried out continuously. A batch still may be used for the separation of relatively small quantities of mixtures into fractions. Such separations are common in the fine organic chemical industry. Batch distillation is discussed in Section 36.1.

34.4 The Number of Trays Required

A binary feed is distilled to give a top product D and a bottom product W, with x_f, x_d, and x_w as the corresponding mole fractions of the more volatile component [1]. See Figure 34.6. The vapor V_t rising from the top plate is condensed to liquid at its boiling point. Part is withdrawn as product and the remainder flows back into the column. A material balance above plate n (indicated by the loop l) gives:

$$V_n = L_{n+1} + D.$$

Expressing this balance on the more volatile component:

$$y_n \cdot V_n = L_{n+1} \cdot x_{n+1} + D \cdot x_d.$$

Thus: $y_n = \dfrac{L_{n+1}}{V_n} \cdot x_{n+1} + \dfrac{D}{V_n} \cdot x_d.$

This equation gives a relation between the composition of the vapor rising to a plate and the composition of the liquid on and leaving the same plate. The plate must be above the feed plate. Since

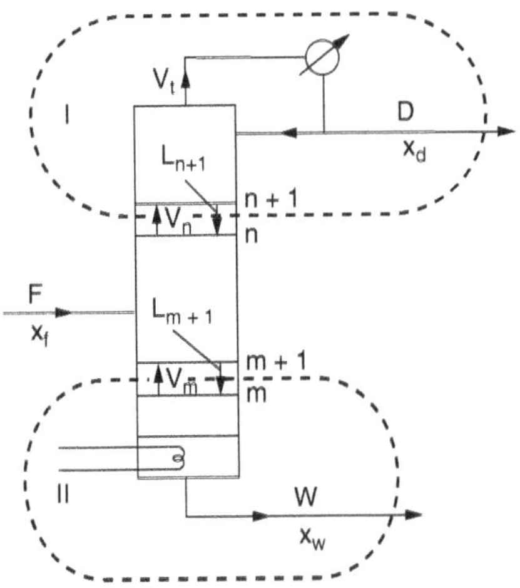

Figure 34.6 Material balances at the top and at the bottom of a continuous distillation column. *Source:* Courtesy of Elsevier, Amsterdam, The Netherlands.

the molar liquid overflow is constant, L_{n+1} can be replaced by L_n. This gives us the general material balance equation for the rectifying section:

$$y_n = \frac{L_n}{V_n} \cdot x_{n+1} + \frac{D}{V_n} \cdot x_d. \tag{34.1}$$

Similarly, taking a material balance for the total streams and for the more volatile component from the bottom to above plate m (indicated by the loop ll in Figure 34.6) and noting that $L_m = L_{m+1}$:

$$L_m = V_m + W,$$

$$y_m \cdot V_m = L_m - W \cdot x_w.$$

$$\text{Thus: } y_m = \frac{L_m}{V_m} \cdot x_{m+1} - \frac{W}{V_m} \cdot x_w. \tag{34.2}$$

This is the general material balance equation for the stripping section. The two Equations 34.1 and 34.2 can be represented by the operating lines in the x/y-diagram. McCabe and Thiele [2] pointed out that, since these equations can be represented by straight lines connecting y_n with x_{n+1} and y_m with x_{m+1}, they can be drawn on the same diagram as the x/y-curve to give a simple graphical solution for the number of stages required. Reference is made to Figure 34.7. First, this figure contains the x/y-curve for the system benzene–toluene. The x/y-curve is found by means of laboratory experiments. The diagonal has also been drawn. The equation of this line is y = x. Two operating lines are shown as well. The operating line for the rectifying section has been drawn as follows. In the accumulator, the benzene content of the arriving vapor is x_d. The benzene content of the liquid in the accumulator is also x_d. Thus, the intersection of the operating line and the diagonal

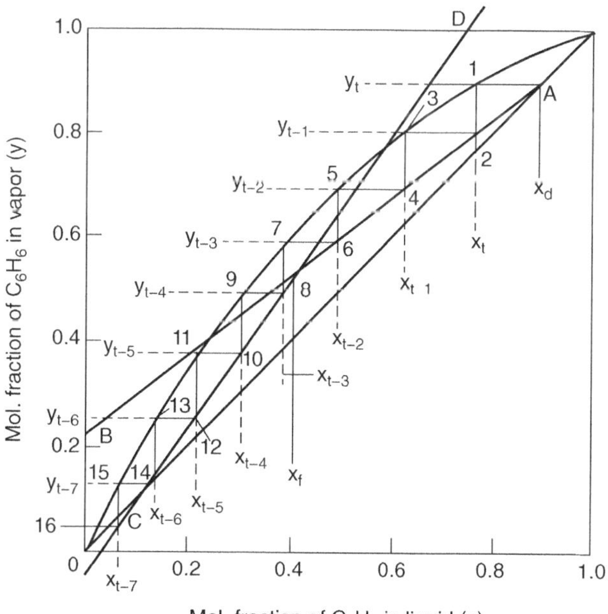

Figure 34.7 Determination of the number of plates by the McCabe–Thiele method. *Source:* Courtesy of Elsevier, Amsterdam, The Netherlands.

is (x_d, x_d). This also follows from Equation 34.1 by realizing that that $x_{n+1} = x_d$ in the accumulator. Then, it follows that $y_n = x_d$. Further, the slope of the line is L_n/V_n. It is also possible to put x_{n+1} in Equation 34.1 equal to zero. Then, it follows that $y_n = (D/V_n)x_d$, giving an easily determined point: $[0, (D/V_n)x_d]$.

The operating line for the stripping section is now dealt with. In Equation 34.2, x_{m+1} is the independent variable and y_m the dependent variable. In the still, x_{m+1} is equal to x_w. According to Equation 34.2, y_m is then equal to x_w because $V_m = L_m - W$. So, the point (x_w, x_w) is a point of the operating line. It is also a point of the diagonal. It is also possible to put x_{m+1} equal to zero, it then follows that $y_m = -(W/V_m)x_w$, giving an easily determined point: $(0, -(W/V_m)x_m)$. The slope of the second operating line is L_m/V_m. Thus, it is possible to also draw that line.

When the two operating lines have been drawn in, the number of stages required can be found by drawing steps between the operating lines and the equilibrium curve starting from point A. This will be explained as follows. Point A indicates the benzene content of both the vapor and the liquid in the accumulator. y_t is equal to x_d because the vapor going to the condenser at the top of the column and the condensed liquid have the same composition. The liquid generating vapor having composition y_t has the composition x_t: point 1. x_t is the composition of the liquid on the top tray and leaving the top tray. y_{t-1} and x_t are related by means of Equation 34.1: point 2. Vapor having composition y_{t-1} was generated by liquid having composition x_{t-1}: point 3. y_{t-2} and x_{t-1} are related by means of Equation 34.1: point 4. It is possible to assess the number of trays required for a separation by drawing steps between the equilibrium line and the operating lines. The number of trays is the number of times the operating lines are met. The number is eight in Figure 34.7.

Figure 34.8 can be read in combination with Figure 34.7. The uneven numbers relate to the fact that liquid on and leaving a plate and vapor leaving a plate are in equilibrium. The even numbers relate to the fact that there is a mass balance relationship between vapor and liquid at any height in the column; these flows are not in equilibrium and do not exchange matter.

The method designed by McCabe and Thiele is one of the most important concepts in chemical engineering and is an invaluable tool for the solution of distillation problems.

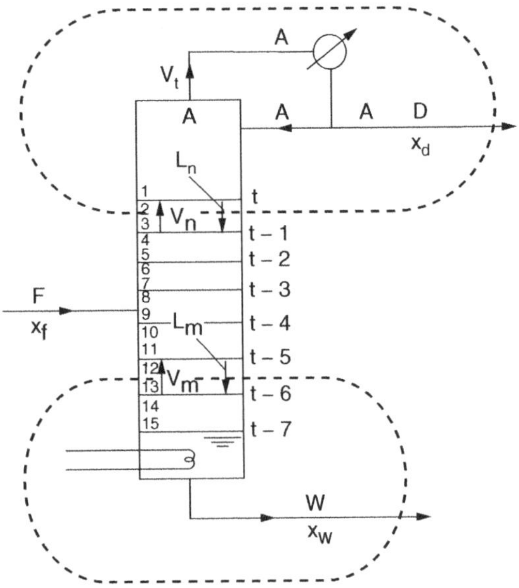

Figure 34.8 Elucidation of Example 34.1.

Example 34.1

A mixture of benzene and toluene containing 40 mol% of benzene is to be separated to give a product of 90 mol% of benzene at the top, and a bottom product with not more than 10 mol% of benzene. The feed is heated so that it enters the column at its boiling point, and the vapor leaving the column is condensed but not cooled. The condensed top product provides both reflux and product. The unit is operated with a reflux ratio of 3 mol per mol of product. The number of theoretical plates and the position of entry for the feed will be determined. Figure 34.7 shows the equilibrium curve at atmospheric pressure.

Considering 100 kmol of feed per h, a total material balance gives:

$$100 = D + W.$$

A balance on benzene gives:

$$0.4 \cdot 100 = 0.9 \cdot D + 0.1 \cdot W.$$

It follows that $W = 62.5$ and $D = 37.5$.

The amount of liquid L_n flowing back into the column is $3D = 112.5$.

The sum of L_n and D is the amount of vapor flowing through the rectifying section of the column: 150.

Thus, the equation of the operating line of the rectifying section is:

$$y_n = \frac{112.5}{150} \cdot x_{n+1} + \frac{0.9 \cdot 37.5}{150}.$$

$$y_n = 0.75 \cdot x_{n+1} + 0.225.$$

Line AB in Figure 34.7 represents this equation. When $x_{n+1} = 0$, $y_n = 0.225$. The slope of the line is 0.75.

Since the feed is all liquid at its boiling point, it will all run down and will combine with the liquid flow from the upper part of the column:

$$L_m = L_n + F,$$

$$L_m = 112.5 + 150 = 212.5.$$

The vapor flow V_m flowing through the stripping section of the column is the difference between L_m and W.

$$V_m = 212.5 - 62.5 = 150.$$

The equation of the operating line of the stripping section is:

$$y_m = \frac{212.5}{150} \cdot x_{m+1} - \frac{0.1 \cdot 62.5}{150}.$$

$$y_m = 1.417 \cdot x_{m+1} - 0.042.$$

Line CD in Figure 34.7 represents this equation. When $x_{m+1} = 0$, $y_m = -0.042$. The slope of the line is 1.417.

Now, starting at point A, the horizontal line to intersect the equilibrium line at point 1 is drawn. Next, the vertical through 1 is dropped to the operating line at point 2. In this way, points 3–6 are obtained. The horizontal line through point 6 is drawn to intersect the equilibrium curve at point 7, and the vertical line is drawn through point 7 to the lower enrichment line at point 8. Points 9–16 are obtained in the same way. The number of stages is counted and assessed as eight. The feed is introduced on the fourth plate from the top.

34.5 The Importance of the Reflux Ratio

The ratio L_n/D of the top overflow to the quantity of product is denoted by R, the reflux ratio. Introducing R into Equation 34.1 results in:

$$y_n = \frac{R}{R+1} \cdot x_{n+1} + \frac{x_d}{R+1}. \tag{34.3}$$

Any change in R will therefore modify the slope of the operating line of the rectifying section. In Figure 34.7, the operating line of the upper column part passes through the point (x_d, x_d). When, for example, R = 3, the slope of the operating line is 0.75. When R is smaller, for example, 2, then the slope decreases to 0.67. The area between the equilibrium line and the operating becomes smaller and more stages will be required to achieve the desired separation. A minimum reflux ratio R_{min} is required to achieve the desired separation. Area must remain between the operating lines and the equilibrium line in order to be able to draw steps. Below the minimum reflux ratio, the desired separation cannot be achieved, however large the number of trays. When R becomes greater than 3, the area between the equilibrium line and the operating line becomes larger and less stages will be required.

On increasing the reflux ratio, the variable costs of the distillation increase because the sum of the top product and the reflux must be transported as vapor through the upper part of the column. The heat required for evaporation is supplied in the reboiler and removed in the top condenser. Figure 34.9 qualitatively illustrates distillation costs. At R_{min}, the variable costs are low, however, the investment is high because many trays are required. When the reflux ratio is high, the variable costs are high because of the large vapor flows and the investment is also high because of the large column diameter required due to the large vapor flows. Thus, there will be an optimum reflux ratio.

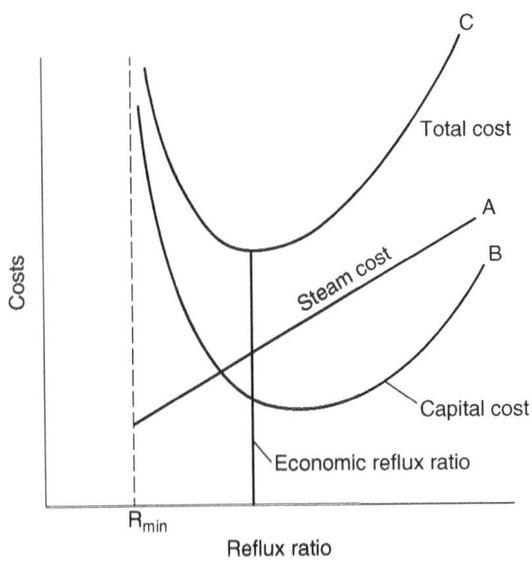

Figure 34.9 Influence of the reflux ratio on capital and operating costs of the still. *Source:* Courtesy of Elsevier, Amsterdam, The Netherlands.

34.6 A Typical Continuous Industrial Distillation

The distillation concerns the separation of i-butane and n-butane [3]. The molecules of both alkanes are apolar, so azeotropy does not occur. The boiling point at atmospheric pressure of i-butane is −11.7 °C and of n-butane −0.5 °C. The industrial distillation is carried out under pressure to avoid having to deal with subzero temperatures. A column pressure of 6.4 barg is reported. The boiling points of the two compounds are relatively close to each other. A difference between the boiling points of, for example, 30 K, is considered to be quite workable. The relatively small difference between the boiling points in this case means that the separation will require many trays and a high reflux ratio.

The industrial distillative separation of i-butane and n-butane occurs at Total's Antwerp refinery in Belgium. See Figure 34.10. The distillation column is called the de-isobutanizer. The subject of the article is an increase of the capacity of the column by installing a new type of tray delivered by the Swiss company Sulzer Chemtec. A capacity increase from 150 to 240 $t \cdot d^{-1}$ is reported. An important step to realize the capacity increase is the replacement of 58 trays by 58 new trays. The article mentions a reflux ratio of 12.5. The preheated feed stream enters the column above tray 29. The feed flow rate is controlled by the level in the feed drum. The reflux rate is controlled by the temperature of tray 10. The i-butane flow is controlled by the level in the reflux drum. The n-butane flow is controlled by the level in the reboiler. Purities of i-butane and n-butane of 97.65 and 98.77% by weight respectively are reported.

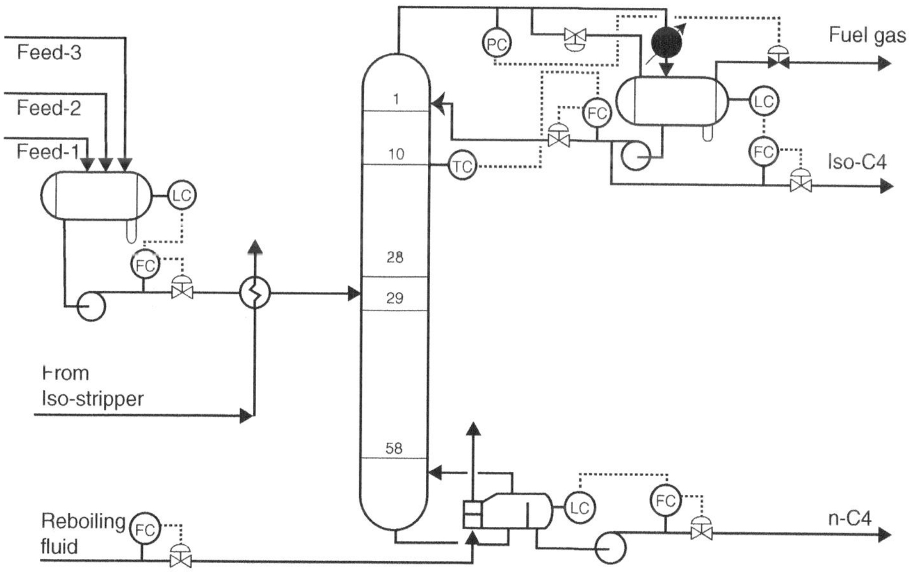

Figure 34.10 The de-isobutanizer at the Total refinery at Antwerp, Belgium. *Source:* Modified from Sulzer Chemtec Ltd, Allschwil, Switzerland.

References

1. Coulson, J. M. and Richardson, J. F. with Backhurst, J. R. and Harker, J. H. (1999). *Chemical Engineering. Particle Technology & Separation Processes* (Vol. 2, pp. 442–459). Oxford, UK: Butterworth-Heinemann.
2. McCabe, W. L. and Thiele, E. W. (1925). Graphical design of fractionating columns. *Industrial & Engineering Chemistry*, 17, 605–611.
3. Mosca, G., van Hemel, G., Tocco, R., and Lestrade, B. (2006). Knowledge is power. *Process Worldwide*, 9 (2), 42–43.

35

Design of Continuous Distillation Columns

35.1 Sieve Tray Columns

The sieve tray is probably the most widely used contacting device found in columns today. Part of a sieve tray is depicted in Figure 35.1. The assessment of the column diameter and the pressure loss per tray will be discussed. A shortcut procedure will be treated [1].

First, the designer has to ensure that the tray can process the maximum gas and liquid flows. That is, a phenomenon called flooding is avoided. Flooding is the condition where the liquid stops flowing down the column. The drag forces exerted by the rising vapor are greater than the gravity force experienced by the liquid. To this end, use is made of the load diagram in Figure 35.2. Tests and dimensional analysis form the basis of the diagram. The diagram will be elucidated by means of a simple model. In the model, the tray is considered at maximum load and the focus is on the pressure drop of the liquid in the downcomer and the pressure drop of the gas across the tray. The two pressure drops are equal. See Figure 35.3. The pressure loss the liquid experiences because of the flow through the nip is:

$$\Delta p_{nip} = c_1 \cdot \rho_l \cdot v_l^2.$$

v_l is the superficial liquid velocity, which means that the reference area is the cross-sectional area of the column. The static head causing the liquid flow is $\rho_l \cdot gH$.

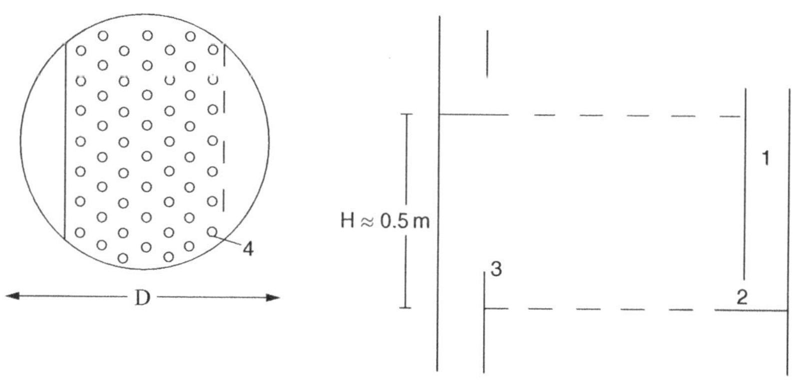

Figure 35.1 Sieve trays in a column: 1, downcomer; 2, nip; 3, weir; 4, holes (≈10 mm). *Source:* Courtesy of Delft Academic Press/VSSD, Delft, The Netherlands.

Introduction to Chemical Engineering, First Edition. C.M. van 't Land.

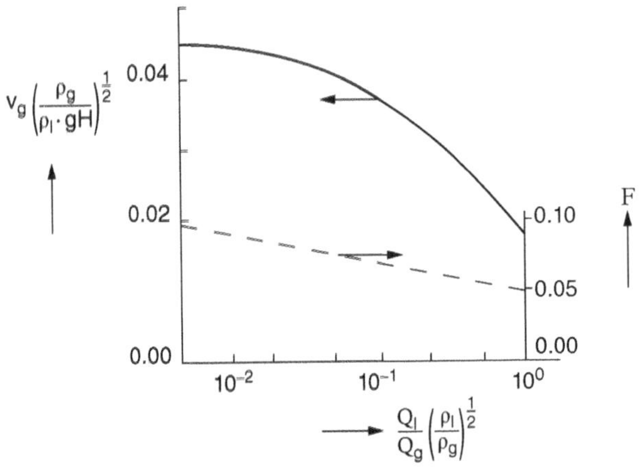

Figure 35.2 Sieve tray load diagram. *Source:* Courtesy of Delft Academic Press/VSSD, Delft, The Netherlands.

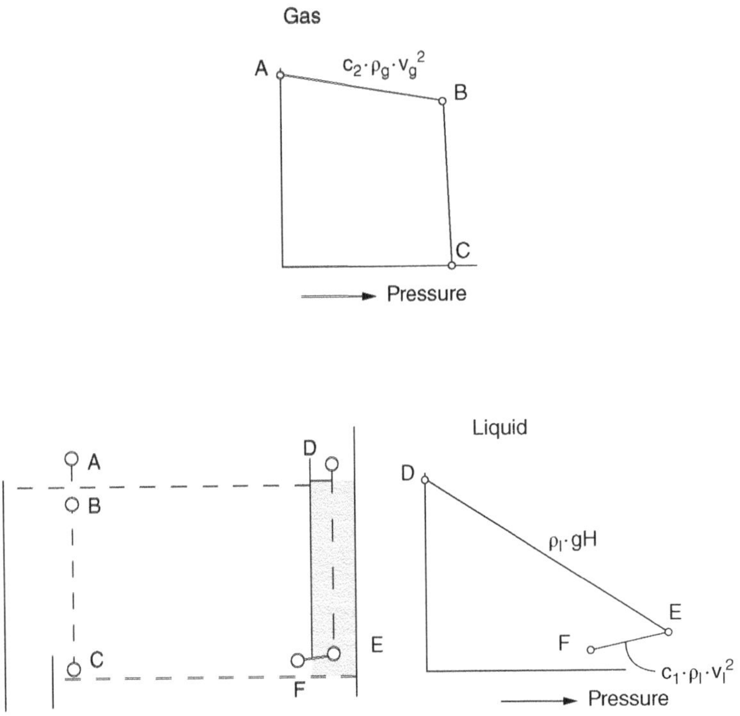

Figure 35.3 Sieve tray liquid and gas pressure drops. *Source:* Courtesy of Delft Academic Press/VSSD, Delft, The Netherlands.

The gas experiences pressure loss due to the flow through the holes and due to the liquid on the tray. The latter pressure loss is small compared with the former pressure loss. The fraction-free area of a sieve tray is smaller than 0.1. Thus, the pressure loss the gas experiences due to the flow through the holes is:

$$\Delta p_{\text{holes}} = c_2 \cdot \rho_g \cdot v_g^2.$$

v_g is the superficial gas velocity, which means that the reference area is the cross-sectional area of the column.

Now, it is possible to state:

$$\rho_l \cdot gH - c_1 \cdot \rho_l \cdot v_l^2 = c_2 \cdot \rho_g \cdot v_g^2.$$

It follows: $c_2 = -c_1 \cdot \frac{\rho_l \cdot v_l^2}{\rho_g \cdot v_g^2} + \frac{\rho_l \cdot gH}{\rho_g \cdot v_g^2}$ and $\frac{\rho_g \cdot v_g^2}{\rho_l \cdot gH} = \left(c_2 + c_1 \cdot \frac{\rho_l \cdot Q_l^2}{\rho_g \cdot Q_g^2}\right)^{-1}$.

Q_l and Q_g are the volumetric liquid and gaseous flows. v_l/v_g has been replaced by Q_l/Q_g, which is allowed because, for both velocities, the reference area is the column cross-sectional area.

The dimensionless groups in the left-hand side and in the right-hand side of the latter equation can also be found in Figure 35.2. The shape of the curve is similar to the shape of the curve predicted by the theory given. However, the theory is not very satisfactory. The constants c_1 and c_2 can be determined independently. On using those values, theory predicts capacities more than twice the capacities actually found.

At distillations, the ratio Q_l/Q_g is strongly pressure-dependent. At low pressures, Q_g is large and vice versa. Q_l is not pressure-dependent. For low pressures, the ratio is small and we are left in the diagram. The right part of the diagram is applicable for high pressures. Furthermore, the fraction free area is relatively large at low-pressure distillations and vice versa.

The shortcut procedure will be given.

1) For H, 0.5 m is chosen. Q_l, Q_g, ρ_l, and ρ_g must be known.
2) The flow parameter Φ is calculated: $\Phi = \frac{Q_l}{Q_g}\sqrt{\frac{\rho_l}{\rho_g}}$.
3) Via Φ, the abcissa in Figure 35.2, a value for $v_g\sqrt{\frac{\rho_g}{\rho_l \cdot gH}}$, the ordinate, is obtained. v_g can be calculated. The design of a tray column is based on 85% of this velocity. Thus, v_g is replaced by $0.85 \cdot v_g$.
4) The column diameter is calculated: $D = \sqrt{\frac{4 \cdot Q_g}{\pi \cdot v_g}}$.
5) F, the fraction free area, is determined by means of the load diagram. F is a function of Φ.

$$\Delta p_g = 1.1 \cdot \rho_g \left(\frac{v_g}{F}\right)^2.$$

5 cm liquid are added, that is, $\rho_l \cdot g \cdot 0.05$.

The design of the downcomers and the weir is not attempted. The turn-down ratio of a well-designed sieve tray column is approximately 0.4.

Example 35.1

The sizing of a sieve tray column processing 100 $kmol \cdot h^{-1}$ of the benzene/toluene mixture in Chapter 34 is attempted. The design is based on the conditions on and above the top tray.

1) First, the gas and liquid flows and their specific masses are calculated. This proceeds as follows. The composition of the liquid on the upper tray can be estimated. The liquid contains 80 mol% of benzene and 20 mol% of toluene. The liquid mixture boils at atmospheric pressure. Now, use is made of Raoult's law.

This law states, for boiling at atmospheric pressure

$$x_A \cdot P_{A0} + x_B \cdot P_{B0} = 101{,}300\ \text{N} \cdot \text{m}^{-2}.$$

In words: the mole fraction of component A times its saturated vapor pressure and the mole fraction of component B times its saturated vapor pressure equal the external pressure. It appears that the left-hand side of this equation equals the right-hand side at 85 °C. So, the top tray temperature is 85 °C.

The specific masses of both benzene and toluene are, at 85 °C, 808 kg · m^{-3}. So, the specific mass of the liquid on the top tray is 808 kg · m^{-3}.

The average molecular weight of the liquid on the top tray is:

$$0.8 \cdot 78.1 + 0.2 \cdot 92.1 = 80.9\ \text{kg} \cdot \text{kmol}^{-1}.$$

The gas specific mass is calculated by using the Universal Gas Law. The average molecular weight of the gas leaving the top tray is: $0.9 \cdot 78.1 + 0.1 \cdot 92.1 = 79.5$ kg · kmol^{-1}. Thus:

$$\rho_g = \frac{MP}{RT} = \frac{79.5 \cdot 101{,}300}{8{,}314 \cdot 358} = 2.71\ \text{kg} \cdot \text{m}^{-3}.$$

$$V_n = 150\ \text{kmol} \cdot \text{h}^{-1}.$$

$$L_n = 112.5\ \text{kmol} \cdot \text{h}^{-1}.$$

$$Q_l = \frac{112.5 \cdot 80.9}{808 \cdot 3{,}600} = 3.13 \cdot 10^{-3}\ \text{m}^3 \cdot \text{s}^{-1}.$$

$$Q_g = \frac{150 \cdot 79.5}{2.71 \cdot 3{,}600} = 1.22\ \text{m}^3 \cdot \text{s}^{-1}.$$

2) $\Phi = \dfrac{3.13 \cdot 10^{-3}}{1.22}\sqrt{\dfrac{808}{2.71}} = 0.0443.$

3) $v_g\sqrt{\dfrac{2.71}{808 \cdot 9.81 \cdot 0.5}} = 0.04 \rightarrow v_g = 1.53\ \text{m} \cdot \text{s}^{-1},$

$$0.85 \cdot 1.53 = 1.30\ \text{m} \cdot \text{s}^{-1}.$$

4) $D = \sqrt{\dfrac{4 \cdot 1.22}{\pi \cdot 1.30}} = 1.09$ m.

5) $F = 0.075.$

$$\Delta p_g = 1.1 \cdot 2.71\left(\frac{1.30}{0.075}\right)^2 = \quad 895.6\ \text{N} \cdot \text{m}^{-2}$$

$$808 \cdot 9.81 \cdot 0.05 = \quad 396.3\ \text{N} \cdot \text{m}^{-2} \; +$$

$$\overline{1{,}291.9\ \text{N} \cdot \text{m}^{-2}}$$

The column is designed with seven sieve trays. The column height is 7 · 0.5 = 3.5 m. The total pressure drop is 9,043 N · m^{-2}.

35.2 Packed Columns

Packed columns are suited for low-pressure operations and for small-diameter columns. Proper liquid distribution is a critical aspect. Therefore, for Raschig rings, the maximum bed height should be equal to 2.5–3.0 times the bed diameter; for saddles, 5–8 times the bed diameter; and for slotted rings, 5–10 times. Total bed height should not exceed 6 m.

Designing a packed column is similar to designing a sieve tray column. Here too, a shortcut procedure for the assessment of the diameter and the pressure drop of a column filled with Raschig rings will be discussed [2]. Raschig rings are small hollow cylinders of which the height equals the diameter. A typical size is 20 mm. In the calculation, a packing factor F_p for the Raschig rings is required. Figure 35.4 gives the packing factor as a function of the ring diameter for columns which are filled at random. The load diagram can be found in Figure 35.5. Here too, the design is based on

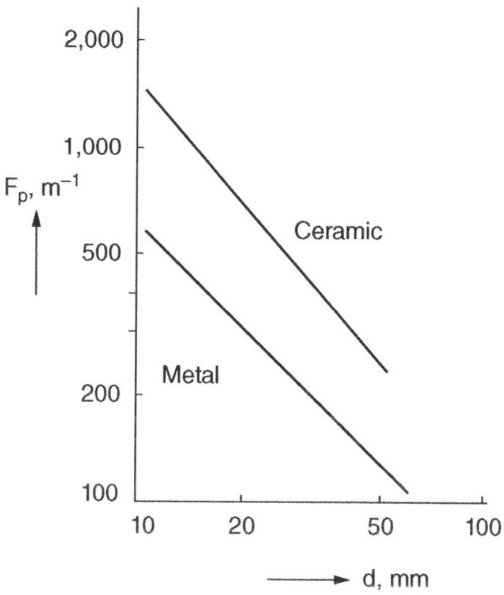

Figure 35.4 Raschig rings packing factor as a function of the nominal ring diameter d. *Source:* Courtesy of Delft Academic Press/VSSD, Delft, The Netherlands.

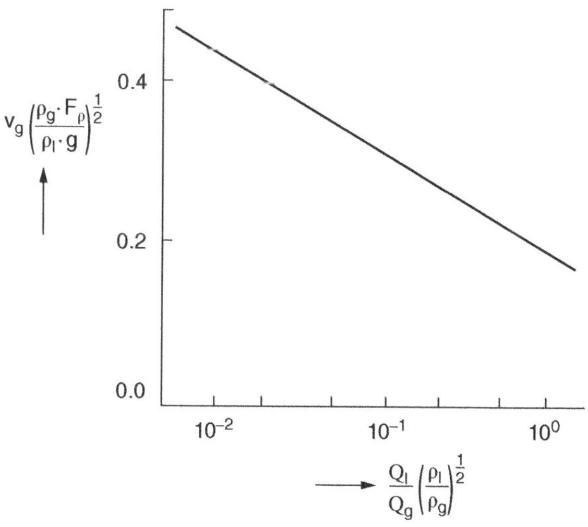

Figure 35.5 Load diagram for a column filled at random with Raschig rings. *Source:* Courtesy of Delft Academic Press/VSSD, Delft, The Netherlands.

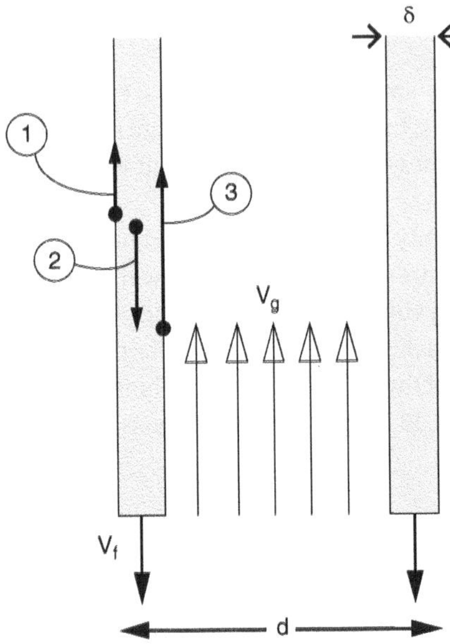

Figure 35.6 Balance of forces for a film in a packed column; 1, shear force wall/liquid; 2, gravity force; 3, shear force liquid/gas. *Source:* Courtesy of Delft Academic Press/VSSD, Delft, The Netherlands.

the desire to avoid flooding. Also in this case, flooding is the condition where the liquid stops flowing down the column. The load diagram for packed columns resembles the load diagram for sieve tray columns as given in Figure 35.2. Compared with that diagram, H has been replaced by $1/F_p$.

The diagram will be elucidated by means of a simple model for the pressure loss. See Figure 35.6. The packing is represented by a series of vertical plates. The distance between the plates is d, the ring diameter. The liquid film having a thickness δ flows down the plates. The average film velocity is v_f. The ratio δ/d is small. The gas flow is countercurrent to the liquid flow and the gas superficial velocity is v_g and $v_g \gg v_f$.

The forces acting on the liquid film per m^2 plate area are:

- the turbulent shear force wall/liquid, that is, $c_1 \cdot \rho_l \cdot v_f^2$;
- the gravity force $\rho_l \cdot g\delta$; and
- the turbulent shear force liquid/gas, that is, $c_2 \cdot \rho g\left(v_f + v_g\right)^2 \approx c_2 \cdot \rho_g \cdot v_g^2$.

The balance of forces is $c_1 \cdot \rho_l \cdot v_f^2 + c_2 \cdot \rho_g \cdot v_g^2 = \rho_l \cdot g\delta$. The balance of forces can also be stated as $c_1' \cdot \rho_l \cdot v_l^2 + c_2 \cdot \rho_g \cdot v_g^2 = \rho_l \cdot g\delta$. In this equation, v_f has been replaced by v_l. v_f is the actual film velocity whereas v_l is the velocity with the cross-sectional area of the column as reference area. This equation is similar to the equation for the sieve tray column. So, a similar load diagram can be constructed.

Now, a closer look will be taken at the first balance of forces in order to arrive at an expression for ε, the volume fraction liquid in the column.

v_f can be replaced by $v_l \cdot \dfrac{d}{2\delta}$ (2 because each plate has two sides).

$c_2 \cdot \rho_g \cdot v_g^2$ can be neglected at small gas loads.

The balance of forces then becomes: $c_1\left(v_l \cdot \dfrac{d}{2\delta}\right)^2 = g\delta$.

The volume fraction liquid ε in the column is: $\varepsilon = \dfrac{2\delta}{d}$.

The latter expression can be expressed explicitly in δ: $\delta = \dfrac{\varepsilon d}{2}$.

Upon substituting the latter expression in the simplified balance of forces, an expression for ε can be obtained: $\varepsilon = (2c_1)^{1/3}\left(\frac{v_l^2}{gd}\right)^{1/3} = c_3\left(\frac{v_l^2}{gd}\right)^{1/3}$.

The latter expression will be used in the calculation.

Before attempting the shortcut procedure, the concept of the HETP (Height Equivalent to a Theoretical Plate) must be discussed. The height of an equivalent equilibrium stage, usually called the height of a theoretical plate (HETP), is the height of packing which will give the same separation as an equilibrium stage. The following values for Pall rings can be used to make an approximate estimate of the bed height required [3].

Size, mm	HETP, m
25 (1″)	0.4–0.5
38 (1.5″)	0.6–0.75
50 (2″)	0.75–1.0

The HETPs for Raschig rings will be larger than those for Pall rings, and the values given above will apply only for Raschig rings at a pressure drop greater than 42 mm wg.

The shortcut procedure follows here.

1) For the HETP, 0.5 m is chosen. Q_l, Q_g, ρ_l, and ρ_g must be known.
2) Figure 35.4 can be used to determine the packing factor.

$$\Phi = \frac{Q_l}{Q_g}\sqrt{\frac{\rho_l}{\rho_g}}.$$

3) The load diagram is consulted. The maximum gas velocity is calculated via the value for the ordinate, $v_g\sqrt{\frac{\rho_g \cdot F_p}{\rho_l \cdot g}}$. The design of packed columns is based on 70% of this velocity. Thus, v_g is replaced by $0.7 \cdot v_g$.
4) The column diameter is calculated: $D = \sqrt{\frac{4 \cdot Q_g}{\pi \cdot v_g}}$.
5) $v_l = \frac{4 \cdot Q_l}{\pi \cdot D^2}$ and $\varepsilon = 1.2\left(\frac{v_l^2}{gd}\right)^{1/3}$.
6) "Dry" pressure loss is $\Delta p_d = 0.8 \cdot \rho_g \cdot v_g^2 \cdot F_p \cdot H$.
"Wet" pressure loss is $\Delta p = \Delta p_d(1 - 3.5 \cdot \varepsilon)^{-3}$.

Example 35.2

A packed column is sized for the duty of Example 35.1 in this chapter. 20-mm ceramic Raschig rings are chosen.

1) $H = 7 \cdot 0.5 = 3.5$ m (7 ideal stages are required).
2) $F_p = 670\ m^{-1}$.
3) $\Phi = 0.0443, v_g\sqrt{\frac{2.71 \cdot 670}{808 \cdot 9.81}} = 0.37 \rightarrow v_g = 0.77\ m \cdot s^{-1}$,
$0.7 \cdot 0.77 = 0.54\ m \cdot s^{-1}$.

4) $D = \sqrt{\dfrac{4 \cdot 1.22}{\pi \cdot 0.54}} = 1.70\ \text{m}.$

5) $v_l = \dfrac{3.13 \cdot 10^{-3} \cdot 4}{\pi \cdot 1.70^2} = 1.4 \cdot 10^{-3}\ \text{m} \cdot \text{s}^{-1},$

$\varepsilon = 1.2\left(\dfrac{1.4 \cdot 10^{-3}}{9.81 \cdot 1{,}000}\right)^{1/3} = 2.6 \cdot 10^{-3}.$

6) $\Delta p_d = 0.80 \cdot 2.71 \cdot 0.54^2 \cdot 670 \cdot 3.5 = 1{,}482\ \text{N} \cdot \text{m}^{-2},$

$\Delta p = 1{,}482(1 - 3.5 \cdot 2.6 \cdot 10^{-3})^{-3} = 1{,}523\ \text{N} \cdot \text{m}^{-2},$

$\Delta p = \dfrac{1{,}523}{9.81 \cdot 1{,}000} = 0.155\ \text{m wg} = 155\ \text{mm wg}.$

Per m packing height: 44 mm wg. This figure is greater than 42 mm $\text{wg} \cdot \text{m}^{-1}$ and hence HETP $\approx$ 0.5 m. The column height is 2.1 times larger than the diameter.

Note

The pressure loss of the packed column is smaller than the pressure loss of the sieve tray column.

References

1. Wesselingh, J. A., and Kleizen, H. H. (1998). *Separation Processes* (pp. 115–119). Delft, The Netherlands: Delft University Press. (in Dutch)
2. Wesselingh, J. A., and Kleizen, H. H. (1998). *Separation Processes* (pp. 119–122). Delft, The Netherlands: Delft University Press. (in Dutch)
3. Sinnott, R. K. (2000). *Coulson & Richardson's Chemical Engineering, Volume 6: Chemical Engineering Design* (p. 593). Oxford, UK: Butterworth-Heinemann.

36

Various Types of Distillation

36.1 Batch Distillation

In batch distillation, the whole of a batch is run into the boiler of the still and, on heating, the vapor is passed into a fractionating column [1]. See Figure 36.1. Since the top product will be relatively rich in the more volatile component (MVC), the liquid remaining in the still will become steadily weaker in this component.

Operation at Constant Reflux Ratio

Suppose a column with several ideal plates is used to separate a mixture of ethanol and water. If the column is operated at constant reflux ratio R, the concentration of the MVC in the top product will continuously fall.

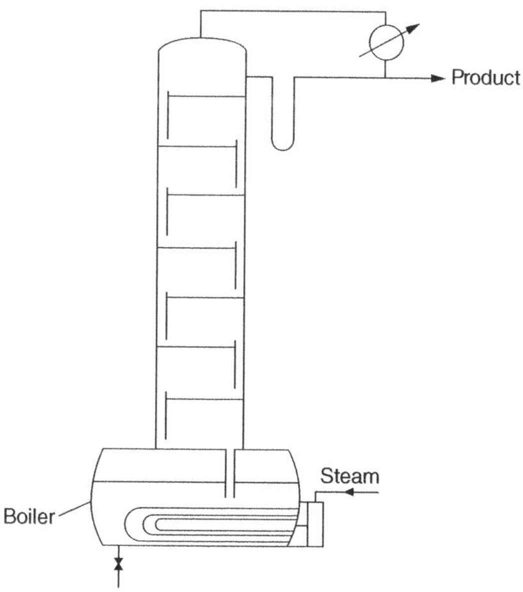

Figure 36.1 Column for batch distillation. *Source:* Courtesy of Elsevier, Amsterdam, The Netherlands.

Introduction to Chemical Engineering, First Edition. C.M. van 't Land.

Let S be the number of kmoles of material in the still and x_s be the mole fraction of the MVC. Suppose an amount dS, containing a mole fraction x_d of the MVC is collected overhead. Rayleigh pointed out that a material balance on the MVC gives:

$$x_d \cdot dS = d(S \cdot x_s).$$

$$x_d \cdot dS = S \cdot dx_s + x_s \cdot dS,$$

$$\int_{S_1}^{S_2} \frac{dS}{S} = \int_{x_{s1}}^{x_{s2}} \frac{dx_s}{x_d - x_s},$$

$$\ln \frac{S_2}{S_1} = \int_{x_{s1}}^{x_{s2}} \frac{dx_s}{x_d - x_s}. \tag{36.1}$$

The right-hand side of this equation can be integrated analytically, numerically, or graphically. In this chapter, the graphical method will be discussed. The latter method encompasses plotting $1/(x_d - x_s)$ versus x_s. This will enable the ratio of the initial to final quantity in the still to be found for any desired change in x_s, and hence the amount of distillate D_b. The heat to be supplied to provide the reflux will be $Q_R = \lambda R \cdot D_b$.

This batch distillation problem can be solved by means of the McCabe–Thiele diagram treated in Chapter 34. An example illustrates this.

Example 36.1

A mixture of ethanol and water with 0.55 mol fraction of ethanol is distilled batchwise by means of a distillation column having four ideal plates (excluding the still) at atmospheric pressure. A constant reflux ratio of 2.1 is used. The distillation is stopped when the ethanol mol fraction in the still is 0.105. The amount of distillate and its composition is calculated. The heat required per kmol of product is also calculated.

Figure 36.2 contains the McCabe–Thiele diagram for this distillation. The curved line is the x/y-diagram for the system ethanol/water. The diagonal has also been drawn. The first step is to check the situation when x_d = 0.74. For this situation, the operating line has been drawn. It can be seen from Equation 34.3 that the operating line passes through the point (0, 0.74/(2.1 + 1)), that is, (0, 0.24). The operating line also passes through the point (0.74, 0.74). Now, five steps can be drawn between the equilibrium line and the operating line to arrive at x_s = 0.105. The meaning of this is as follows. The separation power of this column at R = 2.1 is an upgrade of the concentration of the MVC from 0.105 to 0.74. If we go one step further with this distillation, the ethanol concentration in the still falls below 0.105 and it was a starting point to stop at this concentration. Thus, this is the final situation of this batch distillation.

For an ethanol concentration of 0.78 in the distillate, the aforementioned procedure leads to an ethanol concentration in the still of 0.55. That is the initial situation. The operating line for this situation has also been drawn. The second operating line is parallel to the first operating line because the slope of both lines is R/(R + 1) = 0.68.

Now, it is possible to obtain the following data by taking different x_d-values.

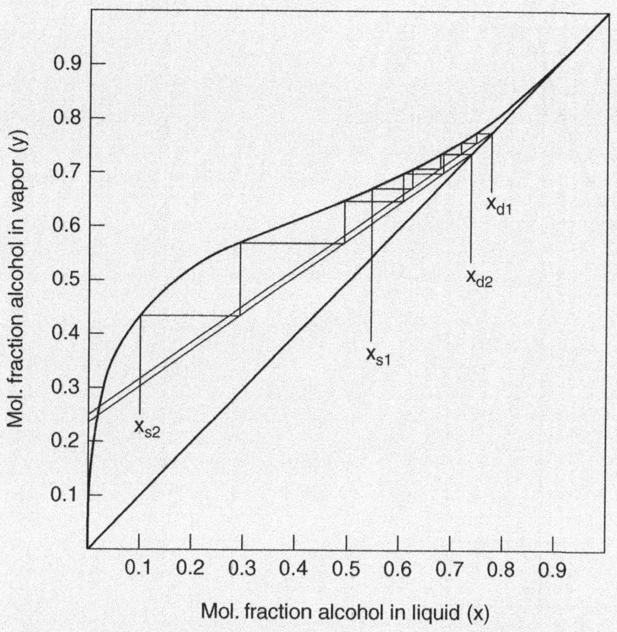

Figure 36.2 Batch distillation–constant reflux ratio. *Source:* Courtesy of Elsevier, Amsterdam, The Netherlands.

x_d	x_s	$x_d - x_s$	$1/(x_d - x_s)$
0.780	0.550	0.230	4.35
0.775	0.500	0.275	3.64
0.770	0.425	0.345	2.90
0.760	0.310	0.450	2.22
0.750	0.225	0.525	1.90
0.740	0.105	0.635	1.57

These data have been plotted in Figure 36.3. It follows:

$$\int_{0.55}^{0.105} \frac{dx_s}{x_d - x_s} = 1.1.$$

It follows from Equation 36.1 that $\ln(S_1/S_2) = 1.1$ and $S_1/S_2 = 3.0$.

Product obtained is $D_b = S_1 - S_2 = 100 - 100/3.0 = 66.7$ kmol.

The initial amount of kmol in the still is 100.

The amount of ethanol in the product is $x_1 \cdot S_1 - x_2 \cdot S_2 = 0.55 \cdot 100 - 0.105 \cdot 33.3 = 51.5$ kmol. Thus, the final composition of the product is 51.5/66.7 = 0.77 mol fraction ethanol. The average latent heat is taken as 40,000 kJ · kmol^{-1}. The heat required to provide the reflux is:

$$\frac{40{,}000 \cdot 2.1 \cdot 66.7}{1{,}000} = 5{,}603 \text{ MJ}.$$

The heat required to provide reflux per kmol of product is:

$$\frac{5{,}603}{66.7} = 84.0 \text{ MJ}.$$

The total amount of heat per kmol of product is 84.0 + 40.0 = 124.0 MJ.

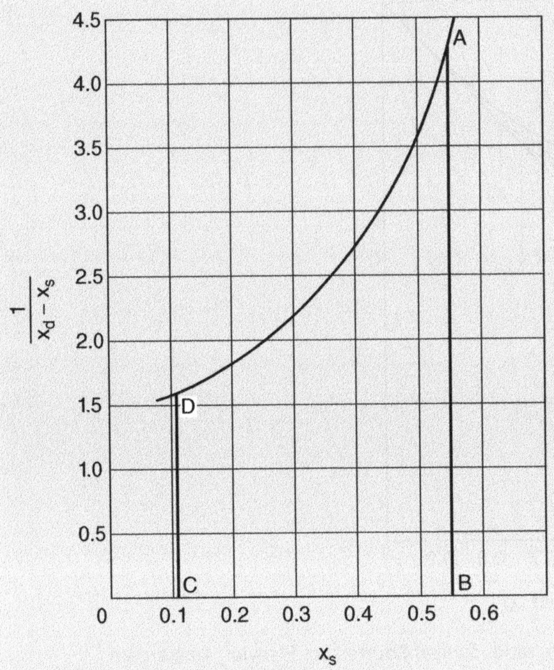

Figure 36.3 Graphical integration for constant reflux ratio. *Source:* Courtesy of Elsevier, Amsterdam, The Netherlands.

Note 36.1

The McCabe–Thiele diagram shows the nonideal behavior of the ethanol/water mixture at concentrations of the MVC exceeding 89 mol%. Concentrations higher than 89 mol% cannot be achieved by straightforward distillation. Further upgrading can be accomplished by, for example, azeotropic distillation. See Section 36.2.

Note 36.2

Azeotropy can be explained by taking a look at the polarity of the molecules of the materials to be separated. In polar molecules, the center of positive electricity does not coincide with the center of negative electricity. In apolar molecules, that coincidence occurs. Generally, "polar molecules and apolar molecules don't like each other." That is demonstrated by the fact that, for example, oil and water do not mix. Let us take, for example, a look at the molecules of water and ethanol. Water consists of polar molecules. The ethanol molecule consists of a polar part, that is, the hydroxyl group. It also contains an apolar part, that is, the alkane part. The result is that, yes, water and ethanol can mix in all ratios. However, there is still some kind of "animosity." That manifests itself at the distillation step. When the ethanol concentration becomes high, that is, 89 mol%, the repulsion in the liquid phase between the two types of molecules becomes that high that the volatility of water is relatively increased. The composition of the vapor is then equal to the composition of the liquid and further distillative separation is not possible. The distillative separation of methanol and water does not have the mentioned complication. Both methanol and water have polar molecules. However, at the distillative separation of propanol-1 and water, the azeotrope occurs at a lower concentration than 89 mol% propanol-1.

Note 36.3

It was already mentioned in Chapter 1 that the molar heat of evaporation of many liquids is approximately 40,000 kJ · kmol^{-1}. The heats of evaporation per kg are different. The situation for ethanol and water is summarized.

	Boiling point, °C	ΔH, kJ · kmol^{-1}	M, kg · kmol^{-1}	ΔH, kJ · kg^{-1}
Ethanol	78.5	38,560	46	838
Water	100	40,644	18	2,258

Operation at Constant Top Product Composition

The reflux ratio R is increased steadily to maintain the desired top product composition. Equation 36.1 is also applicable in this case. The equation can now be solved analytically because x_d is constant. The heat to be supplied to provide the reflux during the complete distillation is given by:

$$Q_R = \lambda \int_0^{L_b} dL_b = \lambda \int_0^{D_b} R \cdot dD_b. \tag{36.2}$$

L_b and D_b are cumulative amounts and not flows in this equation. Again, this batch distillation problem can be solved by means of the McCabe–Thiele diagram. An example illustrates this.

Example 36.2

A mixture of ethanol and water with 0.55 mol fraction of ethanol is distilled to give a top product of 0.75 mol fraction of ethanol. The column has four ideal plates and the distillation is stopped when the reflux ratio has to be increased beyond 4.0. The amount of distillate obtained and the heat required per 100 kmol of initial product are calculated.

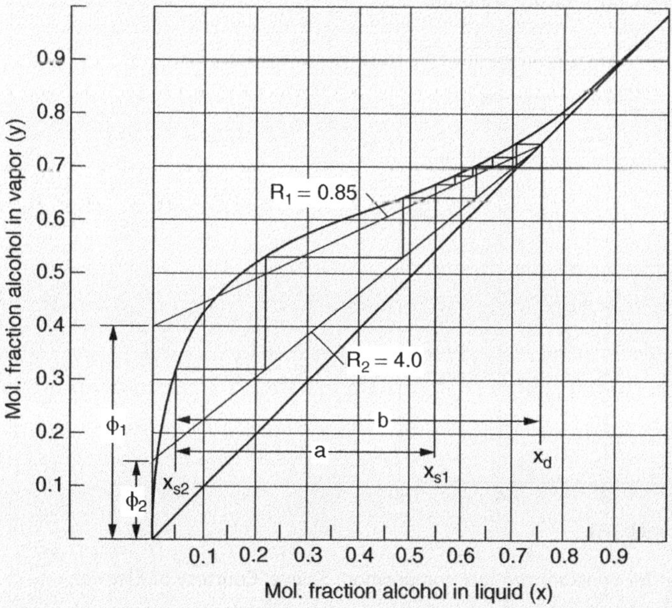

Figure 36.4 Batch distillation—constant product composition. *Source:* Courtesy of Elsevier, Amsterdam, The Netherlands.

Figure 36.4 is the McCabe–Thiele diagram for this distillation. For R = 4.0, the operating line was drawn as follows:

- the point (0.75, 0.75) is on this line, and
- the operating line intersects the ordinate in the point (0, 0.75/(4.0 + 1)), that is, (0, 0.15). See Equation 34.3.

On drawing five steps, $x_s = 0.05$ is found. The meaning of this is as follows. The separation power of this column at R = 4.0 is an upgrade of the concentration of the MVC from 0.05 to 0.75. This is the final situation of the batch distillation of this example.

$$\text{Equation 36.1: } \ln\frac{S_2}{100} = \int_{0.55}^{0.05} \frac{dx_s}{0.75 - x_s} = \ln\frac{0.2}{0.7} \rightarrow S_2 = 28.6 \text{ kmol.}$$

$D_b = 100 - 28.6 = 71.4$ kmol containing 75 mol% of ethanol.

Similarly, D_b-values can be calculated for other R-values. The results of the calculations are as follows:

R	x_s	D_b
0.85	0.55	0
1.0	0.50	20.0
1.5	0.37	47.4
2.0	0.20	63.8
3.0	0.075	70.5
4.0	0.05	71.4

These data have been plotted in Figure 36.5. Now, the integral of Equation 36.2 is solved graphically. The area OABC represents 96 kmol. Taking an average latent heat of evaporation per kmol of 40,000 kJ, the heat to be supplied to provide the reflux is 96 · 40,000/1,000 = 3,840 MJ. The heat to be supplied to provide the reflux per kmol of product is 3,840/71.4 = 53.8 MJ and the total heat per kmol of product is 53.8 + 40.0 = 93.8 MJ.

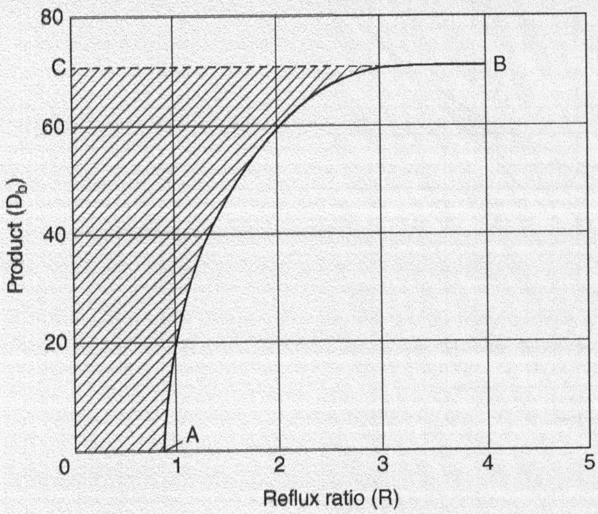

Figure 36.5 Graphical integration for constant product composition. *Source:* Courtesy of Elsevier, Amsterdam, The Netherlands.

Thus, in the previous example, the total heat required per kmol of product is 84.0 + 40.0 = 124.0 MJ and at constant top product composition 93.8 MJ. The product quality is 77 mol% of ethanol in the first example and 75 in the second one.

Review

Example No.	Reflux ratio	Purity distillate, mol%	Yield, %	MJ kmol^{-1} of product
36.1	2.1	77	93.6	124.0
36.2	0.85–4.0	75	97.4	93.8

Going from 75 to 77 mol%, that is, going from 88.5 to 89.5% by weight, reduces the yield and increases the specific energy consumption. It also causes a capacity decrease.

36.2 Azeotropic and Extractive Distillation

Azeotropic Distillation

This type of distillation is applied to separate compounds which form azeotropes. This phenomenon has been discussed in Section 34.1 and it is illustrated in Figure 34.2. The systems acetone/carbon disulphide and acetone/chloroform were given as examples. It is seen that the equilibrium curve crosses the diagonal, indicating the existence of an azeotrope. For the system ethanol/water, the equilibrium curve coincides with the diagonal for ethanol concentrations exceeding 89 mol%, that is, 96% by weight. See Figures 36.2 and 36.4. The ethanol/water mixture containing 89 mol% is also called an azeotropic mixture and cannot be separated by straightforward distillation. The separation of this mixture by means of distillation will now be discussed in some detail.

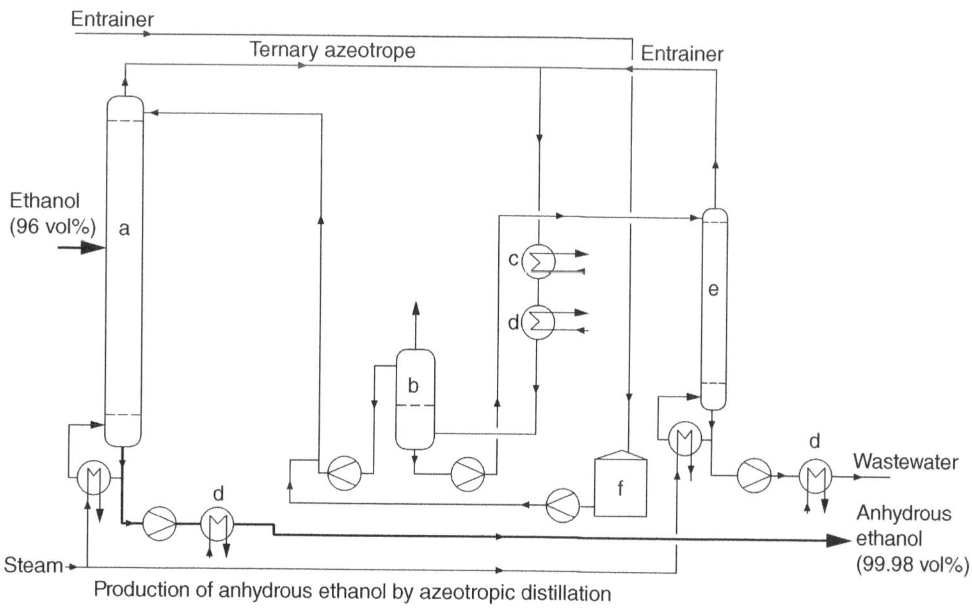

Figure 36.6 Azeotropic distillation for separation of ethanol from water using benzene as entrainer. (a) Dehydrating column; (b) Decanter; (c) Condenser; (d) Cooler; (e) Hydrocarbon stripping column; (f) Entrainer tank.

See Figure 36.6. The system is designed for the production of pure, anhydrous ethanol having a specification of 99.98% by volume. The product contains less than 200 mg · kg^{-1} of water and less than 20 mg · kg^{-1} of other impurities. The dehydrating column and the hydrocarbon stripping column operate at atmospheric pressure. Water is removed from the ethanol/water azeotrope in the form of a ternary azeotrope which is formed by adding an entraining agent such as benzene. Heptane (C_6–C_8 cut) or cyclohexane can also be used. The addition of the relatively nonpolar benzene serves to volatilize water, a highly polar molecule, to a greater extent than ethanol, a moderately polar molecule. The feed, containing 96% by volume of ethanol, enters the dehydrating column near the midpoint. The anhydrous ethanol product collects at the bottom of the tower and is sent through a cooler prior to storage. The ternary azeotrope, having a boiling point of 65 °C, leaves the column as an overhead product which is condensed and then separated into an organic and an aqueous phase in the decanter. The upper organic layer containing the entrainer is returned to the top of the dehydrating column. The lower aqueous layer is pumped to the hydrocarbon stripper, where the hydrocarbon entrainer, the ethanol, and some water are recovered overhead and sent to the condenser/decanter system. Water from the stripper is pumped off as waste. If it contains a substantial amount of ethanol, it may be recycled to the ethanol distillation unit upstream of the azeotropic distillation.

Other azeotropic mixtures can be separated similarly. The third component added to the ternary mixture is sometimes known as entrainer or solvent.

Extractive Distillation

This type of distillation is applied when the equilibrium curve is close to the diagonal, in which case continuous distillation of the mixture to give nearly pure products will require high reflux ratios with corresponding high heat requirements; in addition, it will require a tower having a large cross-section and containing many trays [2].

Also in this case, a third component, termed a solvent, is added. The added solvent is of low volatility and is itself not appreciably vaporized in the fractionator. The effect is to obtain an equilibrium curve which is no longer almost coinciding with the diagonal.

An example of extractive distillation is the separation of toluene, having a boiling point of 110 °C at atmospheric pressure, from paraffin hydrocarbons of approximately the same molecular weight. Such a separation is, in a direct sense, either difficult or even impossible. The separation is practiced in the recovery of toluene from certain petroleum hydrocarbon mixtures. Using i-octane, having a boiling point of 99.5 °C, is an example of using a paraffin hydrocarbon. Figure 36.7 shows that i-octane is the more volatile material in this mixture, but the separation is obviously difficult. In the presence of phenol, having a boiling point of 181.6 °C, however, the difference in volatility between toluene and i-octane increases, so that, with as much as 83 mol% phenol in the liquid, the separation of toluene and i-octane becomes relatively easy. The reason is that there is an affinity between toluene and phenol, both being aromatic compounds, whereas i-octane is an alkane. A flow sheet for a process is in Figure 36.7. 1 and 2 are columns for the separation of i-octane and toluene, respectively.

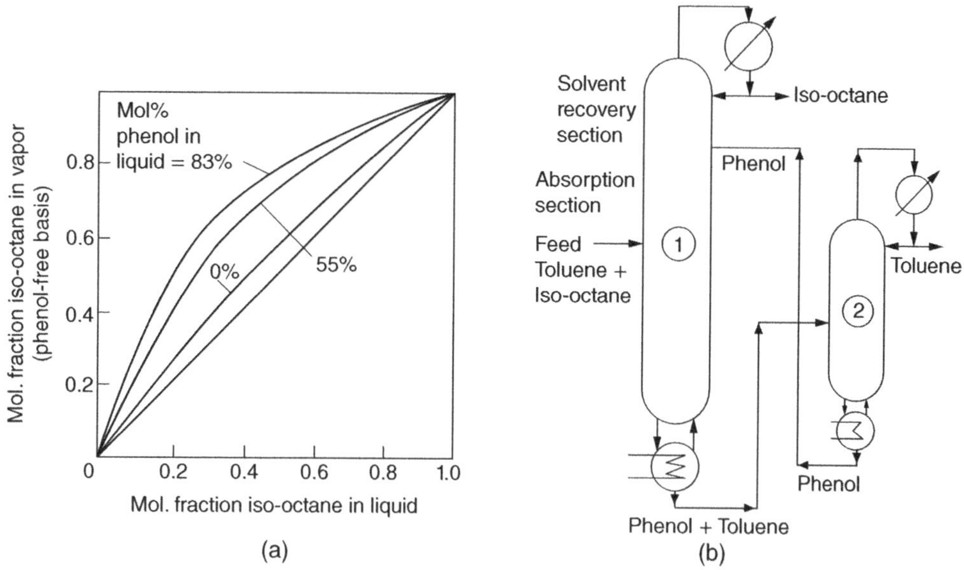

Figure 36.7 Extractive distillation of toluene–i-octane with phenol. *Source:* Courtesy of Elsevier, Amsterdam, The Netherlands.

Extractive distillation is usually a more desirable process than azeotropic distillation since large quantities of solvent must not be vaporized. Furthermore, a greater choice of added component is possible since the process is not dependent upon the accident of azeotropic formation. It cannot be conveniently carried out in batch operation, however.

36.3 Steam Distillation

This type of distillation will be introduced by means of an example [3]. It is required to purify impure nitrobenzene (NB) at atmospheric pressure. The impurities are high boiling components. Live saturated steam is passed through the batch at atmospheric pressure. The steam partly condenses and the heat of condensation is used to evaporate NB. The gaseous mixture is condensed overhead and an NB phase and an aqueous phase are obtained. A still temperature of 99 °C is experimentally found. This means that the sum of the vapor pressures of water and NB is 101,300 $N \cdot m^{-2}$ at 99 °C. The molecular weight of NB is 123 $g \cdot mol^{-1}$ whereas the water molecular weight is 18 $g \cdot mol^{-1}$. The saturated vapor pressure of NB at 99 °C is 2,700 $N \cdot m^{-2}$. It is possible to calculate the mass ratio between the water and the NB collected overhead using these data. The ratio is $((101{,}300 - 2{,}700)/2{,}700) \cdot (18/123) = 5.34$.

Figure 36.8 is a convenient diagram for this type of distillation. It was made by Hausbrand. Its use will be explained as follows. Vapor pressure in $kN \cdot m^{-2}$ is the ordinate and temperature in K is the abcissa. There are three curved lines at the left of the diagram. They are applicable for system pressures of 101,300, 40,000, and 9,300 $N \cdot m^{-2}$. The diagram also contains curves for the saturated vapor pressures of various materials. The system pressure is 101,300 $N \cdot m^{-2}$ in the example concerning the purification of NB. The temperature 99 °C can be found as the abcissa of the intersection of the saturated vapor pressure curve of NB and the curved line for 101,300 $N \cdot m^{-2}$.

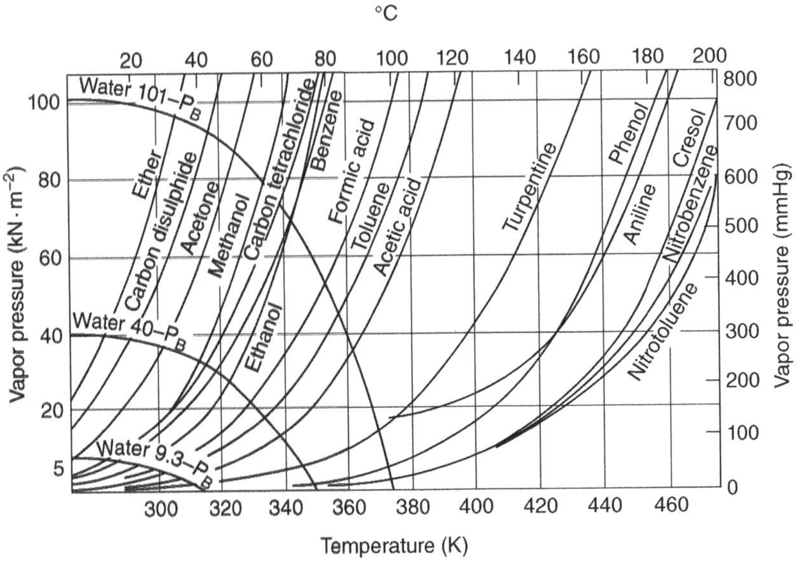

Figure 36.8 Vapor pressure curves for calculations on steam distillation. *Source:* Courtesy of Elsevier, Amsterdam, The Netherlands.

The saturated vapor pressure of NB at 99 °C, that is, 2,700 N · m^{-2}, can be found as the ordinate of the intersection.

Example 36.3

An organic peroxide is synthesized by means of a condensation reaction, that is, a reaction at which water is formed as a by-product. The reactor pressure is 9,300 N · m^{-2}, that is, 948 mm wg. The water produced is evaporated together with toluene. The gaseous flow is condensed overhead. The condensate separates into toluene and water of which toluene is recycled to the reactor. The reactor temperature and the mass ratio water/toluene of the condensate are found and calculated. The molecular weights of toluene and water are 92 and 18 g · mol^{-1}, respectively.

It is possible to read 302 K (29 °C) from Figure 36.8. The saturated vapor pressure of toluene at 29 °C can also be read from the graph, it is 5,000 N · m^{-2}.

$$\text{Mass ratio water/toluene}: \frac{9{,}300 - 5{,}000}{5{,}000} \cdot \frac{18}{92} = 0.17.$$

At this synthesis, toluene is used to remove water from the reactor and to control the reaction.

References

1. Coulson, J. M., and Richardson, J. F., with Backhurst, J. R., and Harker, J. H. (1991). *Chemical Engineering Volume 2, Particle Technology and Separation Processes* (pp. 470–476). Oxford, UK: Butterworth-Heinemann.
2. Coulson, J. M., and Richardson, J. F., with Backhurst, J. R., and Harker, J. H. (1991). *Chemical Engineering Volume 2, Particle Technology and Separation Processes* (pp. 494–497). Oxford, UK: Butterworth-Heinemann.
3. Coulson, J. M., and Richardson, J. F., with Backhurst, J. R., and Harker, J. H. (1991). *Chemical Engineering Volume 2, Particle Technology and Separation Processes* (pp. 497–500). Oxford, UK: Butterworth-Heinemann.

37

Case Studies Distillation

37.1 McCabe–Thiele Diagram

It is required to separate 200 $kmol \cdot h^{-1}$ of a mixture containing 40 mol% of benzene and 60 mol% of toluene into a top product containing 95 mol% of benzene and a bottom stream containing 90 mol% of toluene. q, the mol fraction liquid of the feed, is 0.75. The reflux ratio is 3:1. It is requested to establish the number of theoretical stages required and the tray number at which the feed is added.

Step 1
Obtain an overall balance for the column.

$$200 = D + W,$$

$$200 \cdot 0.4 = 0.95 \cdot D + 0.1 \cdot W,$$

$$D =$$

$$W =$$

Insert the figures into Figure 37.1.

Step 2
Set vapor and liquid flows in the column.

$$L_n = 3D =$$

$$V_n = L_n + D =$$

$$q =$$

$$L_m = L_n + qF =$$

$$V_m = L_m - W =$$

Insert the figures into Figure 37.1.

Step 3
Operating lines

Line	Slope	Point through which it passes
Rectifying line	$L_n/V_n =$	(..., ...)
Stripping line	$L_m/V_m =$	(..., ...)

Draw the operating lines in Figure 37.2.

Introduction to Chemical Engineering, First Edition. C.M. van 't Land.

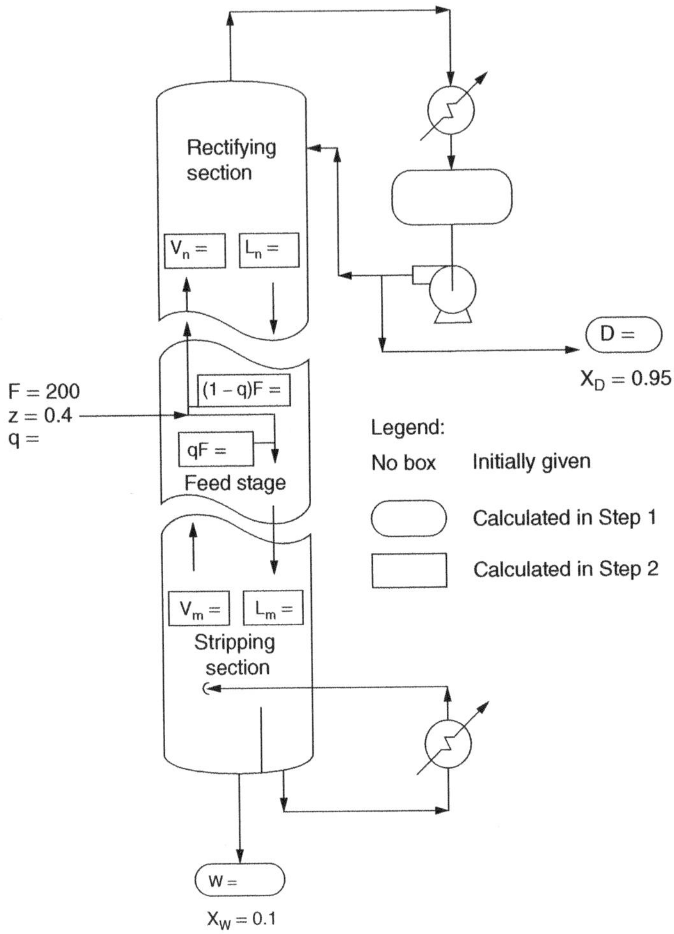

Figure 37.1 Solution to Case Study 37.1: Steps 1 and 2.

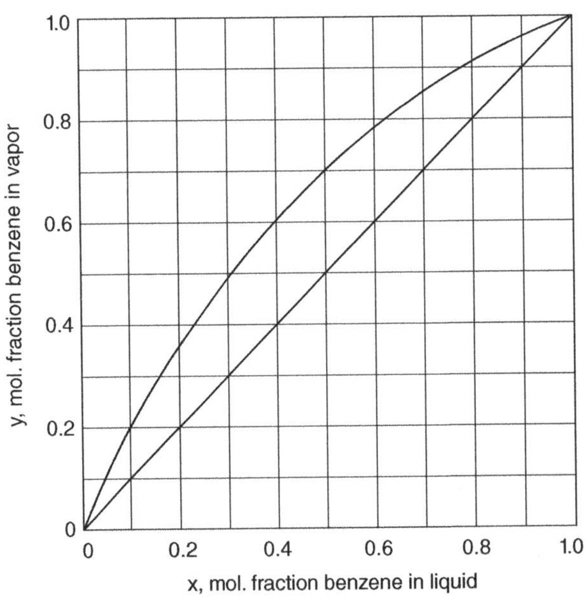

Figure 37.2 Solution to Case Study 37.1: Steps 3 and 4.

Step 4

Step off the stages in the McCabe–Thiele diagram. Start off at the point (0.95, 0.95). Move horizontally to the left until you meet the equilibrium curve. The point of intersection represents the vapor and liquid compositions of stage 1. Then, move vertically down to the operating line. The operating line represents the relationship between the compositions of the vapor approaching a plate and the liquid on and leaving a plate. From there, move left till the equilibrium line is met for the second time, and so on. The number of times the operating line is met is the number of stages.

Number of stages required:
Feed point stage number:

Solution

Step 1

$D = 71.$

$W = 129.$

Step 2

$L_n = 213.$

$V_n = 213 + 71 = 284.$

$q = 0.75.$

$L_m = 213 + 0.75 \cdot 200 = 363.$

$V_m = L_m - W = 363 - 129 = 234.$

Step 3

See Figure 37.3.

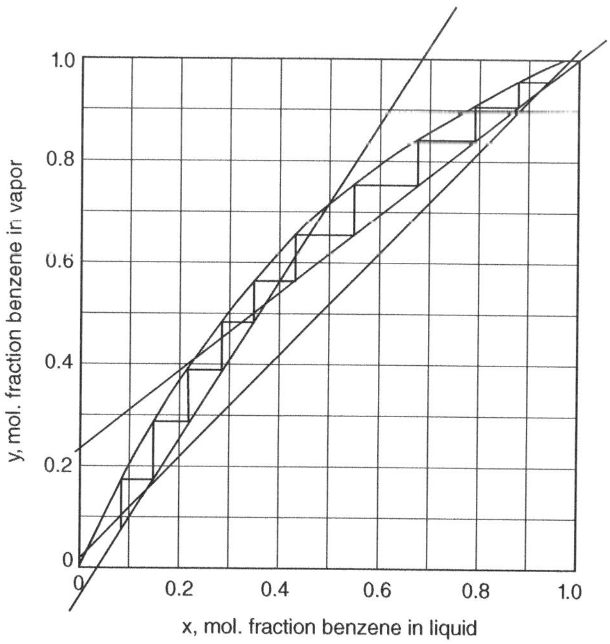

Figure 37.3 Solution to Case Study 37.1: Steps 3 and 4 with drawn lines.

Line	Slope	Point through which it passes
Rectifying line	$L_n/V_n = 0.75$	(0.95, 0.95)
Stripping line	$L_m/V_m = 1.55$	(0.10, 0.10)

Step 4

Number of stages required: 10
Feed point stage number: 6

37.2 Diameter of a Sieve Tray Column and Sieve Tray Pressure Loss

Propane and butane are separated in a sieve tray distillation column [1]. $H = 0.5$ m. The data for the upper tray of the rectifying section are:

$$V_n = 0.1\ \text{kmol} \cdot \text{s}^{-1}.$$

$$L_n = 0.05\ \text{kmol} \cdot \text{s}^{-1}.$$

$$M_g = M_l = 44\ \text{kg} \cdot \text{kmol}^{-1}.$$

$$\rho_l = 500\ \text{kg} \cdot \text{m}^{-3}.$$

$$p = 10^6\ \text{N} \cdot \text{m}^{-2}.$$

$$T = 303\ \text{K}\ (30\ °\text{C}).$$

It is requested to calculate the column diameter and the pressure loss of a sieve tray. $R = 8{,}314\ \text{J} \cdot \text{kmol}^{-1} \cdot \text{K}^{-1}$

$$\rho_g = \frac{M_g \cdot p}{RT} = \qquad \text{kg} \cdot \text{m}^{-3}.$$

$$Q_g = \frac{V_n \cdot M_g}{\rho_g} = \qquad \text{m}^3 \cdot \text{s}^{-1}.$$

$$Q_l = \frac{L_n \cdot M_l}{\rho_l} = \qquad \text{m}^3 \cdot \text{s}^{-1}.$$

$$\Phi = \frac{Q_l}{Q_g}\left(\rho_l/\rho_g\right)^{0.5}$$

Figure 35.2: $v'_g\sqrt{\rho_g/(\rho_l \cdot gH)}$

$$v'_g = \qquad \text{m} \cdot \text{s}^{-1}.$$

$$v_g = 0.85 \cdot v'_g = \qquad \text{m} \cdot \text{s}^{-1}.$$

$$A = \frac{Q_g}{v_g} = \qquad m^2.$$

$$D = \sqrt{4 \cdot Q_g/(\pi \cdot v_g)} = \qquad m.$$

Figure 35.2: F =

$$\Delta p = 1.1 \cdot \rho_g \left(\frac{v_g}{F}\right)^2 + \rho_l \cdot g \cdot 0.05 = \qquad N \cdot m^{-2}.$$

Solution

$$\rho_g = 17.5\ kg \cdot m^{-3}.$$

$$Q_g = 0.25\ m^3 \cdot s^{-1}.$$

$$Q_l = 4.4 \cdot 10^{-3}\ m^3 \cdot s^{-1}.$$

$$\Phi = 0.094.$$

$$v'_g \left(\frac{\rho_g}{\rho_l \cdot gH}\right)^{0.5} = 0.037.$$

$$v'_g = 0.438\ m \cdot s^{-1}.$$

$$v_g = 0.85 \cdot v'_g = 0.372\ m \cdot s^{-1}.$$

$$A = 0.67\ m^2.$$

$$D = 0.93\ m.$$

$$F = 0.065.$$

$$\Delta p = 631 + 245 = 876\ N \cdot m^{-2}.$$

37.3 The Distillation of Wine

The distillation of wine is practiced to obtain, for example, brandy. The vapors escaping from a still are condensed and collected. There is no reflux. The wine initially contains 14 vol% of ethanol and the distillation is stopped when the liquid in the still contains 2.5 vol% of ethanol. It is requested to calculate the amount of distilled liquid obtained from 100 kg of wine, the vol% of ethanol in the overhead product, and the ethanol yield. Figure 37.4 is a graph showing the equilibrium between liquid and vapor for the system ethanol/water. The method of Section 36.1 can be used to calculate the amount of product collected overhead. Figure 37.5 is a graph showing the relationship between vol% and wt% of ethanol/water mixtures. The specific masses of ethanol and water are 790 and 1,000 $kg \cdot m^{-3}$, respectively.

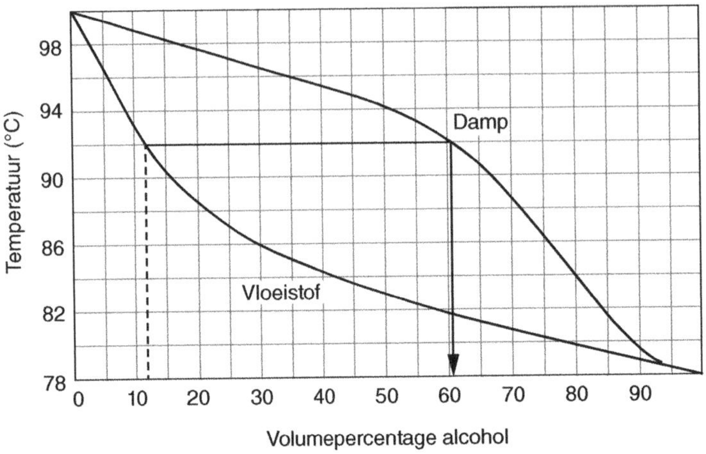

Figure 37.4 Liquid/vapor equilibrium for the system ethanol/water. A translation of the words in the graph follows: Damp, vapor; Temperatuur, temperature; Vloeistof, liquid; Volumepercentage alcohol, volume percentage alcohol.

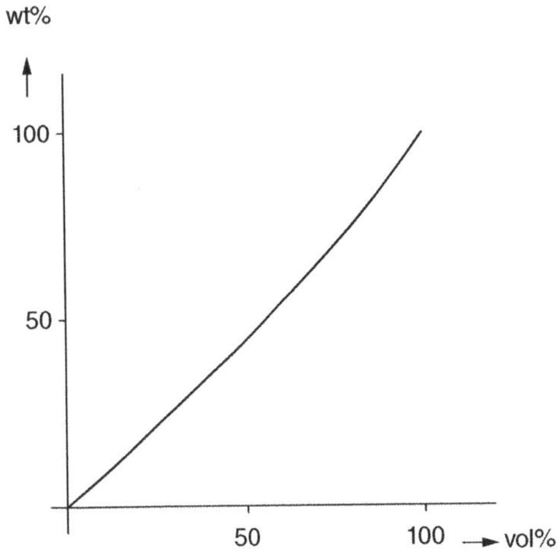

$$y = \frac{0.79 \cdot x}{0.79 \cdot x + 100 \cdot x} \cdot 100.$$

Figure 37.5 Wt% ethanol versus vol% ethanol in ethanol/water mixtures.

Liquid		Condensed Vapor			
vol%	% by weight, x_s	vol%	% by weight, x_d	$x_d - x_s$	$1/(x_d - x_s)$
14					
12					
10					
7.5					
5					
2.5					

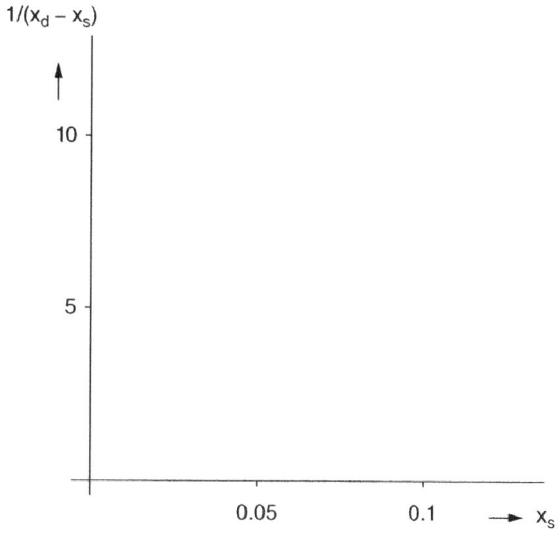

Figure 37.6 $1/(x_d - x_s)$ as a function of x_s without drawn line.

$1/(x_d - x_s)$ can be plotted as a function of x_s in Figure 37.6. The next step is the estimation of the area corresponding with the following integral:

$$\int_{0.11}^{0.02} \frac{dx_s}{x_d - x_s} =$$

$S_1 = 100\ \text{kg},\ \ln \frac{S_1}{S_2} =$

$\frac{S_1}{S_2} =$

$D_b = S_1 - S_2 =$ kg.

Mass of ethanol in overhead product: $x_1 \cdot S_1 - x_2 \cdot S_2 =$ kg.
% by weight of ethanol in overhead product:
Vol% of ethanol in overhead product:
The ethanol yield is:

Solution

Liquid			Condensed Vapor				
vol%	% by weight	x_s	vol%	% by weight	x_d	$x_d - x_s$	$1/(x_d - x_s)$
14	11	0.11	67	61	0.61	0.50	2.0
12	10	0.10	62	56	0.56	0.46	2.2
10	8	0.08	60	54	0.54	0.46	2.2
7.5	6	0.06	48	42	0.42	0.36	2.8
5	4	0.04	38	33	0.33	0.29	3.4
2.5	2	0.02	15	12	0.12	0.10	10.0

See Figure 37.7.

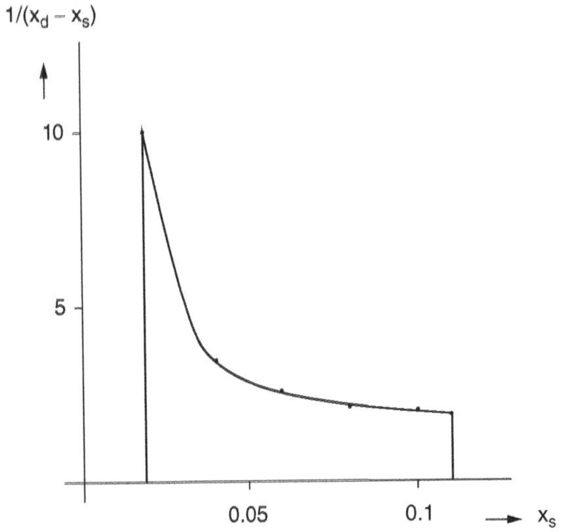

Figure 37.7 $1/(x_d - x_s)$ as a function of x_s with drawn line.

$$\int_{0.11}^{0.02} \frac{dx_s}{x_d - x_s} = 0.29,$$

$$\ln \frac{S_1}{S_2} = 0.29 \rightarrow \frac{S_1}{S_2} = 1.34.$$

$$D_b = S_1 - S_2 = 100 - \frac{100}{1.34} = 25.4.$$

Mass of ethanol in overhead product: $x_1 \cdot S_1 - x_2 \cdot S_2 = 0.11 \cdot 100 - 0.02(100 - 25.4) = 9.5$ kg.
% by weight of ethanol in overhead product: $(9.5/25.4) \cdot 100 = 37.4$.
Vol% of ethanol in overhead product: 43.
Ethanol yield: $(9.5/11.0) \cdot 100 = 86.4\%$.

37.4 Steam Distillation

A convective rotary dryer is used to dry carbon spheres having diameters in the range 1–1.5 mm. The composition of the feed is:

- $100\ kg \cdot h^{-1}$ of carbon,
- $100\ kg \cdot h^{-1}$ of toluene, and
- $100\ kg \cdot h^{-1}$ of water.

The dryer length is 9 m and the dryer diameter is 0.9 m. The rotational speed is several rpm. The residence time of the solid is approximately 20 min. The solid travels in plug flow through the dryer.

Superheated steam having a temperature of 310 °C and being at atmospheric pressure ($101{,}300\ N \cdot m^{-2}$) is used as a strip gas. The product and the superheated steam move countercurrently through the dryer. The dryer is heated indirectly by means of oil burners. Both toluene and water are free moisture, that is, they are not bound to the solid.

The Hausbrand diagram in Section 36.3 can be used.

What is the initial drying temperature?

Which volatile is removed first?

What is the drying temperature when only one of the two volatiles remains?

The spheres leave the dryer at 210–240 °C. Does that make sense?

Solution

Initial drying temperature

Figure 36.8: 83 °C

Volatile removed first

Mass ratio water/toluene in condensate: $(5.63 \cdot 10^3/(4.5 \cdot 10^3)) \cdot (18/92) = 0.245$

Toluene is removed first. See Figure 36.8.

Drying temperature when only water is removed: 100 °C.

Dried product exit temperature.

The product temperature is raised from 100 to 210–240 °C to remove bound moisture.

Reference

1. Wesselingh, J. A., and Kleizen, H. H. (1998). *Separation Processes* (pp. 119, 214). Delft, The Netherlands: Delft University Press. (in Dutch).

Notation IV

A	Column cross-sectional area	m^2
c_1	constant in Sections 35.1 and 35.2	—
c_1'	constant in Section 35.2	—
c_2	constant in Sections 35.1 and 35.2	—
c_3	constant in Section 35.2	—
D	Top product flow	$kmol \cdot h^{-1}$
	Column diameter	m
D_b	Amount of top product	kmol
d	Packing ring diameter	m
F	Feed flow	$kmol \cdot h^{-1}$
	Fraction free area	—
F_p	Packing factor	—
g	Acceleration due to gravity	$m \cdot s^{-2}$
H	Distance between plates	m
	Height of a packed column	m
ΔH	Heat of evaporation	$kJ \cdot kmol^{-1}$
		$kJ \cdot kg^{-1}$
L_b	Amount of liquid fed back to a column	kmol
L_m	Liquid flow in the stripping part	$kmol \cdot h^{-1}$
L_{m+1}	Liquid flow leaving tray m+1	$kmol \cdot h^{-1}$
L_n	Liquid flow in the rectifying part	$kmol \cdot h^{-1}$
		$kmol \cdot s^{-1}$
L_{n+1}	Liquid flow leaving tray n+1	$kmol \cdot h^{-1}$
M	Average molecular weight	$kg \cdot kmol^{-1}$
	Molecular weight	$kg \cdot kmol^{-1}$
M_g	Gas molecular weight	$kg \cdot kmol^{-1}$
M_l	Liquid molecular weight	$kg \cdot kmol^{-1}$
m_v	Amount of vapor	mol
m_l	Amount of liquid	mol
P	Pressure	$N \cdot m^{-2}$
P_{A0}	Saturated vapor pressure of component A	$N \cdot m^{-2}$
P_{B0}	Saturated vapor pressure of component B	$N \cdot m^{-2}$
p	pressure	$N \cdot m^{-2}$
Δp	"Wet" gas pressure loss in a packed column	$N \cdot m^{-2}$

Introduction to Chemical Engineering, First Edition. C.M. van 't Land.

Δp_d	"Dry" gas pressure loss in a packed column	$N \cdot m^{-2}$
Δp_g	Gas pressure loss of a sieve tray	$N \cdot m^{-2}$
Δp_{holes}	Vapor pressure loss due to flow through holes	$N \cdot m^{-2}$
Δp_{nip}	Liquid pressure loss due to flow through a nip	$N \cdot m^{-2}$
Q_g	Gas flow	$m^3 \cdot s^{-1}$
Q_l	Liquid flow	$m^3 \cdot s^{-1}$
Q_R	Amount of heat for reflux	kJ
q	Fraction liquid of the feed	—
R	Reflux flow	$kmol \cdot s^{-1}$
	Reflux ratio	—
	Universal gas constant (8.314 $kJ \cdot kmol^{-1} \cdot K^{-1}$)	
$R_{1,2}$	Reflux ratios Nos. 1 and 2	—
R_{min}	Minimum reflux ratio	—
S	Still content	kmol
$S_{1,2}$	Still contents Nos. 1 and 2	kmol
T	Temperature	°C or K
T'	Temperature	°C or K
$T_{1,2,3}$	Temperature Nos. 1, 2, and 3	°C or K
V_m	Vapor flow in the stripping part	$kmol \cdot h^{-1}$
V_n	Vapor flow in the rectifying part	$kmol \cdot h^{-1}$
		$kmol \cdot s^{-1}$
V_t	Vapor flow leaving a column top	$kmol \cdot h^{-1}$
v_f	Average film velocity	$m \cdot s^{-1}$
v_g	Corrected gas velocity	$m \cdot s^{-1}$
v_g'	Gas velocity based on a column cross-section	$m \cdot s^{-1}$
	Gas velocity	$m \cdot s^{-1}$
v_l	Liquid velocity based on a column cross-section	$m \cdot s^{-1}$
W	Bottom product flow	$kmol \cdot h^{-1}$

The following text is valid for the symbol x with numbers and characters as subscripts. x stands for the dimensionless mole fraction in the liquid phase of the more volatile component of a binary mixture. x also stands for the volume% ethanol in ethanol/water mixtures.

$x_{1,2,3,4,5,6}$	Nos. 1, 2, 3, 4, 5, and 6
x_d	In the top liquid
$x_{d1,d2}$	In the top products Nos. 1 and 2
x_f	In the feed
x_g	In an azeotropic mixture
x_m	In the liquid leaving tray m
x_{m+1}	In the liquid leaving tray m+1
x_{n+1}	In the liquid leaving tray n+1
x_s	In the still content
$x_{s1,s2}$	In the bottom products Nos. 1 and 2
x_t	In the liquid leaving the top tray
x_{t-1}	In the liquid leaving the second tray from the top
x_w	In the bottom

The following text is valid for the symbol x with capitals as subscripts.

x_A	Mol fraction of component A in the liquid phase	—
x_B	Mol fraction of component B in the liquid phase	—

y stands for the dimensionless mole fraction in the vapor phase of the more volatile component of a binary mixture. y also stands for the weight% ethanol in ethanol/water mixtures.

$y_{1,2,3}$	Nos. 1, 2, and 3
y_m	In the vapor leaving tray m
y_n	In the vapor leaving tray n
y_s	In the vapor entering a column at the bottom
y_t	In the vapor leaving the top tray
y_{t-1}	In the vapor leaving the second tray from the top
y_{t-2}	In the vapor leaving the third tray from the top

Greek Symbols

δ	Film thickness	m
ε	Volume fraction liquid in a column	—
λ	Heat of evaporation	$kJ \cdot kmol^{-1}$
ρ_g	Gas specific mass	$kg \cdot m^{-3}$
ρ_l	Liquid specific mass	$kg \cdot m^{-3}$
Φ	Flow parameter	—
$\Phi_{1,2}$	Ordinates Nos. 1 and 2 in Figure 36.4	—

Part V

Liquid Extraction

Part V Content

38

Liquid Extraction – Part 1

38.1 Introduction

Liquid extraction is the separation of the component(s) of a liquid mixture by treatment with a solvent in which one or more of the desired component(s) is preferentially soluble. It is essential that the liquid-mixture feed and solvent are at least partially if not completely immiscible.

Three stages are involved:

- bringing the feed mixture and the solvent into intimate contact,
- separating the feed mixture and the solvent into two phases, and
- recovering the desired component(s) from the solvent.

Contacting and separation can be carried out continuously in stagewise equipment like a mixer/settler unit. See Figure 38.1. Mixer/settler units in a copper solvent extraction plant are depicted in Figure 38.2. Today, about 80% of the copper produced worldwide is obtained from the ore chalcopyrite, that is, $Cu_2S.Fe_2S_3$. Approximately 20% of the copper produced worldwide today is obtained by using a hydrometallurgical process. This process was introduced in the 1960s.

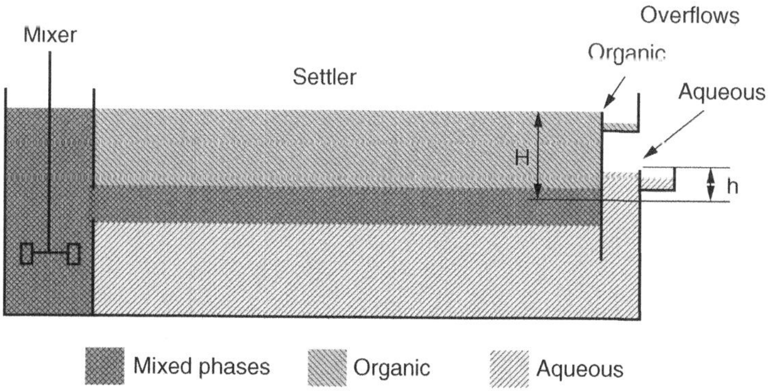

Figure 38.1 A mixer/settler unit. *Source:* Courtesy of Taylor & Francis Group, Abingdon-on-Thames, UK.

Figure 38.2 Copper solvent extraction plant at a mine in Arizona, USA, showing the mixer/settler train with tankhouse in the background. *Source:* Courtesy of Taylor & Francis Group, Abingdon-on-Thames, UK.

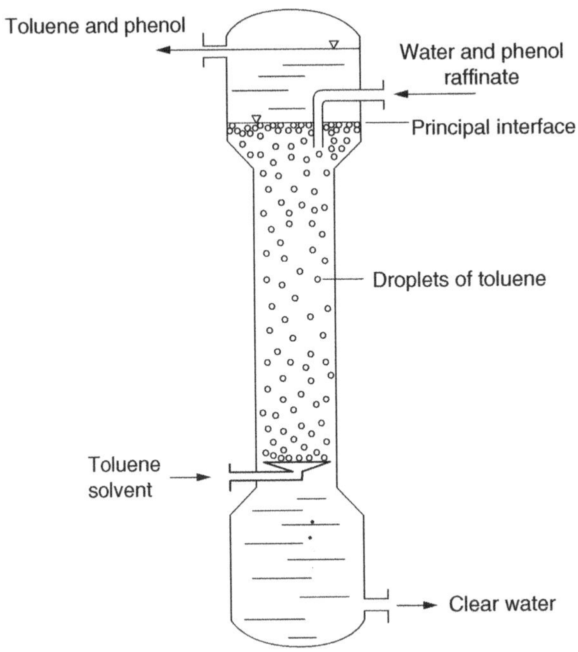

	Density $kg \cdot m^{-3}$	Viscosity $kg \cdot m^{-1} \cdot s^{-1}$	Surface tension $N \cdot m^{-1}$
Water	998	$0.95 \cdot 10^{-3}$	$35.4 \cdot 10^{-3}$
Toluene	866.5	$5.86 \cdot 10^{-3}$	

Figure 38.3 An extraction column. *Source:* Courtesy of Taylor & Francis Group, Abingdon-on-Thames, UK.

The raw material for this process is an oxide-based ore. Milled low-grade ore is treated with diluted sulfuric acid (heap leaching). This ore contains approximately 0.5% by weight of copper. Copper and some undesired components dissolve in the acid. Next, the aqueous solution is contacted with an organic solvent. Copper is transferred selectively from the aqueous solution to the organic solvent. Then, the organic solvent is brought into contact with concentrated sulfuric acid. Finally, copper is obtained from the latter solution by electrowinning. The mixer/settlers depicted in Figure 38.2 are part of a plant in which the hydrometallurgical process is used. Mixer/settlers imply a large holdup and a large area. Therefore, extraction columns are often preferred. Such a piece of equipment (a spray column) is depicted in Figure 38.3. In this figure, the extraction of phenol from coal tar liquors by means of toluene is shown. Figure 38.4 shows various types of stirred columns.

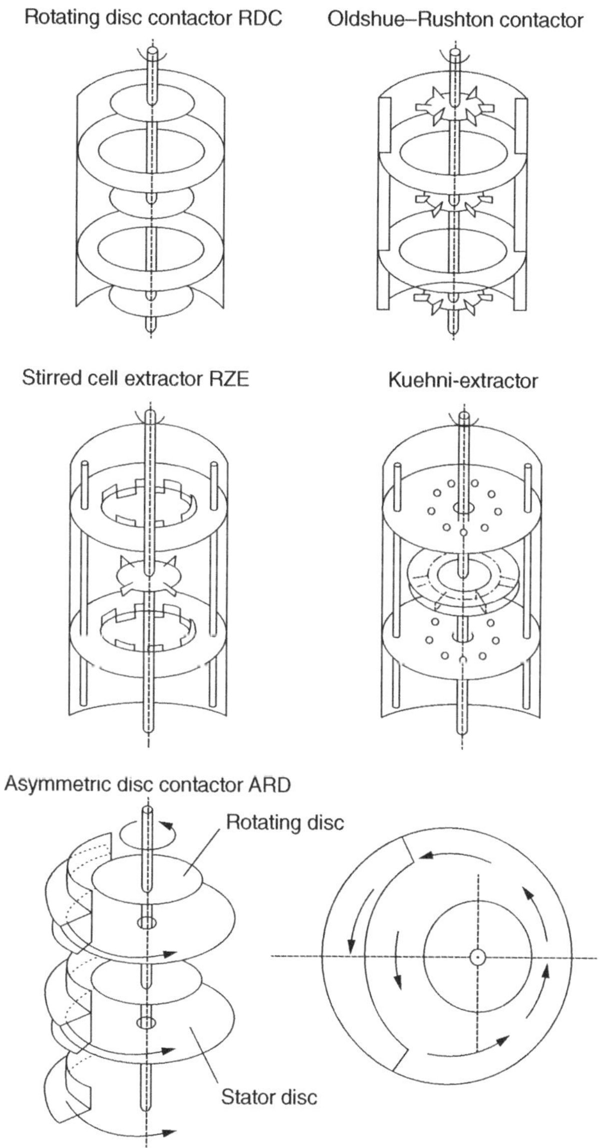

Figure 38.4 Various types of stirred columns. *Source:* Courtesy of Taylor & Francis Group, Abingdon-on-Thames, UK.

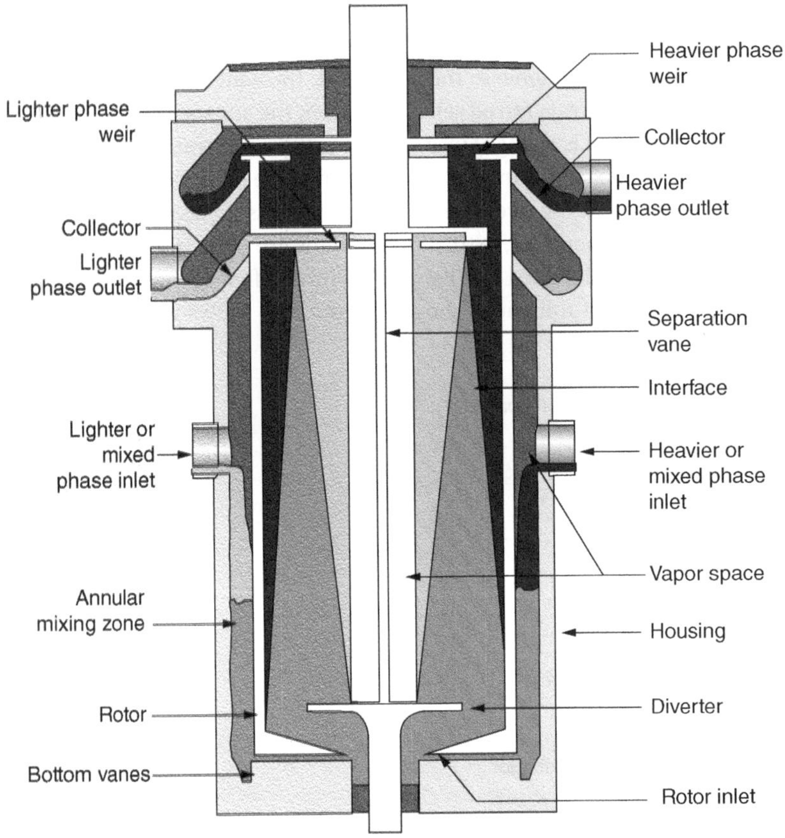

Figure 38.5 A rotating extractor. *Source:* Courtesy of CINC, Brakel, Germany.

Third, there is a variety of rotating devices like the CINC extractor. See Figure 38.5. They are applied when the specific mass difference of the two phases is less than 50 $kg \cdot m^{-3}$. The bonus of rotating equipment over mixer/settlers and columns is the presence of a gravitational field much more powerful than the earth's gravitational field. The trade-off is the greater equipment complexity, like the presence of bearings and seals. The greater complexity is reflected in the price.

The separation of aromatics from paraffin and naphthenic compounds to improve the temperature/viscosity characteristics of lubricating oils is an important application.

The dehydration of acetic acid is shown in Figure 38.6. The wet acid is fed to a countercurrent extraction column and the extract (solvent, acetic acid, and some water) is fed to a distillation column. The top product of the latter column is an azeotropic solvent/water mixture whereas the bottom product is dry acetic acid. The condensed top product separates into an aqueous layer and a solvent layer. The aqueous layer is added to the raffinate, that is, the aqueous phase leaving the extraction column. A small quantity of solvent is stripped from the combined aqueous phases in a second distillation column. It is important that the raffinate contains as little acid as possible, otherwise it will be lost in the effluent.

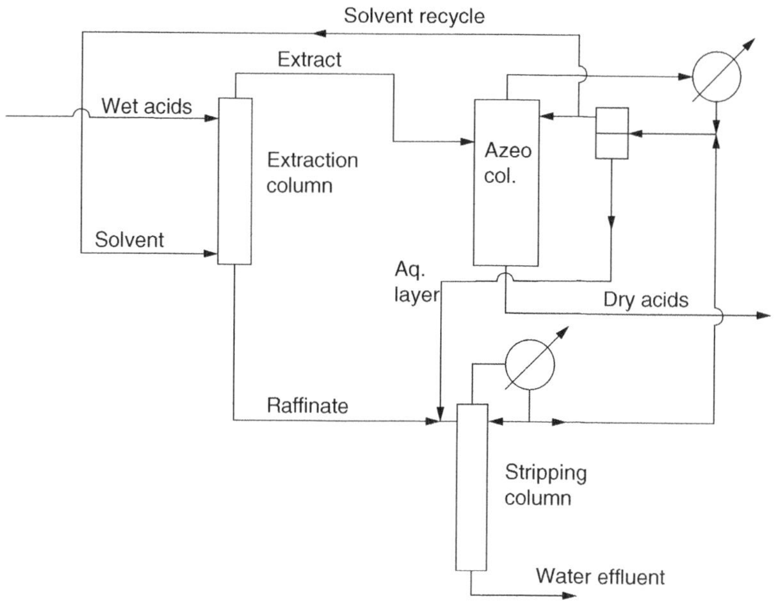

Figure 38.6 The dehydration of acetic acid. *Source:* Courtesy of Taylor & Francis Group, Abingdon-on-Thames, UK.

An alternative for the recovery of acetic acid from aqueous solutions is azeotropic distillation only. The economy of that process is less favorable, both concerning fixed and variable costs [1].

38.2 The Distribution Coefficient

A start is made with the description of an experiment. 3 g of phenol (C_6H_5OH) are dissolved in 1 kg of water. The mass ratio phenol/water is 0.003. The solution is subsequently shaken in a separation funnel with 1 kg of benzene (C_6H_6) for, for example, half a minute at ambient temperature. Next, the aqueous and the organic phases are separated. Analysis shows that the organic phase now contains 2 g of phenol whereas 1 g of phenol stays in the aqueous phase. This means that the mass distribution coefficient of phenol between benzene and water is 2. It is the ratio of the mass ratios phenol/benzene and phenol/water when there is equilibrium between the two phases. It is also possible to define a distribution coefficient based on molar ratios. The molecular weights of phenol, benzene, and water are, respectively, 94, 78, and 18. Based on the aforementioned experimental results, the molar distribution coefficient of phenol between benzene and water is 8.7. For mass balance calculations, mass distribution coefficients or molar distribution coefficients should be used. The volumetric distribution coefficient is useful for equipment sizing. It is the ratio of the concentrations in the two liquid phases. The usual symbol of the distribution coefficient is m. Distillation is almost exclusively described in terms of molar flows and molar ratios. An important reason is the existence of the concept of constant molar overflow. That concept does not play a role in extraction. Therefore, concentrations are sometimes given in $kg \cdot m^{-3}$.

38.3 Calculation of the Number of Theoretical Stages in Extraction Operations

This is the first method to interpret liquid/liquid extraction in extraction columns. An actual extraction column is considered and the question is asked what number of mixer/settlers (stages) would give the same performance as the column. The feed and the solvent flow, like in a column, countercurrently through the series of mixer/settlers. See Figure 38.7. The two flows leaving a stage are in equilibrium. Each circle corresponds with a mixer and a separator. The initial solution of a solute in solvent A is fed to the first unit and leaves that unit as a raffinate. The stream passes through the units and leaves the nth unit. The fresh solvent enters the nth unit and passes in the reverse direction through the units, leaving the installation as an extract.

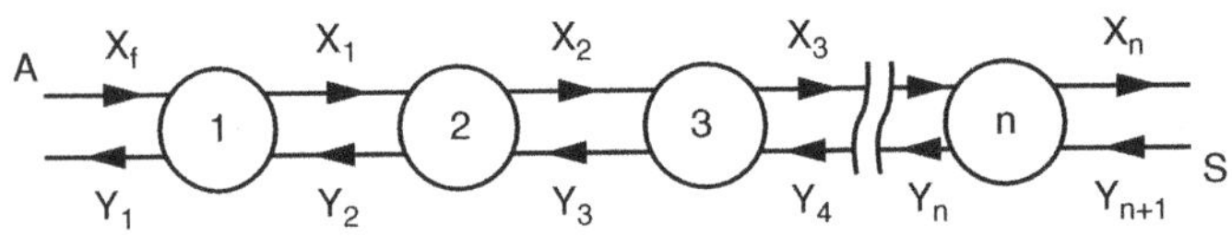

Figure 38.7 Arrangement for multiple-contact extraction in countercurrent flow.

Let X denote the molar ratio of solute to solvent in the raffinate streams, and Y denote the molar ratio of solute to solvent in the extract streams. Since the two solvents are immiscible, the solvent in the raffinate streams remains as A units, and the added solvent in the extract streams as S. The following material balances can be written for the solute.

For the first stage: $A \cdot X_f + S \cdot Y_2 = A \cdot X_1 + S \cdot Y_1$.

For the nth stage: $A \cdot X_{n-1} + S \cdot Y_{n+1} = A \cdot X_n + S \cdot Y_n$.

For the whole installation: $A \cdot X_f + S \cdot Y_{n+1} = A \cdot X_n + S \cdot Y_1$.

The latter equation can be re-written: $Y_{n+1} = \frac{A}{S}(X_n - X_f) + Y_1$. This is the equation of a straight line having a slope A/S, known as the operating line. It gives a relationship between the molar ratios in the two flows between two mixing stages. It is a relationship based on material balances. The two flows are not in equilibrium between two mixing stages. There is an analogy with the derivation of the operating lines for continuous distillation. See Section 34.4 of Part 4, Distillation. It is possible to represent the relationship graphically. The straight line passes through the points (X_f, Y_1) and (X_n, Y_{n+1}). See Figure 38.8. It is also possible to draw the equilibrium line in Figure 38.8. The number of stages required to pass from X_f to X_n is found by drawing steps between the operating line and the equilibrium line. The method was proposed by McCabe and Thiele and was discussed in Section 34.4 of Part 4, Distillation. Four stages are required in the example shown in Figure 38.8.

The attainment of equilibrium in each mixer/settler is a starting point for the calculation scheme of this section. For mixer/settlers and low-viscosity liquids, the equilibrium is approached closely when the average residence time is 1–2 min. The average droplet size is normally in the range 1–3 mm. Smaller droplets lead to larger settlers.

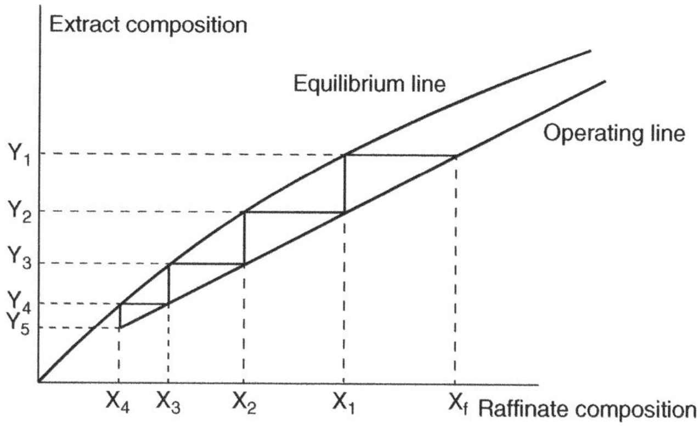

Figure 38.8 Graphical method for determining the number of stages for a process shown in Figure 38.7, using immiscible solvents.

Example 38.1

A glass pulsed packed column is depicted in Figure 38.9. The height of the column is 4 m whereas the diameter is 0.05 m. The column is packed with glass Raschig rings having a height and a diameter of 0.01 m. The column was filled at random with Raschig rings. The pulsation at the bottom of the column served to enhance mass transfer by enlarging the available area. Benzoic acid was extracted from toluene by water. The Raschig rings were preferentially wetted by the aqueous phase. The test temperature was 24 °C. At a typical experiment, the water flow was $0.014\ m^3 \cdot h^{-1}$ and the concentration of benzoic acid in the aqueous phase increased from 0 to 0.01 $kmol \cdot m^{-3}$ ($1.22\ g \cdot l^{-1}$). The toluene flow was approximately $1.56 \cdot 10^{-3}\ m^3 \cdot h^{-1}$ and the concentration decreased from 0.10 ($12.2\ g \cdot l^{-1}$) to 0.01 $kmol \cdot m^{-3}$ ($1.22\ g \cdot l^{-1}$). The volumetric distribution coefficient of benzoic acid between toluene and water is approximately 8. The horizontal piston had a stroke of 8 mm and a frequency of $2\ s^{-1}$.

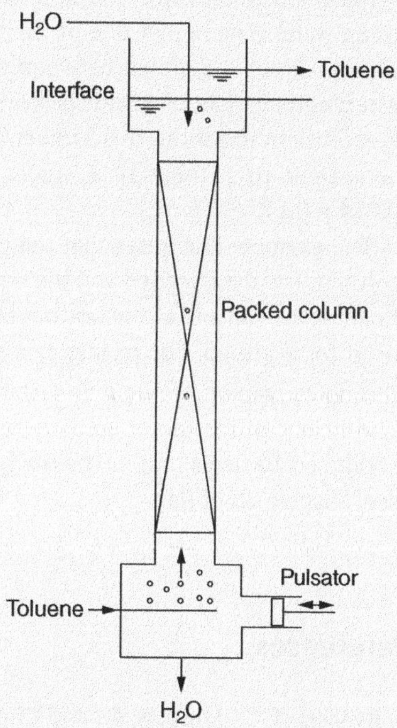

Figure 38.9 A pulsed packed extraction column.

The equilibrium line and the operating line are in Figure 38.10. The equilibrium line is established by means of experiments in a laboratory [2]. The operating line is drawn between two points. These two points are representative for the concentrations in the two phases at the top and at the bottom of the column. Note that, in

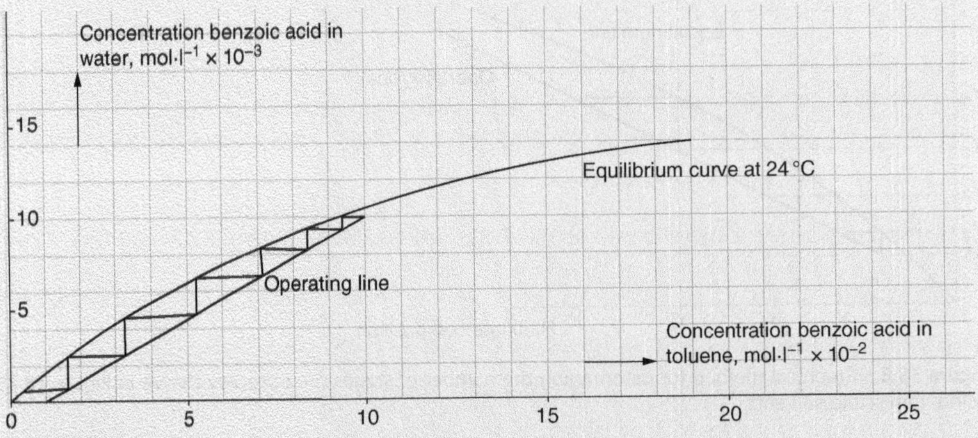

Figure 38.10 Number of stages in a pulsed packed column.

this case, the conditions at the top of the column are represented by the lower of the two points in the graph. At the top of the column, toluene almost depleted from benzoic acid is contacted by clean water. The higher one of the two points represents the conditions at the bottom of the column. Steps are drawn between these two lines and the number of stages is found as 6.5 whereas the height of a stage is 4/6.5 = 0.62 m. This means that the performance of this column is equivalent to a series of 6.5 mixer/settlers. The height of a stage is also indicated as the height equivalent to a theoretical stage (HETS). The slope of the operating line is $1.56 \cdot 10^{-3}/0.014 = 0.11$.

The example illustrates that the method can be used to analyze the behavior of an existing column. The data needed are: the equilibrium line, the flows, and the concentrations in the two flows at the top and at the bottom of the column. The experimental results can, in principle, be used for scale-up. For example, a scale-up factor of 10 would mean 10 times larger flows through a column having a diameter $\sqrt{10}$ times larger than the diameter of the tested column. The height of the larger column could be 4 m. The volume displaced by the pulsation per h should be 10 times larger. The margins to avoid flooding should be wide. Concerning flooding, see Chapter 40 of Part 5.

References

1. Rydberg, J. C., Musikas, C., and Chopin, G. R. (1992). *Principles and Practices of Solvent Extraction* (pp. 329–331). New York, NY: Marcel Dekker.
2. Appel, F. J., and Elgin, J. C. (1937). Countercurrent extraction of benzoic acid between toluene and water: Performance of spray and packed columns. *Industrial and Engineering Chemistry*, 29 (4), 451–459.

39

Liquid Extraction – Part 2

39.1 Calculation of the Number of Transfer Units in Extraction Operations

This is the second method to interpret liquid extraction in extraction columns. An actual extraction column is considered and the number of transfer units (NTU) is calculated. The column height divided by the NTU is the height of a transfer unit (HTU). The method is also suitable for the design of extraction columns. The method resembles the method to design heat exchangers.

The concentration gradients for transfer of a solute from a raffinate to an extract phase are as shown in Figure 39.1. This figure is similar to Figure 13.1 for convective mass transfer in the part on Transport Phenomena.

The transfer through the film on the raffinate side of the interface is brought about by a concentration difference $C_R - C_{Ri}$, and through the film on the extract side by a concentration difference $C_{Ei} - C_E$.

The rate of transfer across these films is:

$$\Phi''_{mol} = k_R(C_R - C_{Ri}), \tag{39.1}$$

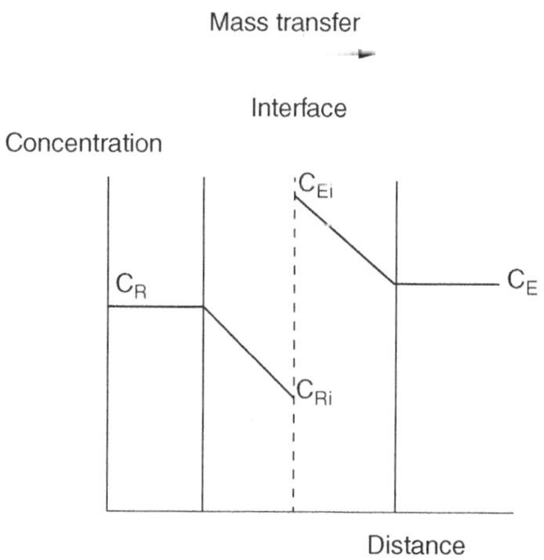

Figure 39.1 Concentration profile near an interface.

and

$$\Phi''_{mol} = k_E(C_{Ei} - C_E). \tag{39.2}$$

k_R and k_E are called partial mass transfer coefficients ($m \cdot s^{-1}$). If the equilibrium curve can be taken as a straight line: $C_{Ei} = m \cdot C_{Ri}$.

The latter equation can be used to eliminate both C_{Ri} and C_{Ei} from Equations 39.1 and 39.2 to obtain one equation that can be written in two different ways, that is,

$$\Phi''_{mol} = K_R\left(C_R - \frac{C_E}{m}\right), \tag{39.3}$$

and

$$\Phi''_{mol} = K_E(m \cdot C_R - C_E), \tag{39.4}$$

in which $\frac{1}{K_R} = \frac{1}{k_R} + \frac{1}{m \cdot k_E}$ and $\frac{1}{K_E} = \frac{1}{k_E} + \frac{m}{k_R}$.

K_R and K_E are called overall mass transfer coefficients. The meaning of these two equations is as follows. On taking a look at Equation 39.3 and Figure 39.1, the question can be asked: given the fact that the concentration in the extract phase is C_E, what should be the concentration in the raffinate phase to have equilibrium between the two phases? The answer is C_E/m. Thus, $C_R - C_E/m$ can be defined as the driving force in the raffinate phase for mass transfer. By the same token, $m \cdot C_R - C_E$ can be defined as the driving force in the extract phase. Note that K_R is a function of both k_R and k_E. K_E is a function of the same two variables as well. Furthermore, $K_R = m \cdot K_E$.

Functioning of a column operating as a countercurrent unit

See Figure 39.2. If Φ'_{vR} and Φ'_{vE} are the volumetric flow rates of raffinate and extract phases per unit column cross-sectional area ($m \cdot s^{-1}$), a is the interfacial surface per unit column volume ($m^2 \cdot m^{-3}$), and Z is the height of packing in the column (m), then, over a small height dZ, a material balance gives:

$$\Phi'_{vR} \cdot dC_R \cdot 1 = \Phi'_{vE} \cdot dC_E \cdot 1 \quad kmol \cdot s^{-1}.$$

It follows from Equation 39.3:

$$-\Phi'_{vR} \cdot dC_R \cdot 1 = K_R \cdot a\left(C_R - \frac{C_E}{m}\right) \cdot dZ \cdot 1 \text{ (per m}^2\text{ cross-sectional area)}.$$

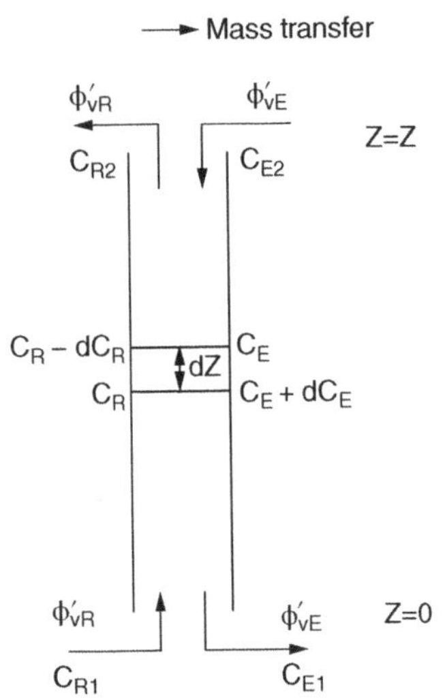

Figure 39.2 Countercurrent flow in a column.

The left-hand side of the equation gives the number of kmol transferred in a column volume having a height dZ and a cross-sectional area of 1 m^2. The minus sign in the left-hand side of the equation follows from the fact that the raffinate phase is depleted from the component transferred between the two phases. The right-hand side of the equation indicates what this kinetically means in terms of $K_R \cdot a$. K_R and a are considered together because the actual area in $m^2 \cdot m^{-3}$ at these processes is unknown.

It follows:

$$-\int_{C_{R1}}^{C_{R2}} \frac{dC_R}{C_R - C_E/m} = \frac{K_R \cdot a}{\Phi'_{vR}} \int_0^Z dZ = \frac{K_R \cdot a}{\Phi'_{vR}} \cdot Z. \qquad (39.5)$$

The integral on the left-hand side of Equation 39.5 is known as the number of overall raffinate film transfer units N_{OR}, and the height of the overall raffinate film transfer unit is:

$$H_{OR} = \frac{Z}{N_{OR}} = \frac{\Phi'_{vR}}{K_R \cdot a}.$$

In a similar manner, we find the height of the overall transfer unit based on concentrations in the extract phase: $H_{OE} = \dfrac{Z}{N_{OE}} = \dfrac{\Phi'_{vE}}{K_E \cdot a}$.

Because $m = K_R/K_E$, it follows: $H_{OR} = \dfrac{\Phi'_{vR}}{m \cdot \Phi'_{vE}} \cdot H_{OE}$.

The group $\Phi'_{vR}/(m \cdot \Phi'_{vE})$ is the ratio of the slope of the operating line and the slope of the equilibrium curve. See Chapter 38. When these two lines are parallel, it follows that H_{OR} and H_{OE} are numerically equal. In deriving the relationships of this chapter, it was assumed that Φ'_{vR} and Φ'_{vE} are constant throughout the column. This will not be true if a large part of the solute is transferred from a concentrated solution to the other phase. It was also assumed that the transfer coefficients are independent of concentration.

To be able to integrate the left-hand side of Equation 39.5, C_E must be expressed as a function of C_R. This can be done by means of a mass balance equation: $\Phi'_{vE}(C_E - C_{E2}) = \Phi'_{vR}(C_R - C_{R2}) \rightarrow C_E = C_{E2} + (\Phi'_{vR}/\Phi'_{vE}).(C_R - C_{R2})$ It is also possible to use a second mass balance equation:

$$\Phi'_{vE}(C_{E1} - C_E) = \Phi'_{vR}(C_{R1} - C_R) \rightarrow C_E = C_{E1} + \frac{\Phi'_{vR}}{\Phi'_{vE}}(C_R - C_{R1}).$$

The result of the integration of Equation 39.5 is:

$$\Phi'_{vR}(C_{R1} - C_{R2}) = K_R \cdot a(\Delta C_R)_m \cdot Z \quad kmol \cdot m^{-2} \cdot s^{-1}. \qquad (39.6)$$

$(\Delta C_R)_m$ is the logarithmic mean of $(C_{R1} - C_{E1})/m$ and $(C_{R2} - C_{E2})/m$.

Equation 39.6 can also be given as:

$$(\Delta C_R)_m = \Phi'_{vR}(C_{R1} - C_{R2}) \frac{1}{K_R \cdot aZ}.$$

In the latter form, the equation is analogous to Ohm's law: Driving force equals flow times resistance (E = iR).

It is also possible to derive the following equation:

$$\Phi'_{vE}(C_{E1} - C_{E2}) = K_E \cdot a(\Delta C_E)_m \cdot Z \quad kmol \cdot m^{-2} \cdot s^{-1}. \qquad (39.7)$$

$(\Delta C_E)_m$ is the logarithmic mean of $m \cdot C_{R1} - C_{E1}$ and $m \cdot C_{R2} - C_{E2}$.

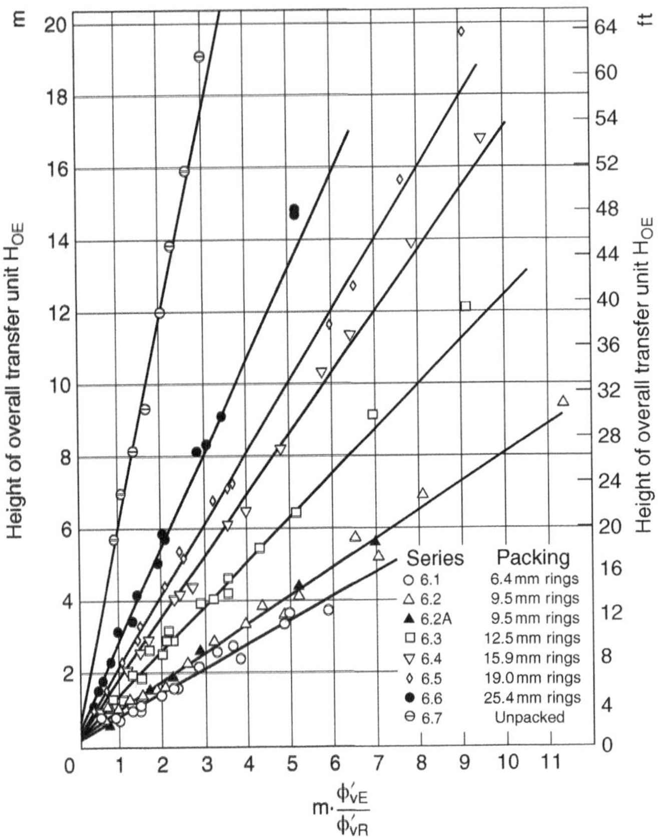

Figure 39.3 Height of the transfer unit H_{OE} as a function of $m \cdot \Phi'_{vE}/\Phi'_{vR}$ for various packings. Transfer of diethylamine from water to dispersed toluene. *Source:* Courtesy of Elsevier, Amsterdam, The Netherlands.

Note: in these equations, we are looking at $1\ m^2$ cross-sectional area.

Equations 39.6 and 39.7 can be used to calculate $K_R \cdot a$ and $K_E \cdot a$ from experimental data. A next step is the calculation of the NTU and the heights of transfer units. Equations 39.6 and 39.7 can, in principle, be used for design purposes. It is then necessary to have valid $K_R \cdot a$- or $K_E \cdot a$-values. Example 39.1 treats a case in which $K_E \cdot a$ and H_{OE} are calculated.

H_{OE} is plotted as a function of $m \cdot \Phi'_{vE}/\Phi'_{vR}$ in Figure 39.3 [1]. The size of the rings in the packed column is the parameter. The extraction is more effective for small rings than for large rings.

Example 39.1

The extraction of benzoic acid from toluene by water as discussed in Chapter 38 will be discussed. The concentrations measured at the top and at the bottom of the column and the water flow rate are shown in Figure 39.4. The column diameter is 0.05 m and the packed height is 4 m.

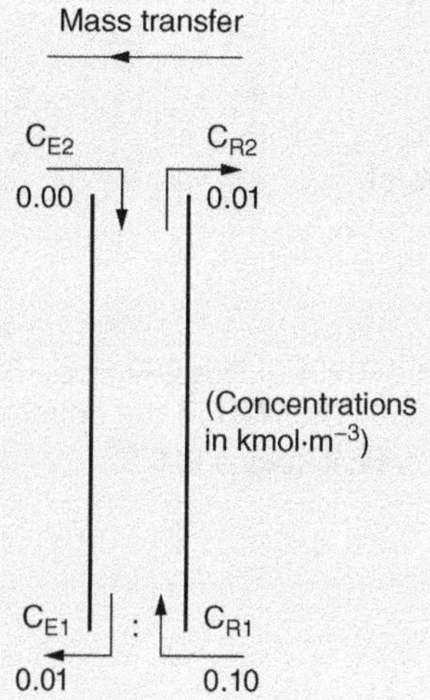

Acid concentration in inlet water phase C_{E2}	: 0.00
Acid concentration in outlet water phase C_{E1}	: 0.01
Flowrate of water phase	: 14 $l \cdot h^{-1}$, i.e., $1.981 \cdot 10^{-3}$ $m^3 \cdot m^{-2} \cdot s^{-1}$
Inlet toluene phase concentration C_{R1}	: 0.10
Outlet toluene phase concentration C_{R2}	: 0.01

Figure 39.4 Extraction process in a pulsed packed column.

The acid transferred to the aqueous phase is:

$$\frac{0.014}{3{,}600}(0.01 - 0.00) = 3.889 \cdot 10^{-8}\ \text{kmol} \cdot \text{s}^{-1}.$$

Per m^2 cross-section: $1.981 \cdot 10^{-5}$ kmol · m · s^{-1}.

The equilibrium relationship for this system is: $\frac{C_{Ei}}{C_{Ri}} - \frac{1}{8} = 0.125$.

The driving force at the top of the column is $(m \cdot C_{R2} - C_{E2}) = 0.125 \cdot 0.01 - 0.00 = 1.25 \cdot 10^{-3}$ kmol · m^{-3}. See Equation 39.4.

The driving force at the bottom of the column is $(m \cdot C_{R1} - C_{E1}) = 0.125 \cdot 0.1 - 0.01 = 2.5 \cdot 10^{-3}$ kmol · m^{-3}.

The logarithmic mean driving force is: $(\Delta C_E)_m = \dfrac{2.5 \cdot 10^{-3} - 1.25 \cdot 10^{-3}}{\ln \dfrac{2.5 \cdot 10^{-3}}{1.25 \cdot 10^{-3}}} = 1.8 \cdot 10^{-3}$ kmol · m^{-3}.

Equation 39.7: $K_E \cdot a = \dfrac{\text{molar transfer rate per } m^2 \text{ cross-section}}{Z(\Delta C_E)_m} = \dfrac{1.981 \cdot 10^{-5}}{4 \cdot 1.8 \cdot 10^{-3}} = 2.75 \cdot 10^{-3}\ s^{-1}$.

$$\Phi'_{vE} = \frac{0.014 \cdot 4}{3{,}600 \cdot \pi \cdot 0.05^2} = 1.981 \cdot 10^{-3}\ m^3 \cdot m^{-2} \cdot s^{-1}.$$

$$H_{OE} = \frac{\Phi'_{vE}}{K_E \cdot a} = \frac{1.981 \cdot 10^{-3}}{2.75 \cdot 10^{-3}} = 0.72\ \text{m}.$$

This the HTU in the extract phase. The NTU in the extract phase is:

$$N_{OE} = \frac{Z}{H_{OE}} = \frac{4}{0.72} = 5.6.$$

The NTU can also be found as the quotient of

$$(C_{E1} - C_{E2}) \text{ and } (\Delta C_E)_m: \quad \frac{0.01 - 0.00}{1.8 \cdot 10^{-3}} = 5.6.$$

This follows from Equation 39.7 and the definitions of H_{OE} and N_{OE}. The values in this example are relatively good because of the pulsation. The NTU in the aqueous phase is 5.6. It was found in Chapter 38 that the number of theoretical stages is 6.5.

Example 39.2

The volume fraction of toluene in the previous example is approximately 0.1. K_E is calculated for the situation that the toluene droplet diameter is 3 mm.

There is 0.1 m^3 toluene per m^3 column volume.

The droplet area is: $$\frac{0.1 \cdot \pi \cdot 0.003^2}{\pi/6(0.003)^3} = 200\ m^2 \cdot m^{-3}.$$

$$K_E = \frac{2.75 \cdot 10^{-3}}{200} = 1.375 \cdot 10^{-5}\ m \cdot s^{-1}.$$

It was shown that the method using transfer units can be used to interpret the behavior of extraction columns. If data like in Figure 39.3 are available, the method can also be used for the design of extraction columns. For example, H_{OE} can be obtained from Figure 39.3 when Φ'_{vE} and Φ'_{vR} are known. Figure 39.3 is applicable for the transfer of diethyl amine from water to dispersed toluene in a packed extraction column. m in the figure is the distribution coefficient for this system.

The data in Figure 39.3 have, regarding mass transfer, the same meaning as a specific Nusselt correlation for heat transfer.

Finally, see the remark made in Section 38.3 concerning scaling up experimental results.

Reference

1. Leibson, I., and Beckmann, R. B. (1953). The effect of packing size and column diameter on mass transfer in liquid–liquid extraction. *Chemical Engineering Progress*, 49, 405–416.

40

Flooding

40.1 General

There are spray columns, baffle-plate columns, packed columns, and agitated columns. Because the droplets of the dispersed phase rise or fall through the continuous phase under the influence of gravity, it will be apparent that there is a limit to the amount of dispersed phase that can pass through the tower for any given flow rate of continuous phase. That limit is called the flooding point. At the flooding point, the light phase enters and leaves the column at the bottom and the heavy phase enters and leaves the column at the top. Mixing the two phases is no longer possible. It is important to be able to predict the conditions at which flooding will occur. It is recommended to stay away from these conditions with ample margins. The diameter of the extraction column should be sized accordingly. The flows to come into contact with each other should be selected carefully.

A semi-emperical approach for the calculation of the flooding point in packed columns can be found in the literature [1, 2]. The forces acting on the separate dispersed and continuous phase channels when coalescence sets in just below the flooding point were considered. The following expression correlates data to within ±20%.

$$1 + 0.835\left(\frac{\rho_d}{\rho_c}\right)^{0.25}\left(\frac{u_d}{u_c}\right)^{0.5} = J\left(\left(\frac{u_c^2 \cdot a_p}{(g \cdot e^3)}\left(\frac{\rho_c}{\Delta\rho}\right)\left(\frac{\sigma}{\sigma_T}\right)^{0.25}\right)\right)^{-0.25}. \tag{40.1}$$

ρ_d: specific mass of the dispersed phase in $kg \cdot m^{-3}$
ρ_c: specific mass of the continuous phase in $kg \cdot m^{-3}$
u_d: linear velocity of the dispersed phase in $m \cdot s^{-1}$
u_c: linear velocity of the continuous phase in $m \cdot s^{-1}$
J: a constant that can be found in Table 40.1, dimensionless
a_p: the packing area per m^3 tower volume in m^{-1}
g: the acceleration due to gravity in $m \cdot s^{-2}$
e: the packing voidage, dimensionless
$\Delta\rho$: the specific mass difference between the two phases in $kg \cdot m^{-3}$
σ: the interfacial tension between the two phases in $N \cdot m^{-1}$
σ_T: the interfacial tension between water and air at 15 °C in $N \cdot m^{-1}$

It can be seen from Equation 40.1 that the right-hand side of the equation is constant for a given system if u_c is constant. On increasing u_d, while u_c remains constant, the left-hand side increases. When the two sides of the equation become equal, flooding occurs.

Introduction to Chemical Engineering, First Edition. C.M. van 't Land.

Table 40.1 Values of the constant J in Equation 40.1.

Type of packing	Nominal packing size, mm					
	6	9	12.5	19	25	57.5
Raschig rings[a]	0.49	0.53	0.57	0.61	0.63	0.63
Lessing rings	—	0.61	0.61	—	—	—
Berl saddles	0.52	—	0.67	0.67	—	—
Spheres	—	—	0.61	—	—	—

[a]For values of the left-hand side of Equation 40.1 < 1.6, the exponent −0.25 on the right-hand side becomes −0.20 and the constant J becomes $1.10 \cdot J^{0.8}$.
Source: Coulson et al. [3].

Dell and Pratt used columns having diameters of 3 and 6 in. The experiments were carried out at ambient temperature and at atmospheric pressure. Ten different liquid/liquid-systems were used. First, seven different solvents (benzene, butyl acetate, dibutyl carbitol, methyl-i-butyl ketone (MIBK), i-octane, dichlorodiethylether, and carbon tetrachloride) were used in conjunction with distilled water. Water was, at almost all tests, the continuous phase. The viscosities of the solvents were in the range 0.50–2.52 cP (20 °C). The interfacial tensions were in the range 0.01–0.045 $N \cdot m^{-1}$. Second, in order to study the effect of variations in physical properties other than interfacial tension, four series of runs were carried out with butyl acetate, using water and three different strengths of glycerol as the continuous phase.

The results of [1, 2] appeared in an adapted form in [3]. The left-hand side of the original equation was already dimensionless; however, the right-hand side was not dimensionless. That side was made dimensionless. The approach in this chapter follows the approach of [3].

Example 40.1

The extraction of benzoic acid from toluene by means of water in a pulsed packed column was discussed in Chapters 38 and 39. At a typical run, 1.4 $l \cdot h^{-1}$ of toluene was dispersed in 14 $l \cdot h^{-1}$ of water. Flooding occurred when it was tried to disperse 8.5 $l \cdot h^{-1}$ of toluene in 14 $l \cdot h^{-1}$ of water. The temperature was 24 °C. These flows are in the range of flows tested in [1, 2]. Both sides of Equation 40.1 are calculated for these two situations.

Note

Dell and Pratt found their results in packed columns and not in pulsed packed columns. Therefore, their equation is not applicable to pulsed packed columns.

Physical Data

$\rho_c = 997\ kg \cdot m^{-3}$.

$\rho_d = 867\ kg \cdot m^{-3}$.

$\sigma_T = 0.073\ N \cdot m^{-1}$ (interfacial tension water/air at 15 °C).

$\sigma = 0.034\ N \cdot m^{-1}$ (see Figure 38.3).

Equipment Data

D = 0.05 m.

1-cm Raschig rings were used, both their height l and diameter d were 1 cm whereas their wall thickness was 1 mm. Cubic packing is assumed, thus

$$a_p = \frac{\pi l\left(2d - 2\cdot 10^{-3}\right) + 2(\pi/4)\left(d^2 - \left(d - 2\cdot 10^{-3}\right)^2\right)}{d^2 l} = 622\ m^{-1}.$$

$$e = 1 - \frac{(\pi/4)\left(d^2 - \left(d - 2\cdot 10^{-3}\right)^2\right) l}{d^2 l} = 0.717.$$

$J = 0.54$ (See Table 40.1).

Operational Data

Typical Run

$u_c = 1.98\cdot 10^{-3}\ m\cdot s^{-1}.$

$u_d = 0.20\cdot 10^{-3}\ m\cdot s^{-1}.$

Flooding

$u_c = 1.98\cdot 10^{-3}\ m\cdot s^{-1}.$

$u_d = 1.20\cdot 10^{-3}\ m\cdot s^{-1}.$

Typical Run

Left-hand side 1.257.

Right-hand side 2.00.

Flooding

Left-hand side 1.628.

Right-hand side 2.11.

According to Equation 40.1, flooding should not yet occur at $\Phi_v = 8.5\ l\cdot h^{-1}$. For flooding to occur, Φ_v should be $26.5\ l\cdot h^{-1}$. However, the experimental results were found in a pulsed packed column whereas Equation 40.1 is applicable to columns that are not pulsed.

References

1. Dell, F. R., and Pratt, H. R. C. (1951). Part 1. Flooding rates for packed columns. *Transactions of the Institution of Chemical Engineers*, 29, 89–109.
2. Dell, F. R., and Pratt, H. C. R. (1951). Liquid–liquid extraction. Part 1. Flooding rates for packed columns. Addendum. *Transactions of the Institution of Chemical Engineers*, 29, 270.
3. Coulson, J. M. and Richardson, J. F. with Backhurst, J. R. and Harker, J. H. (1999). *Chemical Engineering, Volume 2, Particle Technology & Separation Processes* (pp. 613–616). Oxford, UK: Butterworth.

41

The Two Liquids Exchanging a Component Are Partially Miscible

41.1 Triangular Coordinates

An equilateral triangle may be used for representing compositions, use being made of the fact that the sum of the perpendiculars from any point within the triangle to the three sides equals the altitude. The length of the altitude is then allowed to represent 100% composition, and the length of the perpendiculars from any point the percentages of the three compositions. See Figure 41.1.

The apexes of the triangle represent the pure components A, B, and C, respectively. All percentages are percentages by weight. Any point on the side of the triangle represents a binary mixture of the two components at the ends of the side: Thus, point G is a mixture containing 40% of C and 60% of B. A point inside the triangle represents a ternary mixture: Point M thus contains 20% of A, 40% of B, and 40% of C. Points outside the triangle represent imaginary mixtures and have constructional significance only. Use is made of such a point in Section 41.3.

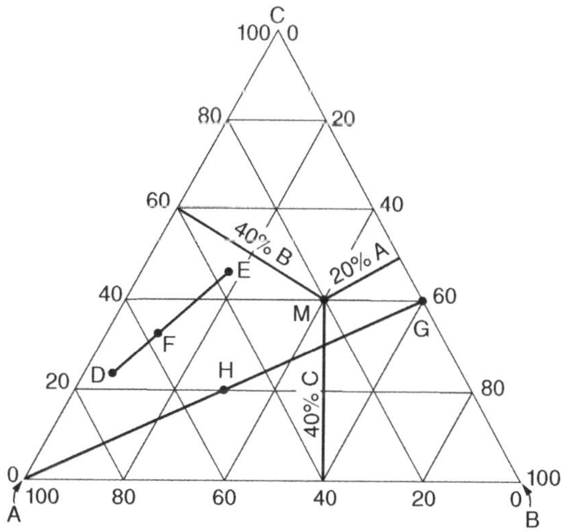

Figure 41.1 Triangular coordinates. *Source:* Courtesy of McGraw-Hill Book Company, Columbus, Ohio, USA.

Introduction to Chemical Engineering, First Edition. C.M. van 't Land.

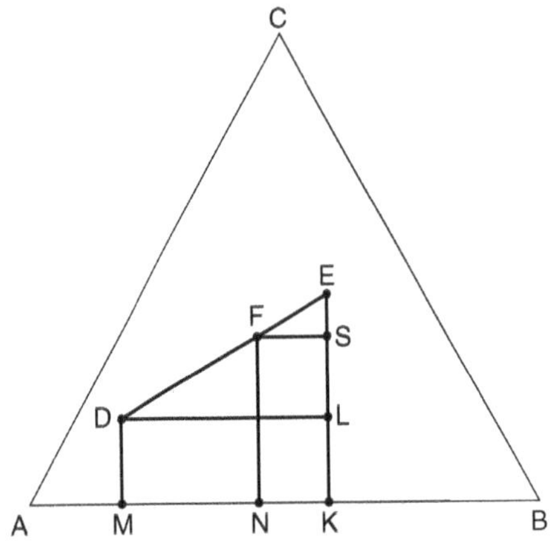

Figure 41.2 The mixture rule. *Source:* Courtesy of McGraw-Hill Book Company, Columbus, Ohio, USA.

Several other characteristics of such plots are significant. If, for example, D kg of the mixture represented by point D, in Figure 41.2, are added to E kg of the mixture at E, the resulting mixture has the composition at F, which is on the straight line DE, such that

$$\frac{E}{D} = \frac{FD}{EF}. \tag{41.1}$$

This is called the Lever rule or Mixture rule. The proof of Equation 41.1 is given with the help of Figure 41.2, where the mixtures at D, E, and F are again shown. A total material balance for the mixture process is $D + E = F$. A material balance for component C is $D \cdot X_{CD} + E \cdot X_{CE} = F \cdot X_{CF}$. Combining the latter two equations results in:

$$\frac{E}{D} = \frac{X_{CF} - X_{CD}}{X_{CE} - X_{CF}}.$$

But $X_{CF} = FN$, $X_{CD} = DM$, and $X_{CE} = EK$, so

$$\frac{E}{D} = \frac{FN - DM}{EK - FN} = \frac{SL}{ES} = \frac{FD}{EF}.$$

Referring again to Figure 41.1, it will be clear from what was described that all points on line AG represent mixtures with constant ratios of C to B with varying amounts of A. The nearer point H is to point A, the richer the mixture is in component A; if all A is removed from mixture H, the binary mixture at G results.

41.2 Formation of One Pair of Partially Miscible Liquids

This most frequently occurring combination is typified by the graph in Figure 41.3. The temperature for such a graph should be given, it is hence also known as an isotherm. In a system of this kind, the liquid pairs A–C and B–C are miscible in all proportions. A and B are partially miscible, and points D and E represent the saturated solutions in the binary system. A typical example is the system benzene (A)–water (B)–ethanol (C). Thus, ethanol could be exchanged between benzene and water by contacting the solutions of ethanol in these two liquids.

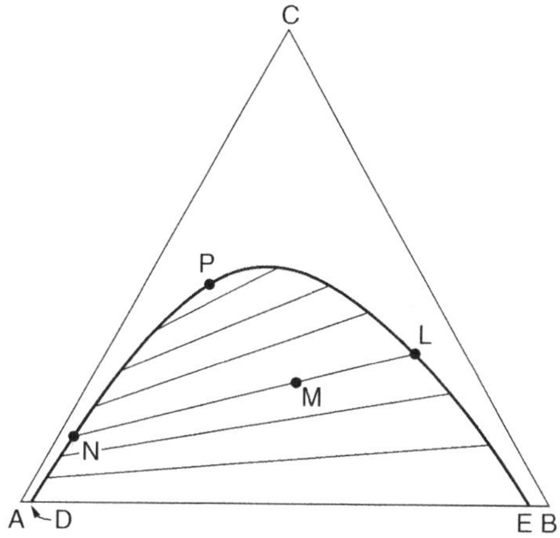

Figure 41.3 Type 1 ternary liquid equilibria. *Source:* Courtesy of McGraw-Hill Book Company, Columbus, Ohio, USA.

All mixtures of the components represented by points outside the area DNPLED in Figure 41.3 are homogeneous single-liquid-phase solutions, whereas mixtures within this area form two insoluble liquid layers. The curve DNPLE represents saturated solutions and is called the solubility or binodal curve.

A mixture of overall composition M will form the two immiscible liquid solutions of composition L and N, respectively, and the point M is therefore on the straight line LN, which is a tie-line. The compositions L and N are found by chemical analysis. All mixtures represented by points of the line LN form two conjugate layers each having a certain composition. The masses of the two layers can be determined from the lengths of the tie-line segments. The area of heterogeneity is to be imagined as filled with an infinite number of tie-lines, only a few of which are shown in Figure 41.3.

Ethanol, when added to a heterogeneous liquid mixture of water and benzene, distributes itself unequally between the two conjugate layers, with a greater concentration in the benzene rich solution than in the water rich solution. As more ethanol is added to such a mixture, the mutual solubility of water and benzene increases. At point P, the plait point, the two branches of the solubility curve merge, ordinarily not at the maximum value of ethanol on the curve.

Example 41.1

100 kg of a homogeneous mixture contains 50 kg of acetone and 50 kg of water [1]. The amount of kg of monochlorobenzene (MCB) required to reduce the acetone content of the aqueous phase to 10% by weight are determined. Figure 41.4 is the triangular diagram for the system acetone/water/MCB.

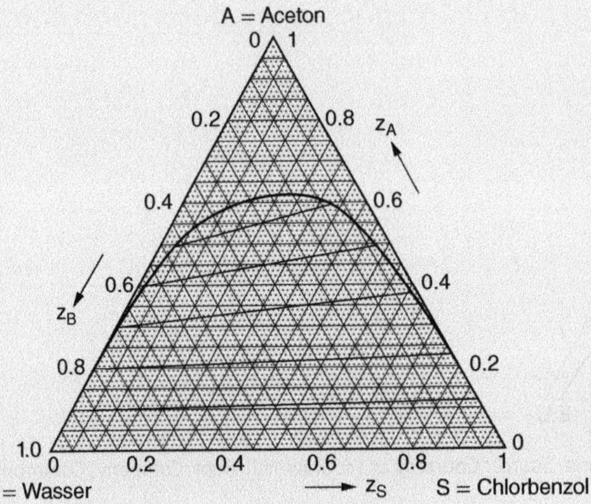

Figure 41.4 Triangular diagram for the system acetone–water–MCB. A translation of the words in the figure follows: Aceton, Acetone; Wasser, Water; Chlorbenzol, Monochlorobenzene.

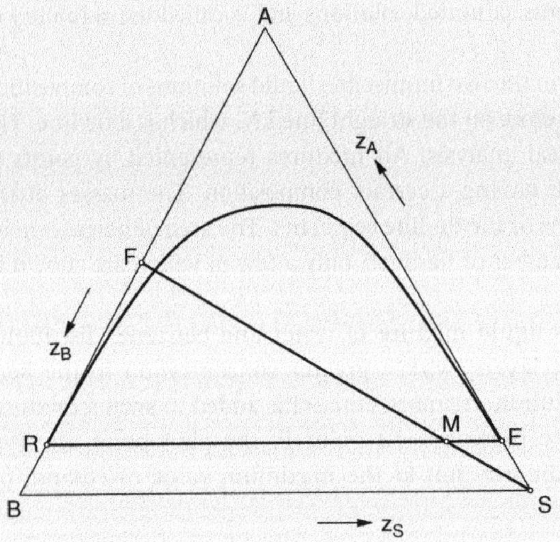

Figure 41.5 Single-stage batchwise extraction of acetone–water with MCB in the triangular diagram.

Figure 41.5 shows the solution. R is the point representing the aqueous phase containing 10% by weight of acetone. The line RE is a tie-line. The line FS is drawn. The two lines intersect at M. The amount of MCB is:

$$100 \cdot \frac{FM}{MS} = 100 \cdot \frac{57}{15} = 380\ \text{kg}.$$

Example 41.2

100 kg · h^{-1} of a 50% by weight solution of acetone in water is extracted continuously in a multistage extraction unit with, in each stage, 25 kg · h^{-1} of pure 1,1,2-trichloroethane [2]. Figure 41.6 shows this setup schematically. Figure 41.7 is the triangular diagram for the system. The number of stages required to reduce the weight fraction acetone in the aqueous phase to less than 0.1 is determined.

Figure 41.8 contains the solution. The unit of the figures is kg · h^{-1}. Note that the symbols F, S, M_1, E_1, and so forth stand for both a composition in the triangular diagram and a mass flow. Note also that, compared to Figure 41.7, 1,1,2-trichloroethane and water have changed places. First, the line SF is drawn. F + S = 25 + 100 = 125. Point M_1 is located:

$$\frac{SM_1}{M_1F} = \frac{100}{25} \text{ and } M_1 = 125.$$

The next step is to draw the tie-line R_1E_1 through M_1. $R_1 + E_1 = 125$, $R_1 = 75$, and $E_1 = 50$. Those figures are found after measuring the lengths of the lines R_1M_1 (representing E_1) and M_1E_1 (representing R_1). The weight fraction acetone in R_1 is 0.35. The fraction is read from the triangular diagram.

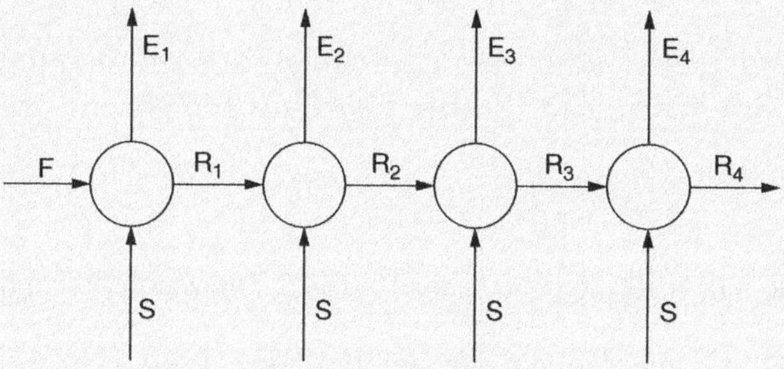

Figure 41.6 Multistage continuous crossflow extraction of acetone–water with 1,1,2-trichloroethane.

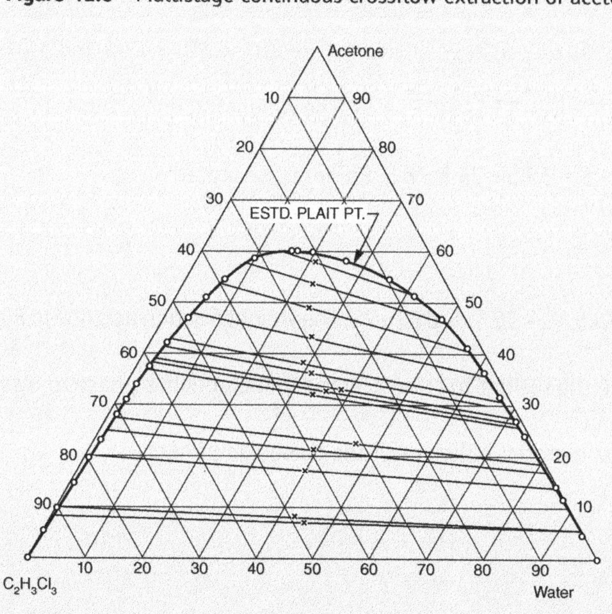

Figure 41.7 Triangular diagram for the system acetone–water-1,1,2-trichloroethane. *Source: Industrial and Engineering Chemistry*, 38 (8): 817–821, Copyright (1946) American Chemical Society.

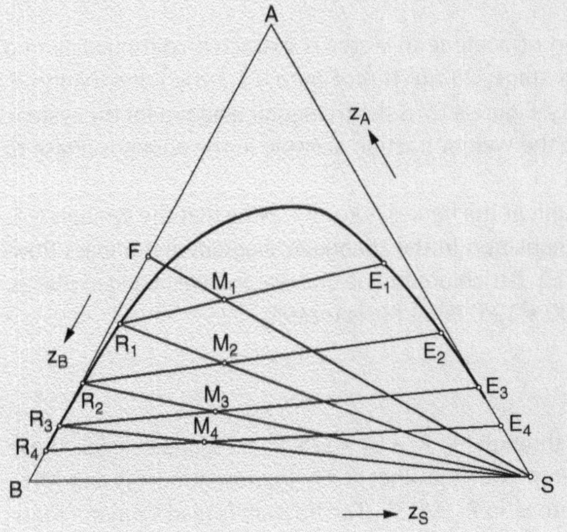

Figure 41.8 Multistage continuous crossflow extraction of acetone–water with 1,1,2-trichloroethane in the triangular diagram.

The next step is to draw the line R_1S. $R_1 + S = 75 + 25 = 100$. Point M_2 is located:

$$\frac{SM_2}{M_2R_1} = \frac{75}{25} \text{ and } M_2 = 100.$$

The tie-line R_2E_2 is then drawn. $R_2 + E_2 = 100$, $R_2 = 60.5$, and $E_2 = 39.5$. The weight fraction acetone in R_2 is 0.22.

The next step is to draw the line R_2S. $R_2 + S = 60.5 + 25 = 85.5$. Point M_3 is located:

$$\frac{SM_3}{M_3R_2} = \frac{60.5}{25} \text{ and } M_3 = 85.5.$$

The tie-line R_3E_3 is drawn. $R_3 + E_3 = 85.5$, $R_3 = 53.5$, and $E_3 = 32$. The weight fraction acetone in R_3 is 0.138.

The line R_3S is then drawn. $R_3 + S = 53.5 + 25 = 78.5$. Point M_4 is located:

$$\frac{SM_4}{M_4R_3} = \frac{53.5}{25} \text{ and } M_4 = 78.5.$$

The tie-line R_4E_4 is drawn. $R_4 + E_4 = 78.5$, $R_4 = 50.5$, and $E_4 = 28$. The weight fraction acetone in R_4 is 0.073.

Four stages are required to reduce the weight fraction acetone in the aqueous phase to less than 0.1.

The reader is advised to check the material balances of the three components.

41.3 Continuous Countercurrent Multiple-contact Extraction

In this type of extraction, a cascade of stages is employed, with feed and solvent entering at opposite ends of the cascade, raffinate and extract solutions flowing countercurrently. Use is made of a triangular diagram [3]. The approach will be discussed by means of an example.

Example 41.3

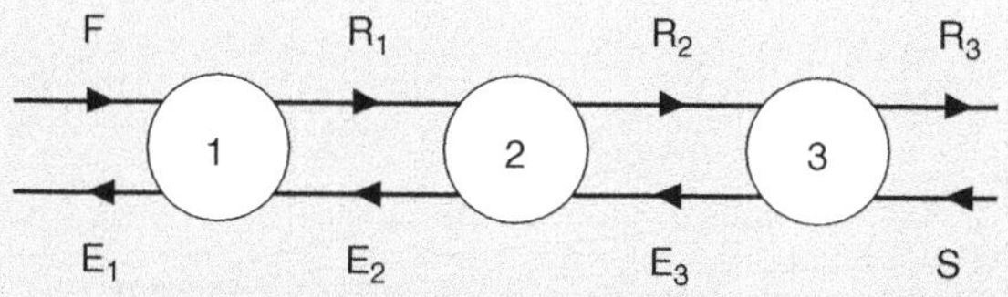

Figure 41.9 Multistage continuous countercurrent extraction.

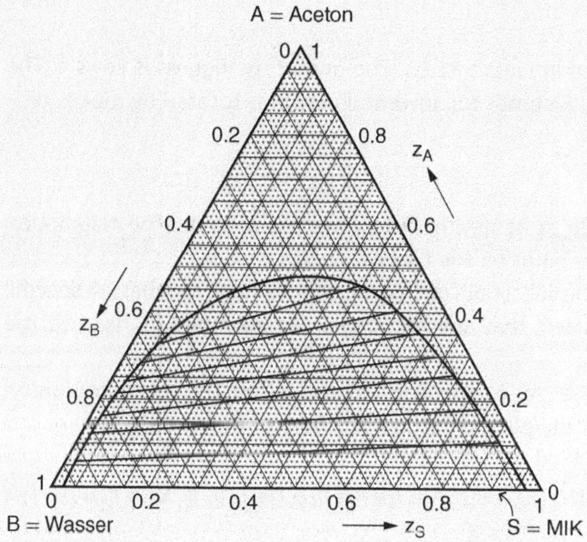

Figure 41.10 Triangular diagram for the system acetone–water–MIBK. A translation of the words in the figure follows: Aceton, Acetone; Wasser, Water; MIK, MIBK.

Acetone is extracted continuously from an aqueous solution by methyl iso-butyl ketone (MIBK) in an extraction system [4]. The two liquid phases move countercurrently through the system. See Figure 41.9. Figure 41.10 depicts the triangular diagram for these three components. The feed is 2.5 kg · s^{-1} aqueous phase containing 49% by weight of acetone. The major part of the acetone is transferred to 2.0 kg · s^{-1} of MIBK entering the system saturated with water. The raffinate leaving the system should contain 2% by weight of acetone maximum. The number of stages to be able to perform this extraction is determined. The symbols F, S, M, E_1, and so forth stand for both a composition in the triangular diagram and a mass flow.

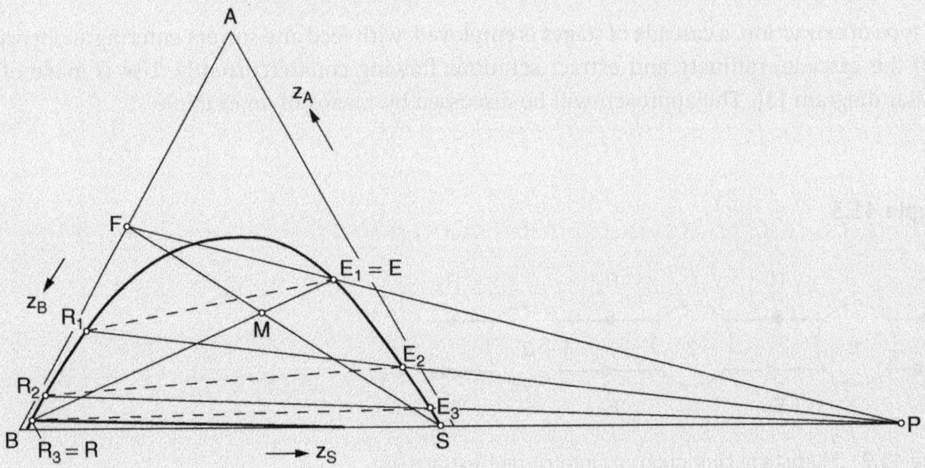

Figure 41.11 Multistage countercurrent extraction of acetone–water with MIBK in the triangular diagram.

The triangular diagram is reproduced in Figure 41.11. The unit of the figures is $kg \cdot s^{-1}$. The line FS is drawn. F stands for feed and S stands for solvent. Point M is located by means of:

$$\frac{MF}{SM} = \frac{2.0}{2.5}.$$

It is assumed that the raffinate leaves the system with 2% by weight of acetone. The assumption will be checked. R_3 is represented by a point on the binodal curve.

It is possibly helpful to realize that the subscripts of E and R refer to stage numbers. A specific subscript indicates that the flow has left that specific stage. For example, E_3 has left the third stage.

It is possibly also helpful to realize that, for example, R_1 stands for both the concentrations in R_1 (R_1 is then a point in the triangular diagram) and the mass flow of R_1 in $kg \cdot s^{-1}$.

Reference is now made to Figure 41.9. The mass balance for the first stage is: $E_2 + F = E_1 + R_1$. The mass balance can be rewritten as: $E_1 - F = E_2 - R_1$.

The mass balance for the second stage is: $E_3 + R_1 = E_2 + R_2$. It can be rewritten as: $E_2 - R_1 = E_3 - R_2$.

The mass balance for the third stage is: $S + R_2 = E_3 + R_3$. It can be rewritten as: $E_3 - R_2 = S - R_3$.

A conclusion is that the difference between the flows between two adjacent stages is constant.

Now, we go back to Figure 41.11. The line R_3ME_1 is drawn. $M = F + S = 2.5 + 2.0 = 4.5$ and $R_3 + E_1 = 4.5$. $R_3 = 1.05$ and $E_1 = 3.45$. The latter two values are found after measuring the lengths of the lines R_3M and ME_1.

The next step is to lengthen the line FE_1 to P. The line FE_1P is called an operating line. The location of P, the operating point, is determined as follows. It is realized that $E_1 = F + P$. So, $P = E_1 - F = 3.45 - 2.5 = 0.95$. Hence, the following ratio is applicable:

$$\frac{E_1P}{FE_1} = \frac{2.5}{0.95}.$$

Then, the tie-line R_1E_1 is drawn. Remember that the tie-lines are found by chemical analysis of laboratory samples. E_1 is in equilibrium with R_1. The line R_1E_2P is drawn. On drawing this line, it is assumed that $E_2 = R_1 + P$. Rewriting: $P = E_2 - R_1$. That assumption is correct because the difference between the flows between two stages is constant.

A second tie-line R_2E_2 is drawn. Next, the line R_2E_3P is drawn. Again, the assumption is $E_3 = R_2 + P$. Rewriting: $P = E_3 - R_2$. The assumption is correct for the already given reason.

A third tie-line R_3E_3 shows that the acetone content of the aqueous phase is approximately 2% by weight.

The planned extraction requires three stages.

The tie-lines represent equilibrium conditions whereas the operating lines are based on mass balances.

The approach has been chosen to treat the theory [3] by discussing an example. The method can also be used in other cases.

References

1. Schlünder, E.-U., and Thurner, F. (1986). *Distillation, Absorption, Extraction* (pp. 167–169). Stuttgart, Germany; New York, NY: Georg Thieme Verlag. (in German)
2. Schlünder, E.-U., and Thurner, F. (1986). *Distillation, Absorption, Extraction* (pp. 171–172). Stuttgart, Germany; New York, NY: Georg Thieme Verlag. (in German)
3. Hunter, T. G., and Nash, A. W. (1934). The application of physico-chemical principles to the design of liquid–liquid contact equipment. Part II. Application of phase-rule graphical methods. *Journal of the Society of Chemical Industry*, 53, 95–102.
4. Schlünder, E.-U., and Thurner, F. (1986). *Distillation, Absorption, Extraction* (pp. 176–178). Stuttgart, Germany; New York, NY: Georg Thieme Verlag. (in German)

42

Case Studies Liquid Extraction

42.1 A Series of Centrifugal Extractors

The extraction of a fermentation product from an aqueous solution by ethyl acetate was described [1]. The latter solvent is only slightly soluble in water. The extraction was performed by means of centrifugal extractors. See Figure 42.1 showing two different sizes. In this type of equipment, the two phases are first brought into contact with each other and then separated. The assumption is that the two liquid phases leaving the equipment are in equilibrium with each other. First, laboratory work was carried out. The next step was the testing of the process in a series of three small centrifugal extractors. The two liquid streams moved countercurrently through the series of extractors. The tests were successful and an industrial extraction unit was built. On scaling-up, the radial acceleration was kept constant at $1{,}000 \cdot g$. The residence time of 14 s in each extractor

Figure 42.1 Two centrifugal extractors. *Source:* Courtesy of CINC, Brakel, Germany.

Introduction to Chemical Engineering, First Edition. C.M. van 't Land.

was also kept constant. Table 42.1 contains results of measurements concerning the industrial unit. [1] contains more results. The requests are:

Table 42.1 Raffinate data on extracting a fermentation product

Extractor no.	Inlet concentration, $g \cdot l^{-1}$	Outlet concentration, $g \cdot l^{-1}$
1	0.94	0.40
2	0.40	0.15
3	0.15	0.02

Raffinate flow: 4,000 $l \cdot h^{-1}$.
Extract flow: 650 $l \cdot h^{-1}$.
Ambient temperature: 20 °C.
Rotational speed of the extractors: 2,700 min^{-1}.
Radial acceleration of the extractors: 1,000 · g.
Source: Vornefeld [1].

- To represent the measurements in the way of Figure 38.7.
- To plot the three points for the equilibrium line and draw that line.
- To plot the four points of the operating line and draw that line.
- To check the actual presence of three extractors in series graphically by drawing steps.
- To determine the distribution coefficient defined as the ratio between the concentrations in $g \cdot l^{-1}$ in the two phases when there is equilibrium.

Solution

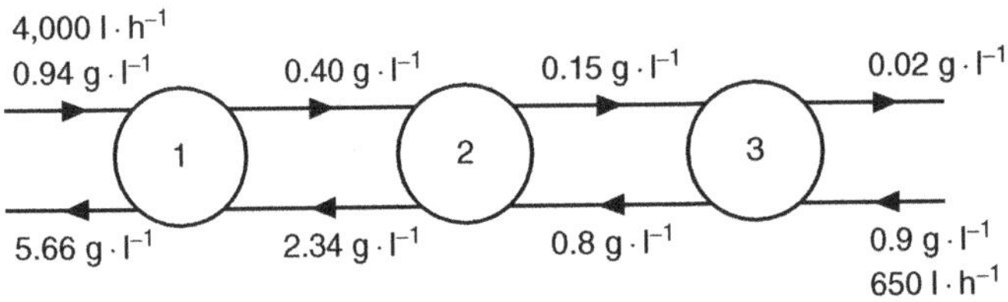

Figure 42.2 A series of centrifugal extractors.

The measurements of Table 42.1 are included in Figure 42.2. The results of the following calculations can also be found in that figure. The material balances for extractors 3, 2, and 1 are:

3. $4 \cdot 10^3(0.15 - 0.02) = 650(x - 0) \rightarrow x = 0.80 \ g \cdot l^{-1}$.
2. $4 \cdot 10^3(0.40 - 0.15) = 650(x - 0.80) \rightarrow x = 2.34 \ g \cdot l^{-1}$.
1. $4 \cdot 10^3(0.94 - 0.40) = 650(x - 2.34) \rightarrow x = 5.66 \ g \cdot l^{-1}$.

Equilibrium line and operating line can be seen in Figure 42.3. The distribution coefficient is approximately 15. The steps are drawn in Figure 42.3.

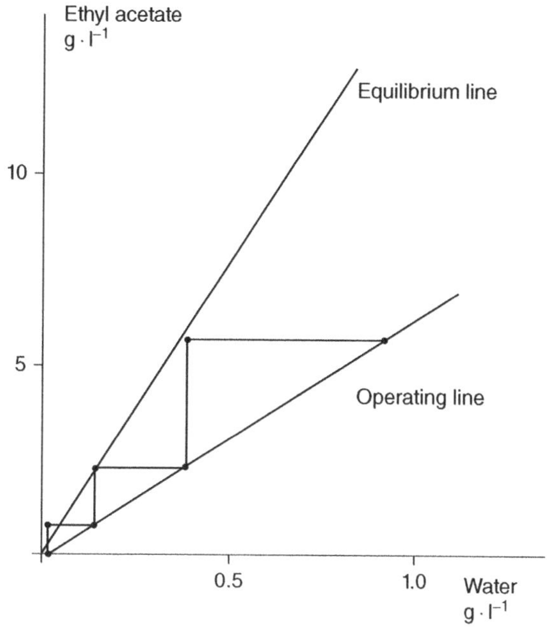

Figure 42.3 Three steps for the extraction of a fermentation product.

42.2 Extraction by Means of An Ionic Liquid

This case concerns the extraction of toluene from heptane by means of ionic liquids. An ionic liquid does not or hardly have a vapor pressure, so there are hardly losses due to the volatility of the extractant. The extractions are carried out in a Rotating Disc Contactor (RDC). The contactor is a column having a height of 6 m and a diameter of 60 mm. The column is divided into sections by annular partitions. A smooth disc rotates in each section. The ionic liquids are the continuous phase and travel from the top to the bottom of the column. Heptane is the dispersed phase and travels from the bottom to the top of the column. Figure 42.4 gives experimental results reported in [2]. It is

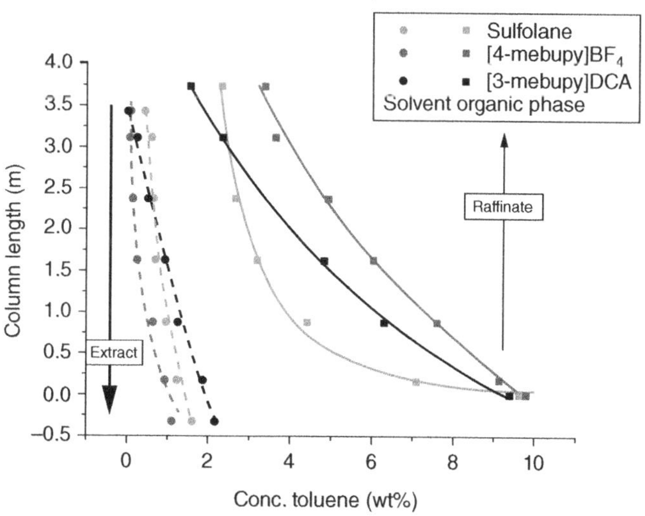

Figure 42.4 Mass transfer in an RDC. *Source:* Courtesy of A.B. de Haan, Delft, The Netherlands.

requested to take a look at the extraction by means of [3-mebupy]DCA. We see two sets of three lines in the figure. The right line of the left set and the middle line of the right set are applicable for this extraction. The distribution coefficient of toluene between the ionic liquid and heptane is 0.415. This means that the ratio of the $kg \cdot kg^{-1}$ concentrations in the two phases is 0.415. The affinity between toluene and heptane is greater than the affinity between toluene and the ionic liquid. It is requested to draw the equilibrium and operating lines in a graph. Next, it is asked to determine the number of mixer/settlers equivalent to the column.

Solution

The equilibrium and operating lines are drawn in Figure 42.5. The slope of the equilibrium line is 0.415. The operating line is drawn between the two points representing the concentrations in $kg \cdot kg^{-1}$ at the top and at the bottom of the column. Next, steps are drawn between the two lines. The number of mixer/settlers equivalent to the column is slightly greater than 2.

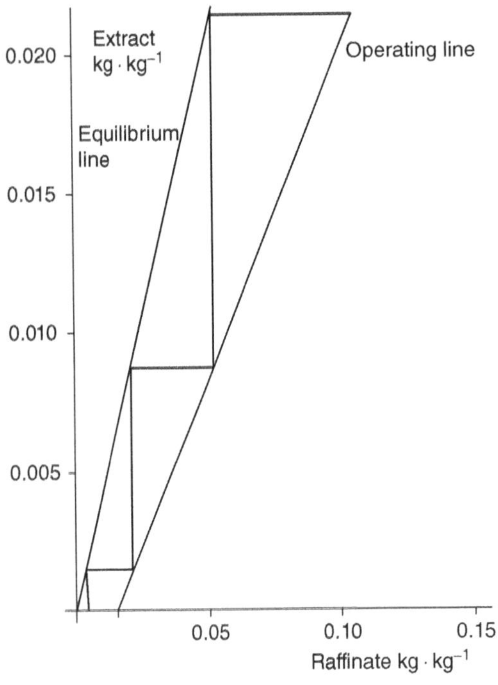

Figure 42.5 Extraction by means of an ionic liquid.

42.3 Overall Transfer Coefficient/Height of a Transfer Unit

It is requested to calculate $K_R \cdot a$, H_{OR}, and N_{OR} from the data given in Example 39.1. In that example, $K_E \cdot a$ and H_{OE} were calculated. See Figure 39.4. The column diameter is 0.05 m and the packed height is 4 m.

The acid transferred to the aqueous phase is:

$$(\quad - \quad) = \quad \text{kmol} \cdot \text{s}^{-1}.$$

Per m^2 cross-section: $\quad \text{kmol} \cdot \text{m}^{-2} \cdot \text{s}^{-1}$.

The equilibrium relationship for this system is:

$$m = \frac{C_{Ei}}{C_{Ri}} = \frac{1}{8} = 0.125.$$

Equation 39.3 reads: $\Phi''_{mol} = K_R\left(C_R - \frac{C_E}{m}\right)$.

So, the driving force is $\left(C_R - \frac{C_E}{m}\right)$.

The driving force at the top of the column is $\Delta C_{R2} = \left(C_{R2} - \frac{C_{E2}}{m}\right) = \quad \text{kmol} \cdot \text{m}^{-3}$. The driving force at the bottom of the column is $\Delta C_{R1} = \left(C_{R1} - \frac{C_{E1}}{m}\right) = \quad \text{kmol} \cdot \text{m}^{-3}$.

The logarithmic mean driving force is $(\Delta C_R)_m = \frac{\quad - \quad}{\ln \quad} = \quad \text{kmol} \cdot \text{m}^{-3}$.

Equation 39.6: $K_R \cdot a = \frac{\text{molar transfer rate per m}^2\text{ cross-section}}{Z(\Delta C_R)_m} = \quad \text{s}^{-1}$. The toluene flow follows from the mass balance equation:

$$\Phi_{vR}(\quad - \quad) = \frac{0.014}{3{,}600}(\quad - \quad)\ \text{kmol} \cdot \text{s}^{-1} \rightarrow \Phi_{vR} = \quad \text{m}^3 \cdot \text{s}^{-1}.$$

Per m^2 cross-section: $\Phi'_{vR} = \quad \text{m}^3 \cdot \text{m}^{-2} \cdot \text{s}^{-1}$.

$$H_{OR} = \frac{\Phi'_{vR}}{K_R \cdot a} = \quad \text{m and } N_{OR} = \frac{Z}{H_{OR}} =$$

Solution

The acid transferred to the aqueous phase is:

$$\frac{0.014}{3.6 \cdot 10^3}(0.01 - 0.00) = 3.889 \cdot 10^{-8}\ \text{kmol} \cdot \text{s}^{-1}.$$

Per m^2 cross-section: $1.981 \cdot 10^{-5}\ \text{kmol} \cdot \text{s}^{-1}$.

The driving force at the top of the column is $\Delta C_{R2} = \left(C_{R2} - \frac{C_{E2}}{m}\right) = 0.01 - \frac{0.00}{0.125} = 0.01\ \text{kmol} \cdot \text{m}^{-3}$.

The driving force at the bottom of the column is $\Delta C_{R1} = \left(C_{R1} - \frac{C_{E1}}{m}\right) = 0.10 - \frac{0.01}{0.125} = 0.02\ \text{kmol} \cdot \text{m}^{-3}$.

The logarithmic mean driving force is $(\Delta C_R)_m = \frac{0.01 - 0.02}{\ln \frac{0.01}{0.02}} = 0.0144\ \text{kmol} \cdot \text{m}^{-3}$.

Equation 39.6: $K_R \cdot a = \frac{\text{molar transfer rate per m}^2\text{ cross-section}}{Z(\Delta C_R)_m} = \frac{1.981 \cdot 10^{-5}}{4 \cdot 0.0144} = 3.44 \cdot 10^{-4}\ \text{s}^{-1}$.

The toluene flow follows from the mass balance equation:

$$\Phi_{vR}(0.1 - 0.01) = \frac{0.014}{3.6 \cdot 10^3}(0.01 - 0.00)\ \text{kmol} \cdot s^{-1},$$

$$\Phi_{vR} = 4.32 \cdot 10^{-7}\ m^3 \cdot s^{-1}.$$

Per m^2 cross-section: $\Phi'_{vR} = 2.2 \cdot 10^{-4}\ m^3 \cdot m^{-2} \cdot s^{-1}$.

$$H_{OR} = \frac{\Phi'_{vR}}{K_R \cdot a} = \frac{2.2 \cdot 10^{-4}}{3.44 \cdot 10^{-4}} = 0.64\ m.$$

$$N_{OR} = \frac{4}{0.64} = 6.25.$$

42.4 Calculation of the Column Height

Acetic acid is extracted from an aqueous solution by benzene in a packed column [3]. The aqueous phase is the continuous phase traveling from the top to the bottom of the column whereas benzene is the dispersed phase traveling in the opposite direction. The temperature is ambient. The diameter of the column is 0.076 m. The inlet concentrations of the extract and the raffinate phase are in Figure 42.6. The specific flow rate of benzene is $\Phi'_{vE} = 1.26 \cdot 10^{-3}\ m^3 \cdot m^{-2} \cdot s^{-1}$. The inlet and desired outlet concentrations of both phases are in Figure 42.6. The equilibrium relationship for this system is

$$\frac{C_{Ei}}{C_{Ri}} = 0.0247 \text{ and } K_E \cdot a = 10^{-3}\ s^{-1}.$$

The column height for this extraction can be calculated.

Equation 39.4 reads: $\Phi''_{kmol} = K_E(m \cdot C_R - C_E)$.

So, the driving force is $(m \cdot C_R - C_E)$.

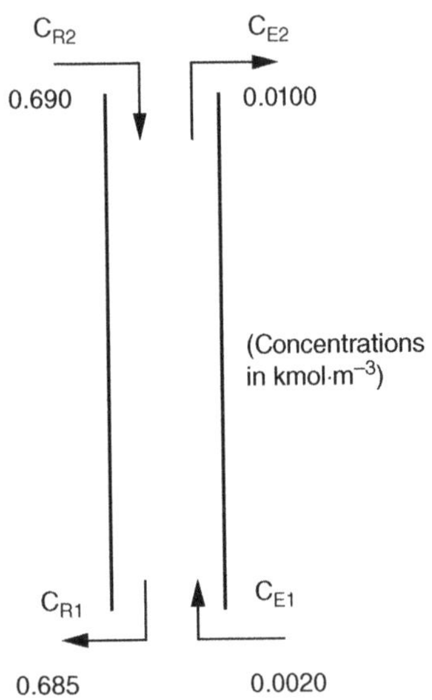

Figure 42.6 Extraction process in a packed column.

The driving force at the top of the column is: $\Delta C_{E2} = (m \cdot C_{R2} - C_{E2}) =$ $\quad$ $kmol \cdot m^{-3}$. The driving force at the bottom of the column is: $\Delta C_{E1} = (m \cdot C_{R1} - C_{E1}) =$ $\quad$ $kmol \cdot m^{-3}$.

The logarithmic mean driving force is $(\Delta C_E)_m = \dfrac{\quad - \quad}{\ln \quad} =$ $\quad$ $kmol \cdot m^{-3}$.

$$H_{OE} = \frac{\Phi'_{vE}}{K_E \cdot a} = \frac{\quad}{\quad} = \quad m.$$

$$N_{OE} = \frac{C_{E2} - C_{E1}}{(\Delta C_E)_m} =$$

$$Z = N_{OE} \cdot H_{OE} = \quad m.$$

Solution

The driving force at the top of the column is: $\Delta C_{E2} = (m \cdot C_{R2} - C_{E2}) = 0.0247 \cdot 0.690 - 0.0100 = 7 \cdot 10^{-3}\ kmol \cdot m^{-3}$.
The driving force at the bottom of the column is: $\Delta C_{E1} = (m \cdot C_{R1} - C_{E1}) = 0.0247 \cdot 0.685 - 2.0 \cdot 10^{-3} = 0.0149\ kmol \cdot m^{-3}$.

The logarithmic mean driving force is: $(\Delta C_E)_m = \dfrac{0.0149 - 7.0 \cdot 10^{-3}}{\ln \dfrac{0.0149}{7.0 \cdot 10^{-3}}} = 0.010\ kmol \cdot m^{-3}$.

$$H_{OE} = \frac{\Phi'_{vE}}{K_E \cdot a} = \frac{1.26 \cdot 10^{-3}}{10^{-3}} = 1.26\ m.$$

$$N_{OE} = \frac{C_{E2} - C_{E1}}{(\Delta C_E)_m} = \frac{0.0100 - 2 \cdot 10^{-3}}{0.010} \quad \text{(see Equation 39.8).}$$

$$Z = 0.8 \cdot 1.26 = 1.01\ m.$$

42.5 Two Partially Miscible Liquids Exchange a Component

The two liquids are water and 1,1,2-trichloroethane. The component exchanged between these two liquids is acetone. Figure 42.7 is the triangular diagram for these three components at ambient temperature [4].

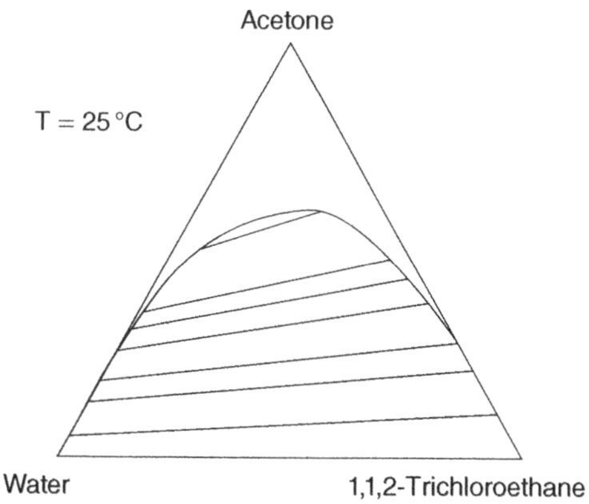

Figure 42.7 Triangular diagram for acetone, water, and 1,1,2-trichloroethane.

The extraction is carried out continuously in a column. The mass flow of pure 1,1,2-trichloroethane entering at the top of the column is 320 $kg \cdot h^{-1}$ (S). The mass flow to be extracted by means of the latter material is 1,000 $kg \cdot h^{-1}$ (F) and contains 45% by weight of acetone. It enters at the bottom of the column. The aqueous flow leaving at the top of the column still contains 10% by weight of acetone. The number of theoretical stages of the column is determined.

The line FS is drawn in the triangular diagram. The point M on line FS is determined by realizing that FM/MS = 320/1,000. The line R_nM is drawn and lengthened until it intersects the binodal curve at E_1. The point R_n is given by the fact that the aqueous phase still contains 10% by weight of acetone on leaving the column.

Determine E_1 and R_n in $kg \cdot h^{-1}$ by realizing that $E_1 + R_n = 1{,}320\ kg \cdot h^{-1}$ and $E_1/R_n = R_nM/ME_1$.

Note that mass flows in $kg \cdot h^{-1}$ and compositions are meant when one capital (for example, F) is used. Line lengths in, for example, mm are meant when two capitals (for example, FE_1) are used. They are used to calculate flows.

$E_1 =$ $\qquad$ $kg \cdot h^{-1}$.

$R_n =$ $\qquad$ $kg \cdot h^{-1}$.

Now, the line E_1F is drawn and lengthened till point P. The location of point P is determined as follows. P stands for $F - E_1 =$ $\qquad$ $kg \cdot h^{-1}$. It is realized that $PF/FE_1 = E_1/(F - E_1)$ and E_1 is known.

Now the tie-line E_1R_1 is drawn. E_1 and R_1 are in equilibrium.

R_1 is the point where the tie-line containing E_1 intersects the binodal curve.

The line PR_1 is drawn and lengthened till it intersects with the binodal curve: E_2. The tie-line E_2R_2 is drawn.

R_2 is the point where the tie-line containing E_2 intersects the binodal curve.

The line PR_2 is drawn and lengthened till the line intersects with the binodal curve: E_3. The tie-line E_3R_3 is drawn.

Continue the construction till R_m is smaller than R_n.

Determine the number of stages.

Solution

See Figure 42.8.

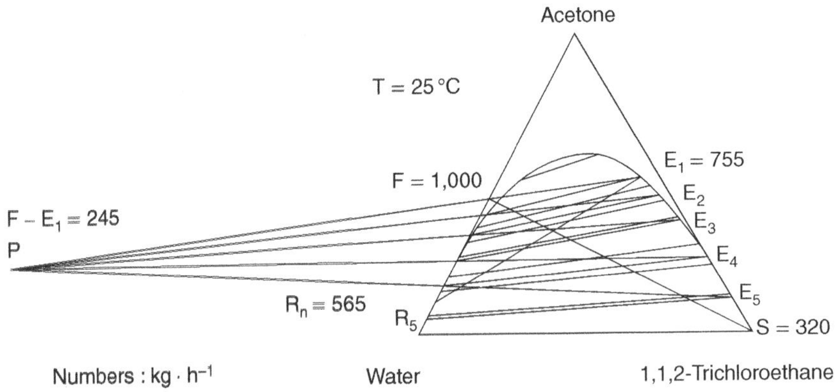

Figure 42.8 Five stages for the extraction of acetone from water by 1,1,2-trichloroethane.

$E_1 = 755\ kg \cdot h^{-1}$.

$R_n = 565\ kg \cdot h^{-1}$.

$F - E_1 = 245\ \ kg \cdot h^{-1}$.

The column contains almost five theoretical stages.

References

1. Vornefeld, M. (2007). Mixing possible. *Process*, 10 (2), 58–60. (in German)
2. Meindersma, G. W., Onink, F., de Haan, A. B., and Hansmeier, A. R. (2009). Improved ionic liquids for the extraction of aromatic hydrocarbons from naphtha. *npt procestechnologie*, 15 (4), 10–11.
3. Coulson, J. M. and Richardson, J. F. with Backhurst, J. R. and Harker, J. H. (1999). *Chemical Engineering, Volume 2, Particle Technology and Separation Processes* (p. 601). Oxford, UK: Butterworth-Heinemann..
4. Treybal, R. E., Weber, L. D., and Daley, J. F. (1946). The system acetone-water-1,1,2-trichloroethane. *Industrial and Engineering Chemistry*, 38(8), 817–821.

Notation V

A	Solvent flow in raffinate	$kmol \cdot s^{-1}$
a	Interfacial surface area	$m^2 \cdot m^{-3}$
a_p	Packing area per m^3 tower volume	m^{-1}
C_E	Concentration in the extract phase	$kmol \cdot m^{-3}$
C_{E1}	Concentration in an extract phase at the bottom of the column	$kmol \cdot m^{-3}$
C_{E2}	Concentration in an extract phase at the top of a column	$kmol \cdot m^{-3}$
C_{Ei}	Concentration in the extract phase at the interface	$kmol \cdot m^{-3}$
C_R	Concentration in the raffinate phase	$kmol \cdot m^{-3}$
C_{R1}	Concentration in the raffinate phase at the bottom of a column	$kmol \cdot m^{-3}$
C_{R2}	Concentration in the raffinate phase at the top of a column	$kmol \cdot m^{-3}$
C_{Ri}	Concentration in the raffinate phase at the interface	$kmol \cdot m^{-3}$
$(\Delta C_E)_m$	Logarithmic mean driving force in the extract phase	$kmol \cdot m^{-3}$
$(\Delta C_R)_m$	Logarithmic mean driving force in the raffinate phase	$kmol \cdot m^{-3}$
D	Amount of a three-component mixture	kg
	Column diameter	m
d	Packing ring diameter	m
E	Amount of extract	kg
	Amount of a three-component mixture	kg
	Electric tension	V
	Extract flow	$kg \cdot h^{-1}$
		$kg \cdot s^{-1}$
$E_{1,2,3,4,5}$	Extract flows leaving stages Nos. 1, 2, 3, 4, and 5	$kg \cdot h^{-1}$
		$kg \cdot s^{-1}$
E_n	Extract flow leaving stage n	$kg \cdot h^{-1}$
E_{n+1}	Extract flow leaving stage n + 1	$kg \cdot h^{-1}$
e	Packing voidage	—
F	Raffinate feed flow	$kg \cdot h^{-1}$
		$kg \cdot s^{-1}$
	Amount of a three-component mixture	kg
g	Acceleration due to gravity	$m \cdot s^{-2}$
H_{OE}	Height of an overall extract film transfer unit	m
H_{OR}	Height of an overall raffinate film transfer unit	m
i	Electric current	A
J	Constant in Table 40.1	—

Introduction to Chemical Engineering, First Edition. C.M. van 't Land.

K_E	Overall mass transfer coefficient based on the extract phase	$m \cdot s^{-1}$
K_R	Overall mass transfer coefficient based on the raffinate phase	$m \cdot s^{-1}$
k_E	Partial mass transfer coefficient in the extract phase	$m \cdot s^{-1}$
k_R	Partial mass transfer coefficient in the raffinate phase	$m \cdot s^{-1}$
l	Packing ring height	m
M	Sum of the extract and the raffinate flow in a stage	kg
		$kg \cdot h^{-1}$
		$kg \cdot s^{-1}$
$M_{1,2,3,4}$	Sum of the extract and raffinate flows leaving stages 1, 2, 3, and 4	$kg \cdot h^{-1}$
		$kg \cdot s^{-1}$
m	Molar distribution coefficient	—
N_{OE}	Number of transfer units based on the extract phase	—
N_{OR}	Number of transfer units based on the raffinate phase	—
P	Difference between the flows between two adjacent stages	$kg \cdot s^{-1}$
R	Amount of raffinate	kg
	Electrical resistance	Ohm
	Raffinate flow	$kg \cdot s^{-1}$
R_m	Raffinate flow	$kg \cdot h^{-1}$
R_n	Raffinate flow leaving stage n	$kg \cdot h^{-1}$
$R_{1,2,3,4,5}$	Raffinate flows leaving stages 1, 2, 3, 4, and 5	$kg \cdot h^{-1}$
		$kg \cdot s^{-1}$
R_{n-1}	Raffinate flow leaving stage n – 1	$kg \cdot h^{-1}$
S	Amount of solvent	kg
	Extract feed flow	$kg \cdot h^{-1}$
		$kg \cdot s^{-1}$
	Solvent flow in an extract stream	$kmol \cdot s^{-1}$
u_c	Linear velocity of the continuous phase	$m \cdot s^{-1}$
u_d	Linear velocity of the dispersed phase	$m \cdot s^{-1}$
X	Molar ratio solute/solvent in a raffinate	—
$X_{1,2,3}$	Molar ratios solute/solvent in the raffinates leaving stages Nos. 1, 2, and 3	—
X_{CD}	Mass fraction of component C in D	—
X_{CE}	Mass fraction of component C in E	—
X_{CF}	Mass fraction of component C in F	—
X_f	Molar ratio solute/solvent in a raffinate feed stream	—
X_n	Molar ratio solute/solvent in the raffinate leaving stage n	—
X_{n-1}	Molar ratio solute/solvent in the raffinate leaving stage n – 1	—
x	Concentration	$g \cdot l^{-1}$
Y	Molar ratio solute/solvent in an extract	—
$Y_{1,2,3,4}$	Molar ratio solute/solvent in the extracts leaving stages Nos. 1, 2, 3, and 4	—
Y_n	Molar ratio solute/solvent in the extract leaving stage n	—
Y_{n+1}	Molar ratio solute/solvent in the extract leaving stage n + 1	—
Z	Packing height in an extraction column	m
z_A	Mass fraction of component A	—
z_B	Mass fraction of component B	—
z_S	Mass fraction solvent	—

Greek Symbols

ρ_c	Specific mass of a continuous phase	$kg \cdot m^{-3}$
ρ_d	Specific mass of a dispersed phase	$kg \cdot m^{-3}$
$\Delta\rho$	Specific mass difference	$kg \cdot m^{-3}$
σ	Interfacial tension in an extraction column	$N \cdot m^{-1}$
σ_T	Interfacial tension air/water	$N \cdot m^{-1}$
Φ''_{mol}	Mass transfer flow	$kmol \cdot s^{-1}$
Φ_v	Flow	$l \cdot h^{-1}$
Φ'_{vE}	Linear velocity of the extract phase	$m \cdot s^{-1}$
Φ_{vR}	Raffinate flow	$m^3 \cdot s^{-1}$
Φ'_{vR}	Linear velocity of the raffinate phase	$m \cdot s^{-1}$

Part VI

Absorption of Gases

Part VI Content

43

Absorption of Gases

43.1 Introduction

The removal of a contaminant from a gaseous stream will be treated in this chapter mainly. A solution of any volatile material exhibits a vapor pressure of the solute above the liquid phase. This vapor pressure increases with solute concentration and liquid temperature. Contaminant removal continues only so long as the partial pressure of that constituent in the gas phase exceeds the vapor pressure of that solute above the liquid phase. Contaminant removal rate is a function of this driving force, a pressure difference, as well as the mass transfer coefficient. The process of physical absorption is illustrated in Figure 43.1. In certain applications where the dissolved solute would exhibit a high vapor pressure above the resultant solution, a reactive chemical is added to the liquid phase. The function of this reactant is to combine rapidly with the absorbed solute so as to eliminate or substantially reduce the vapor pressure of the solute above the liquid phase. In these situations, the mass transfer driving force is maximized. Further, the scrubbing solution can be recirculated while still obtaining a high fume removal efficiency. For example, the use of an alkaline scrubbing liquid for removal of chlorine from a gas stream is common practice. The process of absorption associated with chemical reaction is shown in Figure 43.2. In this figure, the reaction is irreversible.

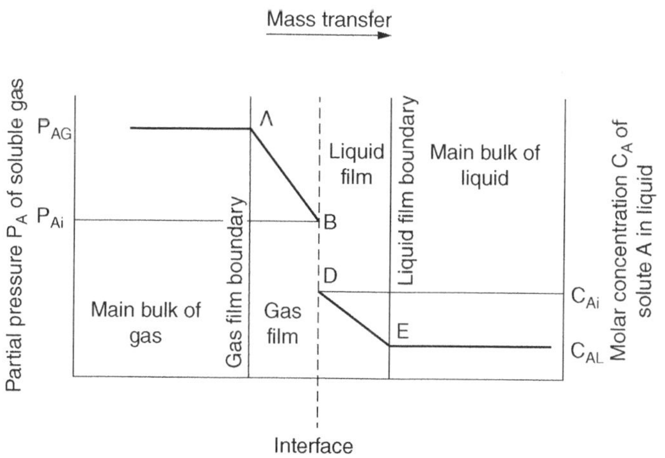

Figure 43.1 Concentration profile for absorbed component A.

Introduction to Chemical Engineering, First Edition. C.M. van 't Land.

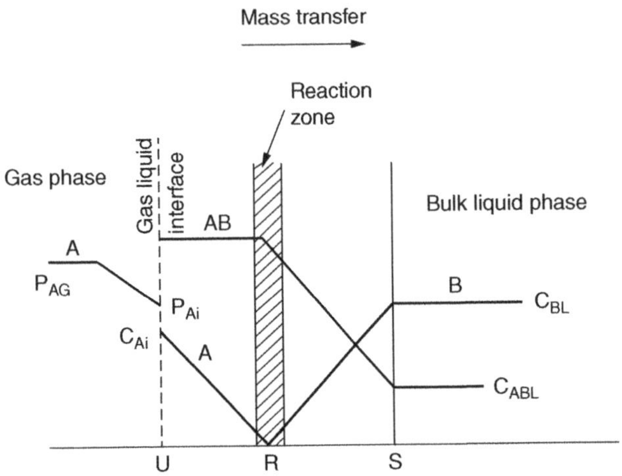

Figure 43.2 Concentration profile for absorption with chemical reaction.

43.2 Determination of the Number of Theoretical Stages at Absorption of Gases

The subject will be treated by taking a look at the drying of natural gas in The Netherlands [1]. Natural gas consists of methane mainly. However, it also contains water vapor that especially in wintertime can lead to the formation of crystals. The crystals consist of a hydrate of methane. The formation of the hydrate can give rise to plugging problems and is thus highly undesirable. Natural gas also contains some H_2S and CO_2. In the presence of water, this may lead to corrosion problems. Hence, for two reasons it is necessary to remove water from natural gas. One way of doing this is absorbing water vapor in di-ethylene glycol (DEG). DEG is a high-boiling ether. Its boiling point is 245 °C at atmospheric pressure and its molecular weight is 106 $g \cdot mol^{-1}$. The formula is $CH_2OH–CH_2–O–CH_2–CH_2OH$. Such a process can be carried out in an absorption column. The natural gas flows through the column from the bottom to the top. DEG flows through the column from the top to the bottom. It is possible to carry out this absorption in a sieve tray column. See Figure 43.3. DEG leaving the absorption column at the bottom is loaded with water. It is pumped to

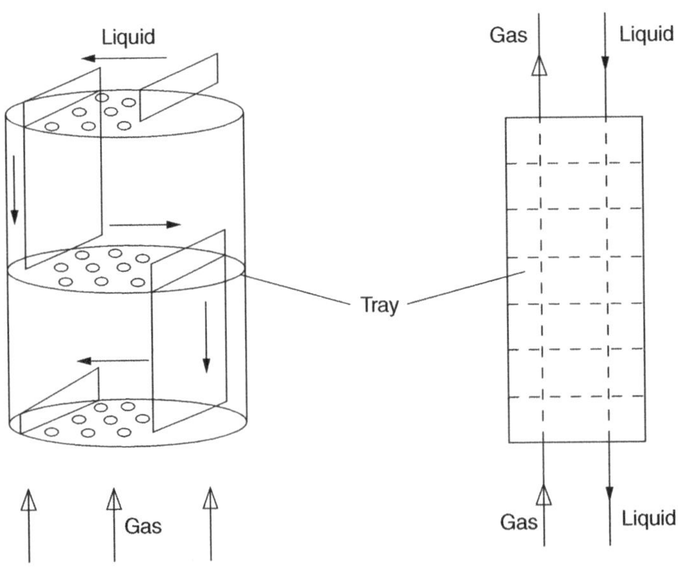

Figure 43.3 Part of a sieve tray column (left) and a full column (right). *Source:* Courtesy of Delft Academic Press/VSSD, Delft, The Netherlands.

the top of a second column, the desorption column. In this piece of equipment, DEG is dried by contacting it with gaseous nitrogen.

It is assumed that gas and liquid leaving a tray are in equilibrium. Figure 43.4 shows the equilibrium stages, flows, and fractions in a typical absorption column. The graphical design of a column for the absorption of water vapor from natural gas by means of DEG will be treated. The design method resembles the design method first proposed by McCabe and Thiele and treated in Chapter 34. In that chapter, the method was discussed generally first and then applied to the separation of benzene and toluene.

The relationship between the water mole fractions in the gaseous and the liquid flows leaving an equilibrium stage is given by the equilibrium line. Note that these two flows are found at different heights in a column. Now, the relationship between the water mole fractions in the gaseous and liquid flows at the same height will be discussed. This relationship can be found via a mass balance. The graphical presentation is called the operating line. In Figure 43.5, it can be seen that the water balance for the upper part of a column is:

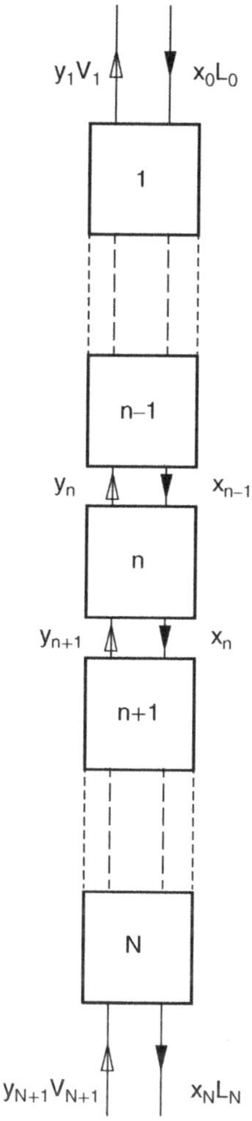

Figure 43.4 Equilibrium stages, flows, and fractions in a typical absorption column. *Source:* Courtesy of Delft Academic Press/VSSD, Delft, The Netherlands.

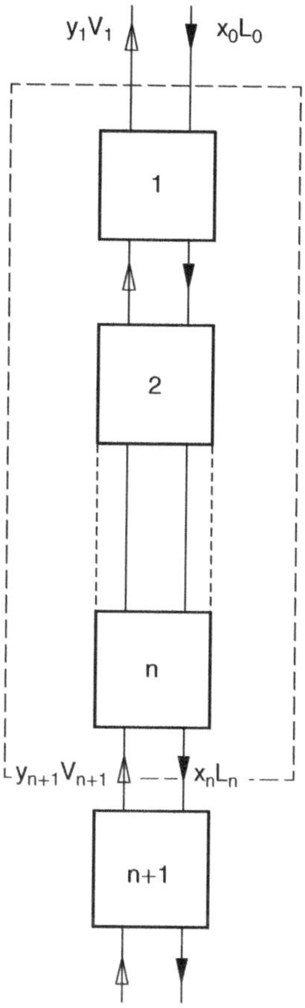

Figure 43.5 The water balance for the upper part of the column. *Source:* Courtesy of Delft Academic Press/VSSD, Delft, The Netherlands.

$$L_0 \cdot x_0 + V_{n+1} \cdot y_{n+1} = V_1 \cdot y_1 + L_n \cdot x_n.$$

L and V are almost constant if x and y are small and hence:

$$V(y_{n+1} - y_1) = L(x_n - x_0).$$

For a point below any stage from the top:

$$V(y - y_1) = L(x - x_0) \rightarrow y = \frac{L}{V} \cdot x - \frac{L}{V} \cdot x_0 + y_1.$$

The latter equation is the equation of the operating line. Its slope is L/V, the designer can assess this ratio. In addition, a point is necessary for the construction of the operating line. For example, (x_0, y_1) is selected. x_0 and y_1 are design specifications: the water contents of the fresh solvent and the purified gas.

The slope of the equilibrium line is given the symbol m = y/x. At 4 MPa and 72 °C, its numerical value is $5.0 \cdot 10^{-3}$. This low value reflects the fact that, in equilibrium, the mole fraction in the vapor is much lower than the mole fraction in the liquid. That is favorable for the removal of water vapor from a gaseous stream. m increases with temperature and is inversely proportional to pressure. The latter statement is explained as follows. On increasing the pressure, the water vapor partial pressure increases. However, the mole fraction remains constant.

In Figure 43.6, the equilibrium line is drawn. An operating line has also been drawn. Its slope is L/V = 0.01, meaning that 100 mol of gas is treated with 1 mol of DEG. The mole fraction of water in the natural gas flow is reduced from $1 \cdot 10^{-3}$ to $2 \cdot 10^{-4}$. That means a reduction with a factor of 5. This purification means that the dew point of natural gas at 4 MPa is decreased from 76 to 4 °C. The mole fractions of water in entering and leaving DEG are, respectively, 0.02 and 0.1. $(0.02, 2 \cdot 10^{-4})$ is a point of the operating line. Note that this point is found low in the graph whereas the point is valid for the top of the column. The background is that both flows contain a minimum of water at the top of the column. The gaseous flow does because it has been purified. The liquid flow does because it has been freed from water in the desorption column. The operating line can now be drawn because we have a point and the slope of the line. By drawing

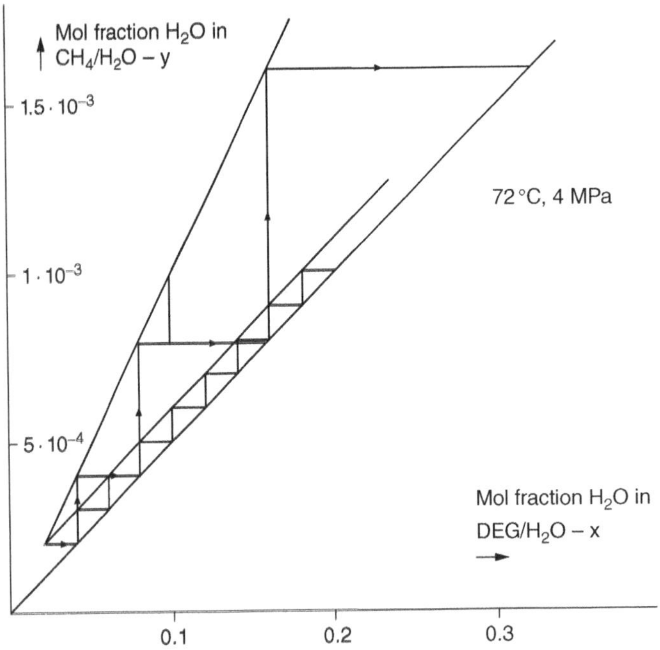

Figure 43.6 Equilibrium line and operating lines.

steps, it is found that 2.25 equilibrium stages are required for this separation. The technique of drawing steps has been explained in Chapter 34.

When we change the ratio L/V to 0.005, we have a situation that 100 mol of gas is treated with 0.5 mol of DEG. Because less DEG is used than in the previous situation, it stands to reason that more theoretical stages are required than for the previous situation. The equilibrium line is not changed for the new situation. However, the operating line changes. A point through which the operating line goes remains $(0.02, 2 \cdot 10^{-4})$. The background is that we do not change the specification of the purified product and that the desorption column is able to also maintain the specification of the cleaned DEG. Note that the water content of the spent DEG rises with a factor of 2: less DEG absorbs the same amount of water. The slope of the operating line decreases with a factor of 2. The new operating line has also been drawn in Figure 43.6. By drawing steps, it becomes clear that the new situation requires nine equilibrium stages.

Generally, more equilibrium stages are required when there is less space between the equilibrium line and the operating line. It is possible to increase the space by increasing the pressure. That affects the equilibrium line. By increasing the pressure with a factor of 2, the slope of the equilibrium line decreases with a factor of 2. The question can be asked why that option is not chosen for the cleaning of natural gas. The answer is that the usual pressure is, with 4 MPa (40 at), already quite high.

In this discussion, m has been assumed to be constant. Then, the equilibrium line is a straight line going through the origin (0, 0). If m is a function of the concentration and the equilibrium line is curved, it is still possible to draw steps between the two lines.

When x and y are relatively large, V and L are not constant. It is recommended to work with the number of impurity moles per mole of inert gas and with the number of impurity moles per mole of inert solvent in that case. The molar flows of inert gas and inert solvent remain constant. The distribution coefficient must be defined accordingly.

A similar purification of natural gas in Rumania was described [2]. Tri-ethylene diglycol (TEG) instead of DEG is used. The gas pressure is between 10 and 30 bar (1–3 MPa). The dew point of the purified natural gas is −15 °C at 40 bar (4 MPa). Winter temperatures in Rumania are, on the average, lower than in The Netherlands.

43.3 Estimation of the Diameter of an Absorption Column for Natural Gas

The diameter of an absorption column is fixed by entrainment considerations. See Figure 43.7. The approximate magnitude of the terminal velocity of droplets having a size of, for example, 1 mm is:

$$v_d = \sqrt{\frac{g \cdot \rho_l \cdot d_p}{\rho_g}} \ \mathrm{m \cdot s^{-1}}. \tag{43.1}$$

$3\ \mathrm{m \cdot s^{-1}}$ is an indication for v_d when $d_p = 0.001$ m, $\rho_l = 1{,}000\ \mathrm{kg \cdot m^{-3}}$, and $\rho_g = 1\ \mathrm{kg \cdot m^{-3}}$.

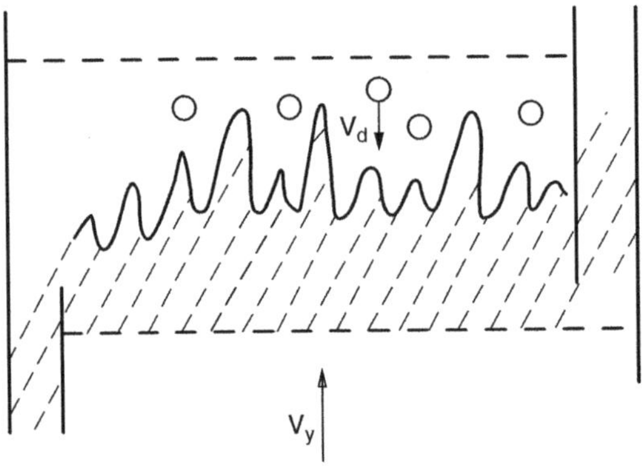

Figure 43.7 Mass transfer on a tray. *Source:* Courtesy of Delft Academic Press/VSSD, Delft, The Netherlands.

Example 43.1

The capacity of a power station is 1,000 MW. The power is raised by the combustion of natural gas. The diameter of an absorption column to remove water from natural gas for this power station is calculated. The column operates at 4 MPa and 72 °C. Further data are: $\rho_L = 1{,}100\ kg \cdot m^{-3}$, $d_p = 0.001$ m, $M(CH_4) = 16\ kg \cdot kmol^{-1}$.

1,000 MW corresponds with 10^6 kJ · s˙.

The efficiency of the power station is 40%. So, the natural gas flow to the power station transfers $2.5 \cdot 10^6\ kJ \cdot s^{-1}$. 1 nm^3 of natural gas provides $3.2 \cdot 10^4$ kJ. The natural gas flow is $\frac{2.5 \cdot 10^6}{3.2 \cdot 10^4} = 78.1\ nm^3 \cdot s^{-1}$.

The natural gas flow is also $\frac{78.1}{22.3} = 3.5\ kmol \cdot s^{-1}$.

$$\rho_g = \frac{Mp}{RT} = \frac{16 \cdot 4 \cdot 10^6}{8{,}314 \cdot 345} = 22.3\ kg \cdot m^{-3}.$$

$$v_d = \sqrt{\frac{9.81 \cdot 1{,}100 \cdot 0.001}{22.3}} = 0.70\ m \cdot s^{-1}.$$

1 kmol occupies 22.3 nm^3 at 0.1 MPa and 0 °C.

Thus, the natural gas flow is: $3.5 \cdot 22.3 \cdot \frac{0.1}{4} \cdot \frac{345}{273} = 2.5\ m^3 \cdot s^{-1}$.

Cross-sectional area: $\frac{2.5}{0.7} = 3.6\ m^2$ and D = 2.15 m.

43.4 The Absorption of Carbon Dioxide

The emission of carbon dioxide (CO_2) is probably the cause of the increase of the average world temperature. The latter increase is the probable cause of a climate change we are experiencing now. The present concentration of CO_2 in the atmosphere is approximately 410 ppm (parts per million) by volume. That corresponds with 0.041% by volume. The annual increase of this number is at present 2 ppm. The four main sources of this emission are power stations, industry, traffic, and house warming. These four categories each cause, in The Netherlands, approximately one-quarter of the emission. At the moment, plans are made to limit or even to avoid this emission. It makes sense to pay attention to the emission of power stations as there is a relatively small number of these installations. An aspect is that the emission of a power station based on natural gas causes approximately half the emission of CO_2 of a power station based on coal and having the same capacity.

Plans were made to absorb CO_2 from the exhaust gas of a power station in The Netherlands based on coal and biomass and having a capacity of 1,100 MW. The idea was to absorb and store underground one quarter of the CO_2 emitted by this power station [3]. That meant the absorption of 187 $t \cdot h^{-1}$. It would mean an absorption column having a height of 40 m and a diameter of 13 m. In this absorption column, CO_2 would be absorbed in mono-ethanol amine (MEA). CO_2 would be resorbed from MEA in a second column having a height of 20 m and a diameter of 7.5 m. These two steps can be built by using proven technology. The resorption would be accomplished by contacting MEA, flowing from the top to the bottom of the column, with nitrogen gas at 100 °C. The conclusion is that a relatively modest separation of CO_2 (25%) emitted by this power plant entails the construction of a plant having an impressive size. Part of the power produced by the power plant is needed to keep this absorption plant functioning. The plant to absorb CO_2 has not been built because of cost reasons. The power plant is operated today without a CO_2 absorption system.

This sequence of facts draws our attention to the issues that will have to be addressed on limiting or even avoiding the emission of CO_2.

43.5 Design of Absorption Columns

The concentration gradients for transfer of a solute from a gaseous phase to a liquid phase are as shown in Figure 43.1. Figure 43.8 depicts the process in a countercurrent absorption tower.

$$G_m = \frac{\text{kmol of inert gas}}{\text{m}^2 \text{ of tower cross-section} \cdot \text{s}}.$$

$$L_m = \frac{\text{kmol of solute-free liquor}}{\text{m}^2 \text{ of tower cross-section} \cdot \text{s}}.$$

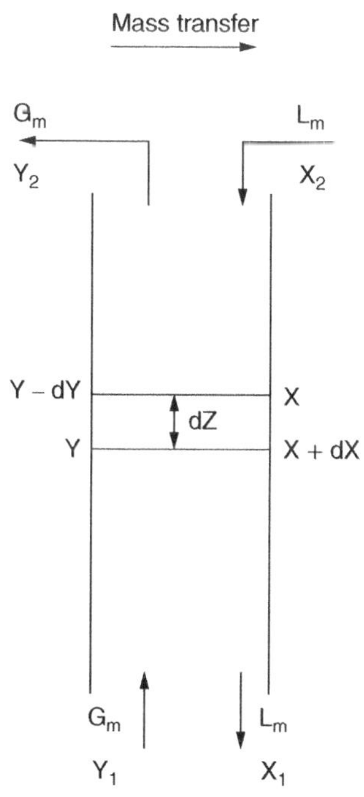

Figure 43.8 Countercurrent absorption tower.

$$X = \frac{\text{kmol of dissolved solute}}{\text{kmol of inert solvent}}.$$

$$Y = \frac{\text{kmol of gaseous solute}}{\text{kmol of inert gas}}.$$

The equation obtained after the integration of the relevant differential equation is:

$$G_m(Y_1 - Y_2) = K_G \cdot a(\Delta Y_G)_m \cdot Z \text{ kmol} \cdot m^{-2} \cdot s^{-1}. \tag{43.2}$$

This equation resembles Equation 39.7. In both equations, the left-hand side represents the amount of kmoles transferred in a column per m^2 column cross-section and per s. $K_G \cdot a$ appears instead of $K_R \cdot a$. $(\Delta Y_G)_m$ is the logarithmic mean driving force for the absorption and replaces $(\Delta C_R)_m$. The gaseous phase is cleaned during the absorption and may be compared to the raffinate phase at extraction whereas the liquid phase absorbs the solute and may thus be compared to the extract phase at liquid extraction. The height of the column appears in both equations. The analogy will be amplified in an example which follows.

The applicability of Equation 43.2 depends on the availability of data concerning $K_G \cdot a$. The latter product is a function of:

- G_m,
- L_m,
- the tower packing,
- the column pressure,
- the column temperature, and
- the physical properties of liquid and gas.

Often, the influence of the physical properties is considered to be minor. Many data concerning $K_G \cdot a$ have been published. See, for example, [4].

Example 43.2

Gas, from a petroleum distillation column, has its concentration of H_2S reduced from 0.03 kmol of H_2S per kmol of inert hydrocarbon gas to 1% of this value by scrubbing with a triethanolamine–water solvent in a countercurrent tower, operating at 27 °C and at atmospheric pressure [4]. See Figure 43.9. The equilibrium relation for the solution may be taken as $m = X_i/Y_i = 0.5$. The solvent enters the tower free of H_2S and leaves containing 0.013 kmol of H_2S/kmol of solvent. The flow of inert gas is 0.015 kmol · $m^{-2} \cdot s^{-1}$. The height and the number of transfer units of the absorption tower are calculated. $K_G \cdot a = 0.04$ kmol · $m^{-3} \cdot s^{-1}$.

Driving force at the top of the column: $Y_2 - \dfrac{X_2}{m} = 3.0 \cdot 10^{-4} - \dfrac{0}{0.5} = 3 \cdot 10^{-4}$.

Driving force at the bottom of the column: $Y_1 - \dfrac{X_1}{m} = 0.03 - \dfrac{0.013}{0.5} = 4 \cdot 10^{-3}$.

The driving force is the actual molar ratio in the gaseous phase (raffinate phase) minus the molar ratio in the gaseous phase that would be in equilibrium with the molar ratio in the liquid phase (extract phase).

Logarithmic mean driving force: $\dfrac{4 \cdot 10^{-3} - 3 \cdot 10^{-4}}{\ln \dfrac{4 \cdot 10^{-3}}{3 \cdot 10^{-4}}} = 1.43 \cdot 10^{-3}$.

Equation 43.2: $0.015(0.03 - 3 \cdot 10^{-4}) = 0.04 \cdot 1.43 \cdot 10^{-3} \cdot Z \text{ kmol} \cdot m^{-2} \cdot s^{-1} \rightarrow Z = 7.8$ m.

Height of a transfer unit: $H_{OG} = \dfrac{G_m}{K_G \cdot a} = \dfrac{0.015}{0.04} = 0.375$ m.

Number of transfer units: $N_{OG} = \dfrac{7.8}{0.375} = 20.7$ (say 21).

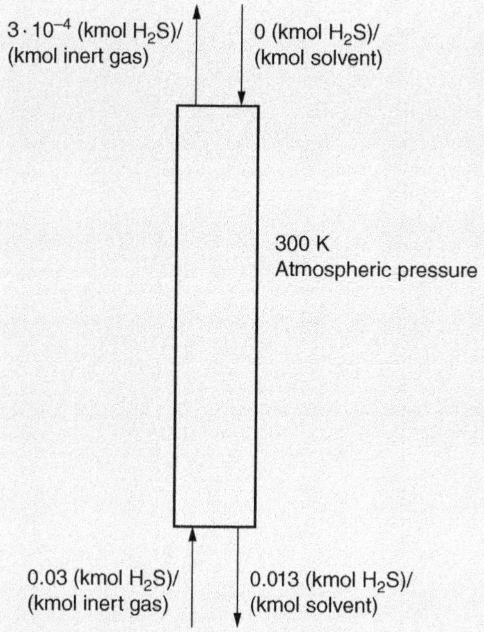

Figure 43.9 Absorption of H_2S from a refinery gas.

References

1. Wesselingh, J. A., and Kleizen, H. H. (1998). *Separation Processes* (pp. 29–35). Delft, The Netherlands: Delft University Press. (in Dutch).
2. Anonymous (2009). The power is in the rest. *Process*, 12 (4), 80–81. (in German).
3. van Baal, M. (2012). CO_2 from the plane of the river Maas to the North Sea. *Technisch Weekblad*, 18 (11), 5. (in Dutch).
4. Coulson, J. M. and Richardson, J. F. with Backhurst, J. R. and Harker, J. H. (1999). *Chemical Engineering, Volume 2, Particle Technology & Separation Processes* (p. 570). Oxford, UK: Butterworth-Heinemann.

Notation VI

a	Area for absorption	$m^2 \cdot m^{-3}$
C_A	Concentration of component A	$kmol \cdot m^{-3}$
C_{Ai}	Concentration of component A at an interface	$kmol \cdot m^{-3}$
C_{AL}	Concentration of component A in a liquid	$kmol \cdot m^{-3}$
C_{ABL}	Concentration of compound AB in a liquid	$kmol \cdot m^{-3}$
C_{BL}	Concentration of component B in a liquid	$kmol \cdot m^{-3}$
$(\Delta C)_m$	Logarithmic mean concentration difference	$kmol \cdot m^{-3}$
d_p	Droplet size	m
G_m	Inert gas flow in a column	$kmol \cdot m^{-2} \cdot s^{-1}$
g	Acceleration due to gravity	$m \cdot s^{-2}$
H_{OG}	Height of a transfer unit	m
K_G	Mass transfer coefficient	$kmol \cdot m^{-3} \cdot s^{-1}$
L	Absorption liquid flow	$kmol \cdot s^{-1}$
L_0	Absorption liquid flow entering a column	$kmol \cdot s^{-1}$
L_N	Absorption liquid flow leaving tray N	$kmol \cdot s^{-1}$
L_n	Absorption liquid flow leaving tray n	$kmol \cdot s^{-1}$
L_{n-1}	Absorption liquid flow leaving tray n−1	$kmol \cdot s^{-1}$
L_m	Absorption liquid flow in a column	$kmol \cdot m^{-2} \cdot s^{-1}$
M	Molecular weight	$kg \cdot kmol^{-1}$
m	Molar distribution coefficient	—
N_{OG}	Number of transfer units	—
P_A	Partial pressure of component A	$N \cdot m^{-2}$
P_{AG}	Partial pressure of component A in a gaseous phase	$N \cdot m^{-2}$
P_{Ai}	Partial pressure of component A at an interface	$N \cdot m^{-2}$
p	Pressure	$N \cdot m^{-2}$
R	Universal gas constant ($8.314\ kJ \cdot kmol^{-1} \cdot K^{-1}$)	
T	Temperature	K
V	Gas flow in a column	$kmol \cdot s^{-1}$
V_N	Gas flow leaving tray N	$kmol \cdot s^{-1}$
V_1	Gas flow leaving a column	$kmol \cdot s^{-1}$
V_{N+1}	Gas flow leaving tray N + 1	$kmol \cdot s^{-1}$
V_{n+1}	Gas flow leaving tray n + 1	$kmol \cdot s^{-1}$
v_d	Terminal droplet velocity	$m \cdot s^{-1}$
v_y	Gas velocity in a column	$m \cdot s^{-1}$

Introduction to Chemical Engineering, First Edition. C.M. van 't Land.

X	Molar ratio dissolved solute/solvent in a column	—
X_1	Molar ratio dissolved solute/solvent at the bottom of a column	—
X_2	Molar ratio dissolved solute/solvent at the top of a column	—
X_i	Molar ratio dissolved solute/solvent at an interface	—
x	Mole fraction water in the absorption liquid	—
x_0	Mole fraction absorbant in L_0	—
x_N	Mole fraction absorbant in L_N	—
x_n	Mole fraction absorbant in L_n	—
x_{n-1}	Mole fraction absorbant in L_{n-1}	—
Y	Molar ratio gaseous solute/inert gas in a column	—
Y_1	Molar ratio gaseous solute/inert gas at the bottom of a column	—
Y_2	Molar ratio gaseous solute/solvent at the top of a column	—
Y_i	Molar ratio gaseous solute/solvent at an interface	—
$(\Delta Y_G)_m$	Logarithmic mean driving force in a gaseous phase	—
y	Mole fraction water in natural gas	—
y_1	Mole fraction water in V_1	—
y_N	Mole fraction absorbant in V_N	—
y_{N+1}	Mole fraction absorbant in V_{N+1}	—
y_n	Mole fraction absorbant in V_n	—
y_{n+1}	Mole fraction absorbant in V_{n+1}	—
Z	Column height	m

Greek Symbols

ρ_g	Specific mass of the gas flow	$kg \cdot m^{-3}$
ρ_l	Specific mass of the absorption liquid	$kg \cdot m^{-3}$

Part VII

Membranes

Part VII Content

44

Membranes—An Introduction

44.1 General

Membranes control the permeation of a compound or of compounds, whereas other compounds are retained. Thus, membranes are used to separate. The feed to the membranes can be liquid or gaseous. If the feed is liquid, it is almost always aqueous. If the feed is gaseous, it consists of compounds having small molecules, for example, at the separation of nitrogen from air. The discovery, which contributed to the transformation of membrane separation from a laboratory to an industrial process, was the development of the Loeb–Sourirajan process [1]. That process, in the early 60s of the previous century, made it possible to make defect-free, high-flux, and anisotropic membranes for reverse osmosis on an industrial scale. Anisotropy means that the result of a measurement carried out depends on the direction in which the measurement is performed. These membranes consisted of an ultrathin, selective surface film on a much thicker but more permeable microporous support, which provides mechanical strength.

Separation by means of membranes is an interesting option because it, unlike unit operations like distillation and drying, does not involve a transfer from the liquid to the vapor phase.

The development of membranes for medical applications occurred independently from the development of industrial membranes. It will be discussed in Section 44.7.

44.2 Membranes

Synthetic membranes are discussed exclusively. The majority of membranes used commercially are polymer-based. Ceramic membranes and dense metal membranes also exist. The term filter is usually limited to structures that separate particles larger than 1–10 μm. Figure 44.1 shows schematically symmetrical and anisotropic membranes. Instead of symmetrical, the word isotropic is also used. Isotropy means that the result of a measurement does not depend on the direction in which the measurement is performed.

Symmetrical Membranes

The pores of microporous membranes differ from the pores in a conventional filter by being extremely small, that is, on the order of 0.01–10 μm in diameter. All particles larger than the largest pores are completely rejected by the membrane. Particles smaller than the largest pores but larger than the smallest pores are partially rejected, according to the pore size distribution of the

Introduction to Chemical Engineering, First Edition. C.M. van 't Land.

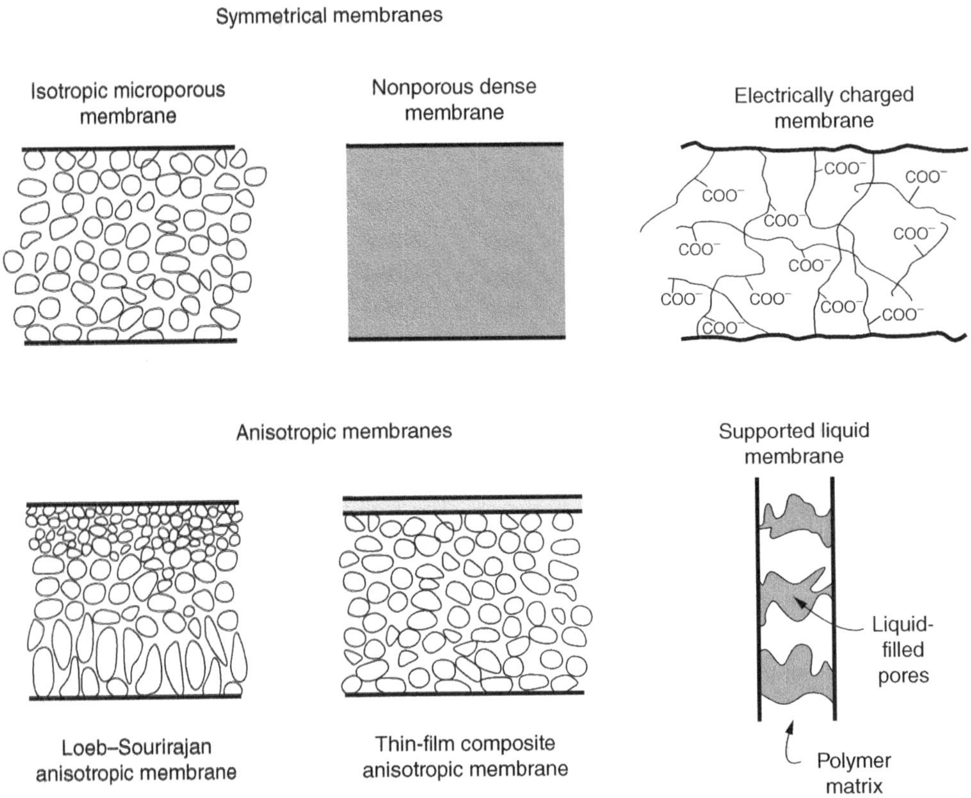

Figure 44.1 Schematic diagrams of the principal types of membrane. *Source:* Membrane Technology and Applications by Baker, R. W. (2012). Reproduced with permission of John Wiley & Sons Inc.

membrane. Nonporous dense membranes consist of a dense film through which permeants are transported by the driving force of a pressure, concentration, or electric potential gradient. Electrically charged membranes are most commonly very finely microporous, with the pore walls carrying fixed positively or negatively charged ions. Anion-exchange membranes carry positively charged ions in their pore walls, whereas cation-exchange membranes carry negatively charged ions. Separation is based mainly on the exclusion of ions having the same charge as the ions of the membrane. Electrolyte solutions are processed by means of electrically charged membranes in electrodialysis. Electrodialysis is discussed in Section 44.4 of this chapter.

Anisotropic Membranes

These membranes consist of an extremely thin surface layer supported on a much thicker, porous substructure. They are called composite membranes when the two layers are made from different polymers. The separation properties and permeation rates of the membrane are determined exclusively by the surface layer; the substructure functions as a mechanical support. Almost all commercial processes use such membranes because of the high fluxes provided by anisotropic membranes. Most membranes for reverse osmosis, gas separation and pervaporation consist of a dense membrane on top of a relatively porous substructure.

44.3 Three Pressure-Driven Membrane Separation Processes for Aqueous Systems

Microfiltration, ultrafiltration, and reverse osmosis are fully developed processes and are mentioned in this section. They will be discussed in some detail in the following three chapters of this part. Figure 44.2 illustrates the ranges of application of these three processes. Basically, microfiltration and ultrafiltration are similar in that the mode of separation is retaining small entities, whereas water molecules pass through a membrane. At ultrafiltration, the pores are smaller than at microfiltration. Microfiltration membranes filter colloidal particles and bacteria having sizes between 0.1 and 10 μm. Dissolved macromolecules like proteins can be separated from aqueous solutions by means of ultrafiltration membranes. Reverse osmosis membranes function differently. The size of the membrane pores varies between approximately 3 and 5 Å, and that is within the range of thermal motion of the polymer chains forming the membrane. For reverse osmosis, the generally accepted mode of transport through the membrane is by solution and diffusion. Solutes permeate the membrane by dissolving in the membrane material and traveling through it driven by a pressure gradient. The main application of reverse osmosis is desalination of seawater and brackish groundwater by permitting water molecules to dissolve into and travel through a membrane while both positively charged sodium ions and negatively charged chloride ions are retained.

The following line of reasoning is given to illustrate that the pressure-driven flow of water through a pore decreases strongly if the pore diameter decreases. That decrease is stronger than

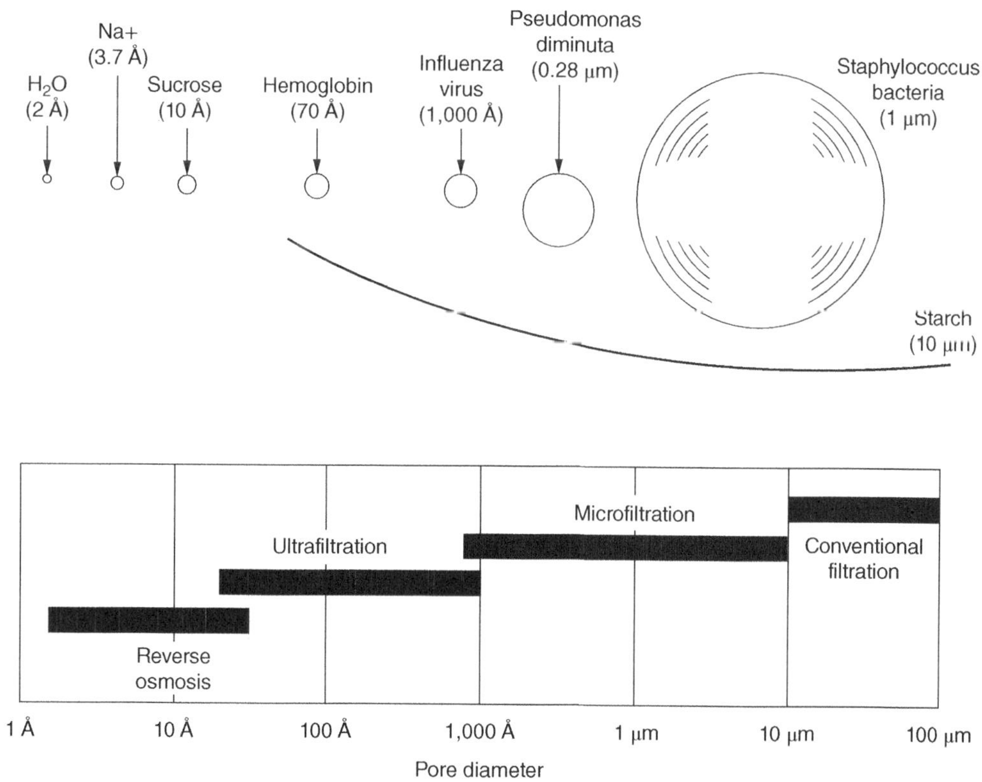

Figure 44.2 The ranges of application of microfiltration, ultrafiltration, and reverse osmosis. *Source:* Membrane Technology and Applications by Baker, R. W. (2012). Reproduced with permission of John Wiley & Sons Inc.

a decrease that would correspond with the area decrease. At both microfiltration and ultrafiltration, the water flow through the pores is laminar. That means that we can use the Hagen–Poiseuille equation. See Equation 4.3. The equation reads:

$$\Phi_V = \frac{\pi \cdot R^4}{8\mu} \cdot \left(-\frac{dp}{dx}\right)\ m^3 \cdot s^{-1}.$$

This general equation can in this case be replaced by:

$$\Phi_V = \frac{\pi \cdot D^4}{128 \cdot \mu l} \cdot \Delta p\ m^3 \cdot s^{-1}.$$

In this equation, Δp is the pressure difference across the pore, μ is the liquid viscosity, and l is the pore length. The flux, or flow per unit membrane area (which has the dimension of velocity, that is, $m \cdot s^{-1}$), is the sum of all flows through the individual pores and so is given by:

$$v = N \cdot \frac{\pi \cdot D^4}{128 \cdot \mu l} \cdot \Delta p\ m \cdot s^{-1}. \tag{44.1}$$

N is the number of pores per m^2 of membrane area. For membranes of equal pore area and fraction free area ε, it is possible to write:

$$\varepsilon = N \cdot \frac{\pi}{4} \cdot D^2 \rightarrow N = \frac{4\varepsilon}{\pi \cdot D^2}.$$

Substituting this expression into Equation 44.1 results in:

$$v = \frac{\varepsilon \cdot D^2}{32 \cdot \mu l} \cdot \Delta p\ m \cdot s^{-1}. \tag{44.2}$$

This equation tells us that for a given combination of fraction free area, viscosity, pore length, and pressure loss, the velocity at which the liquid flows through the membrane in $m \cdot s^{-1}$ (or $l \cdot m^{-2} \cdot h^{-1}$) is proportional to D^2. That means that the flux through a microfiltration membrane can be 10^4 times greater than the flux through an ultrafiltration membrane. On carrying this argument a bit further, the permeance, that is, velocity per unit pressure difference ($v/\Delta p$) of microfiltration membranes is much higher than the permeance of ultrafiltration membranes. This difference affects the operating pressure and the way these membranes are used industrially.

44.4 A Membrane Separation Process for Aqueous Solutions Which Is Driven by an Electrical Potential Difference

The process is electrodialysis and is also a fully developed process. It will be discussed in some detail in Chapter 48. The main application is the desalting of brackish groundwater. The process is also being introduced into the food industry for the deionization of cheese whey. A further application is in pollution control. Figure 44.3 shows schematically the mode of operation of the process for desalting brackish ground water. The electrodialysis stack contains several hundred individual cells. A current passes through the unit from an anode to a cathode. Each cell consists of a pair of anion and cation exchange membranes. The solution flows through the piece of equipment, which resembles a filter press. The sodium and chloride ions pass through the membranes and enter a pick-up solution. Both a demineralized product and a concentrated effluent leave the unit.

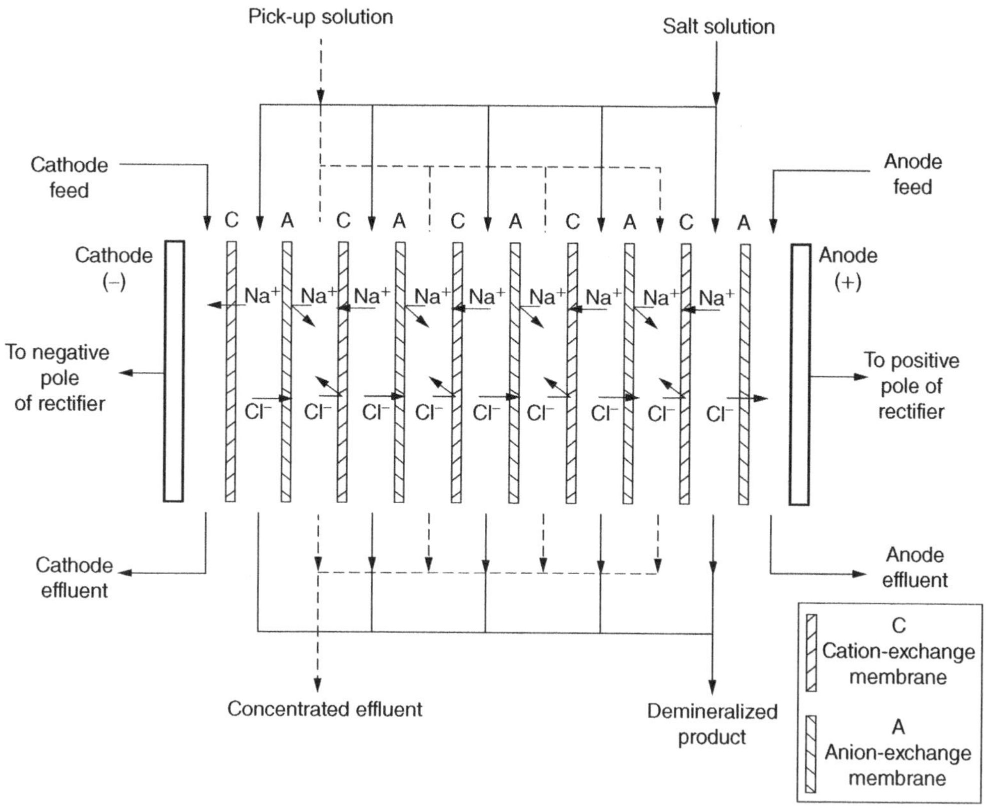

Figure 44.3 Schematic diagram of an electrodialysis process. *Source:* Membrane Technology and Applications by Baker, R. W. (2012). Reproduced with permission of John Wiley & Sons Inc.

44.5 Gas Separation

This separation method is a developing industrial membrane separation process [2]. At least 20 companies worldwide offer systems for various applications. A gaseous mixture at an elevated pressure is passed across the surface of a membrane that is selectively permeable to one component of the feed mixture. Figure 44.4 shows the process schematically. Important current applications

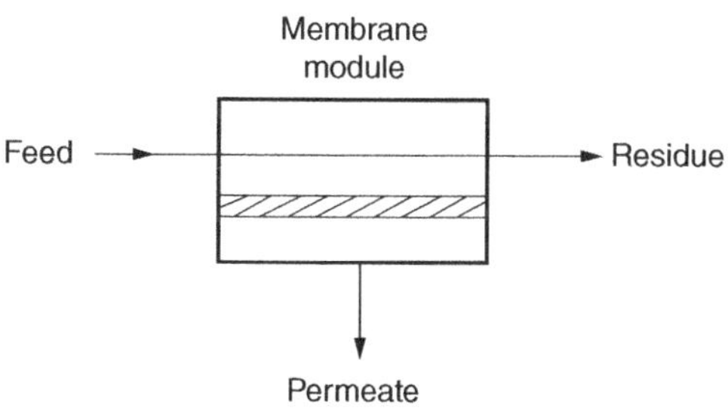

Figure 44.4 Schematic diagram of the basic membrane gas separation process. *Source:* Membrane Technology and Applications by Baker, R. W. (2012). Reproduced with permission of John Wiley & Sons Inc.

are the production of nitrogen from air; the separation of hydrogen from a mixture of nitrogen, argon, and methane in ammonia plants; and the separation of carbon dioxide from natural gas in natural gas operations. There are further applications.

44.6 Pervaporation

This method is also a developing industrial membrane separation process. It is illustrated schematically in Figure 44.5. A liquid mixture is in contact with one side of a membrane. A compound in the mixture permeates through the membrane and is removed in the vapor phase. The vapor pressure of the compound passing through the membrane is low on the permeate side of the membrane because it is condensed by cooling. A typical application is the separation of water from an azeotropic mixture of water and ethanol. Such a mixture contains 95% by weight of ethanol. The separation of water from such a mixture by distillation was discussed in Section 36.2. A second possibility is the application of pervaporation. Water permeates through the membrane. Pervaporation can produce dehydrated ethanol having a purity of more than 99.9% by weight. Further applications are the dehydration of other organic solvents.

44.7 Medical Applications

The artificial kidney, the membrane blood oxygenator, and controlled drug delivery will be discussed shortly. The annual turnover of the first two pieces of equipment is greater than the

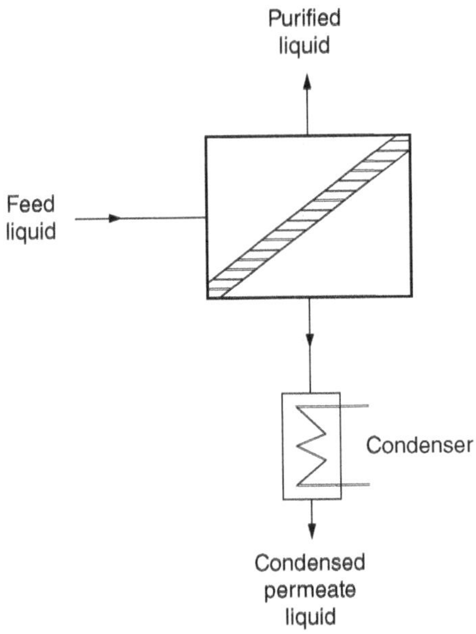

Figure 44.5 Schematic diagram of the basic pervaporation process. *Source:* Membrane Technology and Applications by Baker, R. W. (2012). Reproduced with permission of John Wiley & Sons Inc.

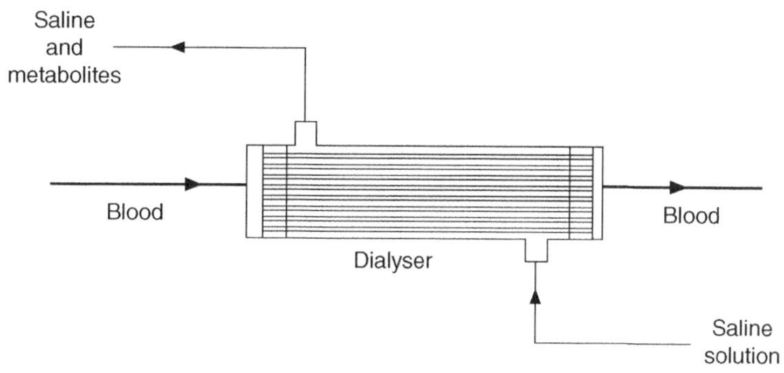

Figure 44.6 Schematic diagram of a hollow-fiber artificial kidney dialyser. *Source:* Membrane Technology and Applications by Baker, R. W. (2012). Reproduced with permission of John Wiley & Sons Inc.

turnover of all industrial membranes [3]. W.J. Kolf demonstrated the first successful operation of an artificial kidney in 1944. Today, most artificial kidneys consist of hollow-fiber membranes formed into modules having an effective area of approximately 1 m^2. See Figure 44.6. Blood passes through the fibers, while isotonic saline flows through the shell. Isotonic means that the concentration of sodium chloride, potassium, and calcium in both liquids is equal, so that these materials do not leave the blood. Metabolites like urea, creatinine, and other metabolites having a low molecular weight diffuse through the fiber wall and are removed with the saline solution. Metabolism is the set of life-sustaining chemical reactions in organisms. Metabolites are the products of these reactions. A treatment requires several hours and must be repeated once or several times a week.

The membrane blood oxygenator is a type of artificial lung. It is used in open-heart surgery. During open-heart surgery, the action of the heart is interrupted. The membrane blood oxygenator then provides oxygen to the blood and removes carbon dioxide. Blood flows through the hollow fibers, while the sweep gas flows through the shell of the device. Oxygen diffuses from the sweep gas to the blood through the wall of the hollow-fiber membrane, while carbon dioxide diffuses in the other direction. Sweep gas consists of oxygen mainly.

Finally, a few words concerning controlled drug delivery. Several systems exist. In membrane diffusion-controlled systems, a drug leaves a device by diffusion through a membrane and is administered to a body.

44.8 Additional Remarks

If the feed to the membrane is liquid, it is almost always aqueous. As far as microfiltration, ultrafiltration, and reverse osmosis are concerned, water is the compound passing through a membrane and pressure is the driving force. At electrodialysis, sodium and chloride ions pass through a membrane, whereas water is retained. An electric potential is the driving force in this case. Molecules having various sizes pass through a membrane at medical applications. Diffusion and convection are then the transport mechanisms.

References

1. Baker, R. W. (2012). *Membrane Technology and Applications* (p. 2). Chichester, UK: Wiley.
2. Baker, R. W. (2012). *Membrane Technology and Applications* (p. 9). Chichester, UK: Wiley.
3. Baker, R. W. (2012). *Membrane Technology and Applications* (p. 3). Chichester, UK: Wiley.

45

Microfiltration

45.1 Introduction

Microfiltration resembles normal filtration. Porous membranes are used to separate particles having sizes between 0.1 and 10 μm from a suspension. The cellulose acetate/cellulose nitrate blend membrane is a widely used microfilter. Other polymeric materials are also used.

Pleated membrane cartridges for large-scale industrial applications were introduced by Gelman in the 1970s. In the 1960s and 1970s, microfiltration became important in biological and pharmaceutical manufacturing. It became important in the processing of air and water at the production of microelectronics in the 1980s. A major part of the manufacture of microfilters is the production of low-cost, single-use, and disposable cartridges for pharmaceutical and electronics processes. In these industries, in most applications of microfiltration, trace amounts are removed from already clean solutions.

Figure 45.1 shows two process designs. The upper one is dead-end or in-line filtration, in which pressure forces the flow through the membrane. Streams containing less than 0.1% by weight of

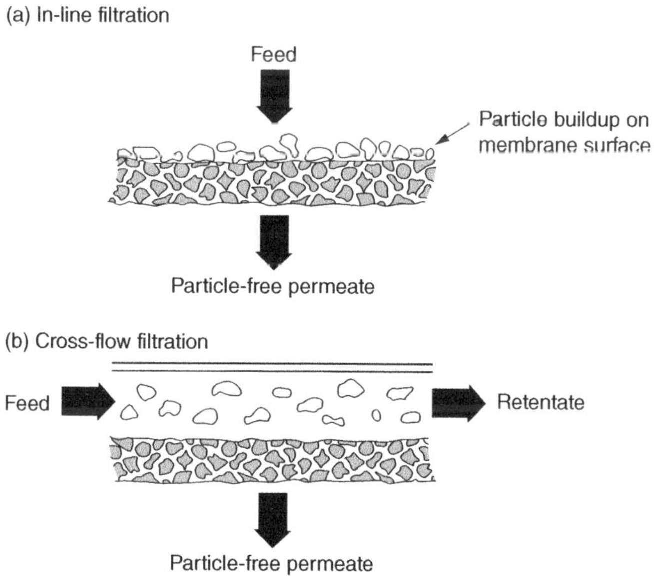

Figure 45.1 Process designs for microfiltration. *Source:* Membrane Technology and Applications by Baker, R. W. (2012). Reproduced with permission of John Wiley & Sons Inc.

Introduction to Chemical Engineering, First Edition. C.M. van 't Land.

solids are almost always treated with in-line filters [1]. Usually, the flow is kept constant. This mode of operation implies that the pressure increases with time because particles accumulate on the membrane surface or in its interior. At some point, the membrane must be replaced. The lower process design in Figure 45.1 is cross-flow filtration. As a rule-of-thumb, this design is adequate when the suspension contains 0.5% by weight of solids [1]. The equipment for cross-flow filtration is more complex than the equipment for in-line filtration and the membrane lifetime is longer than for dead-end filtration.

It is also possible to combine the two designs in an installation. In-line filtration is then followed by cross-flow filtration.

45.2 Membrane Types

The two main types of membrane for microfiltration are screen filters and depth filters. They are shown in Figure 45.2. Screen filters retain particles larger than the pore diameter. Depth filters admit the passage of particles to the interior of the membrane where they are captured.

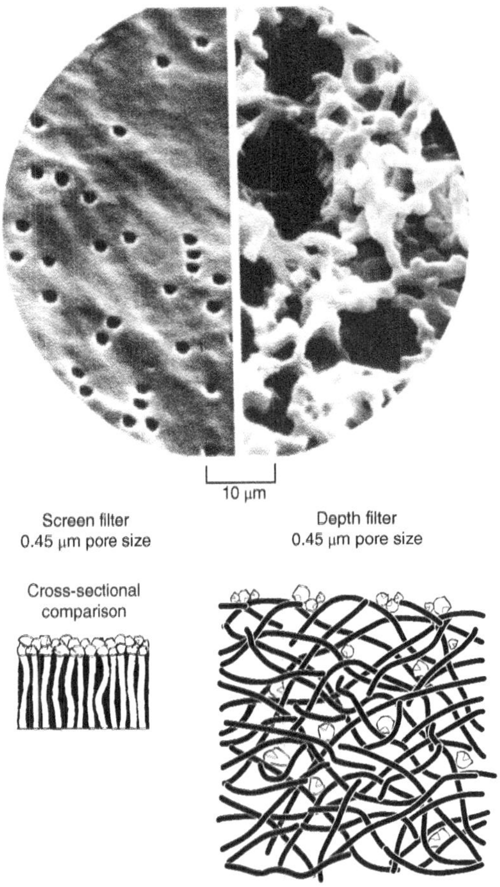

Figure 45.2 0.45-μm screen and depth filters. *Source:* Membrane Technology and Applications by Baker, R. W. (2012). Reproduced with permission of John Wiley & Sons Inc.

Although, in Figure 45.2, the sizes of the openings of the depth filter are larger than the pores of the screen filter, they have nominally the same size. Screen filters rapidly become plugged, whereas depth filters have a relatively large holding capacity. Screen filters are used for cross-flow microfiltration systems. An aspect is that the circulating fluid keeps the filter clean. Depth filters are used for in-line filtration.

45.3 Membrane Characterization

Several tests are used to characterize membrane pore size. The industry has adopted two bacterial challenge tests to measure pore size. At present, most bacterial challenge tests are performed with Pseudomonas diminuta, having an average size of 0.3–0.4 μm, although the size varies significantly with the culture conditions. The log reduction factor (LRV) quantifies the performance of membranes in bacterial challenge tests.

$$LRV = {}^{10}\log\frac{c_f}{c_p}. \tag{45.1}$$

In this equation, c_f is the concentration of particles in the feed suspension and c_p is the concentration in the permeate. It follows that at 99% rejection, c_f/c_p is 100, and the LRV is 2; at 99.9% rejection, the LRV is 3; and so on. An LRV of 7 or 8 is usually required in pharmaceutical and electronic applications.

The usual method to characterize microfiltration membranes is the bubble point test. It is simple, quick, and reliable. Air is fed to a membrane with a layer of liquid on top. It is important that the membrane is fully wetted by the test liquid. Usually, for hydrophylic membranes, water is taken, whereas methanol is used for hydrophobic membranes. The background is that, before the test is started, all pores must be completely filled with liquid. During the test, the air pressure is slowly increased until the first string of air bubbles at the membrane surface is observed. At that point, the prevailing air pressure is called the bubble point pressure. Empirically, it is a characteristic measure of the diameter of the largest pore size. For the bacterium Pseudomonas diminuta, a relationship between the bubble point pressure and the LRV was established for a series of related membranes. Typical bubble point pressures are in the range 3–4 bar [2].

45.4 Filter Construction

Pleated cartridges, first introduced in the early 1970s, are the most important devices used for microfiltration. Figure 45.3 exhibits a disposable cartridge filter of this type. A typical cartridge filter is 25 cm long, has a diameter of 5–6 cm, and contains about 0.3 m^2 of membrane area [3]. The membrane is pleated and then folded around the permeate core in these units. The feed solution, at a pressure in the range 10–120 psi, enters the housing in which the cartridge is placed.

A typical in-line microfiltration unit may consist of a coarse prefilter cartridge before the final polishing filter. The prefilter retains particles in the range 1–10 μm, whereas the polishing filter retains particles as small as 0.2 μm.

The particle-holding capacity of a cartridge filter is between approximately 100 and 300 $g \cdot m^{-2}$ of membrane area.

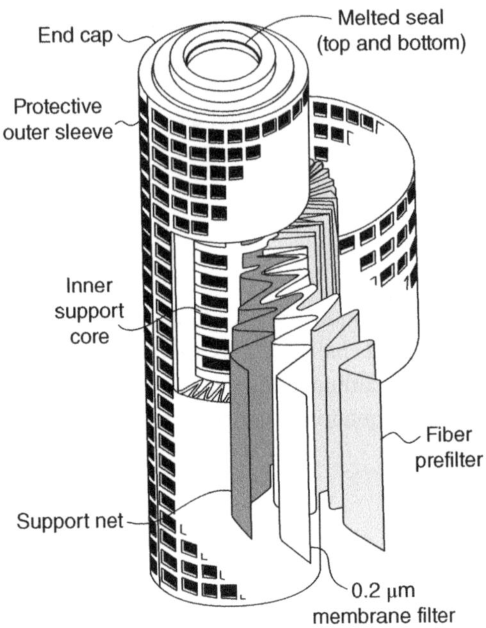

Figure 45.3 Cut-away view of a simple pleated cartridge filter. *Source:* Membrane Technology and Applications by Baker, R. W. (2012). Reproduced with permission of John Wiley & Sons Inc.

Recently, systems were developed which encompass backflushing the membrane intermittently to control membrane fouling.

45.5 Operational Practice

Figure 45.4 shows the results of experiments concerning the rate of fouling of 0.22-μm microfiltration membranes processing dilute biomass suspensions [4]. The pressure to be applied in order to keep the flux in $l \cdot m^{-2} \cdot h^{-1}$ constant is plotted as a function of time in h. For a flux of $30\ l \cdot m^{-2} \cdot h^{-1}$, the pressure increases linearly with time. That means that the cake properties

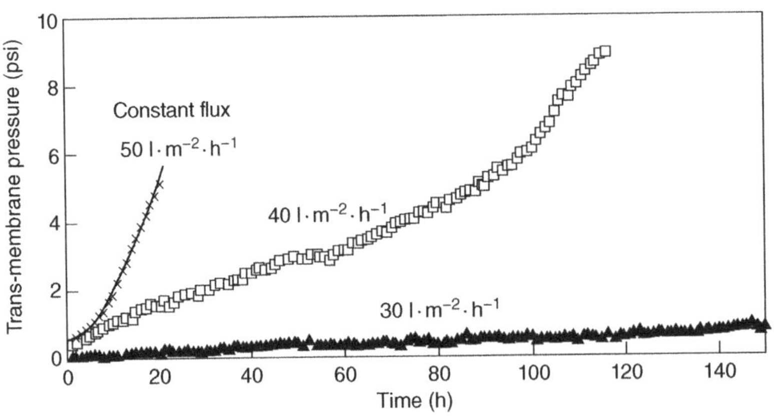

Figure 45.4 The rate of fouling of 0.22-μm microfiltration membranes used to treat biomass suspensions. *Source:* Membrane Technology and Applications by Baker, R. W. (2004). Reproduced with permission of John Wiley & Sons Inc.

do not change, specifically, the cake permeability remains constant. In terms of normal filtration, the cake thickness increases linearly with time and the cake properties remain constant. For a flux of $40\ \mathrm{l \cdot m^{-2} \cdot h^{-1}}$, it strikes in first instance that the pressures required are much higher than the pressures that would be expected on the basis of the results for a flux of $30\ \mathrm{l \cdot m^{-2} \cdot h^{-1}}$. Based on Hagen–Poisseuille's law, the pressure would be only a factor of 4/3 higher at any point in time. See Equation 44.2. Clearly, the higher flux causes different cake properties. Furthermore, the constant cake properties at this higher filtration rate change at a pressure of 6 psi. We see a cake compression. These effects are even more pronounced present at a flux of $50\ \mathrm{l \cdot m^{-2} \cdot h^{-1}}$.

References

1. Baker, R. W. (2012). *Membrane Technology and Applications* (p. 305). Chichester, UK: Wiley.
2. Baker, R. W. (2012). *Membrane Technology and Applications* (p. 312). Chichester, UK: Wiley.
3. Baker, R. W. (2012). *Membrane Technology and Applications* (p. 314). Chichester, UK: Wiley.
4. Baker, R. W. (2004). *Membrane Technology and Applications* (p. 294). Chichester, UK: Wiley.

46

Ultrafiltration

46.1 Introduction

Water and microsolutes can be separated from macromolecules and colloids by means of ultrafiltration. Dissolved sodium chloride is a typical microsolute. Pigment and latex gel particles and proteins can typically be present as macromolecules. Ultrafiltration membranes generally consist of a finely porous surface layer supported on a much more open microporous substrate providing mechanical strength. The average pore diameter of the surface layer is in the range 10^{-3}–0.1 μm. The first successful industrial application was the recovery of electrocoat paint from automobile paint shop rinse water in 1969 [1]. Cheese whey ultrafiltration systems were successfully introduced shortly thereafter. A further application is the use of ultrafiltration membranes in bioreactors, which replace conventional sewage treatment plants. Membrane fouling is the principal problem concerning a wider application of the separation method. The phenomenon is controlled by module design and by cleaning.

46.2 Membrane Characterization

The membranes reject dissolved macromolecules of different sizes. The latter are usually characterized by their molecular weight cut-off, a loosely defined term generally taken to mean the molecular weight of the globular protein molecule that is 90% rejected by the membrane. See Figure 46.1. Retentivity of a membrane depends on the shape of the molecule. Globular protein molecules may be retained, whereas linear molecules having a comparable molecular weight pass through a membrane. The table in Figure 46.1 shows typical results obtained with globular protein molecules and linear polydextran for the same polysulfone. Permeation may also be affected by the pH of the feed solution. That can particularly be the case for solutions containing dissolved polyelectrolytes. For example, ultrafiltration membranes usually reject poly(acrylic acid) at pH-values of 5 and higher, but are permeable at pH-values of 3 and lower. Poly(acrylic acid) is ionized above pH = 5. In the ionized form, the negatively charged carboxyl groups along the polymer backbone repel each other; the polymer coil is then very extended and relatively inflexible [2]. The macromolecule then cannot pass through the pores of the membrane. At pH-values below 3, the carboxyl groups are protonated. The macromolecule is then more flexible and can pass through the pores.

Introduction to Chemical Engineering, First Edition. C.M. van 't Land.

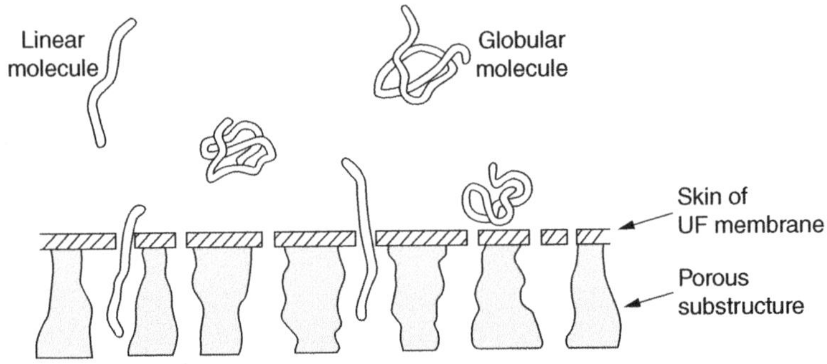

	Globular proteins		Linear polymer
Solute	Pepsin	Cytochrome c	Polydextran
MW (1,000s)	35	13	100
Rejection (%)	90	70	0

Figure 46.1 The ultrafiltration retention of differently shaped molecules. *Source:* Membrane Technology and Applications by Baker, R. W. (2012). Reproduced with permission of John Wiley & Sons Inc.

46.3 Concentration Polarization and Membrane Fouling

Concentration polarization is a key factor determining the performance of ultrafiltration membranes. Concentration polarization is the phenomenon that the concentration of retained compounds in the aqueous system near the membrane surface is higher than their concentration in the bulk of the liquid. The phenomenon can cause membrane fouling as a result of deposition of retained colloidal or macromolecular material on the membrane surface. The flux of pure water through ultrafiltration membranes is usually quite high, for example, $500\ \mathrm{l \cdot m^{-2} \cdot h^{-1}}$. However, a typical initial value of water flux through an ultrafiltration membrane separating colloidal or macromolecular material from water is $50\ \mathrm{l \cdot m^{-2} \cdot h^{-1}}$. This difference is caused by the presence of retained solutes on the surface of the membrane due to concentration polarization. See Figure 46.2.

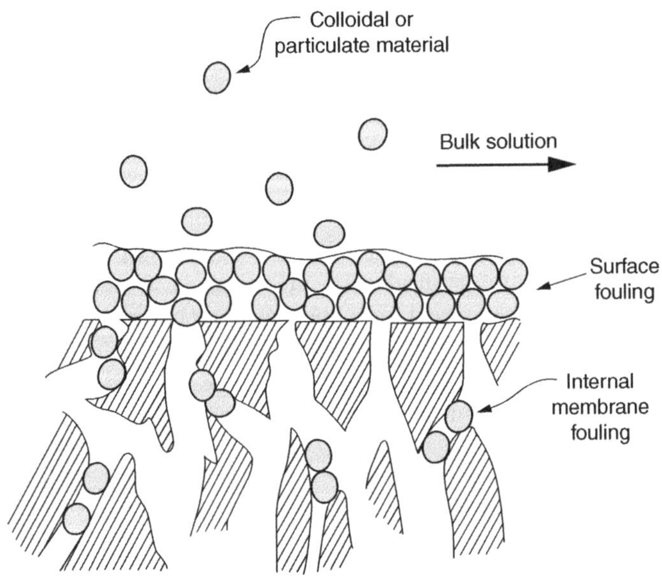

Figure 46.2 Surface fouling and internal fouling. *Source:* Membrane Technology and Applications by Baker, R. W. (2012). Reproduced with permission of John Wiley & Sons Inc.

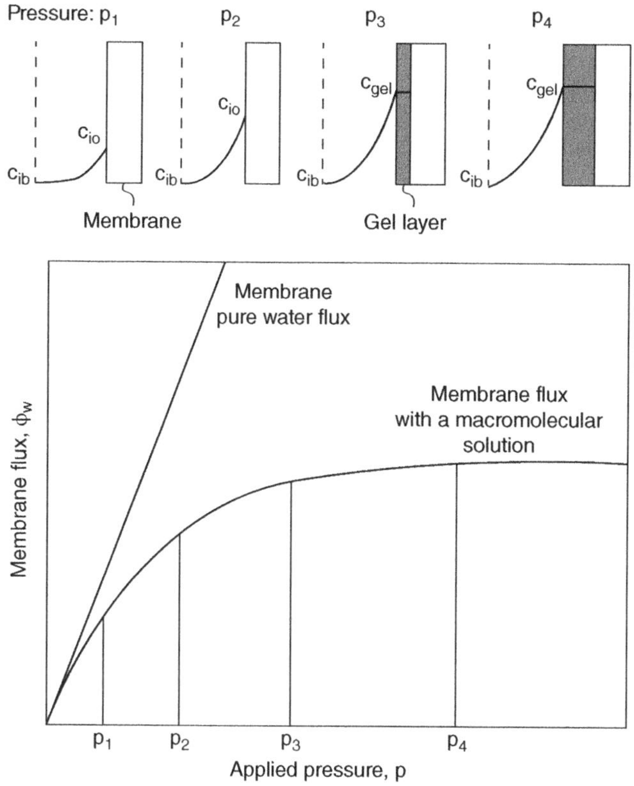

Figure 46.3 The limiting flux. *Source:* Membrane Technology and Applications by Baker, R. W. (2012). Reproduced with permission of John Wiley & Sons Inc.

Consider a solution of macromolecules in an aqueous system. A separation between the macromolecules and the aqueous system is accomplished by means of an ultrafiltration membrane. The flow through the membrane as a function of pressure is shown in Figure 46.3. The upper four small pictures show what happens in the boundary layer on increasing the pressure. At pressure p_1, liquid flows through the membrane while the dissolved macromolecules are retained by the membrane. The concentration of the macromolecules at the membrane surface is higher than the concentration of the macromolecules in the bulk of the solution ($c_{io} > c_{ib}$). The rate of transport of the macromolecules to the membrane equals the diffusion rate of the macromolecules back into the solution. The rate at which the rejected solute molecules diffuse back from the membrane surface into the bulk solution is relatively low. Typically, their concentration at the membrane surface is 20–50 times higher than in the feed solution [3]. The diffusion of the macromolecules back into the solution slows down the transport of the aqueous system (water and ions) through the membrane. The transport of the aqueous system is slowed down even stronger at pressure p_2. At pressure p_3, the concentration of the macromolecules at the membrane is such that the macromolecules no longer stay dissolved but precipitate as a gel. The membrane flux is then called the limiting flux. The phenomenon of the formation of a gel at the membrane surface is caused by concentration polarization. A further pressure increase to p_4 does hardly or not lead to a greater membrane flux because the gel layer gets thicker. The equation describing the transport of the macromolecules or colloids to and from the membrane at the limiting flux is:

$$\Phi_v \cdot c_i = Ð_i \cdot \frac{dc_i}{dx}.$$

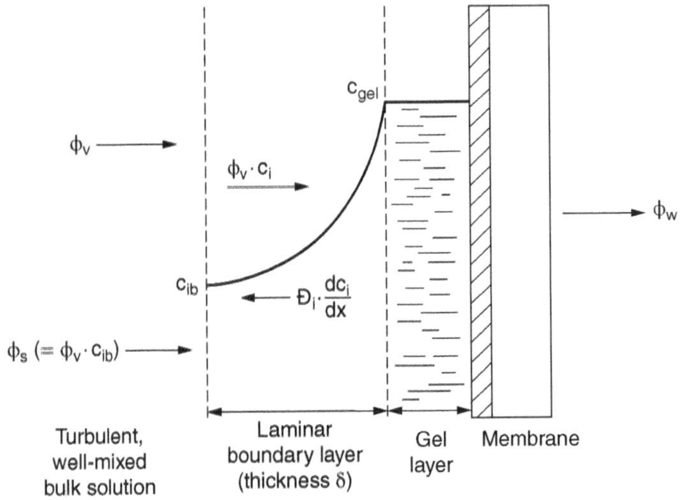

Figure 46.4 Concentration polarization causes the formation of a gel layer. *Source:* Membrane Technology and Applications by Baker, R. W. (2012). Reproduced with permission of John Wiley & Sons Inc.

This is Fick's first law. The integration proceeds as follows.

$$\frac{dc_i}{c_i} = \frac{\Phi_v}{Đ_i} \cdot dx,$$

$$\int_{c_{ib}}^{c_{gel}} \frac{dc_i}{c_i} = \frac{\Phi_v}{Đ_i} \int_0^{\delta} dx,$$

$$\ln \frac{c_{gel}}{c_{ib}} = \frac{\Phi_v \cdot \delta}{Đ_i}. \tag{46.1}$$

See Figure 46.4. In any particular ultrafiltration test, the terms c_{ib}, c_{gel}, $Đ_i$, and δ are fixed because the solution and the operating conditions are fixed. This means that the water flux through the membrane is also fixed and independent from the intrinsic permeability of the membrane. In physical terms, this is because a membrane with a higher intrinsic permeability only causes a thicker gel layer to form on the membrane surface. This lowers the membrane flux until the rate at which solutes are brought toward the membrane surface and the rate at which they are removed are again balanced, as expressed in Equation 46.1.

Going back to Figure 46.3, ultrafiltration membranes are best operated at pressures between p_2 and p_3 at which the gel layer is absent. Operation at pressures such as p_4 leads to formation of thick gel layers, which can consolidate over time, resulting in permanent fouling of the membrane. Equation 46.1 can be rewritten as follows:

$$\Phi_{max} = \frac{Đ_i}{\delta} \left(\ln c_{gel} - \ln c_{ib} \right). \tag{46.2}$$

Φ_{max} is the plateau or limiting flux through the membrane. Figure 46.5 shows the effect of pressure on membrane flux for styrene–butadiene polymer latex solutions in a high-turbulence, thin-channel test. The slopes of the straight lines in the plots in Figure 46.6 are proportional to the term $Đ_i/\delta$. The increase in flux resulting from an increase in the fluid recirculation rate is caused by a decrease of the boundary layer thickness δ.

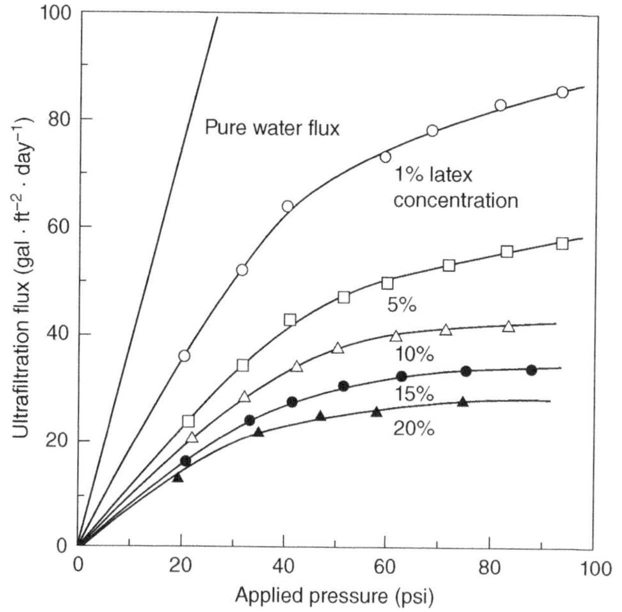

Figure 46.5 The limiting flux is a function of the concentration. *Source:* Membrane Technology and Applications by Baker, R. W. (2012). Reproduced with permission of John Wiley & Sons Inc.

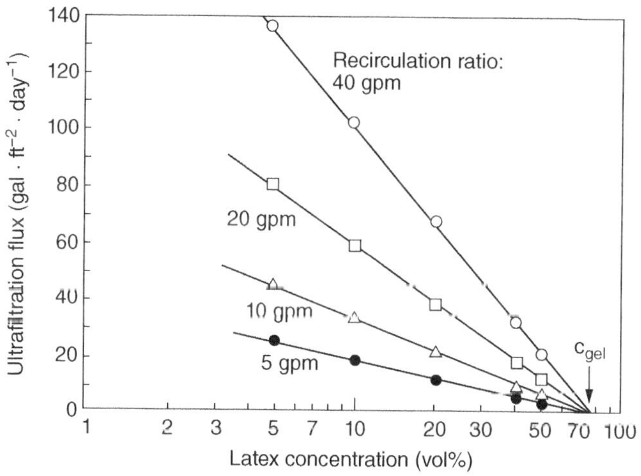

Figure 46.6 A thin laminar boundary layer results from a high recirculation rate. *Source:* Membrane Technology and Applications by Baker, R. W. (2012). Reproduced with permission of John Wiley & Sons Inc.

The occurrence of gel layers is often used to explain the behavior of ultrafiltration membranes. However, not all phenomena can be understood by that model. For example, protein solutions that form tough adherent gels have lower fluxes than pigment and latex solutions forming looser gels. Because the diffusion coefficients of pigment and latex solutions are very low, the opposite would be expected. These observations can be explained by the scouring action of the feed fluid removing undissolved pigment and latex particles from the membrane surface. A further example is the fact that experiments with different ultrafiltration membranes and the same feed solution often yield very different ultrafiltration limiting fluxes.

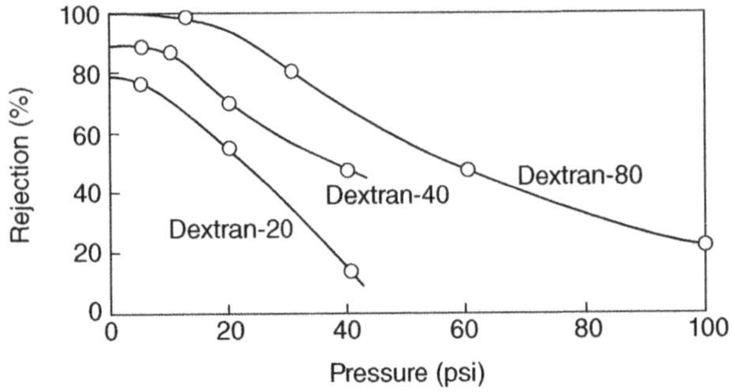

Figure 46.7 Pressure affects retention. *Source:* Membrane Technology and Applications by Baker, R. W. (2012). Reproduced with permission of John Wiley & Sons Inc.

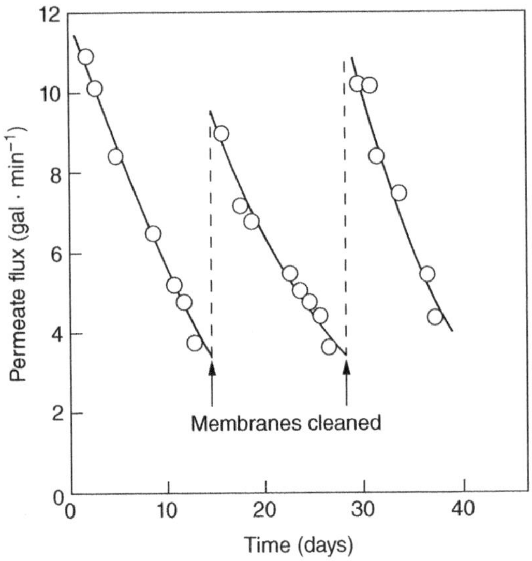

Figure 46.8 Fouling and cleaning. *Source:* Membrane Technology and Applications by Baker, R. W. (2012). Reproduced with permission of John Wiley & Sons Inc.

Figure 46.7 shows that rejection of dextran polysaccharides is a function of pressure. High pressures lead to high concentrations at the membrane surface.

Finally, Figure 46.8 illustrates that for an electrocoat paint latex solution, ultrafiltration rate is a function of time. Gradually, as time goes on, the gel layer is densified and the water flux decreases. It is necessary to clean periodically to maintain high fluxes.

46.4 Membrane Cleaning

To remove the densified gel layer of retained material from the membrane surface, several cleaning methods are used. The circulation of an appropriate cleaning solution through the membrane modules for several hours is the easiest method. Layers consisting of organic polymer colloids or gelatinous materials are usually treated with alkaline solutions followed by hot detergent solutions. When the fouling layer is a proteinaceous gel, enzymatic detergents are particularly effective [4]. Hydrated iron oxide scaling is often a problem because many feed waters contain small amounts of soluble ferrous ion salts. These compounds are oxidized to ferric ion salts by entrained air. An insoluble iron hydroxide gel forms and accumulates on the membrane because ferric ions are insoluble in water. Citric or hydrochloric acid can be used to remove deposits.

Cleaning of ultrafiltration systems for food applications, for instance, for the processing of whey, usually occurs on a daily basis. Systems in use as polishing units for the production of ultrapure water are cleaned, for example, once a month.

Membrane modules are normally stored wet. It is therefore important that the final rinse solution contains, for example, 0.5% by weight formaldehyde to prevent bacterial growth.

Backflushing is practiced to control membrane fouling. It is used to clean capillary and ceramic membrane modules that can permit a flow from the permeate side to the feed side of the membrane without damaging the membrane.

Module lifetimes are rarely longer than 2–3 years. Annual replacement can occur of modules processing cheese whey or electrocoat paint.

46.5 Ultrafiltration Membrane Systems

At the design of ultrafiltration modules, attention must be paid to the need to control concentration polarization and membrane fouling. Tubular modules and plate-and-frame modules were the first commercially successful ultrafiltration systems. Tubular modules resemble shell-and-tube heat exchangers. The feed passes in series through tubes (multiple-pass arrangement). The tubes are plastic or steel support tubes inside which several 0.5- to 1.0-cm-diameter tubes are nested. The feed flows in parallel through the latter tubes, which have porous walls with a membrane on the inside of the tubes. The permeate flows through these walls and is collected in a permeate manifold. Plate-and-frame modules resemble filter presses. The feed mixture is forced across the surface of the membrane. The permeate passes through the membrane. See Figure 46.9. Plate-and-frame modules can be operated at higher pressures than tubular and capillary hollow fiber modules. The pressure can be as high as 10 bar. Tubular modules and plate-and-frame modules are still used for highly fouling solutions. Spiral-wound modules and capillary modules are investment-wise cheaper

Figure 46.9 Horizontal plate-and-frame ultrafiltration system. *Source:* Membrane Technology and Applications by Baker, R. W. (2012). Reproduced with permission of John Wiley & Sons Inc.

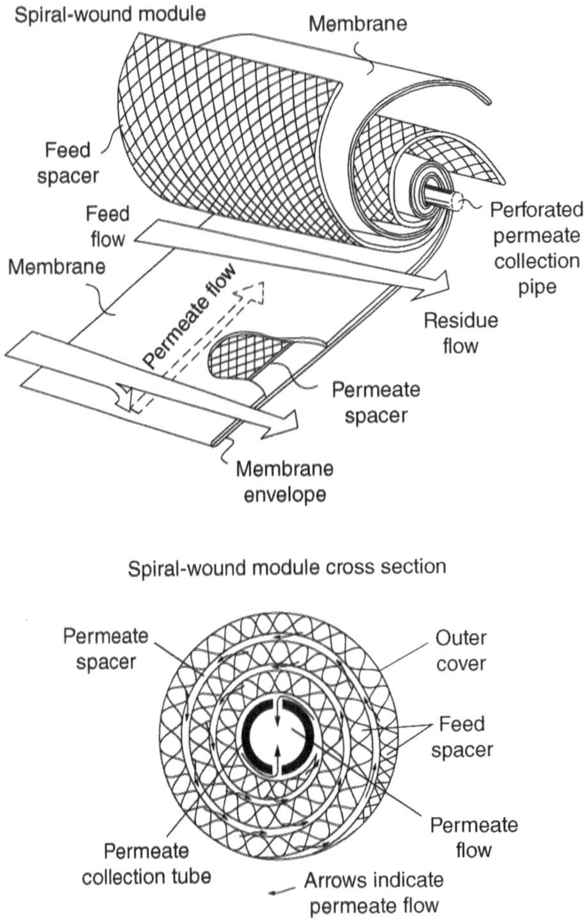

Figure 46.10 Exploded view and cross-section drawings of a spiral-wound module. *Source:* Membrane Technology and Applications by Baker, R. W. (2012). Reproduced with permission of John Wiley & Sons Inc.

and are hence widely used. Figure 46.10 depicts a spiral-wound unit. These units are also used for reverse osmosis for which they were originally developed. Finally, capillary hollow fiber modules contain a number of parallel tubes having a diameter as small as, for example, 0.5 mm. The feed flows in parallel through the tubes. Their design resembles a tube-and-shell heat exchanger with single passes on both the tube and shell sides. The permeate flows through the capillary walls. The fact that these units can be cleaned easily by backflushing is an advantage. To avoid breaking the fibers, the operating pressures are usually limited to 2 bar [5].

46.6 Continuous Systems

Figure 46.11 shows one-, two-, and three-stage designs. Let us assume that the concentration in the feed is 1% by weight and the concentration in the concentrate is 8% by weight in all three designs. First, a few words concerning the one-stage design. We see a recirculation through the unit. As a rule of thumb, the flow through the unit is 5–10 times higher than the feed flow. The purpose of this recirculation is to control the concentration polarization. The flow across the membrane limits the thickness of the boundary layer. Next, a comment regarding the use of multistage units. It is

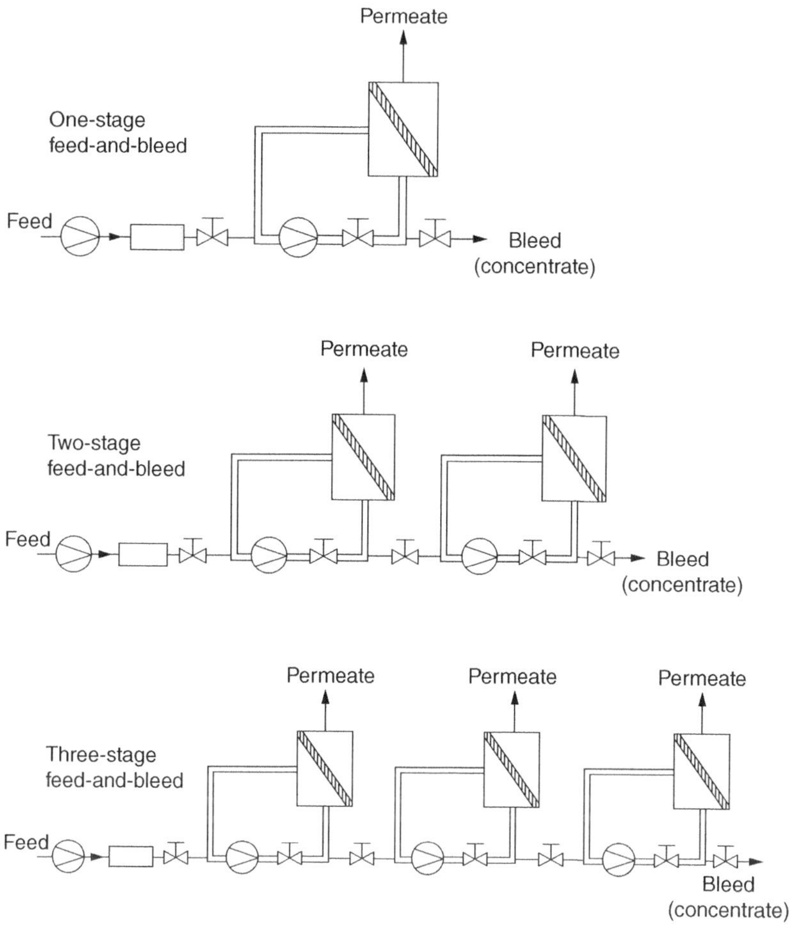

Figure 46.11 One-, two-, and three-stage feed-and-bleed systems. *Source:* Membrane Technology and Applications by Baker, R. W. (2012). Reproduced with permission of John Wiley & Sons Inc.

assumed that the concentrations in the concentrates of the three-stage unit are respectively 2, 4, and 8% by weight. That means that the amount of water to be removed in a stage is half the amount removed in the previous stage. The relatively high recirculation rate in the stages means that the concentrations within a stage are, respectively, 2, 4, and 8% by weight. The stages are considered to be ideally mixed. The concentrations in the feeds to the stages are different and, respectively, 1, 2, and 4% by weight. The areas of the stages are about equal. Thus, the permeate flux in $l \cdot m^{-2} \cdot h^{-1}$ decreases from step to step. That observation is in line with data shown in Figure 46.5, that is, the limiting fluxes decrease as the concentration increases. The effect is that the area of a three-stage unit is typically 40% of the area of a one-stage unit. The limiting flux for a one-stage unit is small indeed. The area decrease is the reason why most large plants have between three and five stages.

46.7 Applications

Figure 46.12 shows the recovery of electropaint in automobile plants. Coach-works of automobiles are immersed in the content of electropaint tanks. A coating is formed over the entire wetted

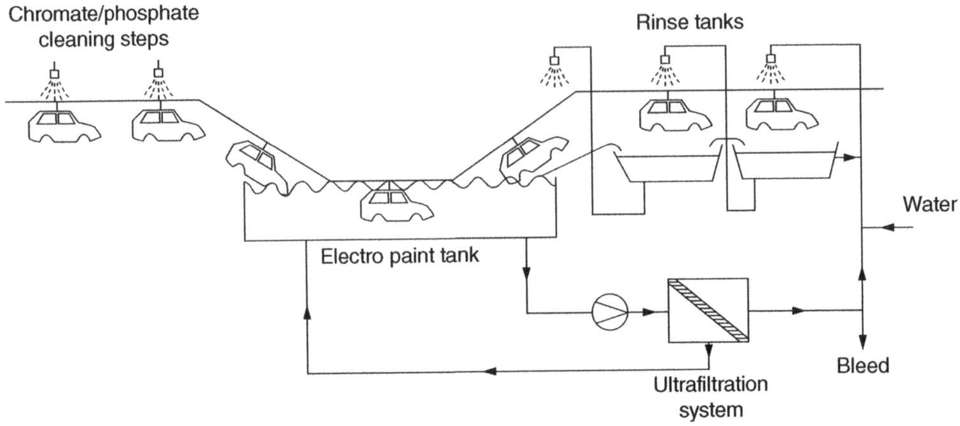

Figure 46.12 An electrocoat paint ultrafiltration system. *Source:* Membrane Technology and Applications by Baker, R. W. (2012). Reproduced with permission of John Wiley & Sons Inc.

surface of the parts. Prior to this tank, the coach-works are cleaned by a solution containing chromate and phosphate ions. These ions are carried over into the paint step. It is necessary to control the concentration of the ions in the content of the electropaint tank because at too high levels of the ions, the stability of the paint solution would be affected. The coach-works are washed after they have left the electropaint tank. The wash is necessary to remove excess paint. The wash water and the coach-works move countercurrently, and the wash water flows into the electropaint tank after the third wash. It is necessary to remove water from the electropaint tank. Ultrafiltration takes care of both the removal of ions and water from this tank. The fluxes of liquid through the membrane are relatively low because of gel formation: 20–30 $\mathrm{l \cdot m^{-2} \cdot h^{-1}}$ [6]. After the washing step, the paint on the coach-works is cured in an oven.

Figure 46.13 concerns the production of cheese. The conventional production is at the left. Milk proteins are precipitated and sent to the cheese fermentation plant. The liquid left is called whey

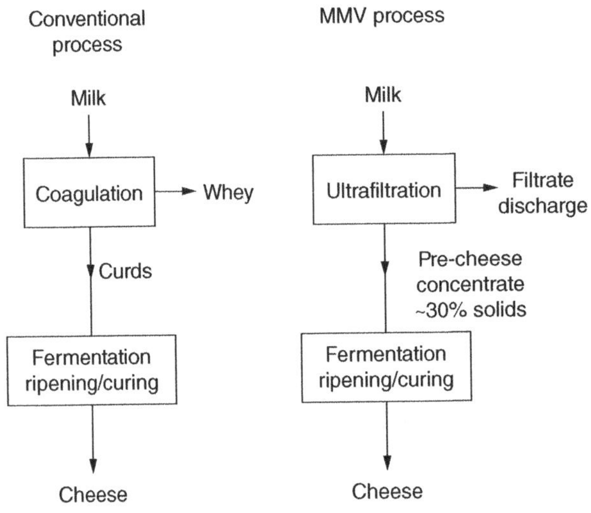

Figure 46.13 Ultrafiltration at cheese production. *Source:* Membrane Technology and Applications by Baker, R. W. (2012). Reproduced with permission of John Wiley & Sons Inc.

and must be disposed of. The production at which ultrafiltration is used is at the right of Figure 46.13. Milk is concentrated by ultrafiltration. The protein content of the milk increases from, for example, 3.3 to 12–17% by weight [7]. When subsequently milk proteins are precipitated, the cheese yield is 10% higher than in the former case. Moreover, the whey contains less solids and can even be used further. These two facts explain the success of ultrafiltration at the cheese manufacture.

References

1. Baker, R. W. (2012). *Membrane Technology and Applications* (p. 253). Chichester, UK: Wiley.
2. Baker, R. W. (2012). *Membrane Technology and Applications* (p. 257). Chichester, UK: Wiley.
3. Baker, R. W. (2012). *Membrane Technology and Applications* (p. 262). Chichester, UK: Wiley.
4. Baker, R. W. (2012). *Membrane Technology and Applications* (p. 271). Chichester, UK: Wiley.
5. Baker, R. W. (2012). *Membrane Technology and Applications* (p. 276). Chichester, UK: Wiley.
6. Baker, R. W. (2012). *Membrane Technology and Applications* (pp. 283–284). Chichester, UK: Wiley.
7. Baker, R. W. (2012). *Membrane Technology and Applications* (p. 285). Chichester, UK: Wiley.

47

Reverse Osmosis

47.1 Osmosis

First, before attempting reverse osmosis, osmosis as a phenomenon is discussed. Figure 47.1 shows the apparatus used by Berkeley and Hartley at Oxford in the United Kingdom. Two concentric tubes can be seen. Water is in the central tube and an aqueous solution of a compound is in the annular space between the two tubes. The compound can be sodium chloride. The wall of the inner tube is a semipermeable membrane. The meaning of this is that, yes, water molecules can move through the wall. However, neither sodium ions nor chloride ions can move through the wall. We see a net water flow through the wall from the central tube to the annular space if the pressures in the central tube and the annular channel are equal. The solution becomes diluted. A possible explanation is based on Figure 47.2. A semipermeable membrane is schematically shown as a wall with holes. Water molecules can move freely through the holes, however, ions cannot. Pressure is nothing else than the bouncing of molecules and ions. The pressure exerted by water molecules is higher in the central tube than in the annular space. Part of the volume in the annular space is occupied by ions that cannot exert pressure. The pressure difference between the two spaces is called the osmotic pressure, and this pressure difference is the driving force for the net transport of water molecules.

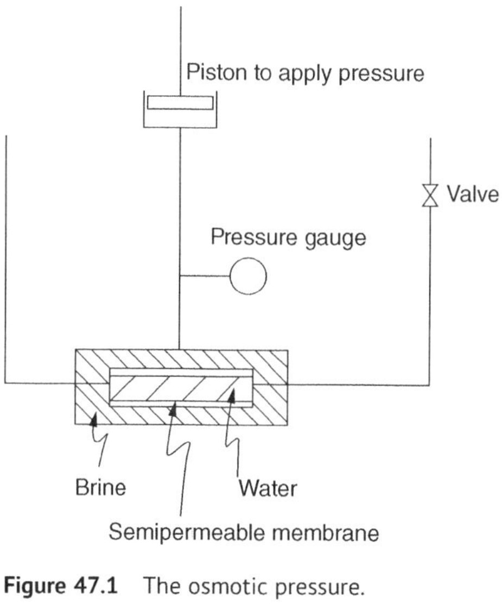

Figure 47.1 The osmotic pressure.

Introduction to Chemical Engineering, First Edition. C.M. van 't Land.

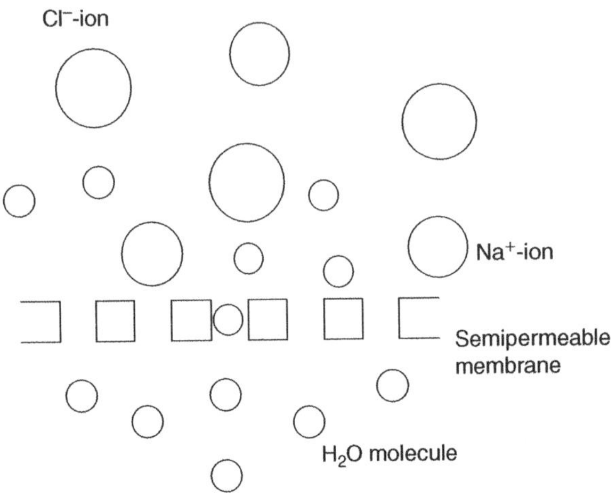

Figure 47.2 A semipermeable membrane as a wall with holes.

The explanation is a simplified one because the model, which is used most frequently for the functioning of a semipermeable wall at reverse osmosis, is the model in which water molecules dissolve in the wall and diffuse through it.

Now reverse osmosis will be dealt with. If the pressure in the annular space is increased, a point is reached at which there is no net transport of water through the wall from the central tube to the annular space. The pressure difference between the central tube and the annular space then equals the osmotic pressure of the solution. If the pressure in the annular space is increased further, there will be net transport of water from the annular space to the central tube. That phenomenon is called reverse osmosis. The driving force for this transport of water is the pressure difference between the pressure applied to the annular space and the osmotic pressure.

47.2 Reverse Osmosis

In reverse osmosis, pressurized water containing dissolved salts contacts the feed side of a membrane and water depleted of salt is withdrawn as a low-pressure permeate. The construction of a large seawater desalination plant at Jiddah in Saudi Arabia in the 1970s was a milestone in reverse osmosis development. This plant was equipped with interfacial composite membranes. A very thin polymer film was formed, by an in-situ chemical reaction, at the surface of a microporous support membrane.

The present industry standard is a fully aromatic interfacial composite membrane based on the reaction of m-phenylene diamine and trimesoyl chloride [1].

The production of municipal drinking water from seawater or brackish groundwater in desert regions and the production of ultrapure industrial water are the main applications.

Interfacial composite membranes are today usually supplied in spiral-wound module form mainly. See Figure 46.10.

47.3 Theoretical Background

The theoretical background will be discussed for the case of the solution of salt (sodium chloride) in water.

The water flux, Φ_{mw}, is linked to the pressure and concentration gradients across the membrane by the empirical equation:

$$\Phi_{mw} = A(\Delta p - \Delta\pi)\ \ \mathrm{kg \cdot m^{-2} \cdot s^{-1}}. \tag{47.1}$$

Δp is the pressure difference between the two sides of the membrane, $\Delta\pi$ is the osmotic pressure differential across the membrane, and A is a constant. Note that this equation is in line with Hagen–Poiseuille's law for laminar flow. See Equation 44.1. At low applied pressure, when $\Delta p < \Delta\pi$, water flows from the dilute to the concentrated salt-solution side of the membrane by normal osmosis. When $\Delta p = \Delta\pi$, flow does not occur. Water flows from the concentrated side to dilute salt-solution side of the membrane when $\Delta p > \Delta\pi$.

The salt flux across a reverse osmosis membrane can empirically be described by the equation:

$$\Phi_{ms} = B\left(c_{jo} - c_{jl}\right)\ \ \mathrm{kg \cdot m^{-2} \cdot s^{-1}}. \tag{47.2}$$

B is the salt permeability constant, and c_{jo} and c_{jl} are, respectively, the salt concentrations on the feed and permeate sides of the membrane. According to the equation, the salt flux is not dependent on the water flux. The transport of salt through the membrane can be described as diffusion through the membrane and is then a function of the salt concentration difference between the two sides of the membrane. See Equation 12.2. The concentration of salt in the permeate solution is usually much smaller than the concentration in the feed, so the former equation can be simplified to:

$$\Phi_{ms} = B \cdot c_{jo}\ \ \mathrm{kg \cdot m^{-2} \cdot s^{-1}}. \tag{47.3}$$

It follows from Equations 47.1 and 47.2 that the membrane becomes more selective as the pressure rises. The salt concentration in the permeate is, according to these two equations, inversely proportional to the difference between the pressure across the membrane and the osmotic pressure difference between the two sides of the membrane. Usually, the selectivity of a membrane is expressed as the salt rejection coefficient R, defined as:

$$R = \left(1 - \frac{c_{jl}}{c_{jo}}\right) \cdot 100\,\%.$$

The salt concentration on the permeate side of the membrane can be related to the membrane flux by the equation:

$$c_{jl} = \frac{\Phi_{ms}}{\Phi_{mw}} \cdot \rho_i\ \ \mathrm{kg \cdot m^{-3}}.$$

In the latter equation, it is implicitly assumed that the water flow through the membrane is equal to the total flow through the membrane ($c_{jl} \cdot \Phi_v = \Phi_{ms}$, $\Phi_{vw} = \Phi_{mw}/\rho_i$). ρ_i is the water specific mass in $\mathrm{kg \cdot m^{-3}}$.

By combining equations, the membrane rejection can be expressed as:

$$R = \left(1 - \frac{\rho_i \cdot B}{A(\Delta p - \Delta\pi)}\right) \cdot 100\,\%. \tag{47.4}$$

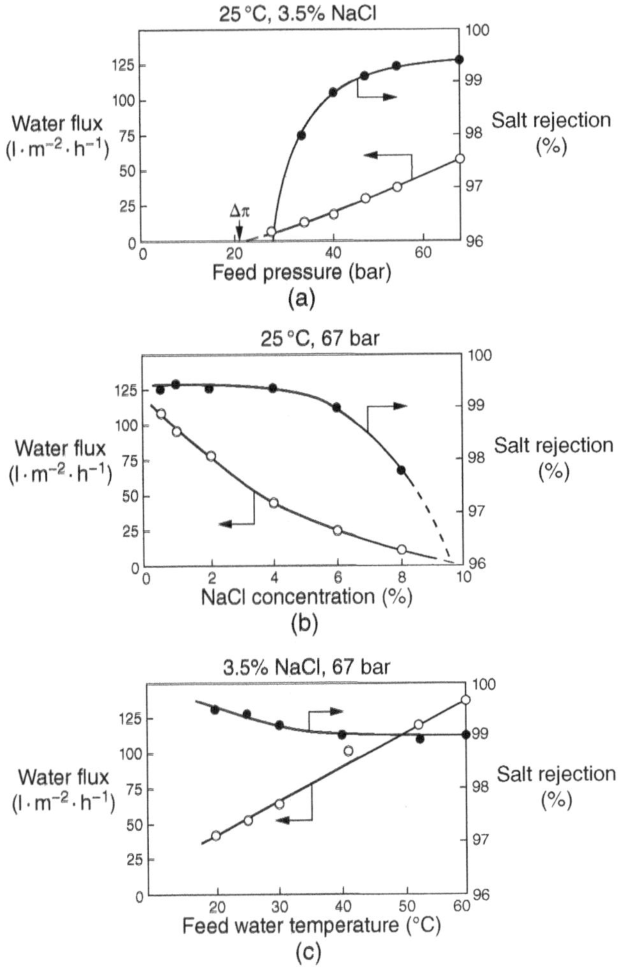

Figure 47.3 Effect of pressure, feed salt concentration, and feed temperature on the properties of good quality seawater desalination membranes (SW-30). *Source:* Membrane Technology and Applications by Baker, R. W. (2012). Reproduced with permission of John Wiley & Sons Inc.

The effects of the most important parameters on membrane water flux and rejection are, for a membrane type SW-30, shown schematically in Figure 47.3 [2].

(a): The graph shows both the water flux in $l \cdot m^{-2} \cdot h^{-1}$ and the salt rejection coefficient as a function of the pressure. Seawater at 25 °C is the concentrate. We see that, yes, the water flux varies linearly with the pressure. For pressures exceeding about 45 bar, the salt rejection coefficient is between 99 and 100%.

(b): The graph shows both the water flux in $l \cdot m^{2} \cdot h^{-1}$ and the salt rejection coefficient as a function of the salt concentration. The pressure is fixed at 67 bar and the temperature is 25 °C. The water flux decreases as the concentration increases because the driving force, that is, $\Delta p - \Delta\pi$, decreases as the concentration increases. The osmotic pressure of the solution increases when the salt concentration in the feed increases. The salt rejection coefficient decreases when the salt concentration in the feed increases.

(c): The graph shows both the water flux in $l \cdot m^{-2} \cdot h^{-1}$ and the salt rejection coefficient as a function of the temperature. The feed is seawater, and the pressure is 67 bar. The flux strongly increases with the temperature. In a way, it can be thought to be caused by a viscosity decrease. The salt rejection coefficient decreases somewhat when the temperature increases.

47.4 Concentration Polarization

Concentration polarization was discussed in Section 46.3. The phenomenon plays an important role at that technique. It also occurs at reverse osmosis, for example, at the desalination of seawater. The reason is as follows. There is a net flow of water through the membrane. This water contains, on approaching the membrane, dissolved salt. The membrane allows the passage of water; however, it does not permit the passage of dissolved salt. The salt diffuses back into the feed. The driving force for this diffusion is a difference between the salt concentrations at the membrane surface and in the bulk of the feed. Concentration polarization at reverse osmosis has two consequences. The first one is that it promotes the passage of salt through the membrane as the salt concentration difference across the membrane increases. See Equation 47.2, which states that the salt flux is a function of the concentration difference across the membrane only. The second one is that it increases the osmotic pressure causing a decrease of the pressure difference across the membrane.

The design of reverse osmosis systems is such that the effect of concentration polarization is limited. An aspect is that the diffusion coefficients of both sodium and chloride ions are much higher than the diffusion coefficients of colloids and macromolecules. A phenomenon like gel formation, discussed in Chapter 46, does not occur at reverse osmosis because the composition of reverse osmosis feeds is different from the composition of ultrafiltration feeds.

47.5 Membrane Specifications

For human consumption, seawater desalination membranes with salt rejections greater than 99.3% are required. The specification is that the permeate should contain less than 500 ppm by weight of salt. With the present membrane qualities, operating pressures between 50 and 60 bar are required.

Reverse osmosis is also used for the desalination of brackish water feeds. Typically, these feeds contain 0.1–0.5% by weight of salt. Operating pressures in the range 10–30 bar are required for salt rejection of approximately 99%.

In the electronics industry, reverse osmosis is at present applied to produce ultrapure water to wash silicon wafers. Municipal drinking water is often the feed. It typically contains 100–200 ppm by weight of dissolved salts, mostly consisting of divalent ions. The specifications are in this case: 98–99% salt rejection but more than 99.5% divalent ion rejection. Operating pressures are usually in the range 8–12 bar.

47.6 Membrane Qualities

The industrial application of reverse osmosis started with the availability of anisotropic cellulose acetate membranes in the 1960s. Anisotropic membranes consist of a very thin surface layer attached to a thicker, porous substructure giving mechanical support. Interfacial composite membranes are significantly better than cellulose acetate membranes and are today the industry standard. Typically, for seawater, the salt rejection is 99.7%, and the water flux is $50\,\text{l}\cdot\text{m}^{-2}\cdot\text{h}^{-1}$ at 35 bar; this is less than half the salt passage of the cellulose acetate membrane and twice the water flux [3]. The only drawback, and a significant one, is the rapid, permanent loss in selectivity that results from exposure to even ppb by weight levels of chlorine or hypochlorite disinfectants.

47.7 Reverse Osmosis Units

Typically, reverse osmosis membranes are modularized. The designs contain a large amount of area in a relatively small volume. Like for ultrafiltration, there are four basic forms for the modules: plate-and-frame, tubular, spiral-wound, and hollow fiber. The plate-and-frame modules resemble filter presses. The tubular modules resemble shell-and-tube heat exchangers with the feed water on the tube side and permeate on the shell side. There is one pass on the shell side, and there are multiple passes on the tube side as the feed passes through the tubes in series. The spiral wound membrane is at present the most common type of module used. It will therefore be described in some detail. Figure 46.10 shows an exploded view of a spiral-wound module. A somewhat more sophisticated design is widely used today. It typically has a diameter of 8 in and a length of 40 in. Such a module has an area of 40 m^2. Figure 47.4 exhibits an RO Municipal System equipped with pressure vessels. They are constructed of fiberglass. Fiberglass is the usual material of construction for the desalination of brackish water and seawater. Stainless steel is the choice for sanitary applications.

It is possible to process the feed in different stages. See Figure 47.5. The first stage consists of four pressure vessels, the second of two, and the last stage of one. Such a set-up is possible because, as the permeate is collected, the feed stream decreases. It is thus possible to maintain a high average feed velocity through the pressure vessels and through the modules contained in those vessels. A high average feed velocity keeps the thickness of the laminar boundary layer within limits.

Finally, the hollow fiber module exists and is also called the capillary module.

Figure 47.4 An RO Municipal System. *Source:* Courtesy of Koch Separation Solutions, Inc. (www.kochseparation.com), Wilmington, Massachusetts, USA and used by the publisher with permission.

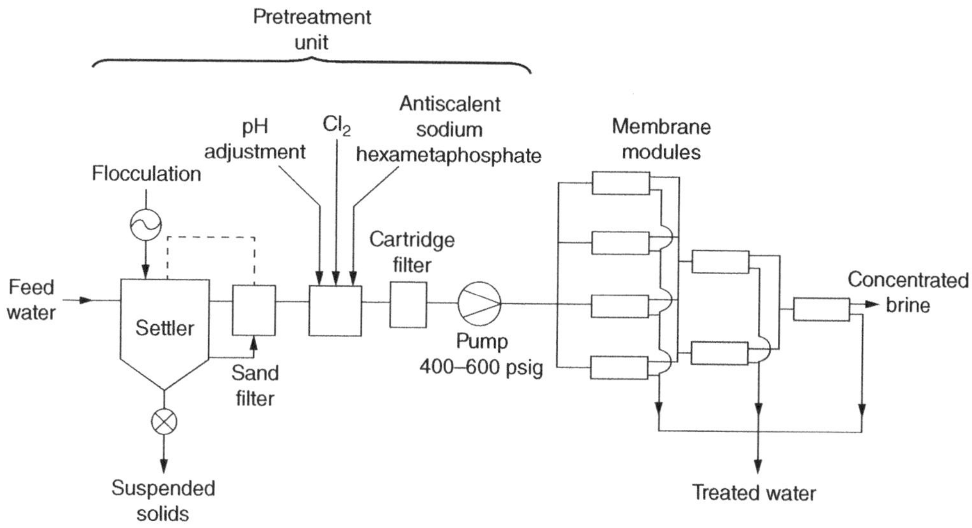

Figure 47.5 Flow schematic of a typical brackish water reverse osmosis plant. The plant contains seven pressure vessels each containing six membrane modules. The pressure vessels are in a "Christmas tree" array to maintain a high feed velocity through the modules. *Source:* Membrane Technology and Applications by Baker, R. W. (2012). Reproduced with permission of John Wiley & Sons Inc.

47.8 Membrane Fouling Control and Cleaning

In general, sources of fouling can be divided into four principal categories: scale, silt, bacteria, and organic. Precipitation of metal salts dissolved in the feed water on the membrane surface causes scale. Fine solid matter suspended in the feed water and retained by the membrane forms silt. Bacteria tend to grow on the membrane surface. Such growth can be prevented via sterilization by chlorination. Finally, organic fouling comprises the accumulation of organic matter like oil or grease on the membrane surface.

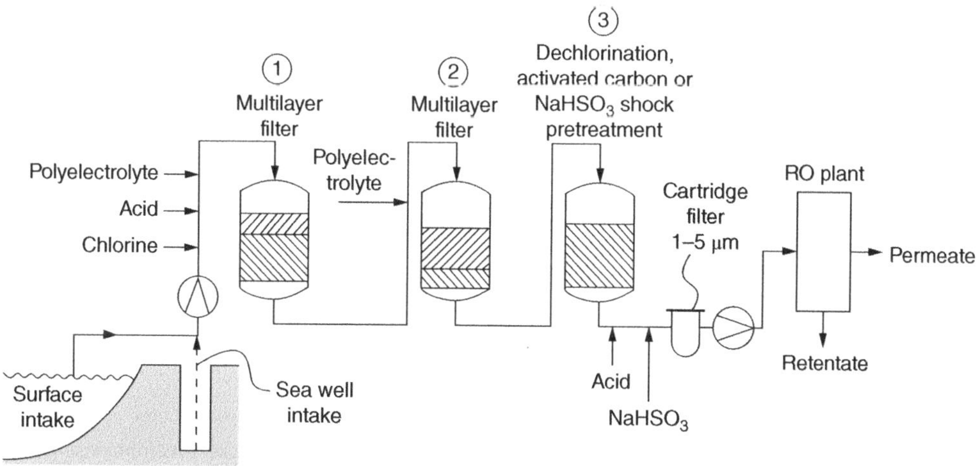

Figure 47.6 Flow scheme showing the pretreatment steps in a typical seawater reverse osmosis system. *Source:* Membrane Technology and Applications by Baker, R. W. (2012). Reproduced with permission of John Wiley & Sons Inc.

Figure 47.6 shows the pretreatment steps in a typical seawater reverse osmosis plant. First, chlorine is added to sterilize the water. The next step is an acidification to bring the pH to 5–6. That causes the precipitation of scale. Next, a polyelectrolyte is added to flocculate suspended matter. The latter is subsequently retained in two multilayer depth filters. Dechlorination, adjustment of the pH, and filtration by a 1–5 μm filter are final steps. Dechlorination is effected by the dosing of sodium bisulfite.

Cleaning is generally done once or twice annually, but more often if the feed is a problem water.

47.9 Applications

Approximately one-half of the reverse osmosis systems currently installed are desalting brackish water or seawater. A further 40% are producing ultrapure water for the electronics, pharmaceutical, and power generation industries [4].

As for desalination, the technology is the best approach for water containing between 1 and 50 $g \cdot l^{-1}$. For higher concentrations, multi-effect evaporation is more attractive than reverse osmosis. Typical reverse osmosis plants processing seawater operate at a recovery rate of maximum 35–45%. The recovery rate expresses the permeate flow as a percentage of the feed flow. The reason for this relatively low value of the recovery is the osmotic pressure. The pressure generated in reverse osmosis plants should be higher than the osmotic pressure to effect separation. Recovery of compression energy from the high-pressure brine is practiced.

There are further typical applications in the treatment of wastewater. Figure 47.7 shows the use of a reverse osmosis system to control nickel loss from rinse water produced in countercurrent electroplating rinse tanks. It concerns a special situation in which the chemical separated from the water is valuable. A typical countercurrent rinse tank produces a waste stream containing

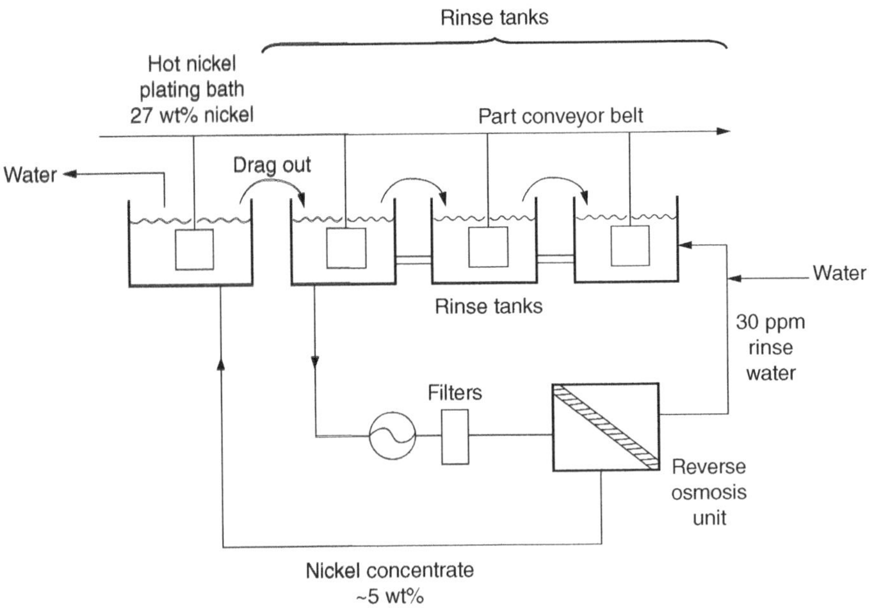

Figure 47.7 Flow scheme showing the use of a reverse osmosis system to control nickel loss from rinse water produced in a countercurrent electroplating rinse tank. *Source:* Membrane Technology and Applications by Baker, R. W. (2012). Reproduced with permission of John Wiley & Sons Inc.

2,000–3,000 ppm by weight nickel; the water is a pollution problem and the valuable material is lost. This is an ideal application for reverse osmosis because the rinse water is at nearly neutral pH. The water balance is taken care of by dosing the water to compensate for evaporation losses from the hot plating tank and the rinse tanks and water removal by the plated parts.

47.10 Nanofiltration Membranes

Nanofiltration membranes are membranes with relatively low sodium chloride rejections and relatively high water permeabilities. They fall in between ultrafiltration membranes and membranes for reverse osmosis. The principal application is in the removal of low levels of contaminants from already clean water.

47.11 Conclusions and Future Directions

As far as desalination is concerned, the water flux has doubled since 1980. Furthermore, the salt permeability has decreased sevenfold in that application [5]. The key short-term technical issue is the limited chlorine resistance of interfacial composite membranes. Three longer-term, related technical issues are fouling resistance, pretreatment, and membrane cleaning.

References

1. Baker, R. W. (2012). *Membrane Technology and Applications* (p. 208). Chichester, UK: Wiley.
2. Baker, R. W. (2012). *Membrane Technology and Applications* (p. 211). Chichester, UK: Wiley.
3. Baker, R. W. (2012). *Membrane Technology and Applications* (p. 217). Chichester, UK: Wiley.
4. Baker, R. W. (2012). *Membrane Technology and Applications* (p. 237). Chichester, UK: Wiley.
5. Baker, R. W. (2012). *Membrane Technology and Applications* (p. 246). Chichester, UK: Wiley.

48

Electrodialysis

48.1 Introduction

Electrodialysis is the most important one of a number of separation processes using ion exchange membranes. At the end of the previous century, desalting of brackish water and the production of boiler feed water were the main applications. At present, a major use, in combination with ion exchange, is the removal of salt from water in ultrapure water plants. The technology is best suited for water containing between 0.05 and 0.2 $g \cdot l^{-1}$ salt. Pressure is the driving force for the membrane separation processes discussed in previous chapters, that is, microfiltration, ultrafiltration, and reverse osmosis. An electric potential difference is the driving force for electrodialysis. See Figure 48.1. It depicts the production of pure water from a diluted salt solution. The technology

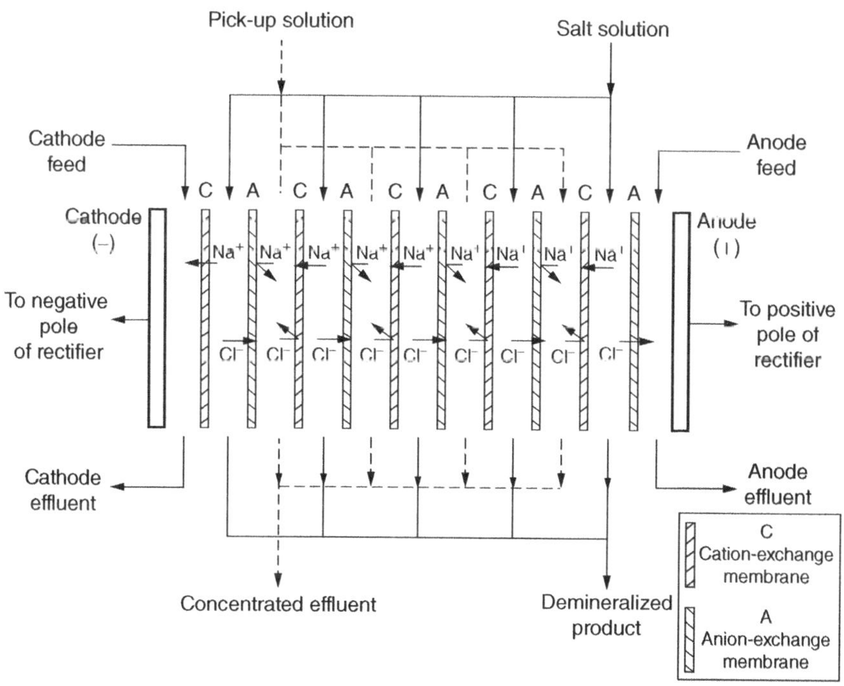

Figure 48.1 Schematic diagram of a plate-and-frame electrodialysis stack. *Source:* Membrane Technology and Applications by Baker, R. W. (2012). Reproduced with permission of John Wiley & Sons Inc.

Introduction to Chemical Engineering, First Edition. C.M. van 't Land.

uses direct current (DC). An anionic membrane with fixed positive groups is freely permeable to negatively charged chloride ions. It excludes positively charged sodium ions. Similarly, a cationic membrane excludes negatively charged ions but allows the passage of positively charged ions. Alternating cation- and anion-permeable membranes are arranged in a stack of up to 100 cell pairs. A cell pair is formed from each set of cationic and anionic membranes. An electric potential is maintained while salt solution is pumped through the cells. The positively charged ions migrate toward the cathode, whereas the negatively charged ions move toward the anode. The net result is that one cell of a pair becomes enriched in ions and the other cell becomes depleted of ions. An electrode rinse solution flows past both anode and cathode.

Modern electrodialysis plants use electrode polarity reversal. This technique comprises the reversal of the polarity of the DC power applied to the electrodes two to four times per hour. Simultaneous with this reversal, the functions of the desalted water and brine compartments are switched. The effect is that freshly precipitated scale is flushed from the membranes before it can attach itself to the membranes. Colloidal particles are also loosened from the membranes.

The separation method can also be used to obtain a concentrated solution. Electrodialysis is used in Japan to produce salt from seawater. That application is shortly discussed in Section 48.7.

48.2 Functioning of Ion-Exchange Membranes

Both the cationic membranes and the anionic membranes mentioned in Section 48.1 are ion exchange membranes. These membranes contain a high concentration of fixed ionic groups, typically 3–4 $meq \cdot g^{-1}$ [1]. Seawater contains approximately 0.6 meq of salt per gram of seawater and an aqueous solution containing 1% by weight of salt contains 0.17 meq of salt per gram of solution. A cationic membrane in contact with a solution containing 1% by weight of salt is permeable to sodium ions but excludes chloride ions almost completely. However, a cationic membrane has a certain small permeability to chloride ions when the salt concentration rises. Qualitatively, a similar statement can be made for anionic membranes.

48.3 Types of Ion Exchange Membranes

It was mentioned that there are anionic and cationic membranes. Both anionic and cationic membranes can be had as homogeneous membranes or as heterogeneous membranes. The charged groups are uniformly distributed through the membrane matrix in homogeneous membranes. Homogeneous membranes consist of polymers to which the active groups were added. For example, the copolymerization of styrene and divinyl benzene results in a cross-linked polymer. In a further stage, it is possible to introduce ionic groups into the polymer. For example, the copolymer of styrene and divinyl benzene is sulfonated to form a cationic membrane. The ionic groups in homogeneous membranes tend to absorb water when placed in water. Most homogeneous ion exchange membranes are highly cross-linked to limit swelling due to this water absorption. Highly cross-linked membranes are brittle when dry. Thus, the membranes are usually stored and handled wet to allow absorbed water to plasticize the membrane. The ion exchange groups are contained in small domains throughout an inert support matrix in heterogeneous membranes. The matrix can be a mixture of two polymers. Charged groups are added to only one of the two polymers. For instance, in a mixture of polystyrene and PVC, polystyrene is sulfonated to contain negatively charged groups. The PVC matrix provides toughness and strength.

48.4 Transport in Electrodialysis Membranes

Figure 48.2 shows a schematic of the concentration and potential gradients in a well-stirred piece of equipment. The salt concentration in alternating compartments increases, whereas the other compartments are simultaneously depleted of salt. The drawing shown implies that the voltage drop caused by the electrical resistance of the apparatus takes place entirely across the ion exchange membrane. In practice, however, the resistance of the membrane is often small in comparison with the resistance of the compartments filled with liquid.

Figure 48.3 shows the concentration gradient of univalent sodium ions next to a cationic membrane. In electrodialysis, the thickness δ of the boundary layer is 20–50 μm [2]. The solution in the

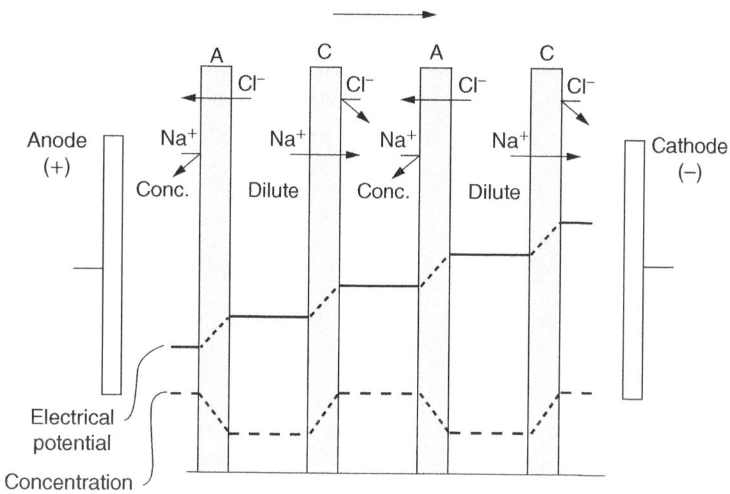

Figure 48.2 Schematic of the concentration and potential gradients in a well-stirred electrodialysis cell. *Source:* Membrane Technology and Applications by Baker, R. W. (2012). Reproduced with permission of John Wiley & Sons Inc.

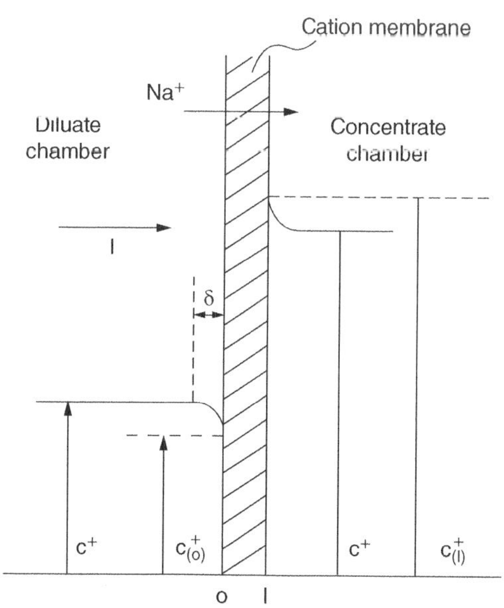

Figure 48.3 The sodium ion concentrations fall in the boundary layer. *Source:* Membrane Technology and Applications by Baker, R. W. (2012). Reproduced with permission of John Wiley & Sons Inc.

diluate chamber next to the membrane becomes increasingly depleted of the permeating ions as the voltage across the stack is increased to increase the flux of ions through the membrane. Ion transport through this ion-depleted aqueous boundary layer generally controls electrodialysis system performance. The maximum transport rate of ions through the boundary layer is reached when the ion concentration in the diluate chamber at the membrane surface is zero. The current density in $mA \cdot cm^{-2}$ at this point is called the limiting current density. Any further increase in voltage difference across the membrane will neither increase transport of ions nor current through the membrane. Normally any extra power is dissipated by side reactions, such as water dissociation into ions. The limiting current density can be determined experimentally by plotting pH of the diluate as a function of the inverse of current. A further possibility is a plot of the electrical resistance of the diluate as a function of the inverse of the current. See Figure 48.4 for a laboratory cell. The graphs are called Cowan-Brown plots. Both the pH and the resistance change strongly when the limiting current density is reached. In industrial practice, it is not easy to determine the limiting current density. The reason is that the boundary layer thickness is not uniform. The limiting current density is also a function of the concentration in the diluate. Large electrodialysis systems contain several electrodialysis stacks in series operating with different current densities in each stack. The background is that the salt concentration in the feed water decreases as salt is removed.

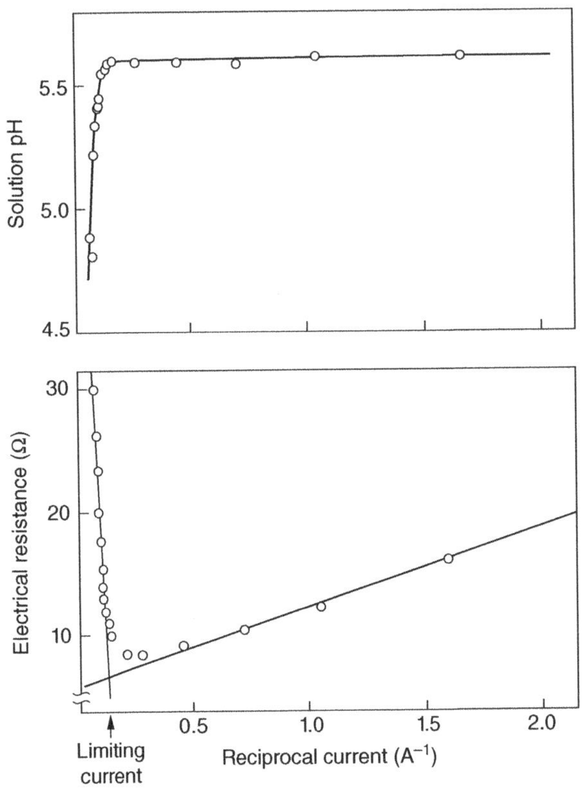

Figure 48.4 Cowan–Brown plots. *Source:* Membrane Technology and Applications by Baker, R. W. (2012). Reproduced with permission of John Wiley & Sons Inc.

48.5 Power Consumption

A distinction is made between the theoretical power consumption and the actual power consumption. The difference between the two is caused mainly by polarization as discussed in Section 48.4. The actual energy consumption is five to ten times greater than the theoretical power consumption. The solutions are circulated at high rates through the stack to control the polarization. A trade-off exists between the power saved due to the increased efficiency of the stack and the power consumed by the pumps. The pumps consume approximately one-quarter to one-half of the total power [3].

48.6 System Design

An electrodialysis plant consists of several elements:

- a feed pretreatment system,
- the membrane stack,
- the power supply and process control unit, and
- the solutions pumping system.

As to feed pretreatment, filtration of feed water is a useful step. It can retain large, charged organic molecules and colloids. Most feed water is sterilized by chlorination to prevent bacterial growth on the membrane. Scaling inhibitors such as sodium hexametaphosphate can be added to control scaling on the membrane by precipitation of sparingly soluble salts such as calcium sulfate. pH adjustment can help to keep salts in the soluble range. Finally, polarity reversal operation can be effective to decrease membrane resistance.

Many plants use a single stack. A stack normally contains 100–200 membrane cell pairs. Each membrane in the stack has an area between 1 and 2 m^2. Using several stacks in series is more efficient for large systems. See Figure 48.5. Three stacks in series are shown. Because the first stack operates on a relatively concentrated feed solution, its current density can be higher than the current density of the third stack. To control the boundary layer thickness and thus the polarization, the solution velocity increases in going from the first stack to the third stack. The number of cell pairs decreases from the first to the third stack, this is known as the taper of the system. It can be noticed from Figure 48.5 that the flows of both solutions through the

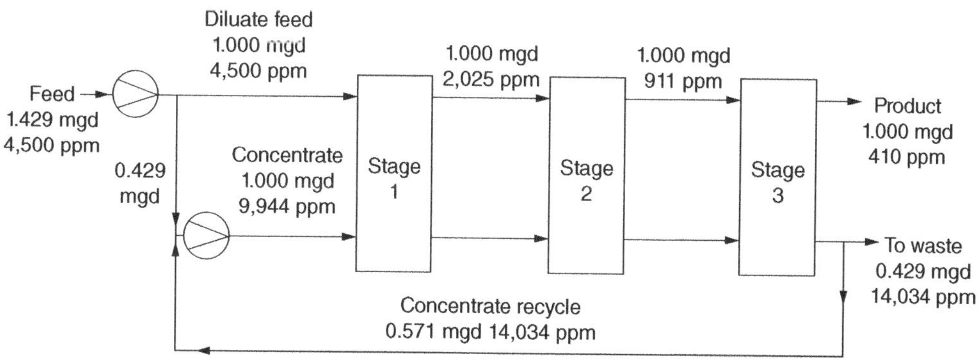

Figure 48.5 Flow scheme of a three-stage electrodialysis plant. *Source:* Membrane Technology and Applications by Baker, R. W. (2012). Reproduced with permission of John Wiley & Sons Inc.

stages are practically constant. The background is that a relatively small amount of impurity is exchanged between the two streams. Regarding this aspect, electrodialysis is different from reverse osmosis. At the latter process, there is not an exchange of matter between two comparable flows but a production of a flow from a feed flow. The produced flow can be, for example, 35% of the feed flow.

The power source delivers amplituding current (AC). AC is converted into direct current (DC) by means of a rectifier. A typical voltage drop across a single cell pair is in the range 1–2 V and a normal current density is 40 mA · cm^{-2}. For a 200-cell-pair stack containing membranes having an area of 1 m^2, the total voltage is approximately 200–400 V and the current about 400 A per stack. Polarity reversal was discussed in Section 48.1.

The pressure drop per stack varies from 1–2 bar for sheet flow cells to as much as 4.5–6 bar for tortuous path cells.

48.7 Applications

The largest application is the desalination of brackish water. For salt contents between 0.05 and 0.25% by weight, electrodialysis is often the best option.

A special application concerns the production of salt from seawater in Japan. See Figure 48.6. Seawater is by means of electrodialysis concentrated from approximately 3.5% by weight of

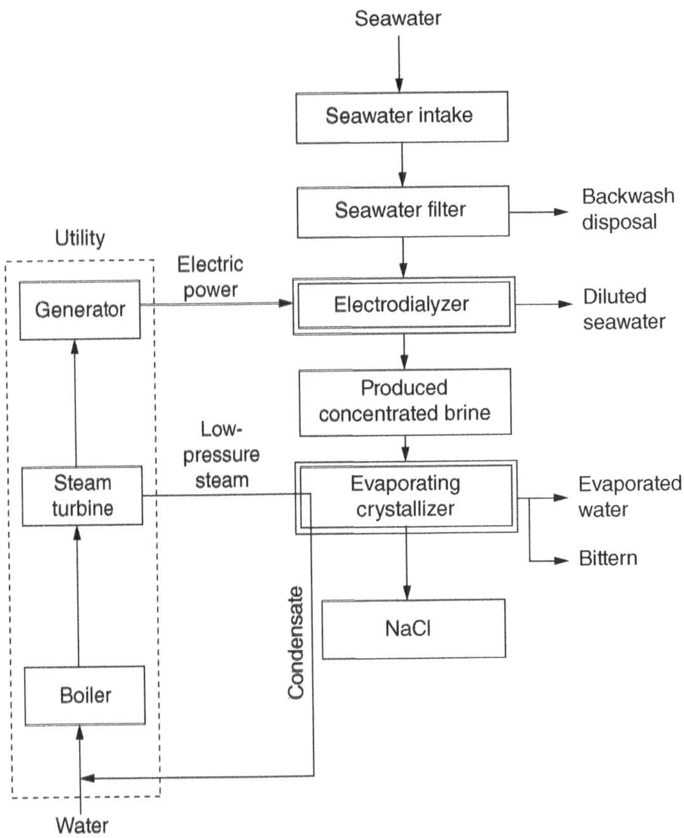

Figure 48.6 Flow scheme of the electrodialysis unit used in a plant producing salt from seawater. *Source:* Membrane Technology and Applications by Baker, R. W. (2012). Reproduced with permission of John Wiley & Sons Inc.

salt to 18–20% by weight. A saturated brine contains about 25% by weight, and that figure is hardly dependent on the temperature between 0 and 150 °C. Water is evaporated from brine in a multiple-effect evaporation/crystallization plant. Salt crystallizes and is separated by means of centrifuges and dried. Electric power for the electrodialysis system is generated in a power station. Steam is raised by means of the combustion of a fuel, steam is passed on to turbines, and turbines drive generators producing electricity. In the power station, low-pressure steam is tapped off from the turbines. That steam has, for example, a pressure of 4.8 bar having a condensation temperature of 150 °C. It is used for the multiple-effect evaporation/crystallization plant. There are several of these salt plants in Japan. Total production is 1.2 million tons per annum, it is subsidized by the Japanese government. Japan does not have other domestic salt supplies [4].

References

1. Baker, R. W. (2012). *Membrane Technology and Applications* (p. 423). Chichester, UK: Wiley.
2. Baker, R. W. (2012). *Membrane Technology and Applications* (p. 429). Chichester, UK: Wiley.
3. Baker, R. W. (2012). *Membrane Technology and Applications* (p. 434). Chichester, UK: Wiley.
4. Baker, R. W. (2012). *Membrane Technology and Applications* (pp. 438–440). Chichester, UK: Wiley.

49

Gas Separation

49.1 Introduction

The basis for industrial membrane gas separation was provided by the development of high-flux anisotropic membranes and large-surface-area membrane modules for reverse osmosis applications. That occurred in the late 1960s and early 1970s. Monsanto, on launching the hydrogen-separating Prism membrane in 1980, was the first company to establish a commercial presence. Especially for the separation of hydrogen from ammonia-plant purge-gas streams, it appeared a viable option. Dow launched Generon, the first commercial membrane system for the separation of nitrogen from air. The latter application of membranes has expanded very rapidly and has now captured more than one-half of the market for nitrogen separation systems from air. Approximately 30,000 nitrogen systems have been installed worldwide today. Gas separation systems are also being used for a wide variety of other, smaller applications.

49.2 Theoretical Background

Gas transport through dense polymer membranes can be described by the following equation:

$$\Phi_{mi} = \frac{Đ(c_{io} - c_{il})}{l} \text{ cm}^3 \text{ (STP) of component i} \cdot \text{cm}^{-2} \cdot \text{s}^{-1}. \quad (49.1)$$

This equation resembles Equation 12.2. The latter equation describes the transport of a compound through a liquid or a gas by diffusion. cm^3 (STP) is a unit of mass comparable to gram or mole. c_{io} and c_{il} are the concentrations of component i at, respectively, the feed side and the product side of the membrane in cm^3 (STP) of component $i \cdot cm^{-3}$ of polymer. l is the membrane thickness in cm.

Henry's law (1803) states that the mass of gas dissolved in a given volume of liquid at a given temperature is proportional to the pressure or partial pressure of the gas. In the case of a gas dissolving in a solid to form a true solution, Henry's law will apply as well. Thus, $c_{io} = K_i \cdot p_{io}$ and $c_{il} = K_i \cdot p_{il}$. In these equations, K_i is the Henry coefficient (termed the sorption coefficient in the remainder of this chapter) in cm^3 (STP) $\cdot cm^{-3} \cdot (cm\ Hg)^{-1}$, and p_{io} and p_{il} are the partial pressures of component i at the respective sides of the membrane in cm Hg. The latter unit is a unit of pressure comparable to, for example, $N \cdot m^{-2}$.

Introduction to Chemical Engineering, First Edition. C.M. van 't Land.

On replacing c_{io} and c_{il} by $K_i \cdot p_{io}$ and $K_i \cdot p_{il}$, respectively, the following equation is obtained:

$$\Phi_{mi} = \frac{Đ_i \cdot K_i(p_{io} - p_{il})}{l} \text{ cm}^3 \text{ (STP) of component i} \cdot \text{cm}^{-2} \cdot \text{s}^{-1}. \tag{49.2}$$

In the latter equation, $Đ_i$ is a measure of the mobility of the gaseous component in the membrane, whereas K_i is a measure of the acceptance of the gaseous component by the membrane. The product $Đ_i \cdot K_i$ can be written as P_i, which is called the membrane permeability, and is a measure of the membrane's ability to permeate the gas.

The permeability of gases through membranes is most commonly measured in Barrer, defined as 10^{-10} cm^3 (STP) $\cdot$ cm^{-1} $\cdot$ s^{-1} $\cdot$ (cm Hg)$^{-1}$ and named after R.M. Barrer, a pioneer in gas permeability measurements. A measure of the ability of a membrane to separate two gases, i and j, is the ratio of their permeabilities, α_{ij}, called the membrane selectivity:

$$\alpha_{ij} = \frac{P_i}{P_j}.$$

The latter equation can also be written as:

$$\alpha_{ij} = \frac{Đ_i}{Đ_j} \cdot \frac{K_i}{K_j}. \tag{49.3}$$

The first ratio, the ratio of the diffusion coefficients, reflects the different sizes of the molecules of the two compounds. It is called the mobility selectivity. The second ratio, the ratio of the sorption coefficients, reflects the relative condensabilities of the two gases. It is called the sorption or solubility selectivity.

The Mobility Selectivity

Because large molecules interact with more segments of a polymer chain than do small molecules, the diffusion coefficient decreases with increasing molecular size in all polymer materials. A conclusion is that the passage of small molecules is always favored over the passage of large ones. A further aspect is that the magnitude of the mobility selectivity term depends strongly on whether the membrane material is above or below the glass transition temperature T_g. The polymer chains are essentially fixed and do not rotate if the material temperature is below T_g. The material is then called a glassy polymer and is tough and rigid. Diffusion through a polymer becomes difficult in that case, especially for relatively large molecules. The segments of the polymer chain have sufficient thermal energy to allow limited rotation around the chain backbone when the membrane temperature is above the glass transition temperature. The mechanical properties of the polymer are changed dramatically, and it becomes a rubber. Figure 49.1 is a log/log-plot of the diffusion coefficient as a function of the Van der Waals molar volume. The latter volume is calculated for the liquid state. The upper line applies for natural rubber and the lower line for PVC. Diffusion coefficients in rubbers decrease much less rapidly with increasing permeate size than diffusion coefficients in glassy materials. For example, the mobility selectivity of natural rubber for nitrogen over pentane is approximately 10. The mobility selectivity of PVC, a rigid, glassy polymer, for nitrogen over pentane is more than 10^5.

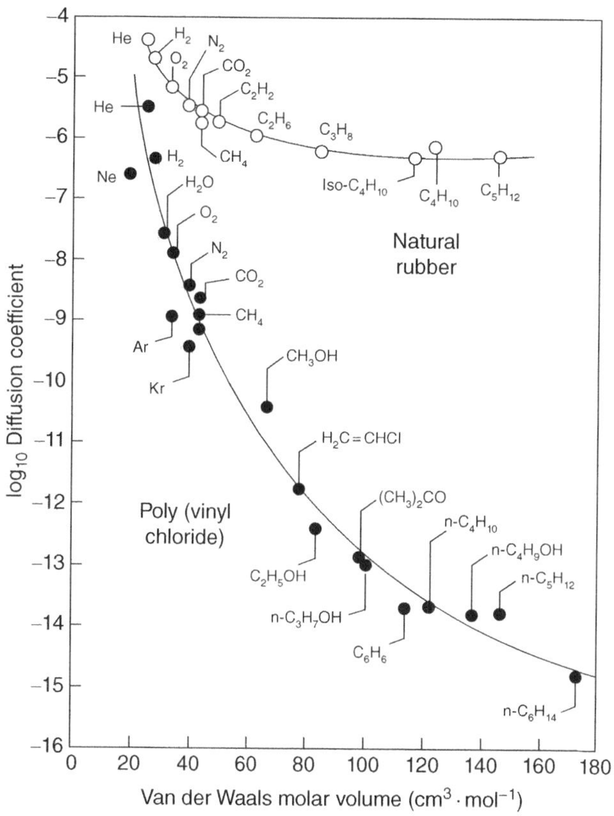

Figure 49.1 Diffusion coefficient as a function of molar volume for a variety of permeants in natural rubber and in PVC, a glassy polymer. *Source:* Membrane Technology and Applications by Baker, R. W. (2012). Reproduced with permission of John Wiley & Sons Inc.

The Sorption Selectivity

See Figure 49.2. The graph gives the sorption coefficient as a function of the Van der Waals molar volume. The curve is valid for natural rubber. The sorption coefficient increases with the size of the molecules. Large molecules are more condensable than small molecules.

The Combination of the Mobility Selectivity and the Sorption Selectivity

See Figure 49.3. It is log/linear-plot of the permeability as a function of the Van der Waals molar volume. The upper line applies for natural rubber, and the lower line is applicable for a glassy polymer. If it is desired to separate, for example, butane from nitrogen, natural rubber would be a good choice as the membrane permeates butane better than nitrogen. If, on the other hand, it is desired to separate nitrogen from propane, polyetherimide would be a good choice as that membrane permeates nitrogen better than butane. An aspect is that, for rubbers, the sorption coefficient varies strongly with molecular size, whereas the diffusion coefficient is only moderately dependent on molecular size. As for glassy polymers, the diffusion coefficient is usually dominant. See also Table 49.1. Silicone rubber permeates propane 13.6 times better than nitrogen. Polyimide permeates nitrogen 40 times better than propane. Figure 49.4 concerns the separation of carbon dioxide from methane with a certain type of membrane. The upper line is a hypothetical line based on

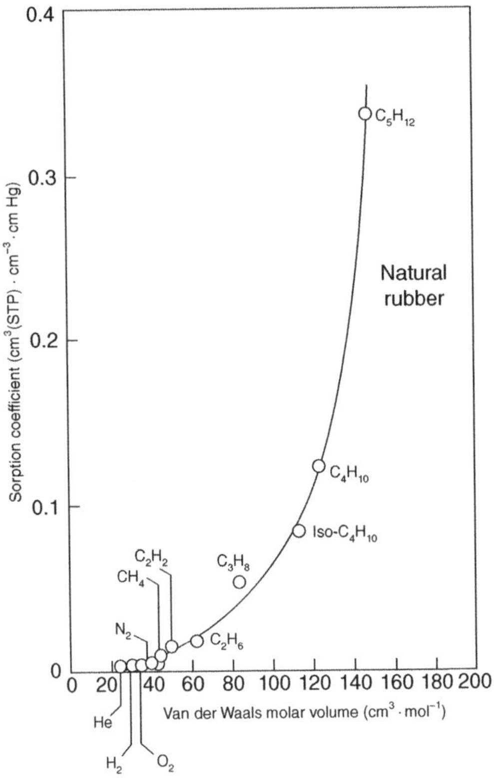

Figure 49.2 Gas sorption coefficient as a function of molar volume for natural rubber. *Source:* Membrane Technology and Applications by Baker, R. W. (2012). Reproduced with permission of John Wiley & Sons Inc.

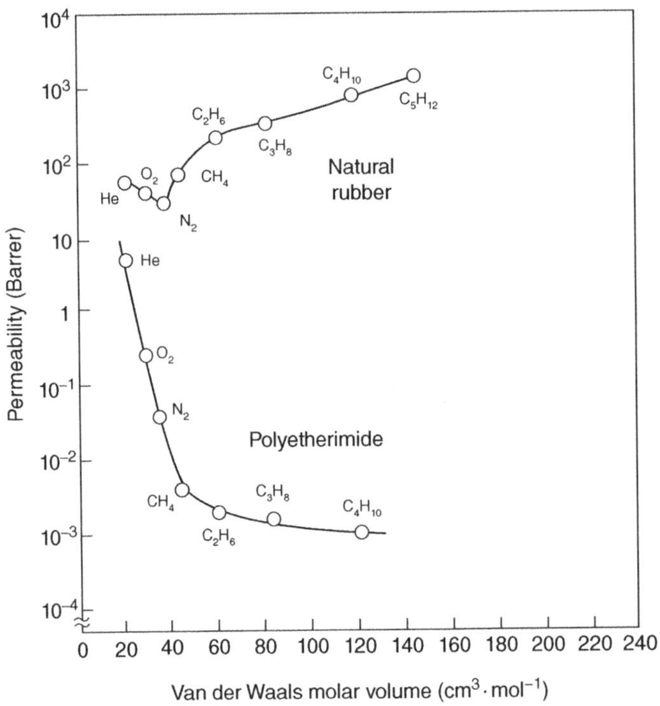

Figure 49.3 Permeability as a function of molar volume for a rubbery and a glassy polymer. *Source:* Membrane Technology and Applications by Baker, R. W. (2012). Reproduced with permission of John Wiley & Sons Inc.

Table 49.1 Permeabilities (Barrer [10^{-10} cm^3 (STP) · cm^{-1} · s^{-1} · (cm Hg)$^{-1}$]) measured with pure gases, at the temperatures given, of widely used polymers.

	Rubbers		Glasses		
Gas	Silicone rubber at 25 °C (T_g = −129 °C)	Natural rubber at 30 °C (T_g = −73 °C)	Cellulose acetate at 25 °C (T_g = 40 − 124 °C)	Polysulfone at 35 °C (T_g = 186 °C)	Polyimide (Ube Industries) at 60 °C (T_g > 250 °C)
H_2	550	41	24	14	50
He	300	31	33	13	40
O_2	500	23	1.6	1.4	3
N_2	250	9.4	0.33	0.25	0.6
CO_2	2,700	153	10	5.6	13
CH_4	800	30	0.36	0.25	0.4
C_2H_6	2,100	—	0.20	—	0.08
C_3H_8	3,400	168	0.13	—	0.015
C_4H_{10}	7,500	—	0.10	—	—

Source: Membrane Technology and Applications by Baker, R. W. (2012). Reproduced with permission of John Wiley & Sons Inc.

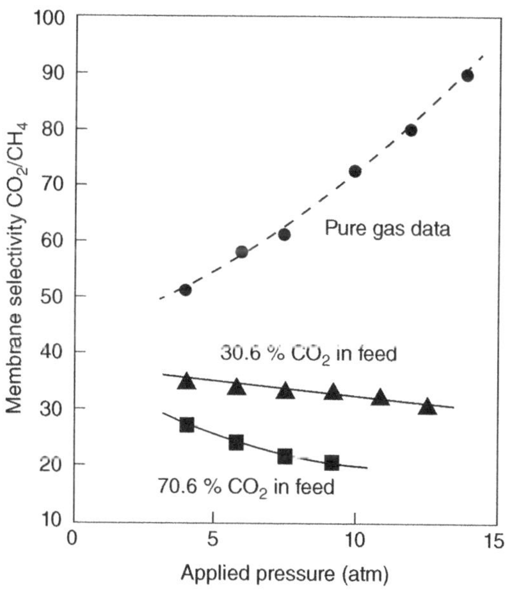

Figure 49.4 The difference between selectivities calculated from pure gas measurements and selectivities measured with gas mixtures can be large. *Source:* Membrane Technology and Applications by Baker, R. W. (2012). Reproduced with permission of John Wiley & Sons Inc.

measured individual permeabilities of the two gases. However, the ratio of the two permeabilities is influenced by the percentage carbon dioxide in the feed. The gases interact with the membrane material. In this case, carbon dioxide is sorbed by the membrane affecting the permeability of methane. Such an effect cannot be noticed at the separation of oxygen from nitrogen.

49.3 Process Design

The three variables, which determine the performance of a membrane separation system, are illustrated in Figure 49.5. These three are the membrane selectivity α, the ratio of feed pressure to permeate pressure φ, and the stage-cut θ. The stage-cut is the permeate flow expressed as a volume fraction of the feed flow. It can be derived that, at zero stage-cut, for a binary system, the volume fraction of component i on the permeate side is a function of α, φ, and the volume fraction of component i on the feed side of the membrane. The equation is [1]:

$$n_{il} = \frac{\varphi}{2}\left(n_{io} + \frac{1}{\varphi} + \frac{1}{\alpha_{ij} - 1} - \sqrt{\left(n_{io} + \frac{1}{\varphi} + \frac{1}{\alpha_{ij} - 1}\right)^2 - \frac{4\alpha_{ij} n_{io}}{(\alpha_{ij} - 1)\varphi}}\right). \tag{49.4}$$

This equation is valid for $\theta = 0$. However, as soon as the permeation starts, and the feed flows along the membrane, the feed volume fraction falls and thus the permeate volume fraction. The equations bear a resemblance to the equation to find the roots of a quadratic equation. The derivation of Equation 49.4 proceeds as follows. The transport equation for component i in the gaseous mixture is:

$$\Phi_{mi} = \frac{P_i(p_{io} - p_{il})}{l} \quad \text{cm}^3\ (\text{STP})\ \text{of component i} \cdot \text{cm}^{-2} \cdot \text{s}^{-1}.$$

The transport equation for component "j" in the gaseous mixture is:

$$\Phi_{mj} = \frac{P_j\left(p_{jo} - p_{jl}\right)}{l} \quad \text{cm}^3\ (\text{STP})\ \text{of component j} \cdot \text{cm}^{-2} \cdot \text{s}^{-1}.$$

Dividing the former equation by the latter equation results in:

$$\frac{\Phi_{mi}}{\Phi_{mj}} = \alpha_{ij} \cdot \frac{p_{io} - p_{il}}{p_{jo} - p_{jl}}. \tag{49.5}$$

The volume fractions of components i and j on the feed side can be expressed as follows:

$$n_{io} = \frac{p_{io}}{p_o} \text{ and } n_{jo} = \frac{p_{jo}}{p_o}.$$

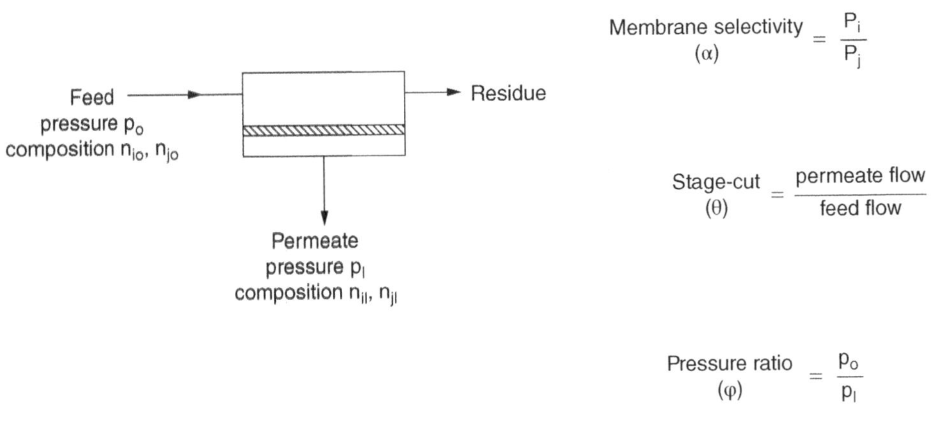

Figure 49.5 Parameters affecting the performance of membrane gas separation systems. *Source:* Membrane Technology and Applications by Baker, R. W. (2012). Reproduced with permission of John Wiley & Sons Inc.

The volume fractions of components i and j on the permeate side can be expressed as follows:

$$n_{il} = \frac{p_{il}}{p_l} \text{ and } n_{jl} = \frac{p_{jl}}{p_l}.$$

Substitution of these four expressions into Equation 49.5 results in:

$$\frac{\Phi_{mi}}{\Phi_{mj}} = \alpha_{ij} \cdot \frac{n_{io} \cdot p_o - n_{il} \cdot p_l}{n_{jo} \cdot p_o - n_{jl} \cdot p_l}.$$

Dividing both the numerator and the denominator of the fraction in the right-hand side of this equation by p_l results in:

$$\frac{\Phi_{mi}}{\Phi_{mj}} = \alpha_{ij} \cdot \frac{n_{io} \cdot \varphi - n_{il}}{n_{jo} \cdot \varphi - n_{jl}}. \tag{49.6}$$

In Equation 49.6, n_{jo} is replaced by $(1 - n_{io})$ and n_{jl} is replaced by $(1 - n_{il})$. From mass balance considerations, we can write:

$$\frac{\Phi_{mi}}{\Phi_{mj}} = \frac{n_{il}}{n_{jl}} = \frac{n_{il}}{1 - n_{il}}.$$

The latter expression is also introduced into Equation 49.6. The equation which results is:

$$\frac{n_{il}}{1 - n_{il}} = \alpha_{ij} \cdot \frac{n_{io} \cdot \varphi - n_{il}}{(1 - n_{io})\varphi - (1 - n_{il})}.$$

The latter equation can be rearranged and be written as a quadratic expression in n_{il}. Equation 49.4 is a root of this expression having a physical meaning.

Example 49.1

A calculation is carried out for the separation of two gases by means of a membrane. The feed mixture contains 10 volume% of component A. Thus, $n_{io} = 0.1$. The permeability of component A through the membrane is 30 times greater than the permeability of component B, that is, $\alpha_{ij} = 30$. The pressure upstream of the membrane is two times the pressure downstream the membrane, that is, $\varphi = 2$. By using Equation 49.4, n_{il} can be calculated. The result is $n_{il} = 0.192$. This result is applicable for a stage-cut of zero, that is, there is no flow through the membrane. As soon as the membrane starts to separate, n_{il} will become smaller than 0.192. The conclusion is that although the permeability of component A is 30 times greater than the permeability of component B, the separation will not be very effective. If φ becomes 5, the situation improves and n_{il} becomes, according to Equation 49.4, 0.405.

49.4 Applications

Air Separation

The production of nitrogen from air is the largest separation process by means of membranes. The membrane materials now used have selectivities of 6–7, providing favorable economics for plants producing $8.5–8.5 \cdot 10^3\ nm^3 \cdot h^{-1}$ of nitrogen [2]. In this range, membranes are the low-cost

Table 49.2 Permeabilities and selectivities of polymers of interest in air separation.

Polymer	Oxygen permeability (Barrer)	Nitrogen permeability (Barrer)	Oxygen/nitrogen selectivity
Poly(1-trimethylsilyl-1-propyne) (PTMSP)	7,600	5,400	1.4
Teflon AF 2400	1,300	760	1.7
Silicone rubber	600	280	2.2
Poly(4-methyl-1-pentene) (TPX)	30	7.1	4.2
Poly(phenylene oxide) (PPO)	16.8	3.8	4.4
Ethyl cellulose	11.2	3.3	3.4
6FDA-DAF (polyimide)	7.9	1.3	6.2
Polysulfone	1.1	0.18	6.2
Polyaramide	3.1	0.46	6.8
Tetrabromo bis polycarbonate	1.4	0.18	7.5

Source: Membrane Technology and Applications by Baker, R. W. (2012). Reproduced with permission of John Wiley & Sons Inc.

process, and most new small nitrogen plants use membrane systems. This is particularly true if the required nitrogen purity is between 95 and 99 vol% nitrogen. Table 49.2 lists the permeabilities and selectivities of some of the materials that are used or have been used for this separation. There is a strong inverse relationship between flux and selectivity. Membranes with selectivities of 6–7 typically have 1% of the permeability of membranes with selectivities of 2–3.

High oxygen/nitrogen selectivity is required for an economical nitrogen production process. Figure 49.6 shows three designs for nitrogen production from air. It is assumed that the percentages are percentages by volume. The upper design is a one-stage design and means a stage-cut of 0.333, as can be calculated from the numbers given. The nitrogen recovery is 41.8%. This design illustrates that it is possible to obtain a residue that is almost pure nitrogen, however, the permeate contains only 31% of oxygen. Other designs could aim for a permeate that is almost pure oxygen, however, in that case, the residue would be impure. Aiming for two pure gases is not possible.

If a selectivity $\alpha_{ij} = 6$ and a pressure ratio $\varphi = 10$ are assumed, it is possible to calculate 55% of oxygen in the permeate by means of Equation 49.4. The stage-cut θ is then 0. So, the oxygen volume fraction in the permeate comes down from 55 to 31% when the stage-cut increases from 0 to 0.333. In making this statement, it has been assumed that, also for the single-step design of Figure 49.6, $\alpha_{ij} = 6$ and $\varphi = 10$.

The other two designs in Figure 49.6 are used to produce purer gas containing > 99% nitrogen. The two-step design saves on membrane area and compressor load. The three-step process shows a further decrease of the membrane area and the compressor load. However, the latter process is limited to large systems in which the savings compensate for the extra complexity and higher maintenance cost of a second compressor.

The oxygen-enriched air, containing about 35% of oxygen, is usually vented [3].

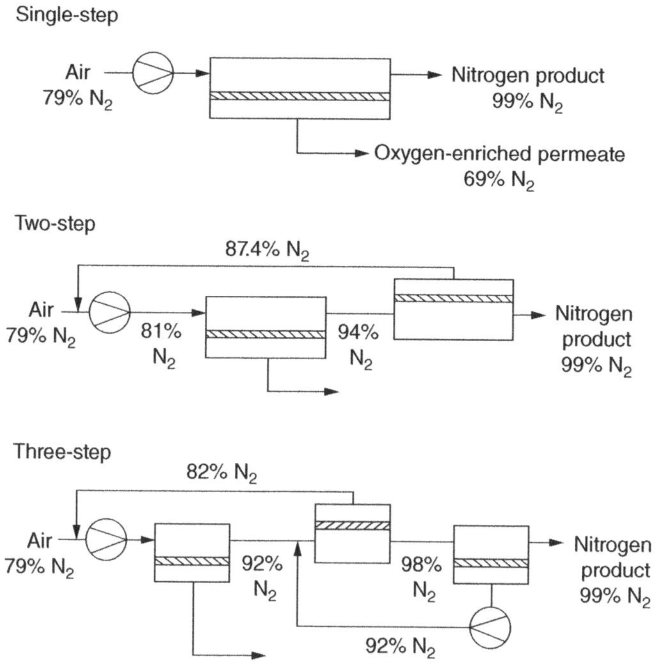

Design	Relative membrane area	Relative compressor HP
One-step	1.0	1.0
Two-step	0.94	0.94
Three-step	0.92	0.92

Figure 49.6 Single-, two-, and three-step designs for nitrogen production from air. *Source:* Membrane Technology and Applications by Baker, R. W. (2012). Reproduced with permission of John Wiley & Sons Inc.

Separation of Hydrogen in Ammonia Plants

The separation of hydrogen from nitrogen in ammonia purge gas streams was the first large-scale commercial application of membrane gas separation. It plays a role in the Haber–Bosch synthesis of ammonia. The synthesis dates back to 1913. The raw materials are nitrogen and hydrogen. See Figure 49.7. Argon enters the high-pressure ammonia reactor as an impurity of the nitrogen stream and methane enters the system as an impurity of the hydrogen stream. Nitrogen and hydrogen react in the reactor to form ammonia, which is removed by condensation in a loop around the reactor. Argon and methane do not participate in the reaction and are not condensed. Hence, a purge flow must leave the reactor to remove the two gases. The purge flow is chosen as to contain 6% of argon and 11% of methane. The purge flow also contains hydrogen and nitrogen, and it is worthwhile to recover hydrogen from this stream by membrane separation. 2–4% of the hydrogen would be lost if the membrane separation would not be present [4]. The separation is assisted by the fact that a hydrogen molecule is small and noncondensable. Compared to all other gases, the molecule is highly permeable. To illustrate this, fluxes through and selectivities of several glassy polymers vis-à-vis three binary systems are summarized in Table 49.3.

Table 49.3 Hydrogen separation membranes.

Membrane (developer)	Selectivity			Hydrogen pressure-normalized flux [10^{-6} cm^3 (STP)/(cm^2·s·cm Hg)]
	H_2/CO	H_2/CH_4	H_2/N_2	
Polyaramide (Medal)	100	>200	>200	—
Polysulfone (Permea)	40	80	80	100
Cellulose acetate (Separex)	30–40	60–80	60–80	200
Polyimide (Ube)	50	100–200	100–200	80–200

Source: Membrane Technology and Applications by Baker, R. W. (2012). Reproduced with permission of John Wiley & Sons Inc.

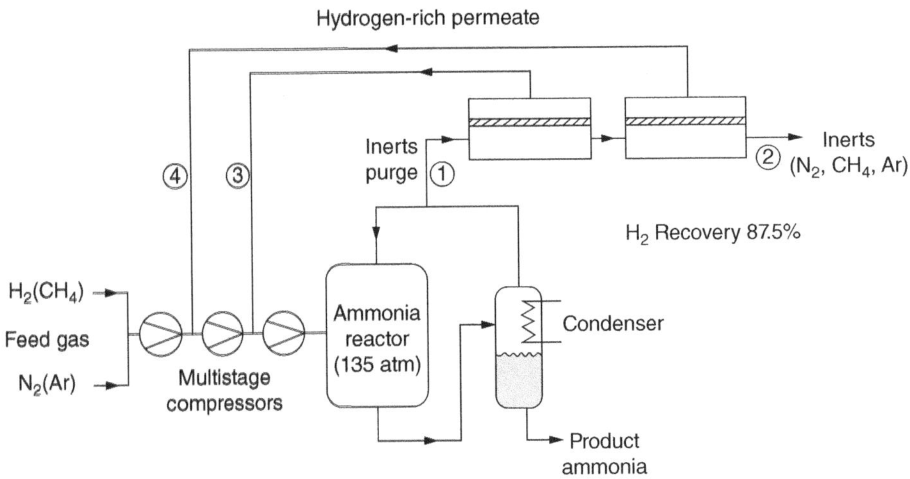

	Stream composition (%)			
	Membrane feed ①	Membrane vent ②	High-pressure permeate ③	Low-pressure permeate ④
Hydrogen	62	21	87.3	84.8
Nitrogen	21	44	7.1	8.4
Methane	11	23	36	4.3
Argon	6	13	2.0	2.5
Pressure (atm)	135	132	70	28
Flow (scfm)	2,000	740	830	430

Figure 49.7 Simplified flow schematic of the PRISM® membrane system to recover hydrogen from an ammonia reactor purge system. *Source:* Membrane Technology and Applications by Baker, R. W. (2012). Reproduced with permission of John Wiley & Sons Inc.

A two-step membrane design is used in the process shown in Figure 49.7. It is chosen to reduce the cost of recompressing the hydrogen permeate stream to the quite high pressure of the ammonia reactor. The pressure ratio of the first membrane separation step is 135/70 = 1.9. The hydrogen content of the permeate of this step is 87.3%, whereas the feed flow contains 62%. The feed flow of the second membrane separation step contains 30% of hydrogen. Now, a pressure ratio of 135/28 = 4.8 is required to recover hydrogen. The two recycle flows to the reactor have approximately the same hydrogen content. Both flows are recompressed to the reactor pressure.

Further applications of gas separation by means of membranes are air drying, natural gas processing, and treatment of petrochemical purge gas.

References

1. Baker, R. W. (2012). *Membrane Technology and Applications* (p. 341). Chichester, UK: Wiley.
2. Baker, R. W. (2012). *Membrane Technology and Applications* (p. 352). Chichester, UK: Wiley.
3. Baker, R. W. (2012). *Membrane Technology and Applications* (p. 356). Chichester, UK: Wiley.
4. Baker, R. W. (2012). *Membrane Technology and Applications* (p. 350). Chichester, UK: Wiley.

50

Case Studies Membranes

50.1 Gel Formation

See Figure 46.6. The gel layer model is applicable for this ultrafiltration. The proof is that straight lines are obtained on plotting the maximum filtration rate as a function of the logarithm of the latex concentration. The line for 10 gpm (gallons per minute) is in focus. It is requested, by using Equation 46.1, to calculate the latex diffusion coefficient in $m^2 \cdot s^{-1}$ for a thickness of the laminary boundary layer of 100 μm. The second request is to estimate the thickness of the laminary boundary layer when the circulation rate is 20 gpm.

1 U.S. gallon is 3.785 l and 1 ft is 0.3048 m.

Solution

Equation 46.1: $\ln \frac{c_{gel}}{c_{ib}} = \frac{\Phi_v \cdot \delta}{Ð_i}$.

Take $c_{gel} = 78$ vol% and, for example, $c_{ib} = 5$ vol%; it then follows that $\Phi_v = 45$ gallon $\cdot$ ft^{-2} $\cdot$ day^{-1}.

$$\ln \frac{78}{5} = \frac{45 \cdot 3.785 \cdot 10^{-3} \cdot 10^{-4}}{0.3048^2 \cdot 24 \cdot 3.6 \cdot 10^3 \cdot Ð_i} \rightarrow Ð_i = 7.724 \cdot 10^{-10}\, m^2 \cdot s^{-1}.$$

This low value for the diffusion coefficient makes sense for latex solutions.

The thickness of the laminar boundary layer for a circulation rate of 20 gpm is approximately $\frac{45}{80} \cdot 100 = 56$ μm.

50.2 Osmotic Pressure

pV = RT is the universal gas law. On using SI-units (Système International), it is possible to state:

p is the pressure in $N \cdot m^{-2}$,
V is the molar volume in $m^3 \cdot kmol^{-1}$,
R is the universal gas constant, that is, $8.314 \cdot 10^3$ $J \cdot kmol^{-1} \cdot K^{-1}$, and
T is the absolute temperature in K.

It is also possible to write: $p = RT/V = \rho RT$. ρ can also be called the concentration in $kmol \cdot m^{-3}$. Van 't Hoff (1852–1911) found that the osmotic pressure of a solution is proportional to the concentration of the solute in $kmol \cdot m^{-3}$ and that the proportionality constant is R, the universal

Introduction to Chemical Engineering, First Edition. C.M. van 't Land.

gas constant. Van 't Hoff was a Dutchman who won, in 1901, the first Nobel Prize for chemistry. The prize was awarded to him because of his work concerning osmotic pressure. He is also known as the founder of stereochemistry.

Thus, π, the osmotic pressure can be calculated by means of the following expression: $\pi = \rho RT$. This expression is called the Van 't Hoff equation. Substantial deviations from his law occur at high concentrations and with macromolecular solutes.

It is requested to calculate the osmotic pressure of an aqueous solution containing 1% by weight of cane sugar at 25 °C. Cane sugar is a disaccharide and has the formula $C_{12}H_{22}O_{11}$. The atomic weights of C, H, and O are, respectively, 12.0, 1.0, and 16.0.

Next, it is requested to calculate the osmotic pressure of seawater at 25 °C. Seawater contains about 3.5% by weight of NaCl. The atomic weights of Na and Cl are, respectively, 23.0 and 35.4. Note that NaCl molecules do not exist. Positive Na-ions and negative Cl-ions are present in an aqueous solution of NaCl.

Solution

Sugar molecular weight: $12 \cdot 12.0 + 22 \cdot 1.0 + 11 \cdot 16.0 = 342.0\ \mathrm{kg \cdot kmol^{-1}}$.
10 kg sugar in $1\ \mathrm{m^3}$, $10/342.0 = 0.029\ \mathrm{kmol \cdot m^{-3}}$, $\pi = 0.029 \cdot 8.314 \cdot 10^3 \cdot 298 = 7.185 \cdot 10^4\ \mathrm{N \cdot m^{-2}} = 0.72$ bar.
NaCl "molecular weight": $23.0 + 35.4 = 58.4\ \mathrm{kg \cdot kmol^{-1}}$.
35 kg NaCl in $1\ \mathrm{m^3}$, $35/58.4 = 0.60\ \mathrm{kmol \cdot m^{-3}}$, $\pi = 2 \cdot 0.60 \cdot 8.314 \cdot 10^3 \cdot 298 = 2.973 \cdot 10^6 = 29.73$ bar.
Note the factor 2 because NaCl "molecules" dissociate into two ions. The measured osmotic pressure of seawater containing 3.5% by weight of NaCl at 25 °C is 23 bar [1]. The discrepancy is caused by the fact that seawater is not a very diluted solution of NaCl in water.

50.3 Membrane Gas Separation

The permeability of methane through a membrane made of polyethylene was measured at various temperatures and pressures [2]. It was found that the permeability is, between 5 and 70 bar, not a function of the pressure. P is a function of the temperature:

T (°C)	P (Barrer)
10	1.8
20	3.5
30	6.3
40	10.1

The transport of methane can be described by an Arrhenius type of equation: $P = P_0 \cdot \exp(-E_{act}/(RT))$ Barrer.

It is requested to calculate the activation energy E_{act} in $\mathrm{kJ \cdot mol^{-1}}$.

$R = 8.314 \cdot 10^3\ \mathrm{J \cdot kmol^{-1} \cdot K^{-1}}$.

Solution

$$\ln P = \ln P_o - \frac{E_{act}}{RT},$$

$$\ln \frac{10.1}{1.8} = \frac{E_{act}}{R}\left(\frac{1}{283} - \frac{1}{313}\right),$$

$$E_{act} = 42.3\,\text{kJ}\cdot\text{mol}^{-1}.$$

References

1. Baker, R. W. (2012). *Membrane Technology and Applications* (p. 210). Chichester, UK: Wiley.
2. Mulder, M. (1996). *Basic Principles of Membrane Separation* (p. 401). Dordrecht, The Netherlands: Kluwer Academic Publishers.

Notation VII

A	Factor in Equation 47.1	$s \cdot m^{-1}$
B	Constant in Equation 48.2	$m \cdot s^{-1}$
c_f	Concentration of particles in a feed suspension	m^{-3}
c_{gel}	Concentration of macromolecules in a gel phase	m^{-3}
c_i	Concentration of macromolecules of compound i	m^{-3}
c_{ib}	Concentration of macromolecules of compound i in the bulk of a liquid	m^{-3}
c_{io}	Concentration of macromolecules of compound i at a membrane surface	m^{-3}
	Concentration of component i at the feed side	$cm^3(STP) \cdot cm^{-3}$
c_{il}	Concentration of component i at the permeate side	$cm^3(STP) \cdot cm^{-3}$
c_{jl}	Salt concentration at the permeate side	$kg \cdot m^{-3}$
c_{jo}	Salt concentration at the feed side	$kg \cdot m^{-3}$
c_p	Concentration of particles in a permeate	m^{-3}
D	Diameter of a flow channel	m
$Đ_i$	Diffusion coefficient of component i	$m^2 \cdot s^{-1}$
		$cm^2 \cdot s^{-1}$
$Đ_j$	Diffusion coefficient of component j	$m^2 \cdot s^{-1}$
		$cm^2 \cdot s^{-1}$
E_{act}	Activation energy	$kJ \cdot kmol^{-1}$
I	Electric current	A
K_i	Henry coefficient of gas i	$cm^3(STP) \cdot cm^{-3} \cdot (cm\ Hg)^{-1}$
K_j	Henry coefficient of gas j	$cm^3(STP) \cdot cm^{-3} \cdot (cm\ Hg)^{-1}$
l	Length of a flow channel	m
	Membrane thickness	m
		cm
N	Number of pores	m^{-2}
n_{il}	Volume fraction of gas i at the permeate side	—
n_{io}	Volume fraction of gas i at the feed side	—
n_{jl}	Volume fraction of gas j at the permeate side	—
n_{jo}	Volume fraction of gas j at the feed side	—
P	Permeability	$cm^3(STP) \cdot cm^{-1} \cdot s^{-1} \cdot (cm\ Hg)^{-1}$
P_i	Permeability of gas i through a membrane	$cm^3(STP) \cdot cm^{-1} \cdot s^{-1} \cdot (cm\ Hg)^{-1}$

Introduction to Chemical Engineering, First Edition. C.M. van 't Land.

P_j	Permeability of gas j through a membrane	$cm^3(STP) \cdot cm^{-1} \cdot s^{-1} \cdot (cm\ Hg)^{-1}$
P_0	Constant in Section 50.3	$10^{-10}\ cm^3(STP) \cdot cm^{-1} \cdot s^{-1} \cdot (cm\ Hg)^{-1}$
p	pressure	$N \cdot m^{-2}$
$p_{1,2,3,4}$	Pressures Nos. 1, 2, 3, and 4	$N \cdot m^{-2}$
p_{il}	Partial pressure of gas i at the permeate side	cm Hg
p_{io}	Partial pressure of gas i at the feed side	cm Hg
p_{jl}	Partial pressure of gas j at the permeate side	cm Hg
p_{jo}	Partial pressure of gas j at the feed side	cm Hg
p_o	Feed pressure	cm Hg
p_l	Permeate pressure	cm Hg
R	Radius of a flow channel	m
	Salt rejection coefficient	—
	Universal gas constant ($8.314 \cdot 10^3\ J \cdot kmol^{-1} \cdot K^{-1}$)	
T_g	Glass temperature	°C or K
V	Molar gas volume	$m^3 \cdot kmol^{-1}$
v	Linear velocity	$m \cdot s^{-1}$
x	Distance	m

Greek Symbols

α	Membrane selectivity	—
α_{ij}	Membrane selectivity for i versus j	—
δ	Diffusion layer thickness	m
ε	Fraction free area of a membrane	—
μ	Dynamic viscosity	$N \cdot s \cdot m^{-2}$
θ	Ratio of permeate flow to feed flow at gas membrane separation (stage-cut)	—
π	Osmotic pressure	$N \cdot m^{-2}$
ρ	Concentration	$kmol \cdot m^{-3}$
ρ_l	Water specific mass	$kg \cdot m^{-3}$
Φ_{max}	Maximum flow	$m^3 \cdot m^{-2} \cdot s^{-1}$
Φ_{ms}	Salt flux through a membrane	$kg \cdot m^{-2} \cdot s^{-1}$
Φ_{mi}	Flux of gas i through a membrane	$cm^3(STP) \cdot cm^{-2} \cdot s^{-1}$
Φ_{mj}	Flux of gas j through a membrane	$cm^3(STP) \cdot cm^{-2} \cdot s^{-1}$
Φ_{mw}	Water flux through a membrane	$kg \cdot m^{-2} \cdot s^{-1}$
Φ_s	Flow of macromolecules	$m^{-2} \cdot s^{-1}$
Φ_v	Flow	$m^3 \cdot m^{-2} \cdot s^{-1}$
		$m^3 \cdot s^{-1}$
Φ_w	Water flow	$m^3 \cdot m^{-2} \cdot s^{-1}$
		$l \cdot m^{-2} \cdot h^{-1}$
Φ_{vw}	Water flow	$m^3 \cdot m^{-2} \cdot s^{-1}$
φ	Ratio of feed pressure to permeate pressure at gas separation	—

Part VIII

Crystallization, Liquid/Solid Separation, and Drying

Part VIII Content

51

Crystallization

51.1 Introduction

Crystallization is the process of producing a solid state of matter characterized by an ordered molecular structure as shown by a reproducible X-ray pattern. The analytical technique to establish such a pattern is called Röntgen diffraction. This chapter will deal with crystallization from a solution.

Substances that can be dissolved in a solvent resulting in solutions containing between 5 and 70% by weight of the solute are called well-soluble substances. Well-soluble substances often crystallize, whereas less-soluble materials often precipitate. The particle size is the basis of the difference between these two steps, for example, industrially, salt having a particle size of 0.5 mm is produced. The reaction/crystallization of calcium citrate by the reaction between aqueous solutions of calcium hydroxide and citric acid produces a much finer material. In the case of crystallization, the supersaturation created is much smaller than the solubility. If the supersaturation is of the same order of magnitude as the solubility, precipitation occurs.

Solubility is discussed in Section 51.2, whereas nucleation and growth are treated in, respectively, Sections 51.3 and 51.4. Several types of industrial crystallizers are dealt with in Section 51.5. A model describing crystallization in a continuous stirred crystallizer with mixed product removal is discussed in Section 51.6. Section 51.7 extends the discussion of the model.

51.2 Solubility

Solid–liquid equilibrium, or the solubility of a chemical compound in a solvent, refers to the maximum concentration of the solute in the solvent at static conditions.

The dependence of solubility on temperature affects the mode of crystallization. Figure 51.1 shows that the solubility of potassium nitrate is strongly influenced by the system temperature but that temperature has a small effect on the solubility of sodium chloride. As a consequence, a reasonable yield of KNO_3-crystals can be obtained by cooling a saturated feed solution. On the other hand, cooling a saturated brine accomplishes little crystallization, and evaporation of water is required to increase the yield of salt crystals.

The production of many high-value chemicals requires maximizing separation from a relatively dilute solution. It is common in such instances to use a combination of methods to reduce solubility in the feed solution. Figure 51.2, for example, illustrates that the addition of methanol to a saturated aqueous solution of l-serine can reduce solubility by more than an order of magnitude.

Introduction to Chemical Engineering, First Edition. C.M. van 't Land.

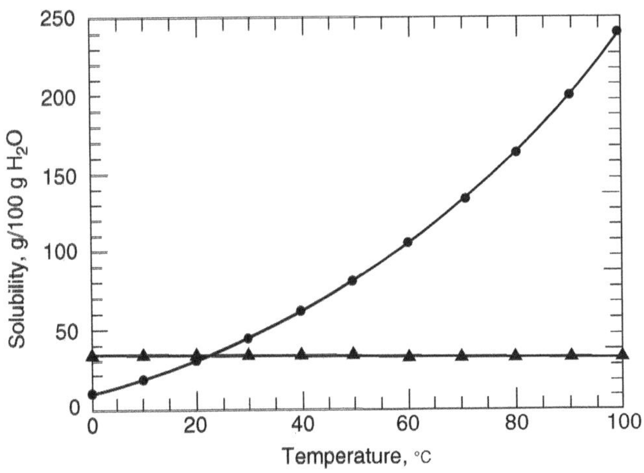

Figure 51.1 Solubilities of NaCl (▲) and KNO_3 (●) in water. *Source*: Processes & Process Engineering/ Crystallization. Kirk-Othmer Encyclopedia of Chemical Technology by Kirk-Othmer (2001). Reproduced with permission of John Wiley & Sons Inc.

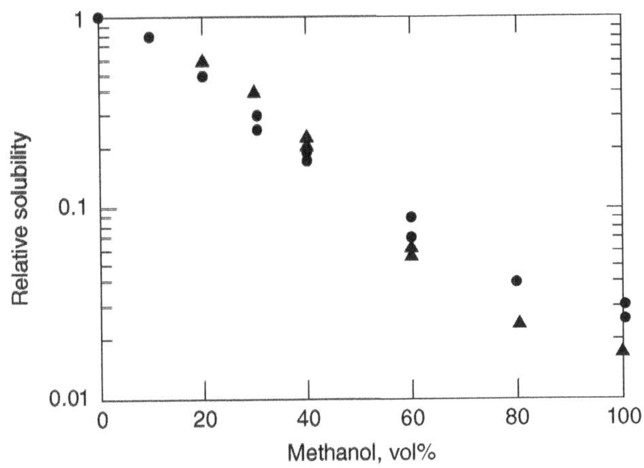

Figure 51.2 Solubilities of L-serine in mixtures of methanol and water. Relative solubility = solubility in methanol–water/solubility in water; ●, 10 °C; ▲, 30 °C. *Source:* Processes & Process Engineering/Crystallization. Kirk-Othmer Encyclopedia of Chemical Technology by Kirk-Othmer (2001). Reproduced with permission of John Wiley & Sons Inc.

51.3 Nucleation

Crystal nucleation is the formation of an ordered solid phase from a liquid or amorphous phase. In primary nucleation, existing crystals do not play a role. This type of nucleation can be homogeneous or heterogeneous. Homogeneous primary nucleation occurs when other solid material is absent. Heterogeneous primary nucleation can start when small different solid particles are present. Primary nucleation can be illustrated by considering a hypothetical experiment in the context of the solubility data in Figure 51.3. Assume that a solution is at a temperature corresponding to point A in the figure. The solution is undersaturated, so any crystals present in the system would dissolve. If the concentration is increased at constant temperature, for example, by evaporation of the solvent, the path followed would cause the solution to reach saturated conditions at point B. Once the concentration becomes greater than at B, the solution is supersaturated and any crystals present in

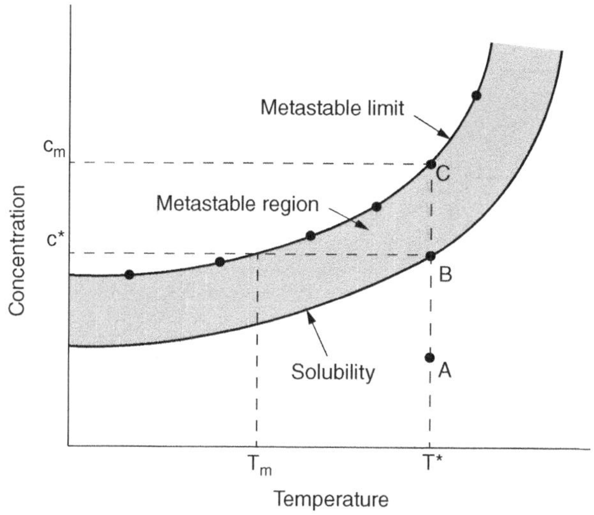

Figure 51.3 Solubility and primary nucleation in a hypothetical experiment. *Source:* Processes & Process Engineering/Crystallization. Kirk-Othmer Encyclopedia of Chemical Technology by Kirk-Othmer (2001). Reproduced with permission of John Wiley & Sons Inc.

the system would grow. However, experience shows that nucleation would not occur until the concentration reaches point C, which defines what is called the metastable limit. If this procedure is repeated at various temperatures, a metastable limit curve could be drawn as shown. The width of the metastable area is larger for homogeneous primary nucleation than for heterogeneous primary nucleation.

Secondary nucleation occurs in the presence of solute crystals. Here we distinguish initial breeding, contact nucleation, and shear breeding. Initial breeding results from immersion of seed crystals in a supersaturated solution. It is thought to be caused by the dislodging of extremely small crystals that were formed on the surface of larger crystals during drying. Contact nucleation results from collisions of crystals with one another, with crystallizer internals, with an agitator, or the impeller of a circulation pump. Shear breeding results when a supersaturated solution flows by a crystal surface and carries with it crystal precursors believed formed in the region of the growing crystal surface. In a study of nucleation of the heptahydrate of magnesium sulfate by shear breeding, it was found that the number of nuclei observed is strongly dependent on the supersaturation at the seed and the severity of shear forces [1].

51.4 Crystal Growth

At least two resistances determine growth kinetics. First, there is the transport of the solute by convection and diffusion from the solution to the crystal surface. Second, we distinguish the integration or incorporation of the crystalline unit or units into the crystal lattice. A simple set of experiments in which the rate of advance of a crystal face is measured can be used to illustrate these two resistances. Data given in Figure 51.4 show the effect of solution velocity at a crystal face at three different conditions. As the solution velocity increases from low values, the growth rates also increase. At about 24 $cm \cdot s^{-1}$, however, the growth rates approach constant values. Such behavior indicates that both bulk mass transfer and surface incorporation are important below 24 $cm \cdot s^{-1}$ but above this velocity, surface incorporation provides the dominant resistance to growth.

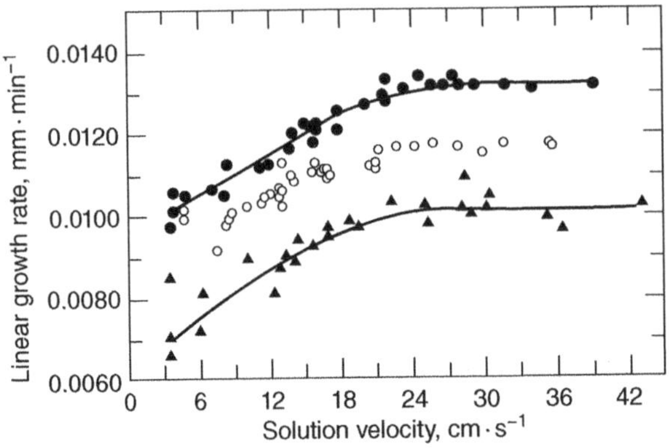

Figure 51.4 Effect of solution velocity on the growth rate of the (110)-face of $MgSO_4.7H_2O$. Concentration of $MgSO_4$: •, 29.02% by weight at 30.8 °C; ο, 28.99% by weight at 30.8 °C; ▴, 18.89% by weight at 31.3 °C. *Source:* Processes & Process Engineering/Crystallization. Kirk-Othmer Encyclopedia of Chemical Technology by Kirk-Othmer (2001). Reproduced with permission of John Wiley & Sons Inc.

51.5 Crystallizers and Crystallizer Operations

Cooling Crystallizers

See Figure 51.5. Cooling crystallizers are used for systems having a steep solubility curve. The heat is removed by indirect heat exchange. Incrustations tend to develop on the walls or other heat exchanging surfaces because those surfaces are cold. A consequence is that batchwise operation is possible, however, continuous operation is not possible.

Vacuum Crystallizers

See Figure 51.6. Vacuum crystallizers are continuous cooling crystallizers in which the cooling is accomplished by the evaporation of a small part of the solvent and not via indirect heat exchange.

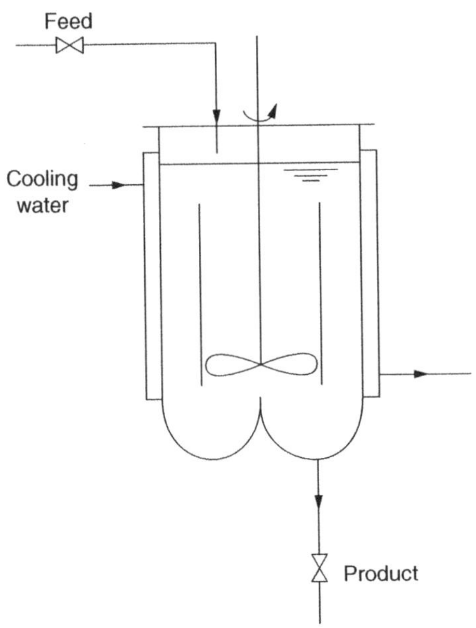

Figure 51.5 Cooling crystallizer.

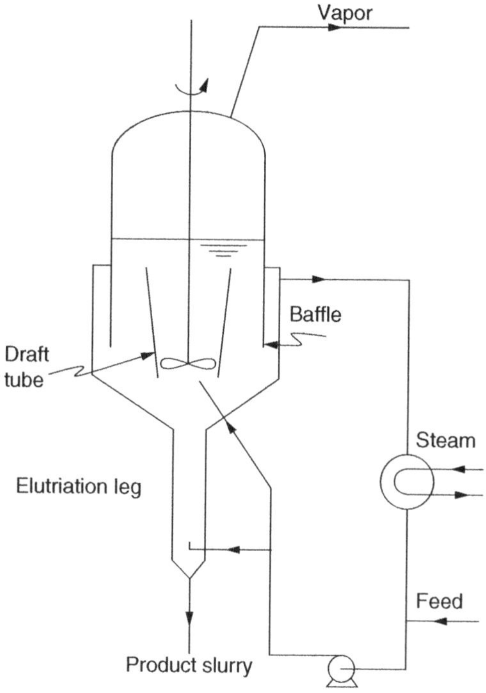

Figure 51.6 Vacuum crystallizer.

An example is the crystallization of potassium chloride from an aqueous solution. First, an explanation of Figure 51.6. The warm feed enters a loop around the crystallizer and the flow in the loop enters the crystallizer in the region of a screw pump. A low pressure is maintained in the vapor head of the crystallizer by means of a vacuum pump. The feed cools down and the product crystallizes. Cooling occurs at the slurry surface in the crystallizer and hence incrustations on metal walls do, in principle, not form. The screw pump transports slurry through a draft tube to the slurry surface. Slurry flows down through the annular space between the draft tube and the baffle. Crystals below a certain size, e.g. 0.5 mm, are entrained by the circulation flow through the crystallizer. However, on becoming larger, they enter the elutriation leg and are removed from the crystallizer in a somewhat concentrated slurry. Part of the flow in the loop around the crystallizer passes in the reverse direction through the elutriation leg. The purpose of this flow is to entrain small crystals back into the crystallizer to acquire more growth. We see in Figure 51.6 a second annular space between the baffle and the crystallizer wall. The velocity in the latter annular space is low to allow crystals to settle. However, very small crystals are entrained and these crystals are re-dissolved in the heat exchanger raising the temperature of the flow a few degrees K.

The contents of the vacuum crystallizer shown in Figure 51.6 are well-mixed. However, the crystallizer cannot be indicated as a Continuous Mixed Suspension Mixed Product Removal (CMSMPR) crystallizer. The reason is that there is classified product removal and nuclei dissolution.

Evaporative Crystallizers

See Figure 51.7. In such a crystallizer, supersaturation is created by evaporation of solvent, usually water. For example, salt is crystallized in these crystallizers. As stated in Section 51.2, the solubility of salt varies hardly with temperature in the range 0–100 °C. The crystallizer depicted in Figure 51.7 consists of three parts connected by lines. We see a vapor separator, a pump, and a heat exchanger. The pump maintains a circulation by removing the crystal slurry out of the vapor separator and

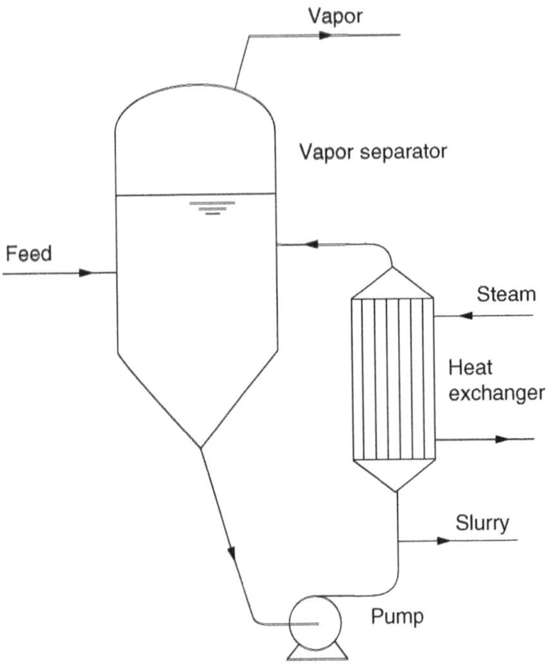

Figure 51.7 Evaporative crystallizer.

passing it, through a heat exchanger, back into the vapor separator. The slurry is heated a few degrees K in the heat exchanger and flashes in the vapor separator. Vapor leaves the vessel at the top. The crystallizer in Figure 51.7 is a continuous mixed suspension mixed product removal (CMSMPR) crystallizer. It is possible to equip these crystallizers, like the vacuum crystallizers, with an elutriation leg leading to classified product removal. Multiple-effect evaporative crystallization plants are built to reduce steam consumption. See Figure 51.8. For example, as a rule-of-thumb, it is possible to evaporate 3 kg of water with 1 kg of steam in a three-effect evaporation plant. The investment increases when the number of effects increases. Table 51.1 contains several figures for the industrial salt production.

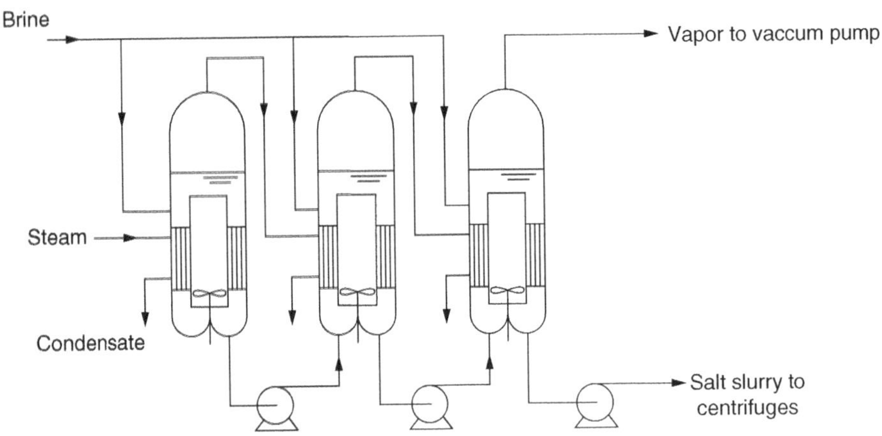

Figure 51.8 Multiple-effect evaporation plant.

Table 51.1 A few figures for the salt production in multiple-effect evaporation plants.

Number of effects	kJ per kg of salt	kg of steam per kg of salt	Steam pressure, barg	Steam condensation temperature, °C
1	7,960			
2				
3	2,510	1.1	0.5	110
4	1,890	0.85	1.5	125
5	1,590	0.7	4.0	150

Reaction Crystallization

In such a process, a chemical reaction is followed by crystallization. The supersaturation is created by a chemical reaction. An example is the production of sodium bicarbonate by the reaction between sodium chloride, ammonia, and carbon dioxide in an aqeous solution in the Solvay process. The solubility of sodium bicarbonate in water is low, and it can be readily crystallized. The next step is calcining the bicarbonate to produce soda.

Salting-Out Crystallization

This type of crystallization uses a third component displacing the solubility of the solute. For example, the addition of ethanol to a saturated brine causes the precipitation of fine salt crystals (2–50 μm).

51.6 The Population Density Balance

The concept of the population density balance helps to understand the behavior of continuous mixed suspension mixed product removal crystallizers (CMSMPR) [2]. The method is applicable to various scales. A particle size distribution of crystals, which is representative for a certain set of operational conditions, is a convenient starting point for the discussion. Plotting the recalculated results of a particle size distribution in a specified way ideally results in a straight line. The nucleation rate and the growth rate of the crystals can be derived from the plot. It is possible to vary the residence time of the flow through the crystallizer, while the other operational conditions, that is, the feed composition, the slurry density, the temperature, and the stirring conditions, are kept constant. For each residence time, a particle size distribution is made and interpreted. Thus, for each residence time, a nucleation rate and a growth rate of the crystals is obtained.

At this point, it is possibly useful to give an example of an industrial crystallization for which it would be possible to make this analysis. See Figure 51.6 depicting a vacuum crystallizer. It would be possible to carry out the mentioned set of measurements for this crystallizer, however, both the elutriation leg and the nuclei dissolution should not be present. The feed composition, the slurry density, the temperature, and the stirring conditions are kept constant. The feed flow through the crystallizer is varied.

A next step could be a variation of the slurry density while keeping the other operational conditions constant. In the case of an industrial vacuum crystallization, this could be done by only varying the feed composition. Finally, it is possible to vary the rotational speed of the pump of an industrial vacuum crystallizer in a single-variable study. It is then important to see to it that the crystallizer contents remain well mixed.

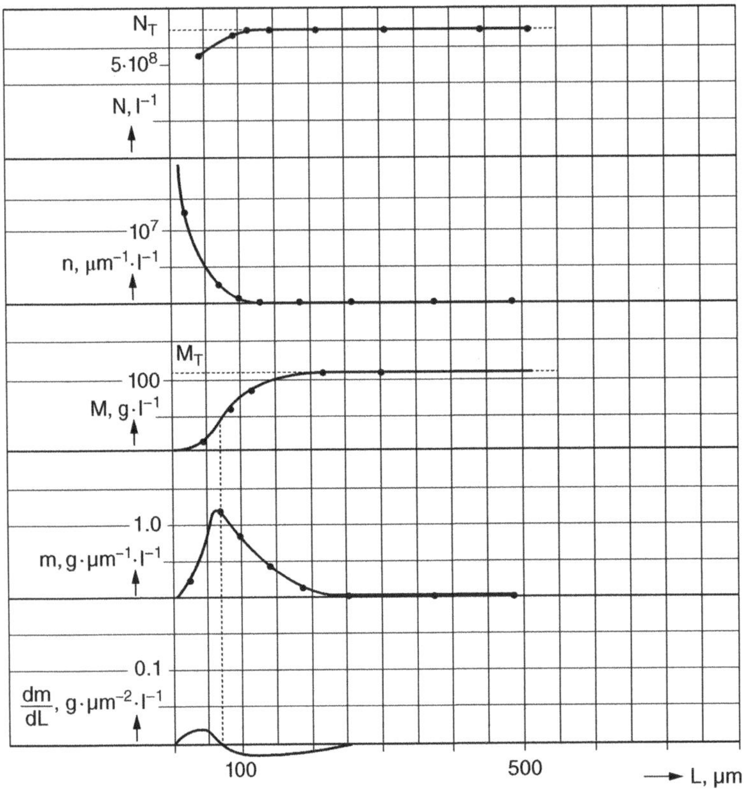

Figure 51.9 Graphical presentation of the figures in the example in Section 51.6.

Not all crystallizers are continuous mixed suspension mixed product removal crystallizers. Still, it is possible to make population density plots. If the plots are not straight lines, explanations to explain this behavior can be looked for. If desired, improvements can be suggested.

Figure 51.9 shows five graphs. The graphs are valid for a crystal size distribution from a CMSMPR-crystallizer [3]. The scale of the crystallization is not specified. The crystal size distribution is reproduced in Table 51.2. The first column from the left contains crystal class boundaries in μm. The next column gives L_{av}, the arithmetically determined average crystal sizes of the classes. The fifth column from the left contains the crystal masses in the classes. The sample mass is 111.85, and 1.30 g of the sample has a size larger than 512 μm. The upper graph in Figure 51.9 contains the cumulative number of crystals as a function of the particle size: N = f(L). This graph was obtained as follows from the fifth column from the left of Table 51.2. The crystals are cubes. The number of particles in a class is obtained by dividing the mass in a class by the mass of a particle in that class. The mass of a particle in a class is the product $\rho_s \cdot k_v \cdot L_{av}^3$ g. k_v is a form factor and is 1 because the particles are cubes. ρ_s, the solid specific mass, is $1.770 \cdot 10^{-12}$ g · μm^{-3}. Thus, $5.70 \cdot 10^8$ particles are smaller than 44 μm and that gives a point of the first graph of Figure 51.9. Further points are obtained from the data in Table 51.2. The slope of the first graph in Figure 51.9 is called the population density n. n is the first derivative of the function N = f (L): n = dN/dL. n indicates how crowded it is at a certain particle size, and hence the name population density. An approximation of n is a certain amount of particles divided by the length of the size range in which the number of particles is present, that is, ΔN/ΔL. n is represented in the seventh column from the left of Table 51.2. It is seen that n decreases strongly with particle size. The explanation of this

Table 51.2 Size distribution from CMSMPR crystallization.

Class boundaries, μm	L_{av}, μm	ΔL, μm	$\rho_s \cdot k_v \cdot L_{av}^3$, g	ΔM, $g \cdot l^{-1}$	ΔN, l^{-1}	n = ΔN/ΔL, $\mu m^{-1} \cdot l^{-1}$	$^{10}\log n$	m = ΔM/ΔL, $g \cdot \mu m^{-1} \cdot l^{-1}$
>512				1.30				
446–512	479	66	$1.95 \cdot 10^{-4}$	0.76	$3.90 \cdot 10^{3}$	59	1.77	0.012
301–446	374	145	$9.26 \cdot 10^{-5}$	1.70	$1.84 \cdot 10^{4}$	127	2.10	0.012
212–301	257	89	$3.00 \cdot 10^{-5}$	3.82	$1.27 \cdot 10^{5}$	$1.43 \cdot 10^{3}$	3.16	0.043
141–212	177	71	$9.82 \cdot 10^{-6}$	9.03	$9.20 \cdot 10^{5}$	$1.30 \cdot 10^{4}$	4.11	0.127
115–141	128	26	$3.71 \cdot 10^{-6}$	12.72	$3.43 \cdot 10^{6}$	$1.32 \cdot 10^{5}$	5.12	0.489
81–115	98	34	$1.67 \cdot 10^{-6}$	27.10	$1.62 \cdot 10^{7}$	$4.77 \cdot 10^{5}$	5.68	0.797
44–81	63	37	$4.42 \cdot 10^{-7}$	44.70	$1.01 \cdot 10^{8}$	$2.73 \cdot 10^{6}$	6.44	1.208
<44	22	44	$1.88 \cdot 10^{-8}$	10.72 +	$5.70 \cdot 10^{8}$	$1.30 \cdot 10^{7}$	7.11	0.244
				111.85				

Source: Modified from Palermo and Larson [4].

phenomenon is as follows. As soon as a particle is born, it starts to grow. As it resides in the crystallizer, it acquires growth. However, this process comes to an end when the particle is removed from the crystallizer. Some particles manage to stay in the crystallizer till they have a size of, for example, 300 μm. But then, on passing the exit stream, they leave. Only a few manage to stay in the crystallizer to grow to a size of, for example, 500 μm. Whether the crystals can stay in the crystallizer or have to leave is governed by the laws of probability because the contents of the crystallizer are ideally mixed. That is why the crystal classes of the larger crystals are less crowded than the crystal classes of the smaller crystals. At this point in time, the mathematical formulation of the phenomenon just discussed is given:

$$n = n_o \cdot e^{-L/(r\tau)}.$$

n is the population density in $\mu m^{-1} \cdot l^{-1}$, n_o is the population density at zero length in $\mu m^{-1} \cdot l^{-1}$, L is the crystal size in μm, r is the growth rate in $\mu m \cdot min^{-1}$, and τ is the residence time in min [1]. The residence time is defined as the quotient of the crystallizer volume in m^3 and the flow through the crystallizer in $m^3 \cdot min^{-1}$. The equation can also be given as:

$$\ln n = \ln n_o - \frac{L}{r\tau} \text{ or:}$$

$$^{10}\log n = {}^{10}\log n_0 - \frac{L}{2.303 \cdot r\tau}. \tag{51.1}$$

The proof of the correctness of Equation 51.1 will be given by drawing a parallel to the behavior of a continuous stirred tank reactor (CSTR). A CSTR is depicted in Figure 51.10. The reactor's volume is V, whereas Φ_V flows through the vessel. At t = 0, an amount of tracer equal to $C_0 \cdot V$ is added. Hence, the initial concentration in the reactor is $C_0 \cdot V/V = C_0$. For example, at t = 0, a small bottle containing red ink is added to the contents of a vessel through which continuously water flows. Initially, the vessel contents are colored red. As time elapses, the red color fades away. The decrease of the concentration C in the vessel as a function of time is depicted in Figure 51.11. A certain C-value is unambiguously coupled to a certain lapse of time. The concentration in the vessel can be expressed as a function of time. The derivation of the formula proceeds as follows.

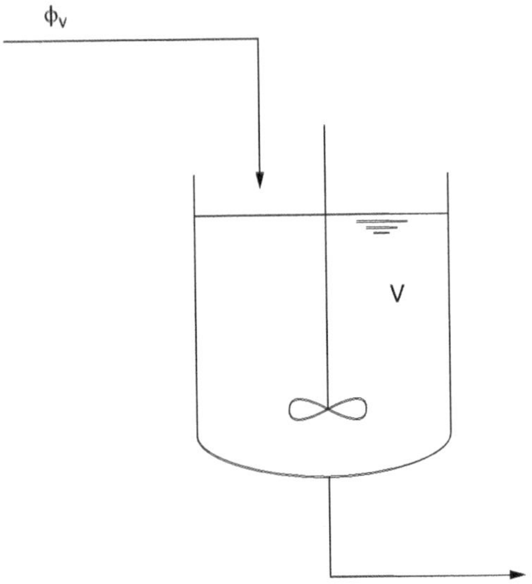

Figure 51.10 A continuous stirred tank reactor (CSTR).

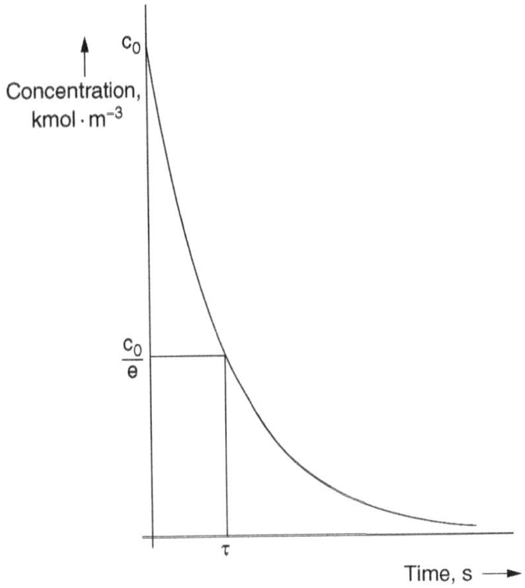

Figure 51.11 The decrease of the concentration in a CSTR on adding a tracer.

The material balance equation is: $-V \cdot dC = \Phi_V \cdot C \cdot dt$. The minus-sign is used because the concentration decreases as time elapses. The integration of this differential equation proceeds as follows.

$$\int_{C=C_0}^{C=C} \frac{dC}{C} = -\int_{t=0}^{t=t} \frac{\Phi_V}{V} \cdot dt,$$

$\frac{V}{\Phi_V}$ is equal to the average residence time τ.

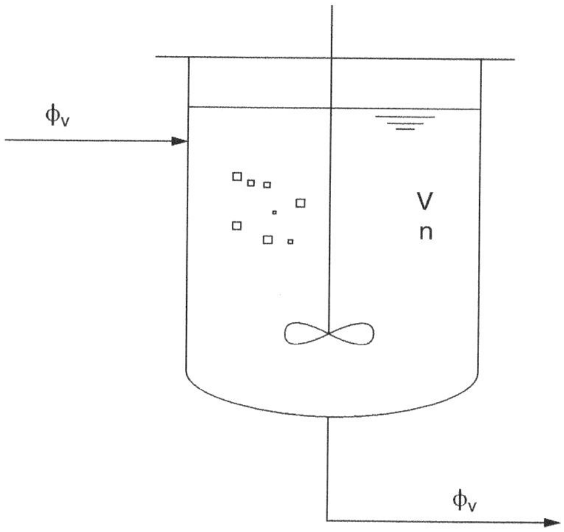

Figure 51.12 A continuous mixed suspension mixed product removal (CMSMPR) crystallizer.

Integration yields $C = C_0 \cdot e^{-t/\tau}$. This equation can also be given as:
$\ln C = \ln C_0 - \frac{t}{\tau}$ or:

$$^{10}\log C = {}^{10}\log C_0 - \frac{t}{2.303 \cdot \tau}. \tag{51.2}$$

Equation 51.2 resembles Equation 51.1. The two equations have the same physical background. In a CSTR, if we have taken the contents of a small bottle of red ink as a tracer, the color of the reactor is unambiguously coupled to a certain time. The red color fades away with time. In a CMSMPR-crystallizer, the size of the crystal is also unambiguously coupled to a certain lapse of time if the growth rate of the crystal is constant. A CMSMPR-crystallizer is depicted in Figure 51.12. The larger the crystal, the longer it has managed to stay inside the crystallizer, and the less colleagues of equal size move along with this crystal inside the crystallizer. Note that $L/r = t$, the time a crystal has resided in the crystallizer. Further conditions for the applicability of Equation 51.1 are:

- the feed does not contain crystals,
- crystals do not agglomerate, and
- crystals do not break.

Example 51.1

The crystal size distribution, which forms the basis of the graphs in Figure 51.9, is given in Table 51.2. $^{10}\log n$ is plotted as a function of L in Figure 51.13. A straight line can be drawn through six points. Two points are not on the straight line. These two points represent $2.1\ g \cdot l^{-1}$, whereas the sample size is $111.85\ g \cdot l^{-1}$. The growth rate can be determined as follows. On progressing from 100 to 200 μm ($\Delta L = 100$ μm), the population density has fallen by a factor of 100. Because $^{10}\log 100 = 2$, the relevant equation is: $\frac{1}{2.303 \cdot r\tau} = \frac{2}{100} \cdot \tau$, the residence time, is 15 min and $r = 1.45\ \mu m \cdot min^{-1}$.

(Continued)

Example 51.1 (Continued)

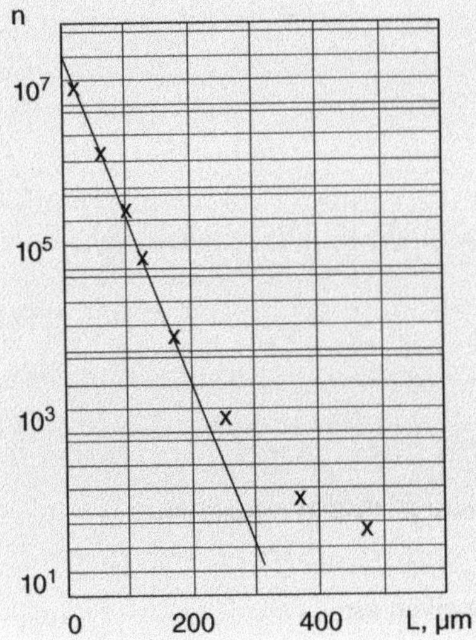

Figure 51.13 $^{10}\log n$ as a function of L.

The rate of growth is in two directions, so that the growth rate of a crystal face is 0.725 μm · min^{-1}. The nucleation rate is calculated as follows. The straight line is extrapolated to the y-axis or $^{10}\log n$-axis. A population density at zero length is obtained from the point where the line intersects the y-axis. It is called n_0 and the unit is μm^{-1} · l^{-1}. A value of $4.8 \cdot 10^7$ is read. The nucleation rate is obtained on multiplying n_0 with r: $4.8 \cdot 10^7 \cdot 1.45 = 6.96 \cdot 10^7$ min^{-1} · l^{-1}.

Both the growth rate and the nucleation rate were obtained by interpreting the population density plot. The values for n_0 and r must also satisfy an equation for the cumulative mass distribution. The reasoning proceeds as follows. The first graph of Figure 51.9 shows the cumulative number of crystals as a function of the crystal size. The second graph contains the derivative of this graph, that is, the population density, as a function of the crystal size. The area between the latter graph and the x-axis, in this case the L-axis, represents the total number of crystals in a liter of slurry:

$$N_T = \int_{L=0}^{L=\infty} n \cdot dL.$$

The expression giving the total mass of crystals per l of slurry is therefore:

$$M_T = k_v \cdot \rho_s \int_{L=0}^{L=\infty} n \cdot L^3 \cdot dL.$$

The integration can be carried out analytically after substituting $n_0 \cdot e^{-L/(r\tau)}$ for n. The result is:

$M_T = 6 \cdot k_v \cdot \rho_s \cdot n_0(r\tau)^4$ g · l^{-1} [4]. The values already found for n_0 and r are used for this expression: $M_T = 6 \cdot 1 \cdot 1.770 \cdot 10^{-12} \cdot 4.8 \cdot 10^7(1.45 \cdot 15)^4 = 114.1$ g · l^{-1}. This value compares well with the actual value of 111.85 g · l^{-1}.

The Dominant Size

The dominant size is the size at which the mass fraction distribution reaches a maximum. See Figure 51.9. The first two graphs were discussed. The third graph shows the cumulative mass of crystals in $g \cdot l^{-1}$ as a function of the crystal size in μm. It can be drawn directly from the data in Table 51.2. The fourth graph shows the mass distribution in $g \cdot \mu m^{-1} \cdot l^{-1}$. The mass distribution is the first derivative of the cumulative mass of crystals in $g \cdot l^{-1}$ as a function of the crystal size in μm. The mass distribution in each crystal class is obtained by dividing the mass in a class in g by the length of the class in μm. The values are in the last column from the left of Table 51.2. It can be read that the mass distribution reaches a maximum when L = 65 μm. Mathematically, the maximum is reached when the derivative of the function of the mass distribution equals zero. The derivative of the latter function is depicted in the fifth graph. The figures were calculated from the fourth graph.

The mass fraction distribution is obtained by dividing the mass distribution by the crystal slurry density in $g \cdot l^{-1}$. The derivation proceeds as follows.

$$dM = k_v \cdot \rho_s \cdot L^3 \cdot dN,$$

$$dM = k_v \cdot \rho_s \cdot L^3 \cdot n \cdot dL,$$

$$\frac{dM}{dL} = k_v \cdot \rho_s \cdot L^3 \cdot n,$$

$$m = k_v \cdot \rho_s \cdot L^3 \cdot n,$$

$$m = k_v \cdot \rho_s \cdot L^3 \cdot n_0 \cdot e^{-L/(r\tau)} \quad g \cdot l^{-1} \cdot \mu m^{-1}.$$

The mass fraction distribution is: $m_n = \dfrac{k_v \cdot \rho_s \cdot L^3 \cdot n_0 \cdot e^{-L/(r\tau)}}{6 \cdot k_v \cdot \rho_s \cdot n_0 (r\tau)^4} = \dfrac{L^3 \cdot e^{-L/(r\tau)}}{6(r\tau)^4} \quad \mu m^{-1}.$

The denominator of the first fraction is the total mass of crystals per l of slurry.

The mass fraction distribution can also be called the normalized mass distribution. Thus:

$$\int_{L=0}^{L=\infty} m \cdot dL = M_T \text{ and } \int_{L=0}^{L=\infty} m_n \cdot dL = 1.$$

The mass fraction distribution reaches a maximum when the first derivative of m and also of m_n is 0. The first derivative of m_n is the second derivative of the normalized cumulative crystal mass M_n as a function of L. The first derivative of the mass fraction distribution is:

$$\frac{1}{6(r\tau)^4}\left(3 \cdot L^2 \cdot e^{-L/(r\tau)} - \frac{L^3}{r\tau} \cdot e^{-L/(r\tau)}\right).$$

This expression is 0 when $3 - \dfrac{L}{r\tau} = 0$. So, $L_D = 3r\tau = 3 \cdot 1.45 \cdot 15 = 65$ μm.

For this calculation, the growth rate derived from Figure 51.13 and the known residence time of 15 min are used. 65 μm can also be read from the fifth graph of Figure 51.9.

51.7 Interpretation of the Results of Population Density Balances

In Section 51.6, a population density balance was made for a given crystal size distribution. It appeared that the crystallizer behavior followed the theory approximately. Six points could be connected by a straight line whereas two points were not on the line. However, these two points only

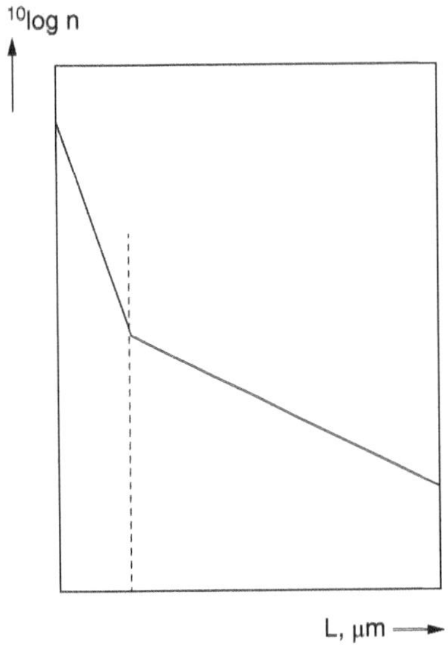

Figure 51.14 Population density plot in case of fines removal.

represented 2.1 $g \cdot l^{-1}$. The sample size was 111.85 $g \cdot l^{-1}$. The interpretation of the two points is that the growth of relatively large crystals is favored with respect to smaller crystals.

Figure 51.14 gives a population density plot for a CMSMPR-crystallizer with fines removal. The life of small crystals is threatened much more than the life of large crystals. Fines removal is practised to obtain relatively large crystals. The net effect of this technique is to force the growth rate to a higher level, producing the same mass on larger average size crystals having less total surface area. Conversely, by plotting $^{10}\log n$ versus L it may become apparent that fines removal is occurring.

Figure 51.15 gives a population density plot for a CMSMPR-crystallizer with classified product removal. The life of large crystals is threatened much more than the life of small crystals. Classified

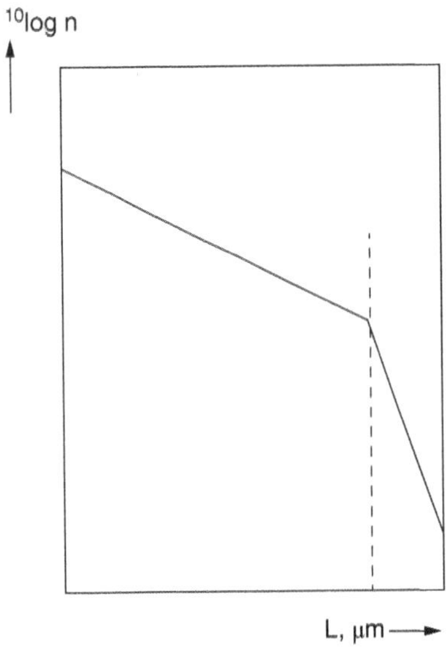

Figure 51.15 Population density plot in case of classified product removal.

product removal is practised to extract particles from a crystallizer when they have reached a certain size. By plotting $^{10}\log n$ as a function of L it may become apparent that there is product classification or abrasion of relatively large crystals.

If, for a certain crystallizer, a series of straight lines is obtained on plotting $^{10}\log n$ as a function of L by varying τ, the residence time, it is possible to go one step further. Use can be made op power law models [5]. The nucleation rate and the growth rate can be expressed as follows:

$n_0 \cdot r = k_1(\Delta C)^b$, and

$$r = k_2(\Delta C)^g.$$

The two variables are written as power law functions of ΔC, the supersaturation. The two expressions can be combined to:

$$n_0 \cdot r = k_3 \cdot r^i.$$

This result can be obtained by writing the equation for the growth rate as $\Delta C = (r/k_2)^{1/g}$ and substituting this outcome into the equation for the nucleation rate. The exponent i equals the quotient of the exponents b and g. It follows:

$$n_0 = k_3 \cdot r^{i-1}. \tag{51.3}$$

Thus, a plot of $^{10}\log n_0$ versus $^{10}\log r$ should give a straight line of slope $(i - 1)$. See Figure 51.16. Furthermore, if the order of the growth process is known, the order of nucleation b can be calculated.

For systems that are believed to follow secondary nucleation mechanisms, the power-law function can be extended to include the slurry density M_T:

$$n_0 = k_4 \cdot r^{i-1} \cdot M_T^j. \tag{51.4}$$

Finally, the power-law relationship can include the agitation and flow:

$$n_0 = k_5 \cdot r^{i-1} \cdot M_T^j \cdot \varepsilon^k. \tag{51.5}$$

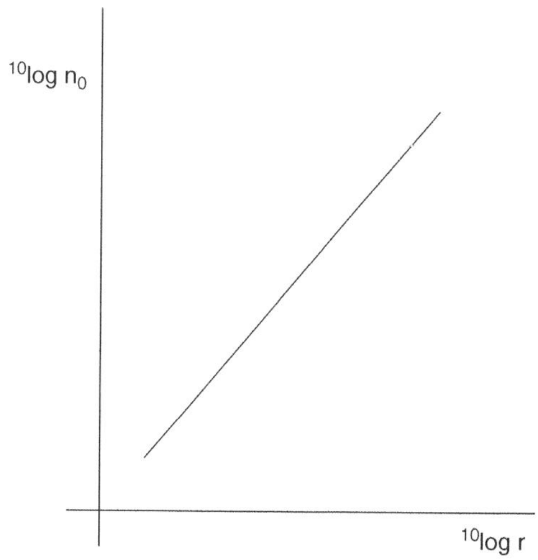

Figure 51.16 $^{10}\log n_0$ as a function of $^{10}\log r$.

References

1. Sung, C. Y., Estrin, J., and Youngquist, G. R. (1973). Secondary nucleation of magnesium sulfate by fluid shear. *American Institute of Chemical Engineering Journal*, 19, 957–962.
2. Randolph, A. D., and Larson, M. A. (1988). *Theory of Particulate Processes*. New York, NY: Academic Press.
3. Palermo, J. A., and Larson, M. A. (1969). *Crystallization from Solutions and Melts* (pp. 1–13). New York, NY: American Institute of Chemical Engineers.
4. Chisholm, J. S. R., and Morris, R. M. (1986). *Mathematical Methods in Physics* (p. 142). Amsterdam, The Netherlands: North-Holland Publishing Company.
5. Kirk-Othmer (2001). *Processes & Process Engineering/Crystallization. Kirk-Othmer Encyclopedia of Chemical Technology*, Wiley Online Library.

52

Liquid/Solid Separation

52.1 Introduction

Liquid/solid separation can be accomplished by means of settlers, hydrocyclones, filters, and centrifuges. Settlers exploit the specific mass difference between the solid and the liquid for the separation of particles from a liquid by gravity. Hydrocyclones can concentrate suspensions by centrifugal settling. Filtration encompasses the separation of solid particles and liquid by means of a filter medium, which retains the solids, whereas the liquid passes through the medium. Flow through the medium is achieved by a static pressure difference. Filtration is treated further in Section 52.2. Liquid/solid separation by means of centrifuges is, like in hydrocyclones, accomplished by means of centrifugal settling. Whereas a hydrocyclone is stationary, centrifuges contain rotating parts. Two subcategories can be distinguished within the category of centrifuges. The first subcategory contains a rotating and a stationary part, whereas there is no screen between the two parts. The separation is accomplished by centrifugal settling. Second, we distinguish centrifugal filters. These centrifuges contain a rotating screen retaining the solids, whereas the liquid passes through the screen. Centrifuges are treated in Section 52.3.

52.2 Filtration

52.2.1 Introduction

Two broad categories can be distinguished in the field of filtration. They will be discussed separately. The two categories are cake filtration and depth filtration. Cake filtration encompasses the passage of a liquid through a porous filter medium, whereas solid particles traveling with the liquid to the filter are retained. The filter medium is relatively thin. The thickness of the cake increases as more liquid is passed through the filter. The pressure exerted on the feed side of the filter must increase if the flow through the filter is kept constant. A characteristic of cake filtration is that the cake formed stays porous to permit the continuous flow through the filter. Cake filtration is discussed in Section 52.2.2. In some instances, cake filtration can be made possible or improved by using filter aids. Filter aids are described in Section 52.2.3.

The second type of filtration process is depth filtration. The suspension to be processed flows through a bed of porous material. The solid particles are trapped within relatively coarse interstices of the bed, whereas relatively clear liquid leaves the bed. Depth filtration is practiced in, for example, sand beds and cartridges. Particles accumulated in a sand bed can be removed by a

Introduction to Chemical Engineering, First Edition. C.M. van 't Land.

cleaning operation. A cartridge filter is usually replaced when the filter material of the cartridge is filled with solid particles. Deep bed filtration is a type of depth filtration and is discussed in Section 52.2.4.

52.2.2 Cake Filtration

Two different flow equations will be derived. The first equation concerns the case of flow through the filter at which the liquid flow through the filter is kept constant. The second equation regards the case of liquid flow through the filter at which the pressure at the feed side of the filter is kept constant. The modified Darcy equation is a starting point for both equations [1]. There are two conditions for the applicability of the latter equation. The first condition is that, as the particles are deposited on the filter cake, they remain in place. The second condition is that the filter cake is incompressible. The modified Darcy equation is:

$$\frac{dV}{A \cdot dt} = \frac{K \cdot \Delta p}{\mu L}. \tag{52.1}$$

The meaning of the symbols in this equation is as follows:

V: liquid volume in m^3,
K: filter cake permeability in m^2,
Δp: pressure drop across the cake in $N \cdot m^{-2}$,
A: filtration area in m^2,
t: time in s,
μ: liquid dynamic viscosity in $N \cdot s \cdot m^{-2}$, and
L: filter cake thickness in m.

The left-hand side of the equation represents the linear velocity of the liquid through the cake. The essence of the equation is that this linear velocity is directly proportional to the pressure and inversely proportional to the dynamic viscosity and the bed height. Darcy formulated his equation after doing experiments. At this point, reference is made to the Hagen–Poiseuille equation discussed in Chapter 4. Equation 4.3 in that chapter reads:

$$\Phi_V = \frac{\pi \cdot R^4}{8\mu}\left(-\frac{dp}{dx}\right).$$

Φ_V corresponds with dV/dt. The modified Darcy equation for a flow channel having a radius R can be written as follows:

$$\frac{dV}{dt} = \frac{AK \cdot \Delta p}{\mu L} = \frac{\pi \cdot R^2 \cdot K \cdot \Delta p}{\mu L}.$$

When $-dp/dx$ can be replaced by $\Delta p/L$, the two equations are identical if $K = R^2/8$. The flow through filter cakes is laminar and the filter cake permeabilty is directly proportional to R^2. This means that the resistance of the filter cake increases when the particle size decreases as the size of a flow channel is related to the particle size.

The cake thickness in Equation 52.1 will be written as a function of the liquid volume passed through the filter. This helps for the next step, the integration of the latter equation. The equation is:

$$L = \frac{VC}{\rho_c \cdot A} \quad \text{m}.$$

C: solids concentration in the suspension in $kg \cdot m^{-3}$, and

ρ_c: bulk density dry filter cake in $kg \cdot m^{-3}$. The implicit assumption is that, on moistening the dry filter cake, the height is not affected.

Equation 52.1 can be given as follows:

$$V \cdot dV = K \cdot A^2 \cdot \rho_c \cdot \frac{\Delta p \cdot dt}{\mu C}. \tag{52.2}$$

The latter equation can be integrated for the case of constant-rate filtration or for the case of constant-pressure filtration.

Constant-Rate Filtration

For this case, the quotient dV/dt becomes V/t and Equation 52.2 can be stated as:

$$\Delta p = \frac{\mu C \cdot V^2}{tK \cdot A^2 \cdot \rho_c} \quad N \cdot m^{-2}.$$

The linear velocity of the liquid through the filter is:

$$v = \frac{V}{At} \quad m \cdot s^{-1}.$$

On substituting the latter expression into the equation for the pressure loss, we obtain:

$$\Delta p = \frac{\mu C \cdot v^2 \cdot t}{K \cdot \rho_c} \quad N \cdot m^{-2}. \tag{52.3}$$

This equation states that the pressure increases linearly with time if the filtration rate is kept constant. That makes sense as the cake thickness increases linearly with time. Thus, a plot of Δp versus t should result in a straight line. However, in practice, these plots tend to have some curvature. Probable reasons are that some particle migration occurs or that the cake exhibits some compressibility. Reference is made to Section 44.5 and Figure 44.4 of Part 7, Membranes. It concerns the filtration of dilute biomass suspensions. The figure shows that, at filtration rates of both 30 and $40\ l \cdot m^{-2} \cdot h^{-1}$, the pressure increases linearly as a function of time. The permeability constant K is not the same for these two filtrations. A constant K-value would imply that the pressure after, for example, 80 h would be, for $40\ l \cdot m^{-2} \cdot h^{-1}$, $(40/30)^2 = 1.78$ times greater than the pressure for $30\ l \cdot m^{-2} \cdot h^{-1}$. The factor is actually approximately 10. A possible explanation is that the particles form different packed beds at different velocities. Still, it is possible to calculate a K-value for each of the two filtrations. The different K-values are possibly caused by the fineness and softness of the particles concerned. These measurements are not typical for many other filtrations.

Equation 52.3 predicts that the pressure varies with the quadrate of the linear velocity of the filtrate. That aspect points to a sensitivity of this type of filtration. If the feed flow increases with, for example, 10%, the pressure increases with 20%. Pressure increases can affect the cake permeability. It is hence good practice to take a margin at the design of the filtration area to keep the filtration velocity relatively low.

The pressure drop caused by equipment piping and filter medium has not yet been discussed. It will add to the filter cake pressure drop. Compared to the resistance of the filter cake, the resistances of the equipment and filter medium are usually small.

Constant-Pressure Filtration

The integration of Equation 52.2 proceeds as follows.

$$\int_{V=0}^{V=V} V \cdot dV = \frac{K \cdot A^2 \cdot \rho_c \cdot \Delta p}{\mu C} \cdot \int_{t=0}^{t=t} dt,$$

$$V^2 = \frac{2K \cdot A^2 \cdot \rho_c \cdot \Delta p \cdot t}{\mu C}. \tag{52.4}$$

An essence of this equation is that we have to multiply the time by four if we want to filter the double volume. A plot of the logarithm of V versus the logarithm of t should result in a straight line having a slope of ½ and an intercept of $(1/2)^{10}\log\,(2K \cdot A^2 \cdot \rho_c \cdot \Delta p/(\mu C))$ with the line t = 1 min.

See Figure 52.1. The drawn line could be measured for a typical constant-pressure filtration. As time elapses and the filter cake builds up, its resistance increases and the resistance of the filter medium, lines, and equipment becomes negligible. The plot of the logarithms of volume of filtrate and time becomes a straight line. This straight line can be extrapolated backward to intersect with the line t = 1 min. The numerical value read on this line can be used to calculate K, the permeability of the filter cake. The calculated K-value can be used for design if it is neither a function of feed pressure nor a function of linear filtrate velocity.

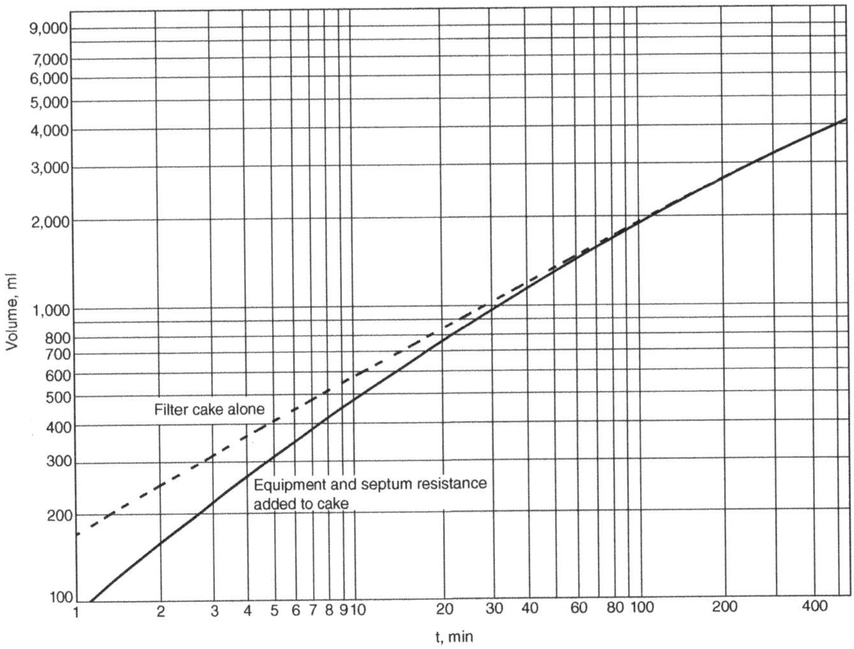

Figure 52.1 Logarithmic plot of collected filtrate as a function of time for constant-pressure filtration in which the effect of equipment and medium resistance is shown. *Source:* Courtesy of McGraw-Hill Book Company, Columbus, Ohio, USA.

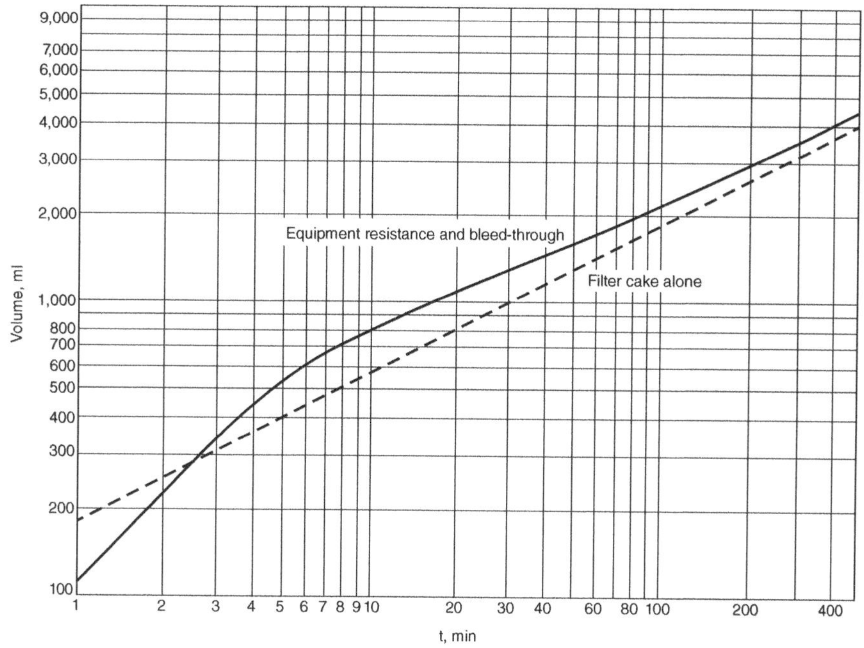

Figure 52.2 Logarithmic plot of collected filtrate as a function of time for constant-pressure filtration in which the effects of equipment resistance and initial bleed-through for constant-pressure filtration are shown. *Source:* Courtesy of McGraw-Hill Book Company, Columbus, Ohio, USA.

In some cases, plotting the logarithms of collected filtrate volume and time results in a line drawn in Figure 52.2. In first instance, too small an amount of filtrate is collected because the resistance of the filter medium, the equipment, and the lines also play a role. More filtrate than theoretically can be expected is collected in the next phase. The cause is bleed of solid material through the filter medium. As the cake thickness increases, small particles are trapped in the filter cake or the filter medium. Thus, the drawn line approaches the theoretical line asymptotically. Again, the theoretical line can be used to determine a K-value.

52.2.3 Filter Aids

Filter aids are fine particles and can be used as body feed or as precoat. Body feed is added to the suspension to be separated. The solids to be separated are retained by the filter medium together with the body feed. The body feed provides strength to the filter cake. It makes the filter cake incompressible or less compressible. In other words, it increases the permeability of the filter cake. Precoats are used for a different purpose. Sometimes, cases are met, in which the direct contact between a filter medium and the filter cake leads to plugging of the filter medium. In such cases, before passing the feed on to the filter, a suspension containing a filter aid is pumped to the filter. The filter cloth retains the filter aid, whereas the suspending liquid passes through the filter cloth. The filter medium is not plugged by the filter aid. The next phase is the pumping of the feed to the filter. The combination of the filter cake retained by the precoat, the precoat, and the filter medium remain permeable. An example is the separation of fine gypsum particles from an aqueous citric acid solution by means of a rotary vacuum filter.

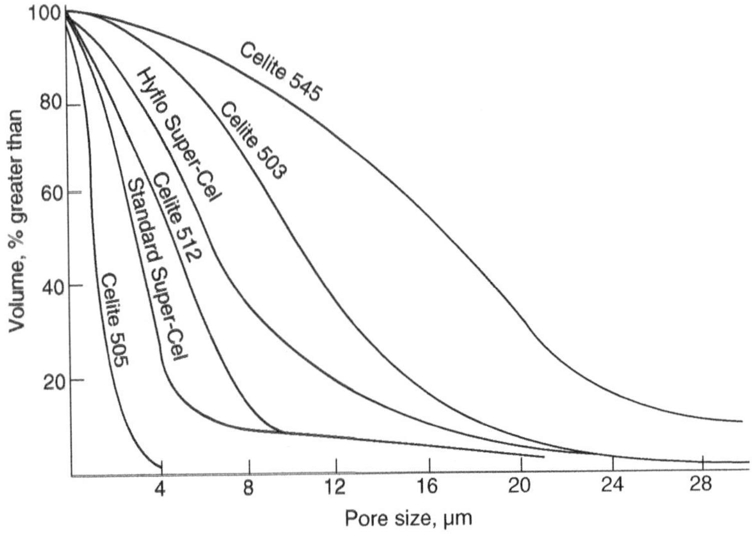

Figure 52.3 Pore size distributions of filter-aid grades (Johns-Manville). Celite, Super-Cel, and Hyflo Super-Cel are trademarks of Johns-Manville for diatomite filter aids. *Source:* Courtesy of McGraw-Hill Book Company, Columbus, Ohio, USA.

The particles of filter aids are rather rigid. Diatomite filter aids are well known. Figure 52.3 contains a graph containing data regarding pore sizes of various filter aids. In a specific case, the selection of a filter aid should be carried out based on experiments. Usually, the optimum filter-aid grade is the one with a median pore size closely matching the median particle size of the turbidity to be removed. The optimum filter-aid dosages must also be established experimentally.

52.2.4 Deep-Bed Filtration

This type of filtration is typically used for the removal of particles smaller than, for example, 1 μm from a liquid. The concentration of the particles in $kg \cdot m^{-3}$ is low. Deep-bed filtration is used for drinking water filtration. A further typical application is the polishing of effluents before their discharge. Commercial filters consist of a vertical closed vessel with a granular bed through which the suspension to be processed passes upward or downward. A packed bed consists of, for example, three layers. First, the liquid flows through a layer of anthracite. The second layer is a sand layer. The anthracite particles are coarser than the sand particles. The third layer consists of gravel. The size of the flow channels in the bed exceeds the size of the particles. The particles are trapped in the bed by, for instance, mechanical interception. The beds are cleaned intermittently by passing water through in a reverse direction.

52.2.5 Filtration Equipment

See Figure 52.4. The left column contains continuous filters. The upper filter of the right column is also a continuous filter. The remaining filters are filters operating batchwise or intermittently continuous. First, the continuous filters are discussed.

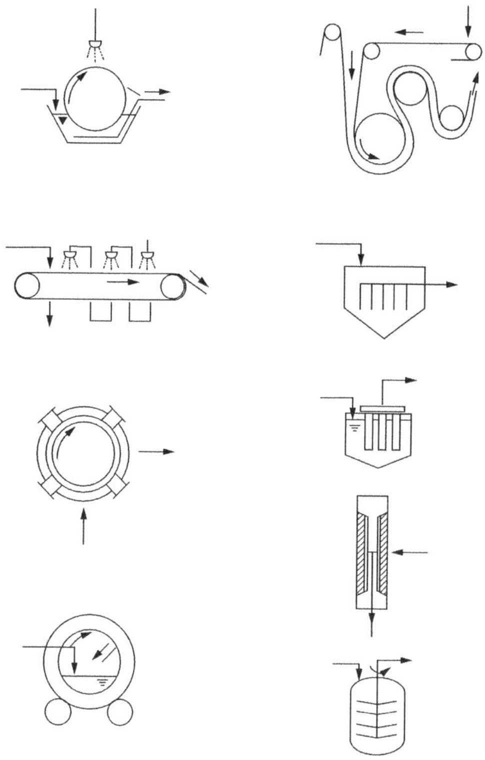

Figure 52.4 Important industrial filters.

Continuous Filtration

Rotary Vacuum Filter

The slurry is pumped into a trough. The contents of the trough are agitated. The drum rotates and as soon as a part of the drum enters the trough, vacuum is created in that part by a vacuum pump, usually a water ring pump. Filtrate passes through the filter cloth, and the filter medium retains particles. As the drum rotates, the filter cake builds up. It is possible to leach the filter cake after the cake has left the trough. The filter cake is removed from the drum by a doctor knife. This filter is a workhorse of industry and sometimes indicated by the name Rovac for rotary vacuum filter. A different mode of operation is as top-feed filter. That mode is selected when the particles are coarse and would settle in a trough. It is possible to apply a precoat before continuous filtration is started. The doctor knife then does not remove the precoat. Scaling-up is based on experiments on a pilot filter. The basis for scaling-up is the amount of product processed in $kg \cdot m^{-2} \cdot h^{-1}$. It is possible to use the theory for constant-pressure filtration in Section 52.2.2 for a calculation of the filtration area if the cake permeability is known. See Equation 52.4. In this equation, t is the time a specific part of the filter cloth resides in the trough. The volume of filtrate obtained in that time can be calculated. A typical application is the separation of fluid cracking catalysts from mother liquor. Fluid cracking catalysts are products used in refineries to convert diesel oil into petrol. The weight average particle size of that material is approximately 80 μm. The particles are rigid. A further typical application is the separation of calcium citrate from an aqueous solution called vinasses at the citric acid production by fermentation. The weight average particle size is about 30 μm, and the particles are relatively rigid.

Belt Filter

The slurry is pumped to the belt of the horizontal filter. The belt is the filter medium. A vacuum is maintained under the filter cloth by a water ring pump. Filtrate passes through the filter medium, which retains the solid material. The filter cake can be leached countercurrently in, for example, three steps. The leached filter cake is removed from the cloth by a doctor knife. The filter medium is cleaned while it travels back to the feed point. This filter also finds many applications. The design can be based on $kg \cdot m^{-2} \cdot h^{-1}$. It is, investmentwise, more expensive than the rotary vacuum filter. Filter cakes can be leached better than on a Rovac. The footprint of a belt filter is larger than the footprint of a rotary vacuum filter. The separation of starch from water is a typical application.

Pressure Filter

A pressure filter can offer advantages when the cake permeability is low. Furthermore, this filter can be applied when the filter should be closed. That is the case when toxic or volatile materials are processed. The four compartments drawn can be used for filtration, leaching, liquid/solid separation, and cake removal. A typical pressure is 3 bar. The design of a filter is based on the results of tests carried out with a pilot filter. The separation of isocyanates from their mother liquor is a typical application. Isocyanates are chemicals produced for the manufacture of polyurethanes, a class of polymers.

Internal Rotary Vacuum Filter

Applied for the separation of metal powders from water.

Belt Filter Press

This continuous filter is used to process slurries containing very fine particles. The feed is pumped to the filter and passed through the filter between two filter media. Gradually, the distance between the two filter media is made smaller, and this causes water to pass through the filter media. The results of small-scale tests can be scaled-up based on $kg \cdot m^{-2} \cdot h^{-1}$. A typical application is the dewatering of a product called EC-Black. EC-Black consists of submicron carbon particles. It is obtained as a by-product at the manufacture of synthesis gas from which it is removed by a water wash. The suspension pumped to the filter contains 1.25% by weight of solids and has the consistency of a paste. The filter cake leaving the belt filter press contains 15% by weight of solids. Dried EC-Black can be sold to the cable industry. It is used as a primary sheathing around copper cables.

Batch Filtration

Leaf Filter

The filter elements are present in a closed vessel. The usually dilute to very dilute suspension is pumped into the vessel. The liquid passes through the cloth, and the solid particles are retained. The pressure differential to maintain a constant filtration rate is measured. The filter is cleaned when the pressure reaches a certain value. The filter was applied for the separation of iron and iron oxide particles from wash water at the manufacture of steel cord. Steel cord is used for the manufacture of certain types of car tyres.

Cartridge Filter

The filter resembles the leaf filter. It operates at atmospheric pressure, and the filter elements are present as cartridges. The separation of magnesium sulfate and an organic peroxide is an application. The magnesium sulfate is used to dry the organic peroxide.

Plate-and-Frame Filter Press

This filter contains many elements in parallel. It is possible to apply high pressures, for example, up to 100 bar. The filter is used to separate very fine particles. A suspension is pumped to the filter, which retains solid material. The liquid passes through the filter. The solid material is removed intermittently from the equipment. The separation of silica and an aqueous solution of sodium sulfate is a typical example.

Funda Filter

This filter can be used as a polishing facility. The suspension is pumped to the filter. The filtrate passes through the filter cloth on the plates and the cake builds up on the plates. Filtration is stopped at the end of a filtration cycle. Cake removal then occurs by rotation of the plates. The cake slides along the wall of the filter to the vessel bottom. A Funda filter played a role in the production of citric acid by fermentation. The reaction between calcium citrate and sulfuric acid is a step in the process. The result is an aqueous solution of citric acid and gypsum particles. The gypsum particles are separated from the solution on a Rovac. However, the solution still contains fine gypsum particles. Those particles are separated from the solution by means of a Funda filter.

52.3 Centrifugation

Table 52.1 shows 10 different types of centrifuges, which are widely used for liquid/solid separation in the process industry. Centrifugal settling is accomplished in the first four machines. The other pieces of equipment are centrifugal filters. Both the particle size and the solid concentration in the feed increase from Machine No. 1 to Machine No. 10. On the other hand, the radial acceleration decreases in that order. The radial acceleration of the decanter lies between 1,000 and 3,000 · g; the radial acceleration of the pusher centrifuge is in the range 200 and 2,000 · g. Machine Nos. 1, 5, and 6 operate batchwise, whereas the other machines function continuously.

Table 52.1 Ten important centrifuges for liquid/solid separation.

No.	Diagram	Name	% by wt of solids	Washing possible	Typical applications
1	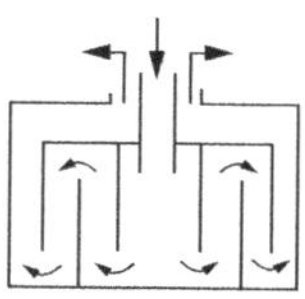	Multiple cylinder clarifier	<0.1	No	Iron oxide particles from water. Clarification of wine.
2	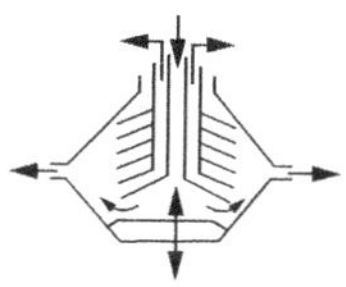	Self-ejecting disc centrifuge	0.1–5	No	Gypsum particles from an unsaturated citric acid solution.
3	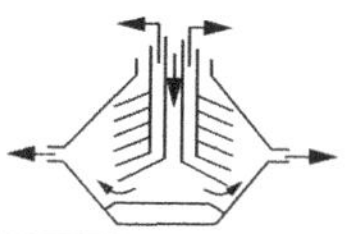	Nozzle centrifuge	0.5–20	Yes	Thickening an effluent containing solid particles (1–100 μm).

(*Continued*)

Table 52.1 (Continued)

No.	Diagram	Name	% by wt of solids	Washing possible	Typical applications
4	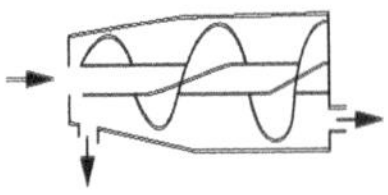	Decanter	2.5–50	Some	Flocculated cyanides from an unsaturated sugar solution. Clarification of effluents after flocculating.
5	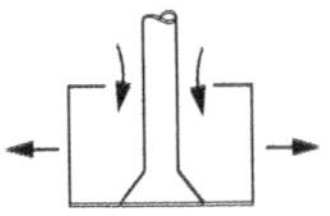	Pendulum centrifuge	5–50	Yes	First quality sugar crystals from mother liquor.
6	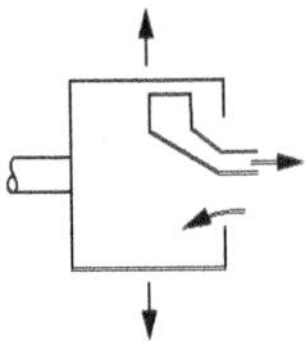	Peeler centrifuge	5–50	Yes	p-Toluene sulfonamide crystals from an aqueous solution.
7	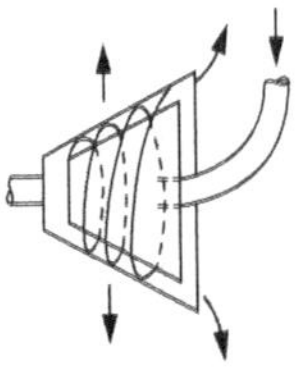	Scroll screen centrifuge	20–75	Yes	Salt (NaCl) crystals from mother liquor (1–2 cP). Citric acid crystals from mother liquor (abt. 40 cP). Capacities up to 50 t h^{-1}.
8	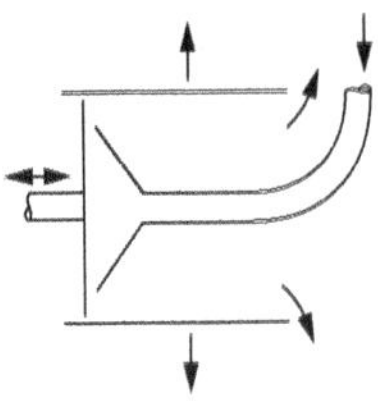	Pusher centrifuge	20–75	Yes	Salt (NaCl) crystals from mother liquor (1–2 cP). Ammonium sulfate crystals from mother liquor. Capacities up to 100 t h^{-1}.
9	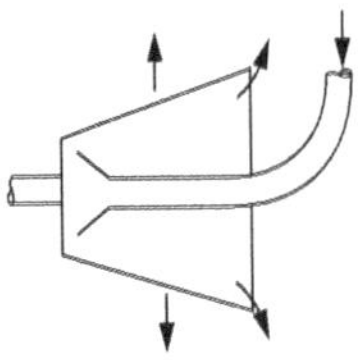	Conical screen centrifuge	20–75	Yes	Second quality sugar crystals from mother liquor. Capacities up to 20 t h^{-1}.
10	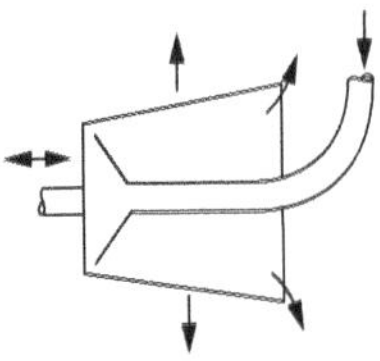	Vibrating screen centrifuge	20–75	Yes	Coal particles (maximum size about 10 mm) from water, max. cap. 300 t h^{-1}.

A typical application of the pusher centrifuge is the separation of vacuum pan salt from mother liquor in salt plants. It is possible to produce 40 $t \cdot h^{-1}$ of salt having a moisture content of 2.5% by weight with a pusher centrifuge. A typical average particle size of this type of salt is 450 μm. The salt is leached in the centrifuge with water to remove sulfate and calcium.

There is a restriction to build large centrifuges. The reason is that, to control the rotor stress, the circumferential velocity must be kept constant on scaling-up. However, on keeping this variable

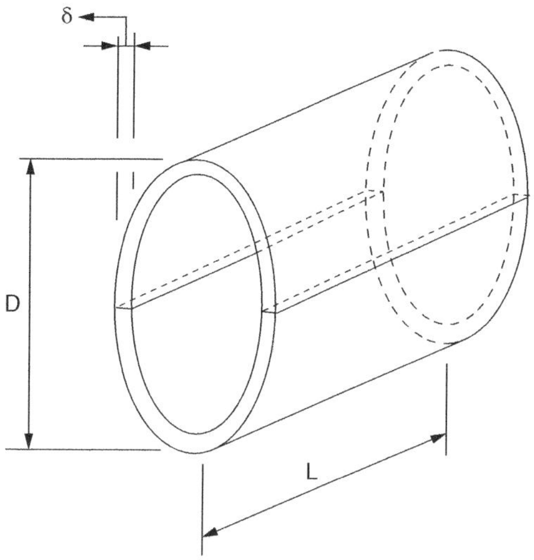

Figure 52.5 Centrifuge bowl.

constant, the radial acceleration decreases on scaling-up. That means that the force to separate the liquid from the solid decreases on scaling-up. See Figure 52.5. There is a tendency to pull, so to speak, the upper half of the bowl from the lower half. The radial acceleration of the rotating bowl is:

$$\frac{v^2}{R} = \frac{\pi^2 \cdot n^2 \cdot D^2 \cdot 2}{D} = 2 \cdot \pi^2 \cdot n^2 \cdot D.$$

The force with which the upper half of the bowl is pulled from the lower half of the bowl is:

$$F_1 = ma = \pi DL\delta\rho 2 \cdot \pi^2 \cdot n^2 \cdot D = 2 \cdot \pi^3 \cdot n^2 \cdot D^2 \cdot L\delta\rho \quad N.$$

In the latter expression, m is the bowl mass in kg and a is the rotor radial acceleration in $m \cdot s^{-2}$. ρ is the specific mass of the bowl material in $kg \cdot m^{-3}$. Note that the expression for the bowl mass is an approximation. The rotor is kept intact by the rotor stress σ in $N \cdot m^{-2}$. The force exerted by the rotor stress is:

$$F_2 = 2L\delta\sigma \quad N.$$

$$F_1 = F_2.$$

$$2 \cdot \pi^3 \cdot n^2 \cdot D^2 \cdot L\delta\rho = 2L\delta\sigma.$$

Writing the latter expression explicit in σ results in:

$$\sigma = \frac{2 \cdot \pi^3 \cdot n^2 \cdot D^2 \cdot L\delta\rho}{2L\delta} = \pi^3 \cdot n^2 \cdot D^2 \cdot \rho \quad N \cdot m^{-2}.$$

The result is that the rotor stress is proportional to $n^2 \cdot D^2$. On scaling-up, it is necessary to keep $n^2 \cdot D^2$ and thus $n \cdot D$ constant. Because the circumferential velocity equals $\pi \cdot n \cdot D$ in $m \cdot s^{-1}$, this implies that the circumferential velocity must be kept constant on scaling-up. The consequence is that the radial acceleration decreases on scaling-up as, for a constant v^2, it is inversely proportional to the bowl radius and thus the bowl diameter. This is the reason that centrifuges with bowl diameters larger than, for example, 2 m are not built. The possibility to dewater properly decreases as the machine diameter increases. Increasing, on scaling-up, the bowl thickness is not an avenue to take as this causes the bowl mass to increase. A consequence is that the production of a vacuum pan salt plant is processed by three or four pusher centrifuges in parallel.

Reference

1. Schweitzer, P. A. (1996). *Handbook of Separation Techniques for Chemical Engineers* (pp. 4-3–4-8). New York, NY: McGraw-Hill.

53

Convective Drying

53.1 Introduction

Drying is defined as the evaporation of a volatile liquid from solids, suspensions, or solutions by heat supply. In many instances, mechanical liquid/solid separation is installed upstream the thermal process. The reason is that such separation is, both investment-wise and concerning variable costs, generally cheaper than thermal processing. The three mechanisms to transfer heat to the stream processed are convection, conduction, and radiation. Convective drying comprises heat transfer by a hot or warm gas, air in many instances, in direct contact with the material dried. The gas cools down, and heat is transferred to the material to be dried where it is utilized as heat for the evaporation of the liquid and the warming of the solid. At conduction drying, also called contact drying, the material to be dried receives the heat for the evaporation of the liquid and the warming of the solid through a metal wall. It is discussed in Chapter 55. Infrared drying and microwave drying fall in the category of radiation drying. Radiation drying is not treated in this book. Concerning drying equipment, there is a great variety. The reasons are:

- the way the liquid is attached or bound to the solid,
- the necessity to transport the material through the dryer in continuous dryers, and
- the necessity to dry safely and preserve the quality of the material to be dried.

The emphasis will be on continuous drying. Water is evaporated in many instances. However, the evaporation of organic liquids, for example, methanol, also occurs.

On drying, free and bound moisture are distinguished. Free moisture can exert the saturated vapor pressure of the liquid, bound moisture cannot. Causes for moisture to be bound are, for instance, its presence as hydrated water in inorganic compounds, or its presence in capillaries.

Air and product can move concurrently, countercurrently, or in both modes through a dryer. The latter mode is also indicated as mixed flow and comprises that in a part of the dryer the mode is concurrent flow and in a different part countercurrent flow. In a fluid bed dryer, cross flow is distinguished.

Four important convective dryers in the chemical industry are discussed in Section 53.2.

Convective drying implies the intensive contact between material to be dried and a drying gas, usually air. The contact occurs in various types of dryers. However, the basic process remains the same. Because it is important to understand the basic process, the Sections 53.3 through 53.5 concern this process. The adiabatic saturation temperature for the system air/water is treated in Sections 53.3 and 53.4, whereas the wet-bulb temperature is discussed in Section 53.5. The Mollier diagram, also known as the psychromatic chart, is dealt with in Section 53.6. Section 53.7 treats the drying of vacuum pan salt in a fluid-bed dryer.

Introduction to Chemical Engineering, First Edition. C.M. van 't Land.

53.2 Four Important Continuous Convective Dryers in the Chemical Industry

Flash Dryer

See Figure 53.1. First, the product to be dried is, in some instances, mixed with product already dried to improve the handling characteristics of the feed. Possibly, 1 kg feed is mixed with 3 kg of dried product. The feed is entrained by a hot or warm air stream in a vertical line. The air velocity is, for instance, $20\ m \cdot s^{-1}$. The residence time of both product and gas is about 1 s. The dry product is separated from the gas stream by means of a cyclone or a filter. Basically, only free moisture can be evaporated because the residence time is short. Cooling the product after drying is not possible. Such cooling is often required to prevent incrustations in a silo. On storing warm, dried product, the air surrounding the particles comes into contact with cold silo walls. That may lead to condensation. This dryer type is a workhorse of industry. The drying of sodium sulfate is a typical example. Two fans appear in Figure 53.1. That setup enables the adjustment of a desired pressure at the feed point. Generally, a slight underpressure is preferred to avoid product being blown out of the dryer.

Spray Dryer

See Figure 53.2. The feed is a suspension or a solution containing, for example, 50% by weight of solids or dissolved material. The feed and a stream of hot or warm air are mixed in the spray dryer chamber. The residence time of both product and gas is about 20 s. The dried product is separated from the gas by means of a filter. Free and some bound moisture can be evaporated because the residence time is short. Spray drying is applied when mechanical liquid/solid separation is difficult or impossible and when the processing step confers good product properties. It is also applied as a shaping step. It is an expensive drying step as the production of 1 kg of product entails the

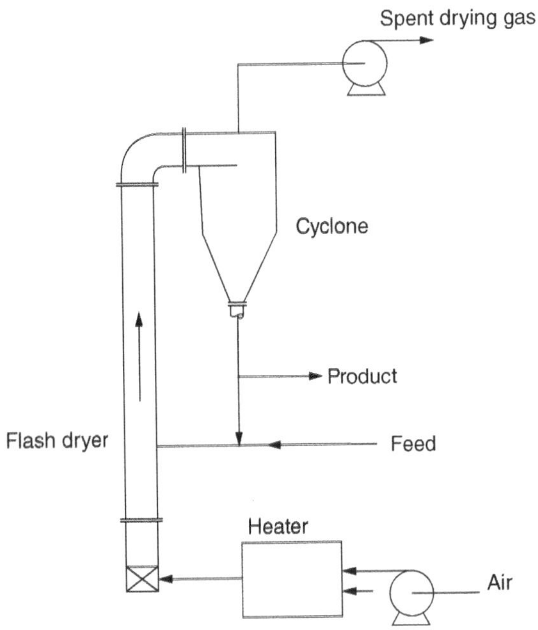

Figure 53.1 Flash dryer.

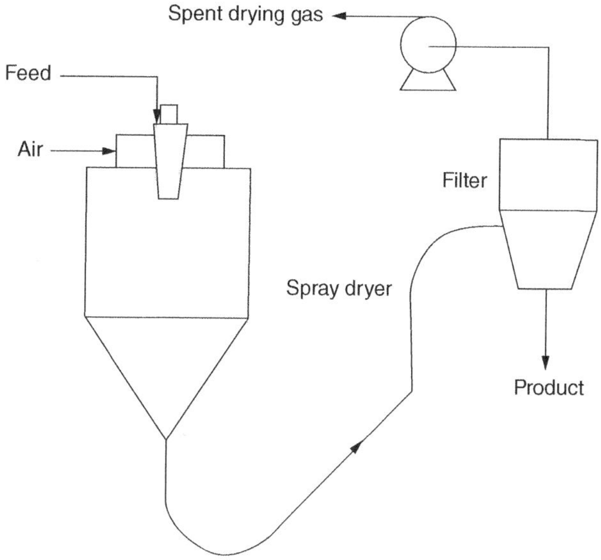

Figure 53.2 Spray dryer.

evaporation of about 1 kg of water and, investment-wise, because of the relatively long residence time of the gas leading to a large dryer. Compare that to the drying of vacuum pan salt in a fluid bed dryer where the removal of 1 kg of water results in the production of about 30 kg of salt and the gas residence time is maybe 2 s. The production of milk powder is a typical example of spray drying.

Convective Rotary Dryer

See Figure 53.3. Besides convective rotary dryers, there are also contact rotary dryers. The feed and hot or warm air are brought into contact in a rotating dryer. The two streams can move concurrently or countercurrently through the dryer. This dryer is a relatively conventional dryer, which is still used in the mining and mineral industry. The residence times of the solids and the gas are, respectively, for example, 20 min and 10 s. It is not common practice to cool the product after drying. The drying of sand is a typical example.

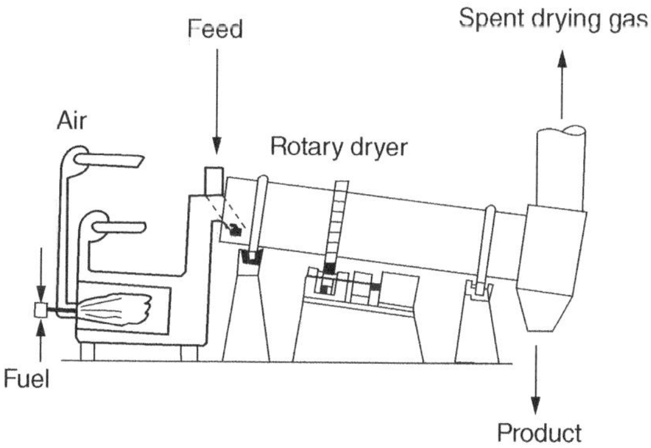

Figure 53.3 Convective rotary dryer.

Fluid-Bed Dryer

See Figure 53.4. A dryer having the character of a continuous stirred tank reactor is at the left, whereas a dryer having a plug flow character is at the right. A few words concerning the first dryer follow. The solids in the bed are fluidized by the hot or warm air stream. The air stream is cooled down by the contact with the product. The moisture content of the bed material equals the moisture content of the product because the bed is ideally mixed. It is not possible to cool after drying. The dryer is applied in the mining and mineral industry. The product fed to the dryer at the right is, in first instance, not fluidized because it is too wet. As it moves from the left to the right, warm air flows through the material and drying occurs. After having moved, for instance, 1 m, the drying has advanced to a degree that fluidization can start. At the end of the drying section, the moisture specification is reached. It is possible to cool the product in a cooling section. The residence time of the product is, for example, 10 min, whereas the gas residence time is about 1 s. The drying of vacuum pan salt, sugar, and citric acid are typical examples.

53.3 A First Example of Convective Drying

On mixing a stream of warm air containing 8 g of water vapor per kg of dry air thoroughly and adiabatically with a lot of water, the air reaches the adiabatic saturation temperature. See for a discussion of the adiabatic saturation temperature Chapter 13. It is assumed that the liquid is already at this temperature. The initial air temperature is 200 °C, and it is assumed that the air cools down to 47 °C. This assumption will be checked. The amount of heat given up by 1 kg of dry air and 8 g of water vapor on cooling is 1.008(200 − 47)1.01 = 156 kJ. The specific heat of air is 1.01 $kJ \cdot kg^{-1} \cdot K^{-1}$. On dividing this amount of heat by the heat of evaporation of water at 47 °C, the amount of water evaporated by 1 kg of air is obtained.

$$\frac{156}{2{,}390} = 0.065 \text{ kg}.$$

This amount of water vapor is added to the amount of water vapor already present in the air, that is, 0.008 kg of water vapor per kg of dry air.

The heat of evaporation of water at 47 °C is 2,390 $kJ \cdot kg^{-1}$. The following survey applies for the gaseous mixture of water and air.

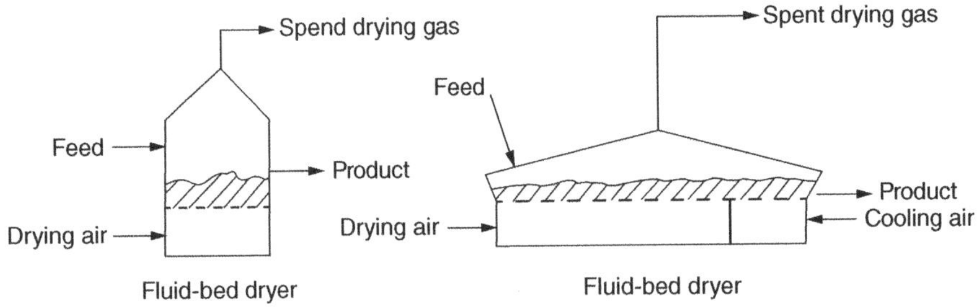

Figure 53.4 Fluid-bed dryer.

	Kg	Mol	Mol%	Partial pressure, $N \cdot m^{-2}$
H_2O	0.073	4.06	10.53	10,667
Air	1.000+	34.48+	89.47+	90,633+
	1.073	38.54	100.00	101,300

Water would have a partial pressure of 10,667 $N \cdot m^{-2}$. The partial pressure of a gaseous component is the pressure the component would have if it would fill the space alone. The partial pressure cannot exceed the saturated vapor pressure. The water saturated vapor pressure at 47 °C is 10,640 $N \cdot m^{-2}$. The assumption of a cooling down of the air to 47 °C was correct.

53.4 The Adiabatic Saturation Temperature

Using the method of the previous section, it is possible to find an adiabatic saturation temperature by trial and error. By using the method to be mentioned in this section, it is also possible to determine the adiabatic saturation temperature. This method also proceeds via trial and error, however, the method is more direct. See Chapter 13 for a discussion of the method. The main reason to discuss this method now is that it enables a comparison with the method to calculate the wet bulb temperature. The wet-bulb temperature is mentioned in Section 53.5. See also the aforementioned chapter. The amount of water evaporated in kg per kg of dry air is:

$$\frac{c_p\left(T_g - T_s\right)}{\Delta H}\ kg \cdot kg^{-1}.$$

c_p is the air specific heat in $J \cdot kg^{-1} \cdot K^{-1}$ and ΔH is the heat of evaporation of water in $J \cdot kg^{-1}$ at T, the arithmetic average of T_g and T_s. This amount of water per kg of dry air can also be represented by the following expression:

$$\frac{p_g^* - p_g}{\rho RT}\ kg \cdot kg^{-1}.$$

An explanation follows. p_g^* is the saturated water vapor pressure at T_s in $N \cdot m^{-2}$, p_g is the partial pressure of water vapor in the drying air, ρ is the air specific mass at atmospheric pressure and at T. The fact that the average temperature is used means that the method is an approximation. The unit of ρ is $kg \cdot m^{-3}$. R is the water gas constant in $J \cdot kg^{-1} \cdot K^{-1}$. That is, the well-known universal gas constant of 8,314 $J \cdot kmol^{-1} \cdot K^{-1}$ divided by 18, the water molecular weight in $kg \cdot kmol^{-1}$. The volume of 1 kg of drying air and a small amount of water vapor is V_1 m^3. For this volume, the universal gas law reads as follows for the water vapor:

$$p_g \cdot V_1 = N_1 \cdot R \cdot T_g\ \ J.$$

N_1 is the amount of water vapor in V_1 before drying in kg. After the drying, the universal gas law reads as follows for the water vapor:

$$p_g^* \cdot V_2 = N_2 \cdot R \cdot T_s\ \ J.$$

V_2 is the volume after drying. N_2 is the amount of water vapor in V_2 in kg after the drying. It follows from the latter two expressions:

$$N_2 - N_1 = \frac{\left(p_g^* - p_g\right)V}{RT} = \frac{p_g^* - p_g}{\rho RT}.$$

V is the arithmetic average of V_1 and V_2.

It follows:

$$\frac{c_p\left(T_g - T_s\right)}{\Delta H} = \frac{p_g^* - p_g}{\rho RT}.$$

This expression can be rewritten as:

$$\frac{p_g^* - p_g}{T_g - T_s} = \frac{RT \cdot c_p \cdot \rho}{\Delta H}. \tag{53.1}$$

The latter equation is the same equation as Equation 13.4 in Chapter 13.

Because of the approximations, the derivation is applicable up to about 200 °C. The background is that dry air of 200 °C can, on cooling down to the adiabatic saturation temperature, accept about 6% by weight of water vapor. The gas physical properties are not strongly affected by the presence of water vapor. Therefore, for the gas physical properties, the air properties are taken. Taking T as the arithmetic average of T_g and T_s also affects the accuracy at temperatures exceeding 200 °C.

Example 53.1

The adiabatic saturation temperature calculated in Section 53.3 is calculated by means of the method of this section.

By trial and error T_s = 49 °C is found.

$T_g = 200\ °C = 473\ K.$

$T_s = 49\ °C = 322\ K.$

$p_g^* = 1.17 \cdot 10^4\ N \cdot m^{-2}.$

$p_g = 1.28 \cdot 10^3\ N \cdot m^{-2}.$

The latter value can be calculated from the presence of 8 g of water vapor per kg of dry air.

$R = 8{,}314/18 = 461.9\ J \cdot kg^{-1} \cdot K^{-1}.$

$T = (200 + 49)/2 = 124.5\ °C = 397.5\ K.$

$c_p = 1.01 \cdot 10^3\ J \cdot kg^{-1} \cdot K^{-1}.$

$\rho = 1.29(273/397.5) = 0.89\ kg \cdot m^{-3}.$

$\Delta H = 2.38 \cdot 10^6\ J \cdot kg^{-1}.$

On using these data in Equation 53.1, it is found:

- left-hand side 69.0, and
- right-hand side 69.3.

Thus, the adiabatic saturation temperature found with the method of this section is 2 K higher than the value found in Section 53.3. The deviation becomes greater for drying gas temperatures higher than 200 °C.

53.5 The Wet-Bulb Temperature

See Chapter 13 for a derivation for the wet-bulb temperature. The aim in this section is to restate the equation derived in that chapter to enable a comparison with the equation derived for the adiabatic saturation temperature.

First, a simple example of a wet-bulb temperature will be given. The reader may think of a visit of the beach at Zandvoort in The Netherlands. It is a sunny but also windy day. On leaving the sea, he walks to his towel. He gets cold because the wind evaporates moisture from his body. His skin tends to adopt the wet-bulb temperature. The evaporation effect does neither impact the temperature nor the composition of the air flow.

The equation derived in Chapter 13:

$$\frac{p_g^* - p_g}{T_g - T_w} = \frac{RT \cdot c_p \cdot \rho \cdot Le^{2/3}}{\Delta H}. \tag{53.2}$$

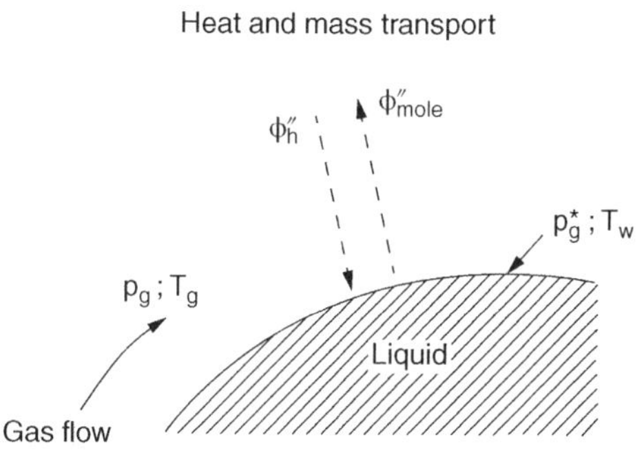

Figure 53.5 Evaporation from a wet surface.

See Figure 53.5. This equation is equal to Equation 53.1 but for the factor $Le^{2/3}$. Because the gaseous mixture air/water vapor contains, up to 200 °C, a small amount of water vapor, the air-specific heat and air-specific mass at T are used in Equation 53.2. The Lewis number is the quotient of the numbers of Schmidt and Prandtl. Both dimensionless numbers contain physical properties of air and water vapor. But for the diffusion coefficient of water vapor in air at T, the air physical properties at T can be used in Sc and Pr. For the combination air/water vapor, $Le^{2/3} \approx 0.95$. The reader can check this for themself. That is the reason that for the combination air/water vapor, the statement can be made that the wet-bulb temperature is approximately equal to the adiabatic saturation temperature. Often, these two temperatures are not distinguished from each other. However, their backgrounds are different. The adiabatic saturation temperature results from mass balance and equilibrium considerations whereas the wet-bulb temperature results from a consideration of the dynamic equilibrium between heat and mass transfer.

Example 53.2

The wet-bulb temperature for the air in Section 53.3 is calculated by using Equation 53.2. The air has a temperature of 200 °C and contains 8 g of water vapor per kg of dry air.

By trial and error, T_w = 48 °C is found.

T_g = 200 °C = 473 K.

T_w = 48 °C = 321 K.

$p_g^* = 1.12 \cdot 10^4\ N \cdot m^{-2}$.

$p_g = 1.28 \cdot 10^3\ N \cdot m^{-2}$.

The latter value follows from the presence of 8 g of water vapor per kg of dry air.

$R = 8{,}314/18 = 461.9\ J \cdot kg^{-1}$.

$T = (200 + 48)/2 = 124\ °C = 397\ K$.

$c_p = 1.01 \cdot 10^3\ J \cdot kg^{-1} \cdot K^{-1}$.

$\rho = 1.29(273/397) = 0.89\ kg \cdot m^{-3}$.

$\Delta H = 2.39 \cdot 10^6\ J \cdot kg^{-1}$.

$Le^{2/3} = 0.95$.

On using these data in Equation 53.2, it is found:

- left-hand side 65.3, and
- right-hand side 65.5.

Thus, the wet-bulb temperature found with the method of this section is 1 K lower than the adiabatic saturation temperature calculated in Example 53.1.

53.6 The Mollier Diagram

Part of the Mollier diagram for the combination air/water is depicted in Figure 53.6. It is also known as the psychrometric chart. Mollier diagrams for other combinations, like air/methanol, also exist. Temperature is on the y-axis. Moisture content of the air in g per kg of dry air and water vapor partial pressure are on the x-axis. The diagram is useful to determine the adiabatic saturation temperature. There are further options. The Mollier diagram is applicable for free moisture only. A discussion of the diagram follows. The batchwise fluid-bed drying of an organic material with air of 40 °C is taken as an example. The organic chemical is insoluble in water. The air attracted by the dryer has a temperature of 10 °C and contains 3.8 g of water vapor per kg of dry air. That is the condition of point A. The partial pressure of the water vapor is 6 mbar. Point A is on the curved line of 50% relative humidity. The relative humidity of air is the percentage of the maximum amount of moisture air can contain at a given temperature. On following the isotherm of 10 °C up to 100% relative humidity, a moisture content of 7.6 g of $g \cdot kg^{-1}$ is found. On drawing a sloped line through A, it is found that the enthalpy in A is 20 $kJ \cdot kg^{-1}$. Note that this value is also applicable for 1 kg of dry air but includes the enthalpy of the water vapor associated with 1 kg of dry air. The enthalpy refers to 0 °C and includes the heat of evaporation of the water vapor.

The air is warmed indirectly in a heat exchanger to 40 °C: point B. The moisture content remains 3.8 $g \cdot kg^{-1}$ as the warming is indirect. The relative humidity decreases to less than 10% because air of 40 °C can contain more moisture than air of 10 °C. The enthalpy is raised to 50 $kJ \cdot kg^{-1}$.

The next phase is the contact between the drying air and the product. The drying process is characterized by a sloped line from point B to point C if the moisture is free moisture, the process is adiabatic, and the contact between air and product is intensive so that equilibrium between moisture and air is reached. The drying process proceeds isenthalpic. The enthalpy remains constant

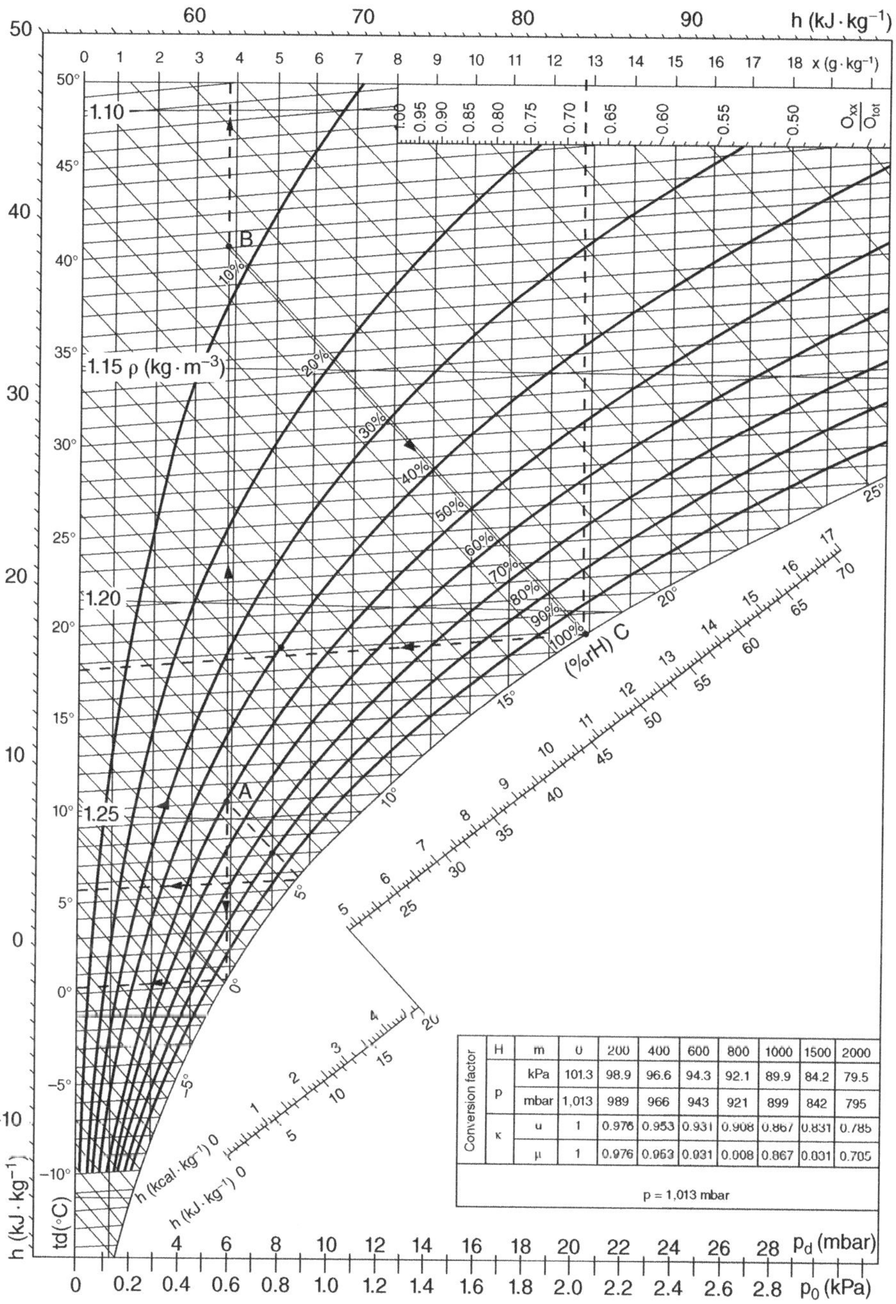

Figure 53.6 Part of the Mollier diagram for air/water.

because, on one hand, the enthalpy decreases because the air cools down, but, on the other hand, it increases because moisture is added to the air. The heat of evaporation is added to the gaseous mixture. The moisture content of the air has increased to 12.8 $g \cdot kg^{-1}$. The adiabatic saturation temperature is 17.8 °C.

The amount of air needed to evaporate a certain amount of free moisture can thus be calculated by consulting the Mollier diagram. The amount of air needed to evaporate bound moisture must be determined experimentally.

The Mollier diagram can also be used to determine the dew point of air. The dew point comprises the following. For example, a room is filled with air of 20 °C and a relative humidity of 50%. When the air in this room is cooled down, condensation will be noticed at a certain temperature. That is the dew point. The air of point A has a dew point of 0.3 °C. See Figure 53.6.

The question can be asked whether it is possible to use air of point A directly for drying purposes. That is possible and the adiabatic saturation temperature of that air is 5.3 °C. The moisture content of the air increases from 3.8 to 5.5 $g \cdot kg^{-1}$. We would need for the drying process more air of 10 °C than air of 40 °C. It is always possible to use air for drying purposes if the relative humidity is less than 100%.

The diagram also contains lines of constant specific mass of the gaseous mixture air/water vapor.

The diagram of Figure 53.6 is for an atmospheric pressure of 1 bar. For drying at Mexico City at an altitude of 2,240 m, we would need a different diagram.

53.7 Drying Vacuum Pan Salt in a Plug Flow Fluid-Bed Dryer

The batchwise drying of an organic chemical was mentioned in Section 53.6. The material is insoluble in water. If the material to be dried is soluble in water, the water is evaporated from a solution. Although the water is free to evaporate, that is, it is not bound to the chemical, it cannot exert the saturated vapor pressure of pure water. The water saturated vapor pressure exerted by the solution is lower than the pure water saturated vapor pressure. Centrifuged vacuum pan salt, having an average particle size of 450 μm, has a moisture content of 2.5% by weight. The moisture surrounding the crystals is a saturated brine containing 25% by weight of dissolved salt. Saturated brine contains 25% by weight of dissolved salt between 0 and 100 °C. The saturated vapor pressure of saturated brine is 75% of the water saturated vapor pressure in the temperature range 0–100 °C. If we would dry centrifuged vacuum pan salt in a continuous plug flow fluid-bed dryer with air of 40 °C, the adiabatic saturation temperature can be found as follows. The sloped line would have to stop at the curved line of 75% relative humidity. The adiabatic saturation temperature would be 21 °C and the moisture content of the drying air would be 11.5 $g \cdot kg^{-1}$.

Vacuum pan salt is dried in plug flow fluid-bed dryers with air of, for instance, 170 °C.

The text of this section is generally applicable for materials which dissolve in water, like soda, sugar, and citric acid.

54

Design of a Flash Dryer

54.1 Introduction

The flash dryer was described in Section 53.2. It is desired to dry an inorganic material having a moisture content of 15% by weight of the mass flow and a temperature of 20 °C. The product is insoluble in water. The dryer delivers 5,000 $kg \cdot h^{-1}$ having a moisture content of 0.1% by weight of the mass flow. The drying air is heated to 600 °C by the direct combustion of Dutch natural gas. There is an empirical equation expressing the gas outlet temperature as a function of the air inlet temperature. The equation is: $TA_{out} = 0.1875 \cdot TA_{in} + 35$ °C [1]. Pilot drying in a small flash dryer was successful. The product specific heat is 0.8 $kJ \cdot kg^{-1} \cdot K^{-1}$. The ambient temperature is 10 °C.

54.2 Design

The mass balance for the product and the water entering with the product in $kg \cdot h^{-1}$ is:

	In	Out
Water	881	5
Product	4,995+	4,995+
	5,876	5,000

Evaporation, 881 – 5 = 876 $kg \cdot h^{-1}$.

Heat balance in $kJ \cdot h^{-1}$.

First, the air outlet temperature and the product outlet temperature are determined. $TA_{out} = 0.1875 \cdot 600 + 35 = 147.5$ °C. The adiabatic saturation temperature is 70 °C. The adiabatic saturation temperature for this design will be discussed in a case study. Experiments have shown that it is necessary to take an air outlet temperature higher than the adiabatic saturation temperature to attain the product moisture specification. Experience learns that the product leaves the dryer cooler than the air, for instance, 30 K cooler. This is because of the short drying time. That means that there is still a driving force for heat transfer at the dryer outlet. $TP_{out} = 147.5 - 30 = 117.5$ °C.

Next, the heat required by the process flow in $kJ \cdot h^{-1}$ is calculated.

Introduction to Chemical Engineering, First Edition. C.M. van 't Land.

Evaporation	
$Q_1 = 876(2{,}500 + 1.9 \cdot 147.5 - 4.2 \cdot 20)$	$= 2{,}361{,}915$
Heat of evaporation at 0 °C 2,500 $kJ \cdot kg^{-1}$, vapor specific heat 1.9 $kJ \cdot kg^{-1} \cdot K^{-1}$, water specific heat 4.2 $kJ \cdot kg^{-1} \cdot K^{-1}$	
Product heating	
$Q_2 = 4{,}995 \cdot 0.8(117.5 - 20)$	$= 389{,}610$
Remaining water heating	
$Q_3 = 5 \cdot 4.2(117.5 - 20)$	$= 2{,}048$ +
	2,753,573

This is the heat exchanged between the air flow and the product in the dryer. It is remarked that the heat spent on product heating is substantial. This heating is required to attain the product moisture specification. Now, the heat to be supplied to the drying gas is calculated. The calculation is:

$$1.2 \cdot \frac{600 - 10}{600 - 147.5} \cdot 2{,}753{,}573 = 4{,}308{,}353 \text{ kJ} \cdot \text{h}^{-1}.$$

1.2 is an empirical factor to allow for heat losses. Atmospheric air is heated to 600 °C and leaves the dryer at 147.5 °C.

Short-term energy consumption per kg evaporated water:

$$\frac{4{,}308{,}353}{876} = 4{,}918 \text{ kJ} \cdot \text{kg}^{-1}.$$

The long-term energy consumption is: $1.5 \cdot 4{,}918 = 7{,}377$ $kJ \cdot kg^{-1}$.

The factor 1.5 is caused by various factors like underloading, air ingress at the feed position, and cleaning [1].

As the water heat of evaporation is about 2,500 $kJ \cdot kg^{-1}$, the energy consumption per kg of water is approximately tripled.

The natural gas consumption under normal, nominal circumstances is:

$$\frac{4{,}308{,}353}{32{,}000} = 134.6 \text{ nm}^3 \cdot \text{h}^{-1}.$$

With a specific mass of 0.8 $kg \cdot m^{-3}$, this is 108 $kg \cdot h^{-1}$.

The lower calorific value of natural gas from Slochteren in The Netherlands is 32,000 $kJ \cdot nm^{-3}$.

The air flow leaving the combustion chamber is:

$$\frac{4{,}308{,}353}{1.05(600 - 10)} = 6{,}955 \text{ kg} \cdot \text{h}^{-1}.$$

The air-specific heat is 1.05 $kJ \cdot kg^{-1} \cdot K^{-1}$.

It can be noticed that the air mass flow is much larger than the natural gas flow. The composition of the gas flow leaving the combustion chamber is, hence, approximately equal to the composition of air.

The main dryer dimensions can now be calculated.

The air flow entering the dryer is: $1.2 \cdot 6{,}955 = 8{,}346$ $kg \cdot h^{-1}$.

The factor 1.2 takes ingress air into account. The specific mass of atmospheric air at 147.5 °C is 0.84 $kg \cdot m^{-3}$. The air flow leaving the dryer is:

$$\frac{8{,}346}{0.84} = 9{,}936 \text{ m}^3 \cdot \text{h}^{-1}.$$

The evaporated water mass flow is 876 $kg \cdot h^{-1}$. The specific mass of water vapor at atmospheric pressure and 147.5 °C is 0.52 $kg \cdot m^{-3}$. The water vapor flow leaving the dryer is:

$$\frac{876}{0.52} = 1{,}685 \text{ m}^3 \cdot \text{h}^{-1}.$$

$$9{,}936 + 1{,}685 = 11{,}621 \text{ m}^3 \cdot \text{h}^{-1}.$$

On taking a gas exit velocity of 10 $m \cdot s^{-1}$, a dryer diameter of 0.64 m can be calculated. A dryer tube length of 12 m appeared adequate in tests. The length of the industrial drying tube will also be 12 m. The air outlet temperature is used for process control.

This design does not include the sizing of the fans.

Several aspects discussed in this section can also be taken into account for the design of other convective dryer types. The reason is that the basic process in all convective dryers is the same. It is the result of an intensive contact between drying air and product and the subsequent transfer of heat and mass.

Reference

1. van 't Land, C.M. (2012). *Drying in the Process Industry* (pp. 120–130). Hoboken, NJ: Wiley.

55

Contact Drying

55.1 Introduction

Contact dryers are indirect heat exchangers. The usual correlations for the design of indirect heat exchangers cannot be used. The reason is that the physical properties of the material to be dried vary during the drying process. So, correlations for the supply of heat to a wet solid do not exist. The usual approach is to carry out experiments on a small scale and to scale up the results. Four different dryer types will be discussed in Sections 55.2 and 55.4 through 55.6. Section 55.3 contains a remark on vacuum drying. The first dryer is operated batchwise, the other three dryers are continuous dryers.

55.2 Scaling Up of a Conical Vacuum Dryer

The conical vacuum dryer is depicted in Figure 55.1. The dryer can be used for sensitive filter or centrifuge cakes. Drying can be carried out at a low temperature. A drawback is that drying times can be quite long, that is, several hours. The dryer is filled with a batch of material. The volume actually filled with wet material is possibly two thirds of the total volume. Next, the screw is activated. We can distinguish two different rotations. First, the screw arm moves along the wall of the dryer. Usual rotational speeds are 1–3 min^{-1}. Second, the screw rotates around its own axis. The usual rotational speeds are in the range 10–50 min^{-1}. The screw rotation transports the material in an upward direction. During discharge, the screw rotates in the opposite direction. The top-driven stirrer without a bottom bearing precludes the penetration of the material being dried into a bearing or packing. The next step is to activate the vacuum pump. Often, a water ring pump is used. The absolute pressure in the dryer is, for example, 10, 20, or 30 mbar. See Section 55.3 for a discussion of this absolute pressure. The heat supply to the jacket is then started. Heat supply can be means of a circulation of warm water. The temperature of the circulating warm water decreases only 1 or 2 K. The reason is the resistance to heat transfer in the dryer. The drying occurs and the temperature of the material is monitored. When the product temperature rises, the free moisture has been removed. The period in which the free moisture is removed is called the first drying period. The product contains the critical moisture content at the end of the first drying period. The removal of bound moisture is the next step. That period is called the second drying period. Usually, the product moisture specification is reached when the product attains a certain temperature. The fact that

Introduction to Chemical Engineering, First Edition. C.M. van 't Land.

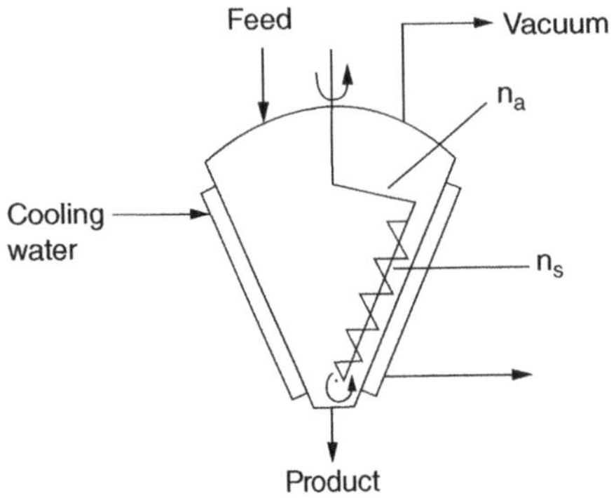

Figure 55.1 Conical vacuum dryer.

the cross-sectional area increases on going from the bottom of the dryer to the top is a bonus to minimize dust entrainment by the leaving vapor. The conical vacuum dryer is a universal piece of equipment as it can also be used for mixing purposes.

Now, the scaling-up experiment and the scaling up will be described.

Pilot plant dryer	Net volume	300 l
Commercial dryer	Net volume	1,000 l

The drying duty concerned a mixture of an organic peroxide and inert fillers. For process safety reasons, a maximum product temperature of 45 °C was prescribed.

Starting Points

- The degree of filling on a volume basis is equal for the two scales,
- the two dryers are geometrically similar,
- the absolute pressure in the dryer is the same on the two scales, and
- the mixing conditions are the same on the two scales.

A series of tests was carried out to determine the sensitivity of the process. The following experiment was selected for scaling up.

- Feed/product

Batch size (wet)	151.2 kg
Liquid	Water
Solubility of the product	Nil
Initial moisture content	8.3% by weight
Final moisture content	0.08% by weight
Critical moisture content	2.0% by weight
Moisture specification	<0.1% by weight
Solid specific heat	$1{,}500\ J \cdot kg^{-1} \cdot K^{-1}$
Feed temperature	15 °C

- Dryer

Wetted area	1.5 m^2
Warmed area	2.1 m^2
n_a	3.5 min^{-1}
n_s	50 min^{-1}
Screw diameter	0.32 m

- Process

Warm water in	46 °C
Warm water out	44 °C
Pressure	20 mbar
Water boiling point at 20 mbar	17.5 °C
Vapor temperature in the first drying period	20.5 °C
Final product temperature	45 °C

- Interpretation of the first drying period

Mass balance, kg

	In	Out
Solids	138.7	138.7
Water	12.5+	2.8+
	151.2	141.5

Evaporation 12.5 – 2.8 = 9.7 kg.
Drying time 144 min.
Product temperature (20.5 + 17.5)/2 = 19 °C [1].
Heat exchanged $9.7 \cdot 2{,}466 = 2.39 \cdot 10^4$ kJ. The heat of evaporation at 19 °C is 2,466 $kJ \cdot kg^{-1}$. It is possible to calculate an average heat transfer coefficient:

$$U = \frac{9.7 \cdot 2{,}466 \cdot 10^3}{(45 - 19)1.5 \cdot 144 \cdot 60} = 71.0\ W \cdot m^{-2} \cdot K^{-1}.$$

The coefficient will initially have been higher and finally have been lower.

- Interpretation of the second drying period

The product temperature rises to 40 °C and the moisture content decreases to 0.15% by weight.
Mass balance, kg

	In	Out
Solids	138.7	138.7
Water	2.8+	0.2+
	141.5	138.9

Evaporation 2.8 – 0.2 = 2.6 kg.
Drying time 160 min.
Heat balance kJ.
See Section 54.2. The amounts of heat for water evaporation, solid warming, and remaining water warming are calculated.

$Q_1 = 2.6(2{,}500 + 1.9 \cdot 30 - 4.2 \cdot 19)$	=	6,441
$Q_2 = 138.7 \cdot 1.5(40 - 19)$	=	4,369
$Q_3 = 0.2 \cdot 4.2(40 - 19)$	=	18+
		10,828

The amount of heat is approximately half the amount of heat exchanged in the first drying period.

- Post drying

The product temperature rose quickly to 45 °C, and the material was kept at this temperature for 30 min. The final moisture content was below 0.1% by weight.
Drying time 45 min

- Scaling up

The batch size of the commercial dryer is (1,000/300)151.2 = 504 kg.
On scaling up, the wall area of this dryer per m^3 decreases. The factor is in this case $(1{,}000/300)^{1/3} = 1.49$.

Times, min	300 l	1,000 l
Charging and homogenizing	30	60
First drying period	144	$1.49 \cdot 144 = 215$
Second drying period	160	$1.49 \cdot 160 = 238$
Post drying	45	45
Transfer	15	30
Spare	15+	30+
	409	618

The fact that, on scaling up, the wall area per m^3 decreases is a well-known drawback of this dryer type. It can also be noticed for other contact dryer types. The effect can sometimes, partly, be compensated by passing heat transfer medium through stirrers or transport screws.
Hourly capacities

$$300\,\mathrm{l}: 20.3\ \mathrm{kg \cdot h^{-1}}.$$

$$1{,}000\,\mathrm{l}: 44.9\ \mathrm{kg \cdot h^{-1}}.$$

The volume of the commercial dryer is more than thrice the volume of the pilot plant dryer. The hourly capacity of the commercial dryer is somewhat greater than twice the hourly capacity of the pilot plant dryer.
The rotational speeds of the screw of the commercial dryer could be varied to attain the same quality of heat transfer as on pilot plant scale. Incrustations did occur neither on pilot plant scale nor on commercial scale. The scaling-up was successful.

55.3 An Additional Remark Concerning Vacuum Drying

Often, for vacuum drying, the vacuum is created by a water ring pump. Then, the absolute pressure in the drying equipment is in the range 30–50 mbar. The reason why the pressure cannot be lower than this range is the fact that the pressure cannot be lower than the saturated vapor pressure of the water in the water ring. The saturated vapor pressure of water at 15 °C is 15 mbar. The pressure loss in the lines could also be 15 mbar, so that the pressure in the dryer is, for example, 30 mbar. Sometimes, the drying time can be shortened by admitting air or nitrogen to the drying chamber. The effect is that the water saturated vapor pressure is lowered. See Figure 55.2. At the left, the situation without the admission of nitrogen can be seen. The vacuum is maintained by a water ring pump. The warm water temperature is 60 °C. The temperature of the water on the product is 36 °C as 60 mbar is the saturated vapor pressure of water at 36 °C. The driving force for heat transfer is 60 – 36 = 24 K. The rate of the water evaporation process is a function of this temperature difference.

At the right, the situation in which nitrogen is admitted. The nitrogen admission rate is adjusted to give a nitrogen partial pressure in the drying chamber of 40 mbar. The water ring pump has ample capacity and maintains the absolute pressure at 60 mbar. The nitrogen admission only leaves 60 – 40 = 20 mbar for water vapor. The product assumes a temperature of 17 °C as the saturated vapor pressure of water at 17 °C is 20 mbar. The driving force for heat transfer is now 60 – 17 = 43 K. That

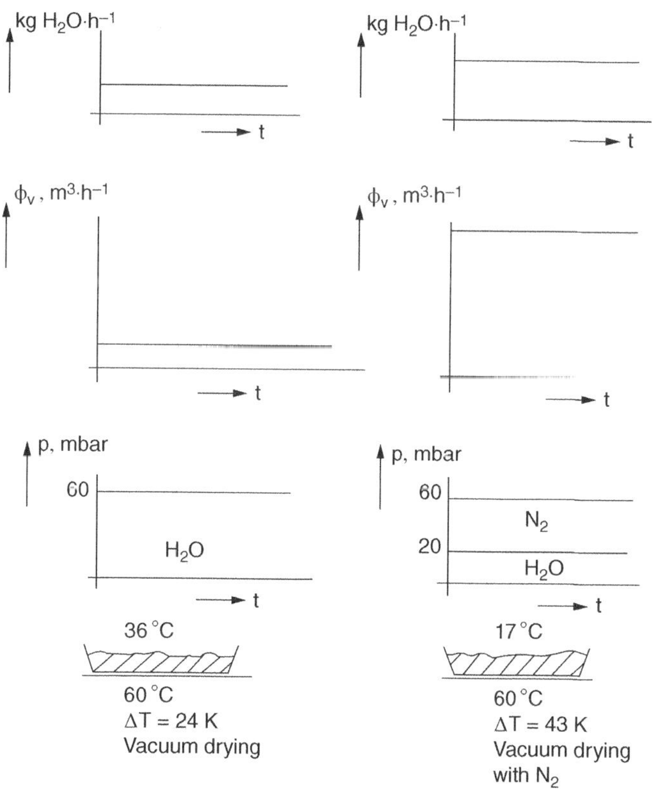

Figure 55.2 Vacuum drying while adding nitrogen gas.

means that the temperature difference is almost doubled. The water evaporation rate is also almost doubled.

This section discusses a simple means to increase the capacity of an existing vacuum drying facility at the cost of increased vacuum pump power consumption. A different possibility would be to buy a different type of vacuum pump, which can maintain a desired vacuum.

55.4 Testing a Small Plate Dryer

See Figure 55.3. The plate dryer is an interesting continuous contact dryer for the chemical industry. A general remark is that this dryer type is suitable for special cases as the dryer is investmentwise not cheap. First, a description of the functioning of the dryer. The centrifuge or filter cake is fed at the top of the dryer. The plates in the dryer are stationary whereas a shaft in the dryer rotates. Heating medium circulates through hollow plates. Two or more arms are attached to the shaft and plows are attached to the arms. The plows transport the material to be dried through the dryer. The slope of the plows can be adjusted to control the layer thickness of the product; 15 mm is a typical layer thickness. The feed enters the dryer centrally. The plows of the two arms above the first plate are

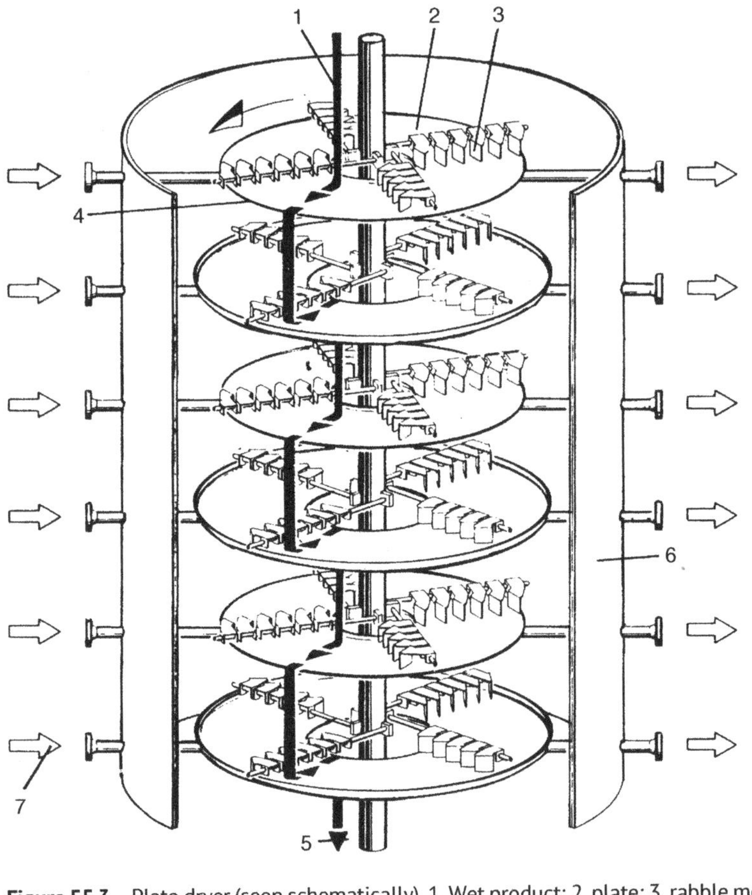

Figure 55.3 Plate dryer (seen schematically). 1, Wet product; 2, plate; 3, rabble mechanism; 4, plate rim; 5, dry product; 6, housing; 7, heating or cooling medium. *Source:* Courtesy of Andritz Gouda BV, Waddinxveen, The Netherlands.

oriented in such a way as to transport material to the periphery of the plate. The wet solid then falls on the second plate. The plows on the arms above the second plate are oriented in such a way as to transport the material to the shaft. On arrival there, the material falls on the third plate. Thus, the material travels in cascade and in plug flow through the dryer. Cooling water circulates through the last set of plates of the dryer. Thus, the warm, dried material can be cooled before storage. An air stream flows from, for example, the right to the left through the dryer to entrain the evaporated moisture. The air velocity is, for example, 0.1 $m \cdot s^{-1}$, and the air is warmed to the temperature of the heating medium in the plates. The air does not substantially transfer heat to the process stream. It is possible to have a nitrogen atmosphere in the dryer when an organic solvent, like methanol, is removed. There is a small overpressure in the dryer in that case to prevent, for process safety reasons, air from entering the dryer. A further possibility is to have a vacuum in the dryer for drying sensitive materials. Special valves are then required for feeding the dryer and for extracting product from the dryer. These valves enable solids to enter and to leave the dryer while air cannot enter the dryer. The main advantage of the dryer is the gentle treatment of the product. There are no high velocities like in a flash dryer, and the material is not fluidized like in a fluid bed dryer.

A test procedure will be described. The test was carried out in a batch dryer. The results can be used for getting an impression of a continuous industrial dryer because the product travels through the industrial dryer in plug flow.

- Feed

Batch size	615 g
Liquid	Water
Solubility of the product	Nil
Initial moisture content	20% by weight
Final moisture content	0% by weight
Temperature	Ambient
Product melting point	146 °C

- Dryer

Area	0.10 m^2
Agitator rotational speed	5 min^{-1}
Number of arms	2

- Process

Warm water temperature	95 °C
Air temperature	95 °C
Pressure	Atmospheric
Air velocity based on the largest cross-sectional area	0.2 $m \cdot s^{-1}$
Layer thickness after drying	10 mm
Final product temperature	85 °C
Drying time	37 min
Product turnover number	$2 \cdot 5 \cdot 37 = 370$

The product turnover number is defined as the product of the number of arms, the rotational speed in min^{-1}, and the drying time in min. It is a dimensionless number. It can, in a way, be seen as the number of contacts between the agitator and a product particle during the drying. The number can be used for scaling-up.

- Mass balance, g

	In	Out
Solids	492	492
Water	123+	0+
	615	492

Evaporation, 123 g.

- Scaling-up to a product mass flow of 350 $kg \cdot h^{-1}$.

The specific capacity of the small dryer is:

$$\frac{0.492}{0.10 \cdot (37/60)} = 7.98\ kg \cdot m^{-2} \cdot h^{-1}.$$

The area for an industrial dryer could be: $350/7.98 = 43.9\ m^2$.

A dryer having an area of 50 m^2 could be a good choice. Cooling the product could also be carried out in this dryer. The dryer is equipped with two arms above each plate for transporting the material across the plate. The rotational speed will be 5 min^{-1}. Such a set-up will lead to a comparable mixing number.

Before selecting an industrial dryer, a continuous pilot dryer should be tested.

55.5 Testing a Continuous Paddle Dryer

See Figure 55.4. A general remark is that this dryer type is considered when convective drying is not feasible. The reason is that this dryer type is investmentwise more expensive and that this cannot be compensated by lower variable costs. First, a description of the functioning of the

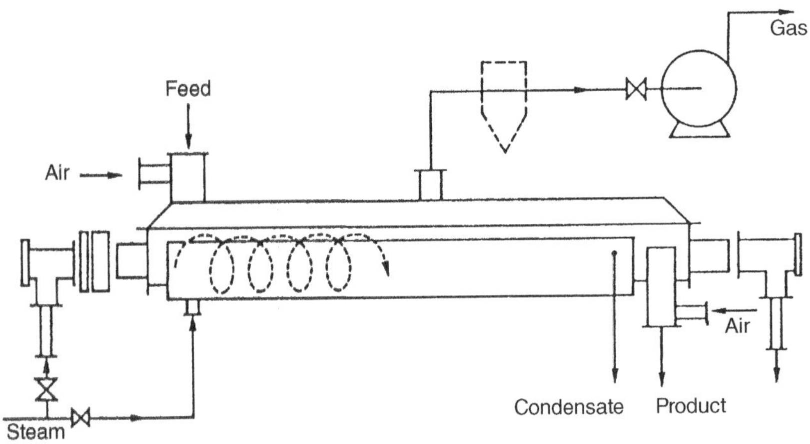

Figure 55.4 Andritz Gouda paddle dryer (schematically). *Source:* Courtesy of Andritz Gouda BV, Waddinxveen, The Netherlands.

dryer will be given. The paddle dryer is a trough equipped with rotating shafts with paddles. Normally, there are two shafts. The rotational speed of the shafts is variable. The paddles do not have a transport function. The dryer has a small slope to enable the transport of the material through the dryer. The product travels through the dryer in plug flow, in Figure 55.4 from the left to the right. The trough is filled with product. The product leaves the dryer at the right, it passes a rim, and falls into a discharge pipe. The height of the rim can be adjusted, usually the paddles are completely covered by product. A heating medium flows through the jacket and through the paddles. Concerning the jacket, the heating medium flows from a stationary line into a stationary jacket and leaves the jacket in the same fashion. As to the paddles, the heating medium flows from a stationary line into rotating parts and vice versa. That implies the necessity of seals. So, this dryer is mechanically more complicated than a convective dryer. A small air or nitrogen flow entrains the evaporated moisture when the dryer functions at atmospheric pressure. If water is evaporated, and if the product is insoluble in water, and the dryer functions at atmospheric pressure, the product quickly attains a temperature of 100 °C. This is different from what happens in a convective dryer in which the product, if free moisture is evaporated, attains the adiabatic saturation temperature. It is possible to have a nitrogen atmosphere in the dryer when an organic solvent, like methanol, is removed. The dryer then functions, for process safety reasons, at a small overpressure to prevent air ingress. A further possibility is to have a vacuum in the dryer for drying sensitive materials. Special valves are then required for feeding the dryer and extracting product from the dryer. These valves enable solids to enter and to leave the dryer, whereas air cannot enter the dryer.

A test procedure will be described. The test considered for scaling up was carried out in a continuous pilot plant dryer.

- Feed

Mass flow	90 $kg \cdot h^{-1}$
Liquid	Water
Solubility of the product	Nil
Initial moisture content	22.5% by weight
Final moisture content	0.2% by weight
Temperature	12 °C
Product melting point	146 °C
Product specific heat	1.0 $kJ \cdot kg^{-1} \cdot K^{-1}$
Bulk density wet	600 $kg \cdot m^{-3}$
Bulk density dry	500 $kg \cdot m^{-3}$

- Dryer

Type	Andritz GMF GPD 1.6W2.5
Jacket area	0.8 m^2
Area of the two shafts	1.7 m^2
Volume	0.07 m^3
Dryer slope	3.5°
Number of shafts	2
Paddle diameter	0.16 m
Shafts rotational speed	30 min^{-1}

- Process

Steam condensation temperature: 125 °C.
A small air flow having a temperature of 110 °C entrains the evaporated moisture.

- Mass balance, $kg \cdot h^{-1}$

	In	Out
Solids	69.75	69.75
Water	20.25+	0.14+
	90.0	69.89

Evaporation, $20.25 - 0.14 = 20.11$.

- Heat balance, $kJ \cdot h^{-1}$

See Section 54.2. The amounts of heat for the evaporation of water, solid heating, and remaining water heating are calculated.

$Q_1 = 20.11(2{,}500 + 1.9 \cdot 100 - 4.2 \cdot 12)$	=	53,082
$Q_2 = 69.75 \cdot 1.0(100 - 12)$	=	6,138
$Q_3 = 0.14 \cdot 4.2(100 - 12)$	=	52+
		59,272

The heat flow is 16.5 kW. It is possible to estimate an average heat transfer coefficient:

$$U = \frac{59{,}272 \cdot 10^3}{2.5 \cdot 3.6 \cdot 10^3(125 - 100)} = 263.4\,W \cdot m^{-2} \cdot K^{-1}.$$

This a rather high heat transfer coefficient. All heat exchanging area is in contact with the product. The actual heat transfer coefficient will be a function of the dryer length, that is, for wet material, we will have a relatively high coefficient and vice versa.

- Paddle data

$$\text{Tip velocity: } v = \frac{\pi \cdot 0.16 \cdot 30}{60} = 0.25\,m \cdot s^{-1},$$

$$Fr = \frac{v^2}{Rg} = \frac{0.25^2}{0.08 \cdot 9.81} = 0.08.$$

The Froude number is the ratio of the radial acceleration of the paddle periphery and the acceleration due to gravity. The trough stays completely filled with material at Froude numbers smaller than 1.

- Scaling up to a dryer product mass flow of $600\,kg \cdot h^{-1}$

The heating area needed is: $2.5 \cdot (600/69.89) = 21.5\,m^2$.
Dryer type GPD 6W25 can be chosen from the Andritz GMF catalogue. The heating area of this dryer is 24.9 m^2. The jacket provides about 8.75 m^2 and the paddles provide approximately 16.25 m^2. The ratio of these two areas is about equal to the ratio of the two areas of the pilot plant dryer.

It is important to pay attention to the ratio of the two heating areas because the heat transfer from the paddles is more efficient than the heat transfer from the jacket. A further aspect is the increase of the residence time of the product in the dryer on scaling up. The volume of the pilot plant dryer is 0.07 m^3, and the residence time is about half an hour. The volume of the industrial dryer is 1.51 m^3, and the residence time will be approximately 75 min. The reason is the decrease of m^2 heating area per m^3 volume on scaling up.

The paddle tip velocity in the pilot plant dryer test run was, for a rotational speed of 30 min^{-1}, 0.25 $m \cdot s^{-1}$. The rotational speed of the shafts of the pilot plant dryer could be varied from 12.5 to 62 min^{-1}. The rotational speed of the shafts of the industrial dryer can be varied from 12 to 20 min^{-1}. The paddle tip velocity can be thus be varied from 0.38 to 0.63 $m \cdot s^{-1}$. It should be possible to have a good heat transfer in the industrial dryer.

Finally, it was decided to stop the work because the dryer, during the test work, started to suffer from incrustations.

55.6 Scale Up of a Thin-Film Dryer

This dryer is called a hybrid dryer because drying proceeds both by convection and by contact. See Figure 55.5. The dryer functions as follows. The product enters at the left side of the dryer. The drying proceeds at atmospheric pressure. A shaft rotates in the center of the dryer. A typical rotational speed for a dryer having a diameter of 0.6 m is 600 min^{-1}. The rotational speed can be

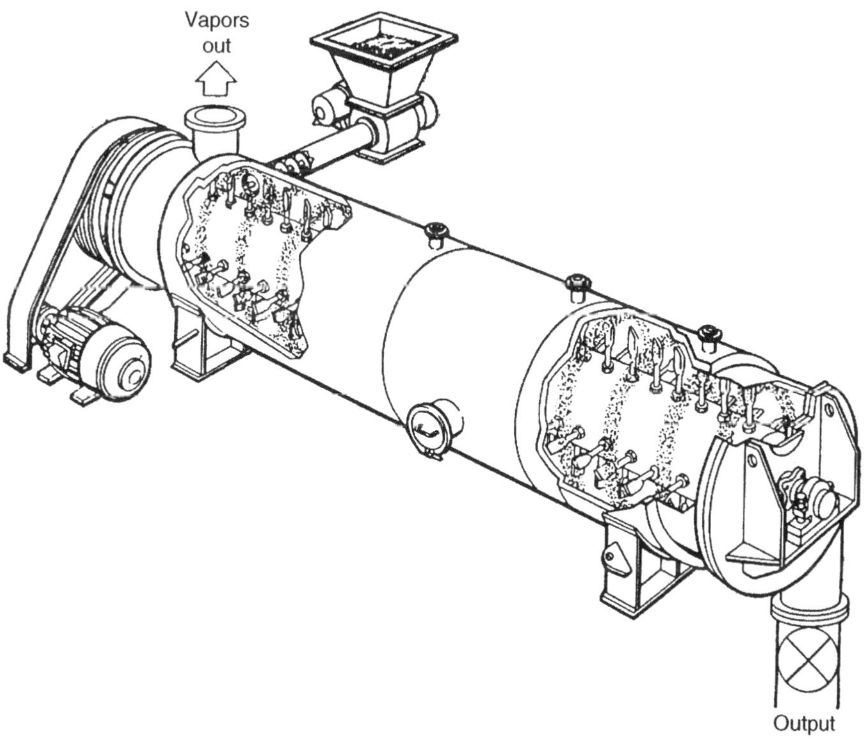

Figure 55.5 Vigorously agitated contact dryer–Solidaire. *Source:* Bepex International LLC, Minneapolis, MN, USA.

adjusted. Arms are attached to the shaft and, at the end of the arms, plows are mounted. The rotational speed of the shaft is high so that the material is concentrated at the wall. The plows transport the material through the dryer. The angle of the plows can be adjusted to vary the residence time of the material in the dryer. Heating medium passes through the jacket. Hot air passes through the dryer concurrently or countercurrently. The product leaves the dryer at the right side. Part of the material falls out of the dryer and part of the material is entrained by the air flow. The latter part is separated from the air by means of cyclones. The dryer is a suitable dryer for materials that tend to form incrustations during the drying process. The vigorous rotation generally prevents incrustations to form.

A test selected for scaling up of the drying of an organic acid will be treated. As a matter of fact, the material had a strong tendency to form incrustations and chunks during the drying process. Both flash drying and fluid-bed drying had failed because of this phenomenon. It appeared that the material could be dried in this hybrid dryer equipped with a stirrer rotating at a high speed.

- Feed

Mass flow	186.55 kg · h^{-1}
Liquid	Water
Solubility of the product	Slightly soluble
Initial moisture content	20.0% by weight
Final moisture content	0.5% by weight
Temperature	20 °C
Temperature sensitivity	Product becomes brown in air at 180 °C
Product-specific heat	1.25 kJ · kg^{-1} · K^{-1}

- Dryer

Make	Hosokawa-Bépex
Jacket area	1.5 m^2
Diameter	0.25 m
Length	2.3 m
Agitator rotational speed	1,300 min^{-1}
Plow position	3

- Process

Product outlet temperature	75 °C
Steam condensation temperature	160 °C
Air mass flow	720 kg · h^{-1}
Air specific heat	1.0 kJ · kg^{-1} · K^{-1}
Air inlet temperature	170 °C
Air outlet temperature	100 °C

- Mass balance, kg · h^{-1}

	In	Out
Solids	149.25	149.25
Water	37.30+	0.75+
	186.55	150.00

Evaporation, 37.30 – 0.75 = 36.55.

- Heat balance, $kJ \cdot h^{-1}$

See Section 54.2. The amounts of heat for the evaporation of water, solid heating, and remaining water heating are calculated.

$Q_1 = 36.55(2{,}500 + 1.9 \cdot 100 - 4.2 \cdot 20)$	=	95,249
$Q_2 = 149.25 \cdot 1.25(75 - 20)$	=	10,261
$Q_3 = 0.75 \cdot 4.2(75 - 20)$	=	173+
		105,683

The heat flow is 29.4 kW. The heat is supplied by both the jacket and the air flow. The air contribution is estimated as follows:

$0.7 \cdot 720 \cdot 1.0(170 - 100) = 35{,}280\ kJ \cdot h^{-1}$, 0.7 is an empirical factor because not all of the heat is used for drying. So, the jacket contribution could be:

$$105{,}683 - 35{,}280 = 70{,}403\ kJ \cdot h^{-1}.$$

It is possible to estimate the heat transfer coefficient of the jacket. The adiabatic saturation temperature of air at 170 °C containing 8 g of water per kg of dry air is 42 °C. Now, the logarithmic mean temperature difference is calculated as follows:

$$(\Delta T)_m = \frac{(160 - 42) - (160 - 75)}{\ln \dfrac{(160 - 42)}{(160 - 75)}} = 100.6\ K.$$

This calculation is but an approximation.

$$U = \frac{70{,}403 \cdot 1{,}000}{3{,}600 \cdot 1.5 \cdot 100.6} = 129.6\ W \cdot m^{-2} \cdot K^{-1}.$$

The calculated value is rather low. The low value is possibly due to the fact that not all area is in contact with product, so that the area actually used for heat transfer is smaller than 1.5 m^2.

- Agitator data

$$\text{Tip velocity: } v = \frac{\pi \cdot 0.25 \cdot 1{,}300}{60} = 17.0\ m \cdot s^{-1}.$$

$$Fr = \frac{v^2}{Rg} = \frac{17.0^2 \cdot 2}{0.25 \cdot 9.81} = 236.$$

The Froude number is the ratio of the radial acceleration of the agitator tip and the acceleration due to gravity. The material travels in a helix from product inlet to product outlet at Froude numbers greater than 10.

A further aspect is that the average particle size of the product is reduced by a factor of approximately four. So, the dryer also functions as a mill. This reduction was not an objective for this and other tests. However, it was not a problem as the customers dissolve the product in water.

- Scaling up

The industrial capacity is 750 $kg \cdot h^{-1}$. So, the scale-up factor is 5. A dryer having an area of $5 \cdot 1.5 = 7.5\ m^2$ would be adequate. The air flow having a temperature of 170 °C could be raised to

$5 \cdot 720 = 3{,}600\ \mathrm{kg \cdot h^{-1}}$. The question was raised whether a dryer having an area of 6.6 m^2 could be chosen. That choice was made. If need be, the air flow could be increased with 25%. An agitator rotational speed of 600 $\mathrm{min^{-1}}$ was selected. The latter speed could be adjusted in the plant. The angle of the plows could also, if need be, be adjusted in the plant. The rotational speed of 600 $\mathrm{min^{-1}}$ corresponds with a tip velocity of 18.8 $\mathrm{m \cdot s^{-1}}$. So, problems concerning incrustations were not expected as this tip velocity is greater than the tip velocity of 17.0 $\mathrm{m \cdot s^{-1}}$ at the test selected for scaling up. The corresponding Froude number is 120 guaranteeing that the material is concentrated at the dryer wall. The scaling up was successful.

Reference

1. Fischer, J. J. (1963). Drying in vacuum tumblers. *Industrial Engineering Chemistry*, 55, 19–24.

56

Case Studies Crystallization, Liquid/Solid Separation, and Drying

56.1 Ultracentrifuges

Ultracentrifuges are used for the manufacture of fuel containing uranium for nuclear power stations. Uranium in uranium ore consists for 99.3% by weight of the isotope U-238 and for 0.7% by weight of the isotope U-235. The fuel used most for nuclear power stations is uranium dioxide (UO_2). Typically, the uranium in the oxide consists of 5% by weight of U-235 and 95% by weight of U-238. So, to manufacture nuclear fuel containing uranium, the ore must be purified, converted, and enriched. Enrichment is the raising of the U-235 content. Worldwide, enrichment is at the moment usually effected by ultracentrifuges. In enrichment plants, uranium is converted into the gaseous uranium hexafluoride. Enrichment by means of a cascade of ultracentrifuges is possible because the molecular mass of the molecule containing U-238 is greater than the molecular mass of the molecule containing U-235. A cascade contains more than 10^3 centrifuges [1]. A rotational speed of $1.6 \cdot 10^3\ \text{min}^{-1}$ and a circumferential velocity of $750\ \text{m} \cdot \text{s}^{-1}$ are typical. It is requested to estimate the centrifuge diameter and the radial acceleration as a multiple of the acceleration due to gravity ($g = 9.81\ \text{m} \cdot \text{s}^{-2}$).

Solution

$$v = \pi D n \rightarrow 750 = \pi D \cdot 1.6 \cdot 10^3 \rightarrow D = 0.15\ \text{m}.$$

$$C = \frac{v^2}{Rg} = \frac{750^2}{0.075 \cdot 9.81} = 7.65 \cdot 10^5.$$

56.2 $Le^{2/3}$

It is stated in Section 53.5 that $Le^{2/3}$ is approximately 0.95. The statement is valid for relatively low concentrations of water vapor in air, atmospheric pressure, and ambient temperatures. It is requested to check the statement. The diffusion coefficient for water vapor in air is given. The other physical properties are for air. Physical properties at atmospheric pressure and 20 and 60 °C are given. The figures for 60 °C are between brackets.

Introduction to Chemical Engineering, First Edition. C.M. van 't Land.

$$\text{Le} = \text{Sc}/\text{Pr},\ \text{Sc} = \nu/\text{Đ},\ \text{Pr} = \nu/a,\ a = \lambda/(c_p \cdot \rho).$$

$$\lambda = 0.0257\ (0.0285)\ \text{W} \cdot \text{m}^{-1} \cdot \text{K}^{-1}.$$

$$c_p = 1{,}007\ (1{,}010)\ \text{J} \cdot \text{kg}^{-1} \cdot \text{K}^{-1}.$$

$$\rho = 1.20\ (1.06)\ \text{kg} \cdot \text{m}^{-3}.$$

$$\text{Đ} = 2.4 \cdot 10^{-5} (3.2 \cdot 10^{-5})\ \text{m}^2 \cdot \text{s}^{-1}.$$

Solution

At 20 °C

$$\text{Le} = \frac{\lambda}{c_p \cdot \rho \text{Đ}} = \frac{0.0257}{1{,}007 \cdot 1.20 \cdot 2.4 \cdot 10^{-5}} = 0.89,$$

$$\text{Le}^{2/3} = 0.93.$$

At 60 °C

$$\text{Le} = \frac{0.0285}{1{,}010 \cdot 1.06 \cdot 3.2 \cdot 10^{-5}} = 0.83,$$

$$\text{Le}^{2/3} = 0.88.$$

The figures for $\text{Le}^{2/3}$ are slightly smaller than 0.95.

56.3 Convective Drying-1

Mexico City is at a height of 2,240 m above sea level. A kilogram of air containing 8 g of water per kg of dry air and having a temperature of 200 °C cools down to the adiabatic saturation temperature. A comparison is made between Mexico City at 2,240 m above sea level and Amsterdam in The Netherlands. Can this kilogram of air absorb more or less water in Mexico City than at Amsterdam? Tip: repeat the calculation in Section 53.3 for the Mexican situation. The atmospheric pressure in Mexico City is 76,742 $\text{N} \cdot \text{m}^{-2}$. The water heat of evaporation is $2.4 \cdot 10^6\ \text{J} \cdot \text{kg}^{-1}$. The water saturated vapor pressure as a function of the temperature can be calculated by means of the Antoine equation:

$^{10}\log p_{sat} = A - \frac{B}{C + T}$. p_{sat} in mm Hg, A = 8.07, B = 1,730.63, C = 233.43, and T in °C. 1 mm Hg = 133 $\text{N} \cdot \text{m}^{-2}$.

Solution

It is assumed that the air cools down to 42 °C. The assumption will be checked. The amount of heat given up by 1 kg of dry air and 8 g of water vapor on cooling is 1(200 – 42)1.01 = 159.6 kJ. The specific heat of air is 1.01 $\text{kJ} \cdot \text{kg}^{-1} \cdot \text{K}^{-1}$. On dividing this amount of heat by the water heat of evaporation, the amount of water evaporated by 1 kg of air is obtained.

$$\frac{159.6}{2{,}400} = 0.067\ \text{kg}.$$

The following survey applies for the gaseous mixture of water and air. The 8 g of water vapor per kg of dry air in the original 1 kg of air are added to the calculated 67 g.

	Kg	Mol	Mol%	Partial pressure, $N \cdot m^{-2}$
H_2O	0.075	4.17	10.87	8,342
Air	0.992+	34.21+	89.13+	68,400+
	1.067	38.38	100.00	76,742

Water would have a partial pressure of 8,342 $N \cdot m^{-2}$. The partial pressure of a gaseous component is the pressure the component would have if it would fill the space alone. The partial pressure cannot exceed the water saturated vapor pressure. The water saturated vapor pressure at 42 °C is calculated as follows:

$$^{10}\log p_{sat} = A - \frac{B}{C + T} = 8.07 - \frac{1{,}730.63}{233.43 + 42} = 1.79 \rightarrow p_{sat} = 62 \text{ mm Hg } (8{,}246 \text{ N} \cdot \text{m}^{-2}).$$

The assumption of the air cooling down to 42 °C was correct because the saturated vapor pressure almost equals the partial pressure. The conclusion is that drying at Mexico City is more effective than drying at Amsterdam in The Netherlands because 1 kg of drying air can, at Mexico City, evaporate more water than at Amsterdam.

56.4 Convective Drying-2

Figure 53.6 of this part contains a Mollier diagram. The diagram contains lines of constant specific mass in $kg \cdot m^{-3}$. These lines intersect with the lines of constant temperature, the isotherms. It is requested to explain this observation.

Solution

The temperature is the ordinate of the diagram, whereas the moisture content is an abcissa. On following an isotherm from the left to the right, the amount of moisture per kg of dry air increases. The pressure is atmospheric. The specific mass in $kg \cdot m^{-3}$ decreases because the molecular weight of water, 18 $g \cdot mol^{-1}$, is smaller than the average molecular weight of air. This average molecular weight is 29 $g \cdot mol^{-1}$. So, an isotherm intersects with the lines of constant specific mass.

56.5 Analysis of a Spray-Drying Operation

Data of a spray-drying operation are reported [2]. The production of a dairy product, possibly milk powder, is described. The heater duty is 1,385 kW, and the moisture content of the leaving air is 53 $g \cdot kg^{-1}$. It is requested to check these data. Assume that 53 $g \cdot kg^{-1}$ means gram of water per kilogram of dry air. Also assume that the air is indirectly heated. Spray drying is described in Section 53.2 and a spray dryer is depicted in Figure 53.2. The data are as follows.

- Product

Product mass flow	1,000 $kg \cdot h^{-1}$
Liquid	Water
Feed moisture content	55% by weight
Product moisture content	3% by weight
Feed temperature	50 °C
Product specific heat	1.5 $kJ \cdot kg^{-1} \cdot K^{-1}$
Product median particle size	120 μm

- Process

Ambient air temperature	20 °C
Drying air inlet temperature	200 °C
Drying air outlet temperature	80 °C
Air moisture content	10 g per kg of dry air
Air specific heat	1.0 $kJ \cdot kg^{-1} \cdot K^{-1}$

- Mass balance, $kg \cdot h^{-1}$

	In	Out
Solids		
Water	______+	______+

Evaporation – =

- Heat balance, $kJ \cdot h^{-1}$

See Section 54.2. The amounts of heat for the evaporation, solids heating, and remaining water heating are calculated.

$Q_1 =$	$(2{,}500 + 1.9 \cdot$	$- 4.2 \cdot$	$) =$	
$Q_2 =$	$\cdot 1.25($ –	$)$	$=$	
$Q_3 =$	$\cdot 4.2($ –	$)$	$=$	______+
Q_{tot1}				

This heat flow is exchanged between air and feed in the spray dryer. The heat flow exchanged in an indirect air heater between condensing steam and air is calculated as follows:

$$Q_{tot2} = 1.2 \cdot \frac{200 - 20}{200 - 80} \cdot \qquad =$$

The latter expression can be understood as follows. Air having a temperature of 20 °C is heated to 200 °C. However, not all of this heat is used for the process flow. The air cools down to 80 °C. The reason is that it is necessary to control this temperature to attain the moisture specification. The factor 1.2 is a loss factor. The heater duty has been obtained in $kJ \cdot h^{-1}$. On dividing by 3,600, we obtain the heater duty in kW:

- Moisture content leaving air

Air mass flow before drying: $\frac{\qquad}{1.0(\qquad - \qquad)} = \qquad kg \cdot h^{-1}$.

The ingress air flow is 10% of this air flow. It is requested to complete the following table. The unit is $kg \cdot h^{-1}$.

	Supply	Ingress air	Evaporation	Off-gas
Air			0	
Water vapor	________+	________+	________+	________+

Note that the calculated air mass flow and the ingress air flow must be split up into air and water vapor. The fourth column is the sum of the other columns. Gram of water per kilogram of dry air in the leaving air:

Solution

- Mass balance, $kg \cdot h^{-1}$

	In	Out
Solids	970	970
Water	1,186+	30+
	2,156	1,000

Evaporation 1,186 – 30 = 1,156.

- Heat balance, $kJ \cdot h^{-1}$

$Q_1 = 1{,}156(2{,}500 + 1.9 \cdot 80 - 4.2 \cdot 50)$	=	2,822,952
$Q_2 = 970 \cdot 1.25(60 - 50)$	=	12,125
$Q_3 = 30 \cdot 4.2(60 - 50)$	=	1,260+
Q_{tot1}		2,836,337

$$Q_{tot2} = 1.2 \cdot \frac{200 - 20}{200 - 80} \cdot 2{,}836{,}337 = 5{,}105{,}407,$$

$$\frac{5{,}105{,}407}{3{,}600} = 1{,}418 \text{ kW}.$$

The latter figure compares well with 1,385 kW [2].

- Moisture content leaving air

Air mass flow before drying: $\dfrac{5{,}105{,}407}{1.0(200 - 20)} = 28{,}363 \text{ kg} \cdot \text{h}^{-1}$.

$kg \cdot h^{-1}$

	Supply	Ingress air	Evaporation	Off-gas
Air	28,083	2,808	0	30,891
Water vapor	280+	28+	1,156+	1,464+
	28,363	2,836	1,156	32,355

$$\frac{1{,}464 \cdot 10^3}{30{,}891} = 47 \text{ g of water vapor per kg of dry air.}$$

The latter figure is slightly smaller than $53 \text{ g} \cdot \text{kg}^{-1}$ [2].

56.6 Estimation of the Size of a Contact Dryer

It is requested to estimate the size of an Andritz Gouda B.V. paddle dryer by calculating the areas required for warming the feed, evaporation of free moisture, and evaporation of bound moisture. This dryer is described in Section 55.5. The dryer is depicted in Figure 55.4. The dryer is heated by steam having a pressure of 3.0 bar and a temperature of 140 °C. The condensation temperature of this steam is 133 °C. To remove the bound moisture, it is necessary to raise the product temperature to 120 °C. The data are as follows.

- Feed

Mass flow	$500\ \text{kg} \cdot \text{h}^{-1}$
Liquid	Water
Solubility of the product	Nil
Initial moisture content	30% by weight
Final moisture content	1.0% by weight
Critical moisture content	3.0% by weight
Temperature	30 °C
Product-specific heat	$1.5\ \text{kJ} \cdot \text{kg}^{-1} \cdot \text{K}^{-1}$

- Water data

Heat of evaporation	$2.23 \cdot 10^3\ \text{kJ} \cdot \text{kg}^{-1}$
Specific heat	$4.2\ \text{kJ} \cdot \text{kg}^{-1} \cdot \text{K}^{-1}$

- Process

Estimated heat transfer coefficients in $\text{kW} \cdot \text{m}^{-2} \cdot \text{K}^{-1}$ are:

Warming the feed	0.1
Removal of free moisture	0.2
Removal of bound moisture	0.05

- Area for warming the feed

$$Q = \frac{1}{3{,}600}(\qquad \cdot 4.2 + \qquad \cdot 1.5)(\qquad - \qquad) = \qquad \text{kW}.$$

$$(\Delta T)_m = \frac{(133 - \qquad) - (133 - \qquad)}{\ln\dfrac{(133 - \qquad)}{(133 - \qquad)}} = \qquad \text{K}.$$

$$A_1 = \frac{Q}{(\Delta T)_m \cdot 0.1} = \qquad \text{m}^2.$$

- Area for the evaporation of free moisture at 100 °C

Mass balance, $kg \cdot h^{-1}$

	In	Out
Solids		
Water	______+	______+

Evaporation − =

$$Q = \frac{\cdot 2.23 \cdot 10^3}{3{,}600} = \quad kW,$$

$$\Delta T = 133 - \quad = \quad K,$$

$$A_2 = \frac{Q}{\Delta T \cdot 0.2} = \quad m^2.$$

- Area for the evaporation of bound moisture

Mass balance, $kg \cdot h^{-1}$

	In	Out
Solids		
Water	______+	______+

Evaporation − =

See Section 54.2. The amounts of heat for the evaporation, solids heating, and remaining water heating are calculated. The heat flows are in $kJ \cdot h^{-1}$.

$Q_1 =$	$\cdot 2.23 \cdot 10^3 +$		$\cdot 4.2(110 - 100)$	=
$Q_2 =$	$\cdot 1.5($	−	)	=
$Q_3 =$	$\cdot 4.2($	−	)	= ______+
Q_{tot}				

$$Q_{tot} = \quad kW.$$

$$(\Delta T)_m = \frac{(133 - \quad) - (133 - \quad)}{\ln \frac{(133 - \quad)}{(133 - \quad)}} = \quad K.$$

$$A_3 = \frac{Q_{tot}}{(\Delta T)_m \cdot 0.05} = \quad m^2.$$

- Dryer area

$$A_{tot} = A_1 + A_2 + A_3 \ m^2.$$

Solution

- Area for warming the feed

$$Q = \frac{1}{3{,}600}(150 \cdot 4.2 + 350 \cdot 1.5)(100 - 30) = 22.5\ \text{kW},$$

$$(\Delta T)_m = \frac{(133 - 30) - (133 - 100)}{\ln((133 - 30)/(133 - 100))} = 61.5\ \text{K},$$

$$A_1 = \frac{22.5}{61.5 \cdot 0.1} = 3.7\ \text{m}^2.$$

- Area for the evaporation of free moisture at 100 °C
  Mass balance, $\text{kg} \cdot \text{h}^{-1}$

	In	Out
Solids	350	350
Water	150+	11+
	500	361

Evaporation 150 – 11 = 139.

$$Q = \frac{139 \cdot 2.23 \cdot 10^3}{3{,}600} = 86.1\ \text{kW},$$

$$\Delta T = 133 - 100 = 33\ \text{K},$$

$$A_2 = \frac{86.1}{33 \cdot 0.2} = 13.0\ \text{m}^2.$$

- Area for the evaporation of bound moisture
  Mass balance, $\text{kg} \cdot \text{h}^{-1}$

	In	Out
Solids	350	350
Water	11+	4+
	361	354

Evaporation 11 – 4 = 7.
Heat flows in $\text{kJ} \cdot \text{h}^{-1}$.

$Q_1 = 7 \cdot 2.23 \cdot 10^3 + 7 \cdot 4.2(110 - 100)$	=	15,904
$Q_2 = 350 \cdot 1.5(120 - 100)$	=	10,500
$Q_3 = 4 \cdot 4.2(120 - 100)$	=	336+
Q_{tot}		26,740

$$Q_{tot} = 7.4\ \text{kW},$$

$$(\Delta T)_m = \frac{(133 - 100) - (133 - 120)}{\ln((133 - 100)/(133 - 120))} = 21.5\ \text{K},$$

$$A_3 = \frac{7.4}{21.5 \cdot 0.05} = 6.9\ \text{m}^2.$$

- Dryer area

$$3.7 + 13.0 + 6.9 = 23.6\ \text{m}^2.$$

Notation VIII

A	Constant in Section 56.3	—
	Filtration area	m^2
A_1	Contact dryer area for feed warming	m^2
A_2	Contact dryer area for the evaporation of free moisture	m^2
A_3	Contact dryer area for the evaporation of bound moisture	m^2
A_{tot}	Total contact dryer area	m^2
a	Acceleration	$m \cdot s^{-2}$
	Radial acceleration of a centrifuge rotor	$m \cdot s^{-2}$
	Thermal diffusivity	$m^2 \cdot s^{-1}$
B	Constant in Section 56.3	—
C	Concentration	$mol \cdot l^{-1}$
	Constant in Section 56.3	—
	Froude number for a centrifuge rotor	—
	Suspension solids concentration	$kg \cdot m^{-3}$
	Tracer concentration	$kmol \cdot m^{-3}$
C_0	Tracer concentration at $t = 0$	$kmol \cdot m^{-3}$
c^*	Concentration	$kmol \cdot m^{-3}$
c_m	Concentration	$kmol \cdot m^{-3}$
c_p	Specific heat	$J \cdot kg^{-1} \cdot K^{-1}$
D	Centrifuge rotor diameter	m
Đ	Diffusion coefficient	$m^2 \cdot s^{-1}$
$F_{1,2}$	Forces Nos. 1 and 2	N
Fr	Froude number	—
g	Acceleration due to gravity	$m \cdot s^{-2}$
ΔH	Water heat of evaporation	$J \cdot kg^{-1}$
		$J \cdot mol^{-1}$
h	Enthalpy	$kJ \cdot kg^{-1}$
		$kcal \cdot kg^{-1}$
K	Filter cake permeability	m^2
$k_{1,2,3,4,5}$	Constants in Section 51.7	—
k_v	Crystal form factor	—

Introduction to Chemical Engineering, First Edition. C.M. van 't Land.

L	Centrifuge rotor length	m
	Crystal size	μm
	Filter cake thickness	m
L_{av}	Arithmetically averaged crystal size	μm
L_D	Dominant particle size	μm
Le	Lewis number	—
M	Cumulative mass of crystals	$g \cdot l^{-1}$
M_T	Slurry density	$g \cdot l^{-1}$
m	Crystal mass distribution	$g \cdot l^{-1} \cdot \mu m^{-1}$
	Centrifuge rotor mass	kg
m_n	Crystal mass fraction distribution	—
N	Cumulative number of crystals	l^{-1}
N_1	Amount of water vapor in V_1	kg
N_2	Amount of water vapor in V_2	kg
N_T	Total cumulative number of crystals	l^{-1}
n	Crystal population density	$\mu m^{-1} \cdot l^{-1}$
	Centrifuge rotor rotational speed	min^{-1} s^{-1}
n_a	Mixer arm rotational speed	min^{-1}
n_s	Mixer rotational speed	min^{-1}
n_0	Crystal population density at $L = 0$	$\mu m^{-1} \cdot l^{-1}$
Pr	Prandtl number	—
p	Pressure	$N \cdot m^{-2}$ mbar
p_d	Pressure	mbar
p_g	Water vapor partial pressure	$N \cdot m^{-2}$
p_0	Pressure	mbar
p_g^*	Saturated water vapor pressure	$N \cdot m^{-2}$
p_{sat}	Saturated vapor pressure	mm Hg
Q	Heat flow	kW
Q_1	Heat flow required for evaporation	$kJ \cdot h^{-1}$
Q_2	Heat flow required for solid heating	$kJ \cdot h^{-1}$
Q_3	Heat flow required for heating remaining water	$kJ \cdot h^{-1}$
Q_{tot}	Heat flow in a contact dryer	$kJ \cdot h^{-1}$ or kW
Q_{tot1}	Net heat flow at drying	$kJ \cdot h^{-1}$
Q_{tot2}	Heat flow required for drying	$kJ \cdot h^{-1}$
R	Dryer paddle or agitator radius	m
	Centrifuge rotor radius	m
	Flow channel radius	m
	Universal gas constant ($8.314 \cdot 10^3\ J \cdot kmol^{-1} \cdot K^{-1}$)	
	Gas constant for water ($461.9\ J \cdot kg^{-1} \cdot K^{-1}$)	
r	Crystal growth rate	$\mu m \cdot min^{-1}$
Sc	Schmidt number	—
T	Temperature	°C
	Arithmetic average of T_g and T_s	°C or K
	Arithmetic average of T_g and T_w	°C or K
T^*	Temperature	°C

TA_{in}	Temperature of air entering a dryer	°C
TA_{out}	Temperature of air leaving a dryer	°C
T_g	Air temperature	°C or K
T_m	Temperature	°C
TP_{out}	Temperature of product leaving a dryer	°C
T_s	Adiabatic saturation temperature	°C or K
T_w	Wet bulb temperature	°C or K
$(\Delta T)_m$	Logarithmic mean temperature difference	K
t	Time	min
t_d	Temperature	°C
U	Overall heat transfer coefficient	$W \cdot m^{-2} \cdot K^{-1}$
V	Arithmetical average of V_1 and V_2	$m^3 \cdot kg^{-1}$
	Liquid volume	m^3
	Reactor volume	m^3
V_1	Volume of drying air before drying	$m^3 \cdot kg^{-1}$
V_2	Volume of drying air after drying	$m^3 \cdot kg^{-1}$
v	Centrifuge rotor circumferential velocity	$m \cdot s^{-1}$
	Dryer paddle tip velocity	$m \cdot s^{-1}$
	Linear liquid velocity	$m \cdot s^{-1}$
x	Air moisture content	$g \cdot kg^{-1}$
	Distance	m

Greek Symbols

α	Heat transfer coefficient	$W \cdot m^{-2} \cdot K^{-1}$
δ	Centrifuge rotor thickness	m
ε	Agitator power input	$W \cdot l^{-1}$
λ	Thermal conductivity	$W \cdot m^{-1} \cdot K^{-1}$
μ	Liquid dynamic viscosity	$N \cdot s \cdot m^{-2}$
ν	Kinematic viscosity	$m^2 \cdot s^{-1}$
ρ	Air-specific mass	$kg \cdot m^{-3}$
	Centrifuge rotor material specific mass	$kg \cdot m^{-3}$
ρ_c	Dry filter cake bulk density	$kg \cdot m^{-3}$
ρ_s	Solid-specific mass	$kg \cdot m^{-3}$
σ	Centrifuge rotor stress	$N \cdot m^{-2}$
τ	Residence time (V/Φ_v)	min
ϕ''_h	Heat flux at convective drying	$W \cdot m^{-2}$
ϕ''_{mole}	Water molar flux at convective drying	$kmol \cdot m^{-2} \cdot s^{-1}$
Φ_v	Cake filtration flow	$m^3 \cdot s^{-1}$
	Flow through a crystallizer or reactor	$m^3 \cdot min^{-1}$
	Gas flow	$m^3 \cdot h^{-1}$

Part IX

Gas/Solid Separation

Part IX Content

57

Introduction

Gas/solid separation is met at various places in the process industry, for example, at convective drying. Convective drying implies intimate contact between the gas used for drying and the material being dried. The drying gas is usually air; however, convective drying with warm nitrogen can also be met. Used air must be returned to the atmosphere. It then must satisfy certain specifications, for instance, the solid contents in $mg \cdot nm^{-3}$. The pieces of equipment for the separation of entrained solid particles from gases treated in this part are cyclones, filters, scrubbers, and electrostatic precipitators. An adage in the process industry for separating dust from gases is: "When it is dry, keep it dry." Cyclones, filters, and electrostatic separators satisfy this requirement. Scrubbers do not. On using scrubbers for separating particulate material from gases, a solid–gas separation task is converted into a solid/liquid separation task.

Cyclones are capable of operating with high inlet dust levels, for example, $200\ g \cdot m^{-3}$, and it is common to install them upstream of such other collectors as bag filters and scrubbers. They are suitable for the collection of particles larger than 5–10 μm.

Because of their screening effect, fabric filters will generally ensure a high collection efficiency (greater than 99%) even on submicrometer-sized dusts. Their main limitation is the heat sensitivity of the filter media. Natural fibers can be used up to temperatures of approximately 80 °C, whereas synthetic fibers can be used up to 150 °C. Glass fiber will withstand continuous operation up to 260 °C.

Concerning scrubbers, low-energy scrubbers are used for soluble particles, and high-energy types can separate insoluble particles from an air stream. The power required for the latter task increases strongly when the particle size decreases.

Electrostatic precipitators are used for the separation of inorganic, incombustible particles from large gas flows. The application at the drying of marl, a raw material in cement manufacture, is an example. Marl is dried convectively using high air temperatures, in the range of 400–500 °C. Installing filters is a risk, as the hot air might come into direct contact with the fabric. The performance of electrostatic precipitators is almost as good as the performance of filters.

Cyclones, filters, scrubbers, and electrostatic precipitators are available in packaged assemblies.

It is important to keep the temperature of the processed air streams 20 K above their dew point to avoid condensation if the dust is to be handled dry. If condensation occurs, cyclones can become plugged by incrustations, and filter fabrics are easily clogged by caking material and are then difficult to clean.

Introduction to Chemical Engineering, First Edition. C.M. van 't Land.

58

Cyclones

58.1 Introduction

A cyclone is a device that accomplishes a solid/gas separation by imparting a spinning motion to the two-phase system. The particles move outward to the wall of the cyclone because of the centrifugal force and subsequently travel downward to enter a discharge hopper. The gas that is freed from solids leaves at the top through a central outlet. Figure 58.1 shows a Stairmand cyclone with all the necessary dimensions [1]. The Stairmand cyclone is a high-efficiency, medium-throughput cyclone. The dust-containing gas enters the cyclone body tangentially and moves downward in a vortex motion. The flow direction reverses at the apex of the cone, and the cleaned gas travels upward in a second vortex.

It is stated sometimes that a cyclone is a cheap centrifuge, and there is certainly an element of truth in that statement. It must, however, be realized that cyclones generate much lower radial acceleration than centrifuges. For example, a cyclone with a diameter of 1 m and an inlet velocity of 20 $m \cdot s^{-1}$ is characterized by a radial acceleration (expressed as a multiple of the acceleration due to gravity, i.e., 9.81 $m \cdot s^{-2}$) of $20^2/(0.5 \cdot 9.81) = 81.5$. This acceleration is considerable but not spectacular, since peeler centrifuges operate with acceleration figures in the range 500–1,000. A cyclone with a diameter of 2 m and the same inlet velocity has a radial acceleration of 40.75. Consequently, large cyclones are less efficient than small ones if, on scaling up, the inlet velocity is kept constant. The pressure drop is, in principle, equal for small and large cyclones if their geometric ratios and the inlet and outlet velocities are equal. That is the reason that we almost never see a cyclone having a diameter larger than 2 m. Large gas flows are processed by a series of two or four cyclones in parallel.

Cyclones are cheap devices for dust collection, and their main limitation is their poor efficiency in collecting particles smaller than about 5 μm. Cyclones can process air streams with high dust loadings.

58.2 Sizing and Process Data

Cyclones typically have diameters in the range 0.1–2 m. Cyclones can be sized using the rule of thumb that the normal flow rate in $m^3 \cdot s^{-1}$ equals $1.5 \cdot D_c^2$ for a high-efficiency, medium throughput pattern. D_c is the diameter of the cylindrical part of the cyclone. The factor becomes 4.5 for a medium-efficiency, high-throughput pattern. For the applicability of the rule of thumb, the entrance velocity is approximately 15 $m \cdot s^{-1}$ in both types. An entrance velocity in excess of

Introduction to Chemical Engineering, First Edition. C.M. van 't Land.

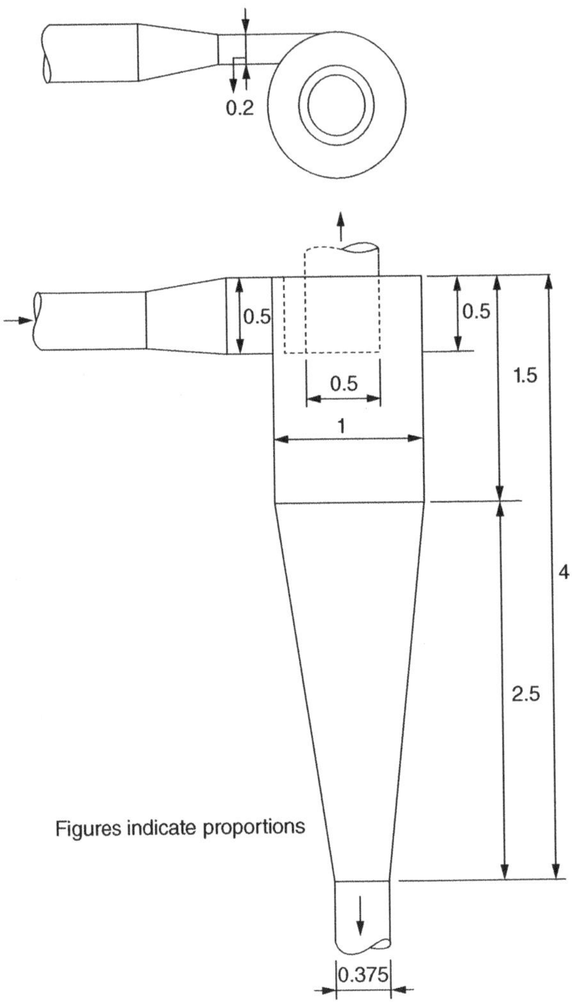

Figure 58.1 Stairmand cyclone for gas/solid separations. Figures indicate proportions.

$30\ m \cdot s^{-1}$ causes turbulence, which, in turn, leads to bypassing and re-entrainment of separated particles and hence a decrease in efficiency [2].

Pressure drops of between 50 and 150 mm water gauge are normal. Koch and Licht [1] recommend Shepherd and Lapple's approach to estimate the pressure drop in a cyclone. First, the dimensionless number of hydraulic heads is calculated:

$$N_H = K\left(\frac{ab}{D_e^2}\right). \tag{58.1}$$

N_H can be compared to the dimensionless friction loss factor as discussed in Chapter 5. See Equation 5.9. K is 7.5 if a neutral inlet vane is present and 16 if an inlet vane is absent. a and b are, respectively, the height and the width of the inlet duct, and D_e is the diameter of the outlet duct. Second, the pressure drop is calculated:

$$\Delta p = N_H \cdot 0.5 \cdot \rho_g \cdot v_i^2 \quad N \cdot m^{-2}. \tag{58.2}$$

$0.5 \cdot \rho_g \cdot v_i^2$ is the velocity head of the entering air.

Example 58.1

The pressure loss of a cyclone is calculated. The cyclone has the geometric ratios of Figure 58.1. An inlet vane is absent, $\rho_g = 1.0\ kg \cdot m^{-3}$, and $v_i = 15\ m \cdot s^{-1}$.

$$N_H = 16 \cdot \frac{ab}{D_c^2} = 16 \cdot 0.5 \cdot 0.2/0.5^2 = 6.4.$$

$\Delta p = 6.4 \cdot 0.5 \cdot 1.0 \cdot 15^2 = 720\ N \cdot m^{-2}$ (73.5 mm water gauge).

Koch and Licht [1] recommend relationships for the calculation of collection efficiency. A potential source for poor cyclone performance is air ingress into the dust outlet of the cyclone. A further aspect is the height of the cyclone. The ratio between the height of the cyclone and the diameter of the cylindrical part is 4 in Figure 58.1. The collection efficiency can be raised if the ratio is increased to, for example, 6. However, the capital cost increases as well. A typical value for the maximum solids content of air processed by a cyclone is $20\ mg \cdot nm^{-3}$. Cyclones are applied widely. Their construction is simple, and the energy requirements are low. They can withstand high temperatures if their material of construction is a metal. Cyclones can be cleaned efficiently. It has occurred that a cyclone became plugged with particulate material and hence must the supporting structure be able to support the filled cyclone.

Explosion relief vents, which are located in the cylindrical part of the cyclone, are used when processing explosive dusts. A possible discharge should neither harm people nor equipment. Units should be grounded to avoid electrostatic charge buildup.

References

1. Koch, W. H., and Licht, W. (1977). New design approach boosts cyclone efficiency. *Chemical Engineering*, 84, 80–88.
2. Muir, D. M. (1985). *A User Guide to Dust and Fume Control* (pp. 37–48). Rugby, UK: The Institution of Chemical Engineers.

59

Fabric Filters

59.1 Introduction

Fabric filters are devices that accomplish gas/solid separation by means of a fabric, with the gas passing through a cloth and the particulate material being retained. Fabric filters can separate particles as small as 0.1 μm, and the treated gas usually contains less than 10 $mg \cdot nm^{-3}$ of solids. Today's trend is to even less than 5 $mg \cdot nm^{-3}$. The structural enclosure containing the bag-shaped filters is termed a *baghouse*. The reverse-pulse filter is for large gas flows today state of the art. Figure 59.1 shows such a filter schematically, and Figure 59.2 shows such a filter as part of a spray-drying installation. The reverse-pulse filter will be described. The dust-laden air enters via an inlet pipe and flows to the bags, where the dust accumulates outside the bags and the gas passes through the dust layer and the cloth and leaves the filter through the outlet pipe. The bags are supported by wire cages, which have a cylindrical cross-section. Venturi nozzles are located in the clean-gas outlets of the bags. For cleaning, jets of high-velocity air are directed through the venturi nozzles and into the bags. The pulses of air expand into the bags and dislodge the collected dust.

The gas velocity is reduced from the value in the inlet pipe (e.g., 15 $m \cdot s^{-1}$) to the low fabric approach velocity (e.g., 1 $m \cdot min^{-1}$). Subsequently, the gas velocity increases to, for example, 15 $m \cdot s^{-1}$ in the outlet pipe. Often, filters are provided with internals, which assist in an orderly decrease of the velocity in the inlet pipe to the fabric approach velocity. With abrasive dusts, this is important to avoid fabric wear. The fabric approach velocity is low to enable bag cleaning and prevent bag plugging. The consequence is a large filtration area.

59.2 Fabrics

The dust layer may be completely removed from the surface of the fabric by the cleaning action of the pulse. Hence, the fabric itself must serve as the principal filter medium for at least a substantial part of the filtration cycle. Woven fabrics are unsuitable for such service, and felts of various types are used. Felt ensures that a good collection efficiency is also maintained during the first phase of dust layer formation.

Felted fabrics consist of fibers that are compressed under high pressure. The commonly used range of filtration velocities for felted fabrics is 1.2 – 2.5 $m \cdot min^{-1}$ [1]. As an alternative for felt, it is possible to use a polyester fabric coated with a Teflon membrane. The membrane allows passage of gas and retains solid material at the surface.

Introduction to Chemical Engineering, First Edition. C.M. van 't Land.

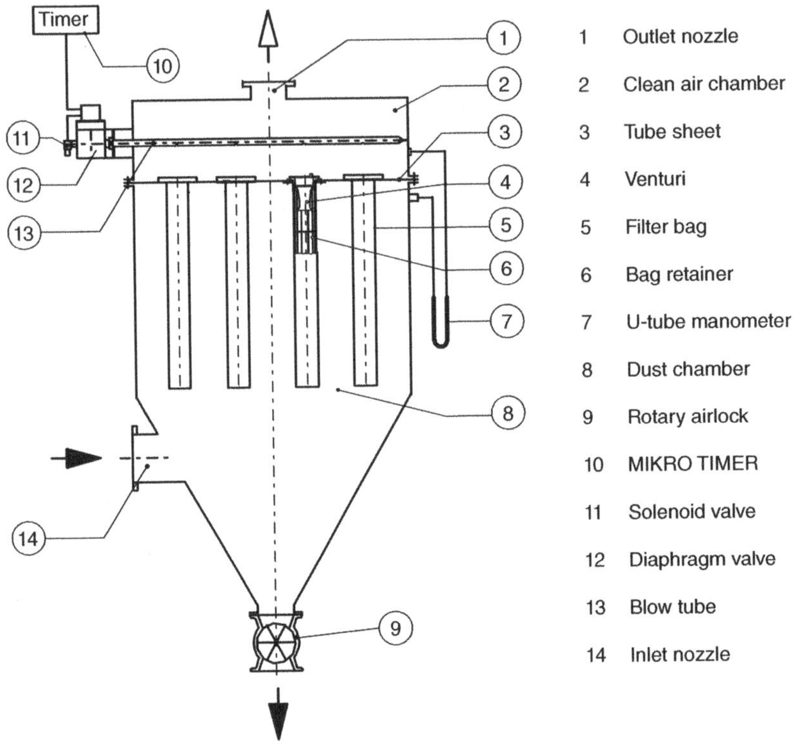

Figure 59.1 Sectional view of a reverse-pulse fabric filter. *Source:* Courtesy of Mikropul GmbH, Cologne, Germany.

Figure 59.2 Reverse-pulse filter to separate silica particles from the gases leaving a spray dryer, having an area of 1,370 m^2. *Source:* Courtesy of Mikropul GmbH, Cologne, Germany.

59.3 Baghouse Construction and Operation

The cross-section of the baghouse can be circular or rectangular. For relatively heavy dusts, the gas enters the filter in the lower part to enable direct settling of part of the dust. For relatively light dusts, the gas enters the filter in the upper part to enable uniform fabric charging. Typical bag diameters are 70, 120, and 160 mm. The maximum length of the bags is 8 m, as the jets of high-velocity air must also be able to reach the bottom parts of the bags. The filter bags are mounted in the supporting plate by means of a bayonet locking.

The cleaning air pulses typically last 80 ms. It is common to clean a row of bags at a time. Each row (or pair of rows) of filter elements is cleaned in turn, while the gas or air continues to flow through the remaining elements. A typical time between cleaning by means of air pulses is 15 s. The air pressure is usually 6 bar, but it is possible, if need be, to use lower pressures. Normal filter pressure losses are in the range 80–150 mm water gauge. Initially, the cloth is the filter medium, however, the surface layer generally becomes the dominating filter medium once it has formed. The gas flow through the cloth and the dust layer is laminar and hence Darcy's equation is applicable. This equation is treated in Section 52.2.2. The consequence is that the pressure drop across the fabric itself is proportional to the velocity, and the pressure loss, which is due to flow through the dust deposit, is proportional to both the velocity and the dust layer thickness. Despite the air pulses, the filter pressure loss increases with time, and eventually it is necessary to wash the filter bags. Ultimately, it will be necessary to replace the filter bags. Table 59.1 contains the maximum recommended operating temperatures for common fabrics. Glass can be used at a relatively high temperature; however, the finish limits the maximum temperature range. Fabric filters are not recommended if there is a possibility that the filter can come into contact with hot gas. If the dust can cause a dust explosion, baghouses must be safeguarded against them. Thorough bonding is then necessary, and metal- or carbon-cured fibers are often included into the fabrics. Explosion panels that relieve directly or through short lines to the atmosphere are normal features of baghouses. A relief should neither affect personnel nor equipment. A dust explosion can be followed by a fire because considerable dust holdup in the baghouse can occur.

The resistance against acids and alkalis of the filter fabrics should be checked. *Perry's Handbook* [1] provides much information concerning the properties (including mechanical) of fibers for

Table 59.1 Maximum recommended operating temperatures for fabrics.

Fabric	Temperature (°C)
Cotton	80
Wool	90
Nylon	90
Polypropylene	90
Polyacrylonitrile	125
Polyester/Teflon	140
Nomex (m-aramid)	230
Glass fiber	260
Ceramic fiber and sintered metal	650

fabrics. To avoid condensation in the filter, it is important that the temperature of the gas leaving the filter is above its dew point. See Section 53.6 for a definition of the dew point. Insulation of the filter is recommended to avoid condensation.

Reference

1. Perry, R. H. and Green, D. W. (2008). *Perry's Chemical Engineers' Handbook* (pp. 17-46–17-51). New York, NY: McGraw-Hill.

60

Scrubbers

60.1 Introduction

A scrubber is a device that uses a liquid (normally, water or an aqueous solution of a mineral acid or an alkali) to separate solid particles from a gas. If the solid material is soluble in the liquid, the aqueous solution can be processed. If the solid material is insoluble in the liquid, we have converted a solid–gas separation task into a solid–liquid separation one. Scrubbers are used in the chemical industry for several reasons, one of which is that the gaseous flows in that industry are usually relatively modest (e.g., 50,000 $m^3 \cdot h^{-1}$). Also, there is considerable processing of liquids and suspensions in the industry.

Good contact must be obtained between the liquid and the gaseous phase: to this end, the scrubbing liquid is dispersed in a jet or spray or spread in a film over the internal surfaces of the scrubber. The three relevant basic particle collection mechanisms in all scrubbers are centrifugal deposition, inertial impaction, and the Brownian movement.

The centrifugal deposition mechanism comprises spinning out of particles by centrifugal force caused by a change in gas flow direction. This is the mechanism prevailing in cyclones and is effective for particles down to approximately 5 μm.

Inertial impaction occurs when a gas stream flows around a small object and the particles suspended in the gas stream continue to move toward the object and eventually are trapped. This mechanism is effective for particles as small as 0.5 μm.

Actually, there is no fundamental difference between these two collection mechanisms. The first is concerned with large-scale flow-direction changes, and the second concerns small-scale changes. The two collection methods are only slightly dependent on the residence time. Hence, multiple-stage contacting has no real advantage over single-stage contacting for the first two collection methods.

Brownian diffusion becomes important for particles that are smaller than 0.3 μm. It comprises the movement at random of particles due to collisions with gas molecules. This mechanism becomes quite effective for very small particles, and the performance depends on the residence time.

For a given particulate, there is a relationship between the collection efficiency of a scrubber and the energy dissipated in the gas–liquid contacting process [1]. It appears that the collection efficiency depends very little on the way the power is applied and is relatively independent of scrubber geometry.

The gaseous flow entering a scrubber contains particles as small as 0.5 μm, for example. The gaseous flow leaving the scrubber often contains droplets of, for example, 100 μm. The separation of these droplets from the gaseous flow is relatively simple and can be achieved by a cyclone or a chevron demister. See Figure 60.1.

There is no risk of a dust explosion or a fire if water or an aqueous solution is used as the scrubbing liquid.

Introduction to Chemical Engineering, First Edition. C.M. van 't Land.

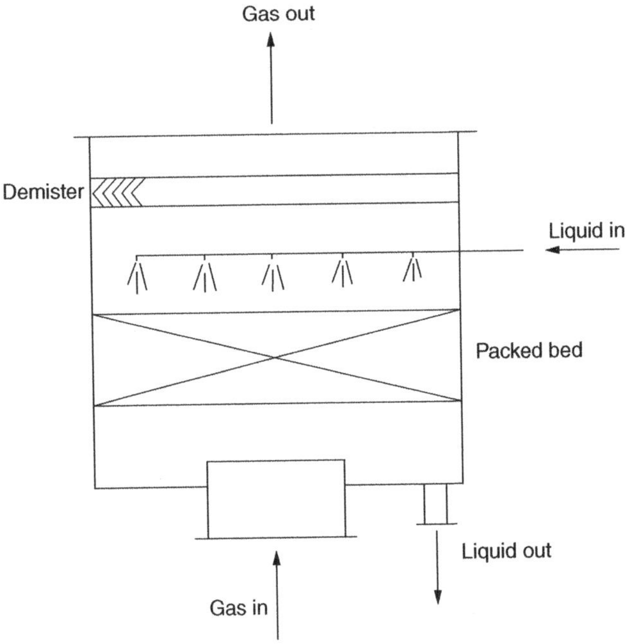

Figure 60.1 Packed-bed scrubber.

A distinction is often made between low-, medium-, and high-energy scrubbers. High-energy scrubbers have pressure drops exceeding 500 mm water gauge and can collect particles as small as 0.5 μm; however, their running costs are high. Low-energy types have pressure drops of up to 100 mm water gauge [2].

It is not clear whether wettable particles are more readily collected than non-wettable particles or whether the use of wetting agents promotes collection. Normal liquid/gas ratio are between 0.1 and 10 m^3 of liquid per 1,000 m^3 of gas. Solid concentrations in the recycle scrubbing liquid can be as high as 10% by weight [2].

The scrubber types most frequently applied in the chemical industry are packed bed, venturi, and mechanical. These three are considered next in some detail.

60.2 Packed-Bed Scrubbers

Figure 60.1 is the schematic of a tower scrubber. Normally, packed-bed scrubbers are used for the removal of soluble particles since high loadings of an insoluble dust can cause plugging. The scrubbers may be packed with, for example, Raschig rings or Berl saddles, and gas–liquid contacting is performed in a countercurrent manner. Both Raschig rings and Berl saddles are well-known packings for distillation columns. Typical superficial gas velocities are on the order of 1 $m \cdot s^{-1}$, and droplet entrainment is prevented by means of a chevron demister.

The removal efficiency depends on the kinetic energy possessed by the particle, which is $0.5 \cdot m \cdot v^2$. Thus, the gas velocity must be increased by a factor of 2.15 if a particle of 3 μm must be captured with the same efficiency as a particle of 5 μm. Because pressure drop is proportional to the square of the gas velocity, the scrubber pressure drop will increase by a factor of 4.62 (2.15^2) to be able to capture the 3-μm particle. In a countercurrent packed-bed scrubber, the power applied (as reflected by pressure drop) is limited by hydraulic flooding of the packing [3]. As a rule of thumb, for a 3-μm particle, 75–80% removal efficiency can be obtained [4].

Packed-bed scrubbers are, for example, used in salt plants. The centrifuged vacuum pan salt, having a weight average particle size of 450 μm, is dried in fluid-bed dryers. See Section 53.2. A typical production capacity of a fluid-bed dryer is 20 $t \cdot h^{-1}$. The design of the fluid-bed dryers is such that 5% by weight of the throughput is entrained by the spent drying air. The first step of gas–solid separation is separating by means of cyclones. The second step is by means of packed-bed scrubbers. Water is used as the liquid for the scrubbers. Dilute brine is recycled to the salt plant.

60.3 Venturi Scrubbers

See Figure 60.2. A venturi scrubber in its simplest form consists of a constriction in the duct carrying the gas that increases the velocity to a value in the range of 60–100 $m \cdot s^{-1}$. The width of the constriction is adjustable by means of a servomotor to enable adjustment of the pressure drop. Scrubbing liquid is introduced at or upstream of the constriction. The liquid is atomized, and the high relative velocity between the liquid droplets and the dust particles leads to efficient collection of even fine particles. The liquid droplets are then separated from the gas in a cyclonic separator. In the constriction, a swirling motion is imparted to the gas flow. This high-energy scrubber is the most efficient type of scrubber if the gas flow through the venturi can be maintained. The

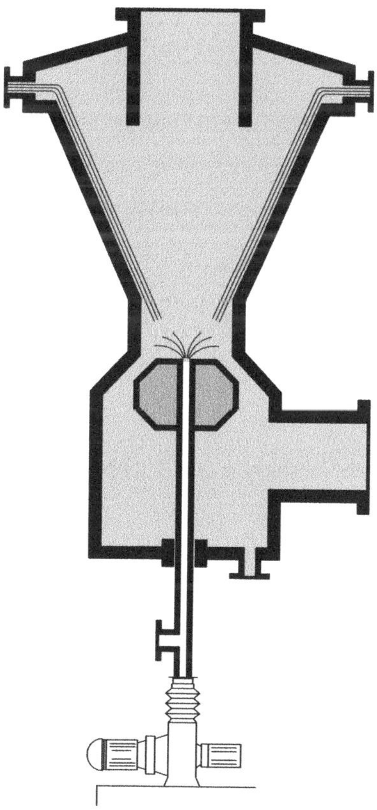

Figure 60.2 Venturi scrubber (seen schematically). *Source:* Courtesy of TREMA Verfahrenstechnik GmbH, Kemnath, Germany.

pressure drop is in the range 300–2,000 mm water gauge. Particles as small as 0.1 μm can be removed with approximately 50% efficiency. Figure 60.3 shows a venturi scrubber in a chemical plant. An example of an installed venturi scrubber is described in [1]. It concerns a scrubber in a spray-drying plant; 26,000 m^3 of spent drying gas leaving the spray dryer per h, having a temperature of 80 °C, are in first instance passed through a cyclone. The drying gas is nitrogen. The second cleaning step comprises a venturi scrubber. Spray drying is described in Section 53.2.

Figure 60.3 Venturi scrubber in a chemical plant. *Source:* TREMA Verfahrenstechnik GmbH, Kemnath, Germany.

60.4 Mechanical Scrubbers

Figure 60.4 shows a mechanical scrubber schematically. This scrubber type incorporates one or two rotors in the wet collection zone. The rotors produce a fine spray that contacts the gas. The liquid droplets are then separated from the gas in a cyclonic separator. The liquid flows down along the wall of the separator, while the gas leaves the device at the top. The circumferential velocity of the rotors is in the range 60–70 $m \cdot s^{-1}$. This scrubber can remove particles of 0.2 μm with

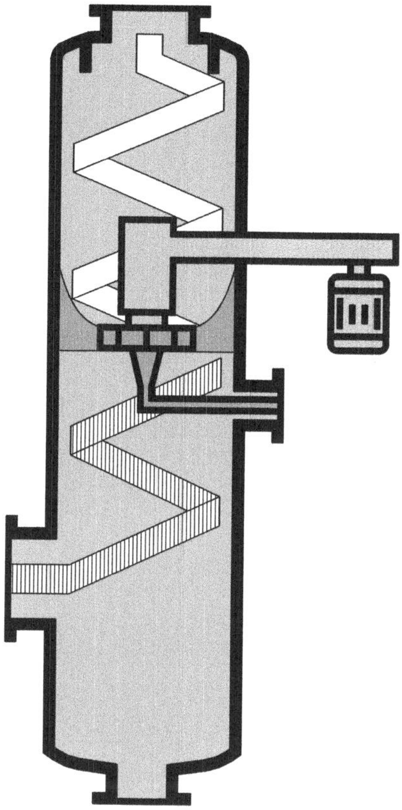

Figure 60.4 Mechanical scrubber (seen schematically). *Source:* Courtesy of TREMA Verfahrenstechnik GmbH, Kemnath, Germany.

approximately 50% efficiency. The pressure loss is in the range 40–100 mm water gauge. The energy consumption per 1,000 m^3 of gas is comparable to the energy consumption of the venturi scrubber. The investment is higher than for a venturi scrubber. An advantage over the venturi scrubber is that the collection efficiency does not depend on the gas velocity. An aspect to be kept in mind vis-à-vis the venturi scrubber is that the mechanical scrubber contains rotating parts.

References

1. Holzer, K. (1979). Wet scrubbing of particles and aerosols. *Chemie Ingenieur Technik*, 51, 200–207. (in German)
2. Muir, D. M. (1985). *A User Guide to Dust and Fume Control* (pp. 49–60). Rugby, UK: The Institution of Chemical Engineers.
3. Strigle, R. F. (1987). *Random Packings and Packed Towers* (pp. 31–32). Houston, TX: Gulf Publishing.
4. Muir, D. M. (1985). *A User Guide to Dust and Fume Control* (pp. 37–48). Rugby, UK: The Institution of Chemical Engineers.

61

Electrostatic Precipitators

61.1 Introduction

Electrostatic precipitators can achieve a separation between particulates and a gas stream. To this end, the gas stream is directed through a number of grounded channels in which high-voltage discharge electrodes are centrally positioned at regular intervals. The process is illustrated in Figure 61.1. Uncharged particles entering the channels are charged by the negative discharge electrodes or ionized gas and then move toward the grounded walls of the channel. The term *electrostatic* is not entirely correct, as a direct current is always flowing from the collecting plates, through the channel filled with gas and to the electrodes.

The pressure drop to move the gas through the separator is very low, which compensates partially for the relatively high capital cost of the equipment.

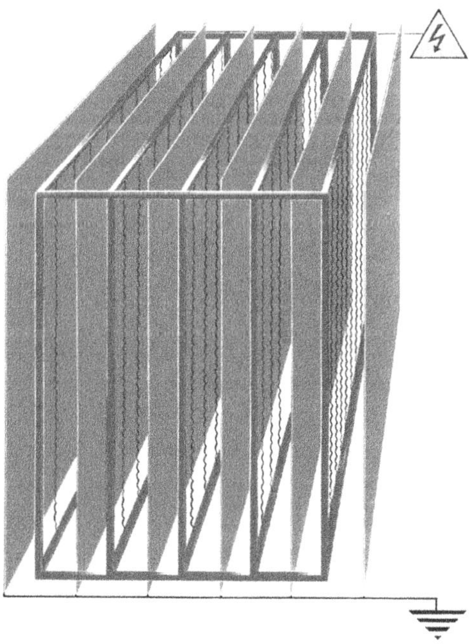

Figure 61.1 Electrostatic precipitator (seen schematically). *Source:* Courtesy of FLSmidth A/S, Valby, Denmark.

Introduction to Chemical Engineering, First Edition. C.M. van 't Land.

Electrostatic precipitators, also called *electrofilters*, are suitable for very large gas flows (for instance, $10^6\ m^3 \cdot h^{-1}$). They effectively remove particles as small as 1 μm and can clean gases down to 20 $mg \cdot nm^{-3}$ and, in some cases, even down to 10 $mg \cdot nm^{-3}$. Electrofilters are unsuitable for combustible materials. They are sometimes employed in convective drying systems in which high (for instance, up to 600 °C) drying-air temperatures are used. Filters are then less suitable, as there is a risk that the hot gas enters the filter and melts or ignites the fabric.

A typical average tension of the discharge electrode is −55 kV; the tension varies sinussoidally from −47 to −63 kV.

61.2 Principle of Operation

Figure 61.1 shows a perspective view of an electrostatic precipitator. The gases pass horizontally through a number of ducts. Two parallel rows of vertically mounted collecting plates form a duct in which a number of discharge electrodes are suspended vertically between the plates. A strong electrical field between the discharge electrodes and the grounded collecting plates is created by the high negative tension applied to the discharge electrodes. The field strength is maximum near the discharge electrodes. Electrical breakdown of the gas takes place on raising the tension. The breakdown, called *corona*, appears as a bluish glow around the discharge electrode. The precipitation mechanism is as follows.

1) The gas always contains free electrons. These electrons rapidly move away from the discharge electrode, collide with gaseous molecules, and remove additional electrons from these molecules.
2) At a certain distance from the electrode, the free electrons are captured by gaseous molecules, which become negatively charged and also move away from the discharge electrode.
3) The next step is that negatively charged gaseous molecules (ions) are captured by dust particles, which become negatively charged and move toward the collecting plates. These particles form a dust layer on the plate surface, which is dislodged by rapping.
4) The dust particles are discharged by the collecting plates and the charge flows to ground.
5) The positively charged molecules (ions) move toward the discharge electrode and are discharged. The discharge electrode is rapped as well.

61.3 Process Data

The velocity of the gases passing through the ducts is typically 1 $m \cdot s^{-1}$. The pressure of the separation is atmospheric. The maximum temperature of the separation process is in the range 400–500 °C. The process loses efficiency at higher temperatures.

A typical dust resistivity is $10^9\ \Omega \cdot cm$. When the resistivity is relatively low, particles reaching the collection plates lose their charge relatively easily and the dust can be re-entrained. At high levels of dust resistivity, an insulating layer is formed on the collecting plates, which prevents normal corona discharge, and this can lead to electrical instability and decreased collection efficiency.

A transformer-rectifier set converts the available low-voltage alternating-current supply to a high-voltage direct-current supply for the discharge electrodes. If the primary tension into the transformer is gradually increased, the high tension also increases until a maximum is achieved. A further increase in the primary tension also increases the sparkover, and the high tension falls. The automatic control system keeps the primary tension continuously at the optimum value.

Scaling up experiments are usually carried out on a 1/16 to 1/8 scale [1].

61.4 Construction

Sometimes, the precipitator unit is divided into two or more separate chambers to allow one chamber to be off-line for inspection and maintenance. Each chamber consists of several (for

Figure 61.2 Electrostatic precipitator in a chemical plant. *Source:* Courtesy of FLSmidth, Valby, Denmark.

instance, three) fields in series. Each field operates independently, with a separately optimized energy supply. The reason for this set-up is that the gas flows in plug flow through the fields and the dust content decreases rapidly.

Typical collecting plate spacings are 300 and 400 mm. A typical field length is 3.5 m and a typical plate height is 12 m. The plates are usually made of mild steel, but stainless steel can be used where corrosion or abrasion resistance is required. Figure 61.2 shows an electrostatic precipitator installed in a plant.

Example 61.1

This example describes the use of electrostatic precipitators in the drying of marl. Dried marl is a raw material used to make cement. It is a mineral obtained by mining and typically contains 10% by weight of water. The solid material consists of 90–95% by weight calcium carbonate and 10–5% by weight sand or clay. Marl is dried in large direct-heat rotary dryers by using hot air having a temperature of 450 °C, for example. The separation of entrained particles from the gases leaving the dryer is achieved by means of electrostatic precipitators. This equipment is preferred over fabric filters because fabric cannot endure high temperatures.

Example 61.2

This example concerns the use of electrostatic precipitators in a waste incineration plant. In such a plant, the gases leaving the incinerator are cleaned carefully before they are passed on to the atmosphere. First, the gases are used to heat water in heat exchangers. Steam is raised from the hot water and the steam is used to generate electricity. The gases have a temperature of 245 °C after they have passed the heat exchangers. The first step in cleaning the gases is a passage through an electrostatic separator to separate fly ash. The second step in cleaning the gases is their passage through a spray dryer in which a suspension of gypsum particles in an aqueous solution of calcium chloride is spray dried. The suspension originates from a downstream cleaning step of the gases. On processing the suspension, fine dry gypsum particles and dry calcium chloride particles are produced. 10% by weight of the particles falls into the spray dryer cone and 90% by weight is entrained by the gas. The gases, having a temperature of 170 °C, pass through a second electrofilter, separating the particulate material from the gaseous flow. Electrofilters are selected for these applications because of their ability to process hot gases and their efficiency.

Reference

1. Muir, D. M. (1985). *A User Guide to Dust and Fume Control* (pp. 61–72). Rugby, UK: The Institution of Chemical Engineers.

Notation IX

a	Height of a cyclone inlet duct	m
b	Width of a cyclone inlet duct	m
D_c	Diameter of the cylindrical part of a cyclone	m
D_e	Diameter of a cyclone outlet duct	m
K	Constant in Equation 58.1	—
m	Mass	kg
N_H	Number of hydraulic heads	—
p	Pressure	$N \cdot m^{-2}$
v	Velocity	$m \cdot s^{-1}$
v_i	Linear velocity of entering air	$m \cdot s^{-1}$

Greek Symbols

ρ_g	Air-specific mass	$kg \cdot m^{-3}$

Introduction to Chemical Engineering, First Edition. C.M. van 't Land.

Index

Printed and bound by CPI Group (UK) Ltd, Croydon, CR0 4YY
17/09/2026
14964322-0008